Lecture Notes in Computer Science 14432

Founding Editors

Gerhard Goos
Juris Hartmanis

T0177964

The series Lecture Notes in Computer Science (LNCS), including its subseries Lecture Notes in Artificial Intelligence (LNAI) and Lecture Notes in Bioinformatics (LNBI), has established itself as a medium for the publication of new developments in computer science and information technology research, teaching, and education.

LNCS enjoys close cooperation with the computer science R & D community, the series counts many renowned academics among its volume editors and paper authors, and collaborates with prestigious societies. Its mission is to serve this international community by providing an invaluable service, mainly focused on the publication of conference and workshop proceedings and postproceedings. LNCS commenced publication in 1973.

Qingshan Liu · Hanzi Wang · Zhanyu Ma ·
Weishi Zheng · Hongbin Zha · Xilin Chen ·
Liang Wang · Rongrong Ji
Editors

Pattern Recognition and Computer Vision

6th Chinese Conference, PRCV 2023
Xiamen, China, October 13–15, 2023
Proceedings, Part VIII

Springer

Editors
Qingshan Liu (iD)
Nanjing University of Information Science
and Technology
Nanjing, China

Zhanyu Ma (iD)
Beijing University of Posts
and Telecommunications
Beijing, China

Hongbin Zha (iD)
Peking University
Beijing, China

Liang Wang
Chinese Academy of Sciences
Beijing, China

Hanzi Wang (iD)
Xiamen University
Xiamen, China

Weishi Zheng (iD)
Sun Yat-sen University
Guangzhou, China

Xilin Chen (iD)
Chinese Academy of Sciences
Beijing, China

Rongrong Ji (iD)
Xiamen University
Xiamen, China

ISSN 0302-9743   ISSN 1611-3349 (electronic)
Lecture Notes in Computer Science
ISBN 978-981-99-8542-5   ISBN 978-981-99-8543-2 (eBook)
https://doi.org/10.1007/978-981-99-8543-2

Preface

Welcome to the proceedings of the Sixth Chinese Conference on Pattern Recognition and Computer Vision (PRCV 2023), held in Xiamen, China.

PRCV is formed from the combination of two distinguished conferences: CCPR (Chinese Conference on Pattern Recognition) and CCCV (Chinese Conference on Computer Vision). Both have consistently been the top-tier conference in the fields of pattern recognition and computer vision within China's academic field. Recognizing the intertwined nature of these disciplines and their overlapping communities, the union into PRCV aims to reinforce the prominence of the Chinese academic sector in these foundational areas of artificial intelligence and enhance academic exchanges. Accordingly, PRCV is jointly sponsored by China's leading academic institutions: the Chinese Association for Artificial Intelligence (CAAI), the China Computer Federation (CCF), the Chinese Association of Automation (CAA), and the China Society of Image and Graphics (CSIG).

PRCV's mission is to serve as a comprehensive platform for dialogues among researchers from both academia and industry. While its primary focus is to encourage academic exchange, it also places emphasis on fostering ties between academia and industry. With the objective of keeping abreast of leading academic innovations and showcasing the most recent research breakthroughs, pioneering thoughts, and advanced techniques in pattern recognition and computer vision, esteemed international and domestic experts have been invited to present keynote speeches, introducing the most recent developments in these fields.

PRCV 2023 was hosted by Xiamen University. From our call for papers, we received 1420 full submissions. Each paper underwent rigorous reviews by at least three experts, either from our dedicated Program Committee or from other qualified researchers in the field. After thorough evaluations, 522 papers were selected for the conference, comprising 32 oral presentations and 490 posters, giving an acceptance rate of 37.46%. The proceedings of PRCV 2023 are proudly published by Springer.

Our heartfelt gratitude goes out to our keynote speakers: Zongben Xu from Xi'an Jiaotong University, Yanning Zhang of Northwestern Polytechnical University, Shutao Li of Hunan University, Shi-Min Hu of Tsinghua University, and Tiejun Huang from Peking University.

We give sincere appreciation to all the authors of submitted papers, the members of the Program Committee, the reviewers, and the Organizing Committee. Their combined efforts have been instrumental in the success of this conference. A special acknowledgment goes to our sponsors and the organizers of various special forums; their support made the conference a success. We also express our thanks to Springer for taking on the publication and to the staff of Springer Asia for their meticulous coordination efforts.

We hope these proceedings will be both enlightening and enjoyable for all readers.

October 2023

Qingshan Liu
Hanzi Wang
Zhanyu Ma
Weishi Zheng
Hongbin Zha
Xilin Chen
Liang Wang
Rongrong Ji

Organization

General Chairs

Hongbin Zha   Peking University, China
Xilin Chen   Institute of Computing Technology, Chinese Academy of Sciences, China
Liang Wang   Institute of Automation, Chinese Academy of Sciences, China
Rongrong Ji   Xiamen University, China

Program Chairs

Qingshan Liu   Nanjing University of Information Science and Technology, China
Hanzi Wang   Xiamen University, China
Zhanyu Ma   Beijing University of Posts and Telecommunications, China
Weishi Zheng   Sun Yat-sen University, China

Organizing Committee Chairs

Mingming Cheng   Nankai University, China
Cheng Wang   Xiamen University, China
Yue Gao   Tsinghua University, China
Mingliang Xu   Zhengzhou University, China
Liujuan Cao   Xiamen University, China

Publicity Chairs

Yanyun Qu   Xiamen University, China
Wei Jia   Hefei University of Technology, China

Local Arrangement Chairs

Xiaoshuai Sun	Xiamen University, China
Yan Yan	Xiamen University, China
Longbiao Chen	Xiamen University, China

International Liaison Chairs

Jingyi Yu	ShanghaiTech University, China
Jiwen Lu	Tsinghua University, China

Tutorial Chairs

Xi Li	Zhejiang University, China
Wangmeng Zuo	Harbin Institute of Technology, China
Jie Chen	Peking University, China

Thematic Forum Chairs

Xiaopeng Hong	Harbin Institute of Technology, China
Zhaoxiang Zhang	Institute of Automation, Chinese Academy of Sciences, China
Xinghao Ding	Xiamen University, China

Doctoral Forum Chairs

Shengping Zhang	Harbin Institute of Technology, China
Zhou Zhao	Zhejiang University, China

Publication Chair

Chenglu Wen	Xiamen University, China

Sponsorship Chair

Yiyi Zhou	Xiamen University, China

Exhibition Chairs

Bineng Zhong	Guangxi Normal University, China
Rushi Lan	Guilin University of Electronic Technology, China
Zhiming Luo	Xiamen University, China

Program Committee

Baiying Lei	Shenzhen University, China
Changxin Gao	Huazhong University of Science and Technology, China
Chen Gong	Nanjing University of Science and Technology, China
Chuanxian Ren	Sun Yat-Sen University, China
Dong Liu	University of Science and Technology of China, China
Dong Wang	Dalian University of Technology, China
Haimiao Hu	Beihang University, China
Hang Su	Tsinghua University, China
Hui Yuan	School of Control Science and Engineering, Shandong University, China
Jie Qin	Nanjing University of Aeronautics and Astronautics, China
Jufeng Yang	Nankai University, China
Lifang Wu	Beijing University of Technology, China
Linlin Shen	Shenzhen University, China
Nannan Wang	Xidian University, China
Qianqian Xu	Key Laboratory of Intelligent Information Processing, Institute of Computing Technology, Chinese Academy of Sciences, China
Quan Zhou	Nanjing University of Posts and Telecommunications, China
Si Liu	Beihang University, China
Xi Li	Zhejiang University, China
Xiaojun Wu	Jiangnan University, China
Zhenyu He	Harbin Institute of Technology (Shenzhen), China
Zhonghong Ou	Beijing University of Posts and Telecommunications, China

Contents – Part VIII

Neural Network and Deep Learning I

A Quantum-Based Attention Mechanism in Scene Text Detection

Hao Wu[1], Jun Zhou[1], Qiong Zhang[1], Yang Lei[1], Kun Yu[1,2], Wenbo An[1], and Juntao Zhang[1(✉)] (ID)

[1] Institute of System Engineering, AMS, Beijing, China
zhangjt0902@hust.edu.cn
[2] University of Electronic Science and Technology of China, Chengdu, China

Abstract. Attention mechanisms have provided benefits in very many visual tasks, e.g. image classification, object detection, semantic segmentation. However, few attention modules have been proposed specifically for scene text detection. We propose an attention mechanism based on Quantum-State-based Mapping (QSM) that enhances channel and spatial attention, introduces higher-order representations, and mixes contextual information. Our approach includes two attention modules: Quantum-based Convolutional Attention Module (QCAM), a plug-and-play module applicable to pre-trained text detection models; Adaptive Channel Information Transfer Module (ACTM), which replaces feature pyramids and complex networks of DBNet++ with a 35.9% reduction in FLOPs. In CNN-based methods, our QCAM achieves state-of-the-art performance on three benchmarks. Remarkably, when compared to the Transformer-based methods such as FSG, our QCAM remains competitive in F-measure on all benchmarks. Notably, QCAM has a 29.5% reduction in parameters compared to FSG, resulting in a balance between detection accuracy and efficiency. ACTM significantly improves F-measure over DBNet++ on three benchmarks, providing an alternative to feature pyramids in scene text detection. The codes, models and training logs are available at https://github.com/yws-wxs/QCAM.

Keywords: Scene text detection · Quantum mechanics · Attention mechanism

1 Introduction

Text in natural scenes usually carries notable semantic information. In recent years, with the development of deep learning methods in the field of computer vision, especially benefiting from the development of object detection, semantic segmentation, and instance segmentation, scene text detection has also received significant improvements. However, scene text detection is still challenging due to the complexity of natural scenes, multi-scale, multi-oriented, multi-lingual, and arbitrary shapes of scene text.

H. Wu and J. Zhou—Contributed equally to this work.

Q. Liu et al. (Eds.): PRCV 2023, LNCS 14432, pp. 3–14, 2024.
https://doi.org/10.1007/978-981-99-8543-2_1

Recent scene text detection methods are mainly based on deep learning methods and can be broadly divided into two categories [28]: regression-based methods and segmentation-based methods. Segmentation-based methods determine whether each pixel in the image belongs to the text, which have a natural advantage over regression-based methods for detecting arbitrarily shaped text, as exemplified by DBNet [10] and DBNet++ [11]. DBNet predicts both the text region probability map and the text boundary region threshold map, and then adaptively sets the binarization threshold using an approximate binarization function named Differentiable Binarization (DB) to finally obtain the text region binarization map. DBNet makes the binarization process in the segmentation network trainable, simplifying the post-processing procedures and thus increasing the inference speed while achieving state-of-the-art performance on five standard scene text benchmarks. DBNet++ adds the Adaptive Scale Fusion (ASF) module to DBNet. The main function of the ASF module is to add a spatial attention mechanism to the feature pyramids fusion process, which further improves the performance and scale robustness of the original DBNet model.

However, we note that the ASF module only uses spatial attention mechanisms and needs to be pre-trained with the original network, thus it is not plug-and-play. A Quantum-State-based Mapping (QSM) for machine learning is proposed in [26], and the application of QSM to MLP-Mixer [19] architecture, which only stacks multi-layer perceptrons (MLPs), performs well in the image classification domain. [26] does not discuss how QSM could be adapted to CNN architectures and does not apply QSM in the field of scene text detection.

In this paper, we propose two new attention modules based on QSM, the first one is named Quantum-based Convolutional Attention Module (QCAM), which can be directly inserted into the text region probability map branch of the pre-trained DBNet++ model. QCAM improves the performance of the original model on three scene text detection benchmarks without extra data.

Current approaches to scene text detection typically use ResNet-50 [7] as the backbone network, followed by individually designed post-processing procedures to complete the text detection task. To improve scale robustness, many methods, such as PSENet [23] and DBNet++, use feature pyramids [12] to fuse multiple level features of the backbone network. We denote the backbone output features with downsample rates of 2, 4, 8, 16, 32 as C1, C2, C3, C4, C5, respectively. We question the necessity of feature pyramids in the text detection domain, arguing that the C5 feature already contains enough information for the text detection task. To verify this assumption, we propose another new attention module named Adaptive Channel Information Transfer Module (ACTM), which is simple and effective, using only the C5 feature. We use a single ACTM to replace the feature pyramids and the ASF module of DBNet++. In a fair comparison, our experiments show that DBNet++ with ACTM outperforms the original model on three scene text detection benchmarks.

The main contributions of this paper can be summarized as follows:

We propose two novel attention modules, QCAM and ACTM. QCAM is a lightweight and general module that can theoretically be applied to any pre-trained segmentation-based text detection models without extra data. ACTM transfers the channel information to the spatial dimension and then recalibrates the spatial features. Our experiments demonstrate that both QCAM and ACTM consistently improve the F-measure on three scene text detection datasets compared to the DBNet++ model. We confirm that an attention module using only the C5 feature layer is an alternative to feature pyramids in scene text detection.

2   Related Work

2.1   Attention Mechanism

SCA-CNN [2] points out that, in deep neural networks, different channels in different feature maps generally represent different objects. Channel attention adaptively recalibrates the weights of each channel, thereby determining *what to pay attention to*. Spatial attention adjusts the selection of spatial regions, determining *where to pay attention*. CBAM [25] incorporates the advantages of both spatial and channel attention, selecting significant objects and regions through sequentially stacked channel and spatial attention. Inspired by CBAM, we propose QCAM and ACTM.

2.2   Revisit Quantum-State-based Mapping

QSM uses wave functions describing microscopic particle systems as mappings. By QSM, original inputs or features extracted by neural networks are processed as quantum states to train wave function parameters. The wave function is a fundamental concept in quantum mechanics, which describes the behavior of matter at the microscopic scale. Specifically, the wave function is a complex exponential function that represents the probability amplitude of a particle at a given point of space and time, and the square modulus of it denotes the probability of the particle's existence at that point of space and time. In the context of a one-dimensional particle confined in a square potential well of infinite depth, the wave function has a simple form, and the time-dependent Schrödinger equation can be solved explicitly. The time-dependent Schrödinger equation is given by:

$$
\begin{cases}
\Psi(x,t) = \sum_{n=1}^{\infty} c_n \sqrt{\frac{2}{a}} sin(\frac{n\pi}{a}x) e^{-iE_n t/\hbar} \\
E_n = \frac{n^2\pi^2\hbar^2}{2ma^2} \quad (n = 1, 2, 3, ...)
\end{cases}
\tag{1}
$$

where a is the well width and m is the particle mass, which can be regarded as a constant in this paper. Although the wave function $\Psi(x,t)$ in Eq. 1 is dependent on t, the probability density $|\Psi(x,t)|^2$ is independent of t:

$$
|\Psi(x)|^2 = \frac{2}{a} \sum_{n=1}^{\infty} c_n^2 sin^2(\frac{n\pi}{a}x)
\tag{2}
$$

where $|\Psi(x)|^2 \in [0,1]$. Specifically, assuming that the feature vector of the neural network after pooling is $X \in R^{1 \times d}$, relating the state of the particle to the feature vector, we get QSM: $X \to |\Psi(X)|^2$, which maps the feature vector of the original space to the probability space.

In our view, taking channel attention as an example, QSM fundamentally maps the global spatial one-dimensional vector to a probability space of higher dimensions. This mapping is followed by superposition into a one-dimensional vector through the principle of quantum superposition. By preserving the dimensionality of the feature vector and including more high-order global channel information, the mapping enhances the model's representation capabilities.

3 Approach

In this section, we first present two sub-modules of our proposed attention modules, the QSM-based Channel Attention (QCA) module and the QSM-based Spatial Attention (QSA) module. Next, we present QCAM, which is seamlessly integrated into an existing network architecture, specifically the DBNet++ model, which serves as the base module for our experiments. The resulting structure after embedding QCAM is illustrated in Fig. 2. We then replace the original feature pyramids and the ASF module of DBNet++ with the proposed ACTM to verify the necessity of the feature pyramids. The overall network architecture after this replacement is shown in Fig. 3.

3.1 QSM-Based Channel Attention (QCA) Module and QSM-Based Spatial Attention (QSA) Module

The QCA module, illustrated in Fig. 1(a), takes as input a feature map $X \in R^{C \times H \times W}$, where C, H, and W denote the number of channels, height, and width respectively. The module first performs global average pooling to extract global spatial information and obtain $X_{gap} \in R^{C \times 1 \times 1}$. The QSM and a fully-connected layer are then applied to capture channel-wise relationships, followed by a sigmoid non-linear layer to obtain the final attention vector. Each channel of the input feature is scaled by multiplying the corresponding element in the attention vector to obtain the feature map $QCA(X) \in R^{C \times H \times W}$. The process is expressed as:

$$QCA(X) = \left[\sigma \left(FC \left(QSM(AvgPool(X)) + AvgPool(X) \right) \right) \right] \odot X \quad (3)$$

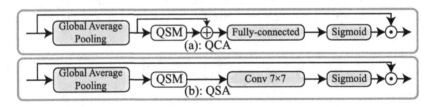

Fig. 1. QSM-based Channel Attention (QCA) module and QSM-based Spatial Attention (QSA) module.

where σ denotes the sigmoid function, FC denotes the fully-connected operation, $AvgPool$ denotes global average pooling, and $\odot$ denotes element-wise multiplication, where the corresponding value of the attention vector is broadcast (copied) during multiplication. The QSM uses the probability density $|\Psi(x,t)|^2$ for the mapping, where $\Psi(x,t)$ in this paper is the wave function of the one-dimensional infinite deep square potential well. The QSM is defined as:

$$
\begin{cases}
QSM(X_{gap}) = \dfrac{2}{a} \displaystyle\sum_{n=1}^{N} b_n sin^2(\dfrac{n\pi}{a} X_{gap}) \\
a = max(|X_{gap}|)
\end{cases}
\tag{4}
$$

where b_n is obtained through training the neural network, and since this paper does not involve the discussion of hyperparameters taking values, N is set to 10 unless otherwise specified.

The QSA module, illustrated in Fig. 1(b), takes as input a feature map $X \in R^{C \times H \times W}$, and performs global average pooling along the channel axis to extract global channel information, obtaining $X_{gap} \in R^{1 \times H \times W}$. The module then captures important spatial information using the QSM and a 7×7 convolutional layer with a large receptive field. A sigmoid non-linear layer is used to obtain the attention vector, and each spatial location of the input feature is scaled by multiplying the corresponding element in the attention vector to obtain the feature map $QSA(X) \in R^{C \times H \times W}$. The process is expressed as:

$$
QSA(X) = \left[\sigma \left(Conv \left(QSM(AvgPool(X)) \right) \right) \right] \odot X
\tag{5}
$$

where $Conv$ denotes the 7×7 convolution, $AvgPool$ denotes the global average pooling along the channel axis, and QSM is the same as in Eq. 4.

3.2 Quantum-Based Convolutional Attention Module (QCAM)

As shown in Fig. 2, our proposed QCAM joins in parallel the branch of DBNet++ that predicts the probability map of text region. The structure of QCAM is depicted in Fig. 2(a). QCAM includes QCA and QSA modules, applied sequentially to recalibrate the channel and spatial information of the original input feature map. Additionally, a transposed convolution is added after the QCA and QSA modules, respectively. Given an input feature map $F \in R^{C \times \frac{H}{4} \times \frac{W}{4}}$, the QCAM outputs $QCAM(F) \in R^{1 \times H \times W}$, which can be expressed as:

$$
QCAM(F) = ConvTr \left(QSA \left(ConvTr(QCA(F)) \right) \right)
\tag{6}
$$

where the $ConvTr$ denotes the transposed convolution.

We add the QCA module first and then the QSA module to better recalibrate the rich channel information in the input feature map before recalibrating its spatial features. Almost all segmentation-based scene text detection methods

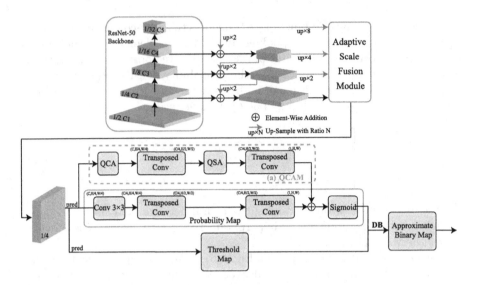

Fig. 2. The DBNet++ architecture with QCAM.

have branches for predicting the probability map of text region, and since QCAM incorporates transposed convolution, by adjusting its parameters, the height and width of the feature map can be enlarged in any ratio, so theoretically QCAM can be applied to any segmentation-based scene text detection method.

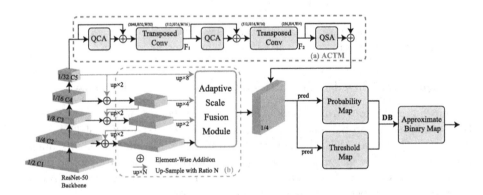

Fig. 3. The DBNet++ architecture with ACTM.

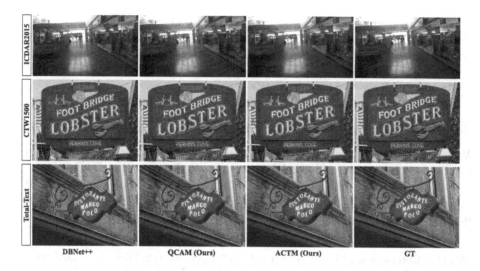

Fig. 4. Qualitative comparisons with DBNet++ on three datasets

3.3 Adaptive Channel Information Transfer Module (ACTM)

As shown in Fig. 3, a simple ACTM (Fig. 3(a)) can replace the feature pyramids plus ASF module (Fig. 3(b)) of DBNet++. The details of the ACTM are shown in Fig. 3(a). The ACTM consists of QCA, QCA, and QSA attention modules applied sequentially, with residual connections added to each attention module. The transposed convolution is inserted between the attention modules. We use the ResNet-50 backbone output feature $F_{c5} \in R^{2048 \times \frac{H}{32} \times \frac{W}{32}}$ as input to the ACTM, where H and W are the height and width of the original image input to the backbone network. To match the structure of DBNet++, the output feature map of the ACTM is set to $ACTM(F_{c5}) \in R^{256 \times \frac{H}{4} \times \frac{W}{4}}$, with two intermediate features $F_1 \in R^{512 \times \frac{H}{16} \times \frac{W}{16}}$ and $F_2 \in R^{256 \times \frac{H}{4} \times \frac{W}{4}}$ as shown in Fig. 3(a). The expression for ACTM can be represented by the following equation:

$$\begin{cases} F_1 = ConvTr\big(QCA\,(F_{c5}) + F_{c5}\big) \\ F_2 = ConvTr\big(QCA\,(F_1) + F_1\big) \\ ACTM\,(F_{c5}) = QSA\,(F_2) + F_2 \end{cases} \tag{7}$$

4 Experiments

4.1 Implementation Details

Our methods are implemented in MMOCR [9]. We use the config files provided by benchmark-DBNet++. To evaluate the plug-and-play nature of QCAM, we add it to the pre-trained DBNet++ model and fine-tune it directly on standard datasets. As removing the FPN and ASF module and adding ACTM represents a significant change to the model, we re-pretrain the modified model on the SynthText [6] dataset before fine-tuning it on other datasets, ensuring a fair comparison.

All the experiments are conducted on a server equipped with one RTX A6000 GPU and two Xeon Gold 6230R CPUs.

4.2   Performance Comparison

We evaluate the performance of our proposed QCAM and ACTM for scene text detection on three standard benchmarks: CTW1500 [16], Total-Text [4], and ICDAR2015 [8]. We compare our methods with previous state-of-the-art approaches, including a Transformer-based method FSG [18]. Qualitative and quantitative comparisons are shown in Fig. 4 and Table 1, respectively.

Table 1. Comparison with related methods on CTW1500, Total-Text and CDAR2015, where "DBNet++†" denotes duplicated DBNet++, "*" denotes with Transformer [20] structure. "P", "R", and "F" represent Precision, Recall, and F-measure, respectively.

Methods	Paper	CTW1500			Total-Text			ICDAR2015		
		P(%)	R(%)	F(%)	P(%)	R(%)	F(%)	P(%)	R(%)	F(%)
DB [10]	AAAI'20	86.9	80.2	83.4	87.1	82.5	84.7	91.8	83.2	87.3
Boundary [22]	AAAI'20	–	–	–	85.2	83.5	84.3	88.1	82.2	85.0
DRRG [27]	CVPR'20	85.9	83.0	84.5	86.5	84.9	85.7	88.5	84.7	86.6
ContourNet [24]	CVPR'20	83.7	84.1	83.9	86.9	83.9	85.4	87.6	**86.1**	86.9
TextRay [21]	MM'20	82.8	80.4	81.6	83.5	77.9	80.6	–	–	–
ABCNet [15]	CVPR'20	84.4	78.5	81.4	87.9	81.3	84.5	–	–	–
FCENet [28]	CVPR'21	87.6	83.4	85.5	89.3	82.5	85.8	90.1	82.6	86.2
I3CL [5]	IJCV'22	87.4	**84.5**	**85.9**	89.2	83.7	86.3	–	–	–
DBNet++† [11]	TPAMI'22	87.9	82.8	85.3	88.9	83.2	86.0	90.9	83.9	87.3
QCAM	Ours	**89.6**	81.6	85.4	**90.6**	85.0	**87.7**	**92.0**	83.6	**87.6**
ACTM	Ours	86.5	84.3	85.4	89.4	**85.6**	87.5	90.0	85.0	87.4
FSG* [18]	CVPR'22	88.1	82.4	85.2	90.7	85.7	88.1	91.1	86.7	88.8

Directly comparing CNN-based and Transformer-based methods is unfair due to their quite different settings. Besides Transformer-based methods, QCAM achieves best precision and F-measure on Total-Text and ICDAR2015 datasets, while on the CTW1500 dataset, QCAM achieves the best precision and competitive F-measure. Furthermore, both QCAM and ACTM improve F-measure over DBNet++ on all three benchmarks. Notably, our QCAM achieves a significant 1.7% improvement in F-measure, and ACTM achieves a significant 1.5% improvement in F-measure over DBNet++ on Total-Text. Findings suggest that DBNet++ feature pyramids are inefficient, and ACTM provides a more effective alternative.

Even when compared to the FSG method, QCAM remains competitive in terms of F-measure on all three benchmarks, while still achieving the best precision on CTW1500 and ICDAR2015 datasets. Specifically, QCAM achieves a 1.5% improvement of precision over FSG on CTW1500 and a 0.9% improvement of precision over FSG on ICDAR2015. Notably, FSG method has 38.3M parameters [18], whereas our QCAM method has only 27.0M parameters, resulting in a 29.5% reduction compared to FSG. We achieve a tradeoff between detection accuracy and efficiency.

4.3   Ablation Study

In this subsection, we conduct ablation experiments on the ACTM to investigate the effect of different attention module sequences and the necessity of the QSM for scene text detection. As discussed earlier, the ACTM is designed to sequentially apply channel, channel, and spatial attention modules to recalibrate the feature and improve text detection performance. Our view is that you need to know *what to pay attention to* when channel information is abundant, and *where to pay attention* when spatial information is abundant.

To verify the effectiveness of the attention module sequences, we compare four stacked attention module approaches: (1) QCA, QCA, QSA (CCS); (2) QSA, QCA, QCA (SCC); (3) QCA, QSA, QCA (CSC); (4) QSA, QSA, QCA (SSC). Additionally, we conduct an ablation study by removing all QSM layers in the ACTM to test their impact on text detection performance. The experiments are performed on the ICDAR2015 dataset, and the results are presented in Table 2.

Table 2. Ablation experiments on ICDAR2015. "Baseline*" denotes the result of reimplemented DBNet++. For FLOPs, both sides of input images are set to 640.

Sequences	QSM	P(%)	R(%)	F(%)	Parameters	FLOPs
Baseline*		90.41	84.02	87.10	26.91 M	50.66 G
CCS	√	90.01	**84.98**	**87.42**	35.46 M	32.43 G
SCC	√	90.16	84.74	87.37	31.33 M	32.43 G
CSC	√	89.94	83.10	86.39	35.26 M	32.44 G
SSC	√	**90.75**	84.11	87.31	31.07 M	32.43 G
CCS	–	89.85	84.83	87.27	35.46 M	32.43 G

The results demonstrate that all four stacked approaches, except CSC, improve the F-measure over the baseline. Specifically, CCS achieves the highest improvement in F-measure and recall, while SSC achieves the highest precision. This suggests that spatial attention helps to better localize the scene text region, improving precision, while channel attention helps to discover more scene text, improving recall. Furthermore, the results validate the importance of the proposed attention module sequencing principle in the ACTM.

The ablation study on QSM shows that the F-measure of the ACTM without QSM is 0.17% higher than the baseline, demonstrating the validity of the overall structure of the ACTM. However, removing QSM from the ACTM reduces the F-measure value by 0.15% compared to the original value, indicating that QSM has a positive effect on feature recalibration. This effect is achieved with only a small increase in overhead, as ACTM involves just three QSM layers with a total of 30 parameters.

5   Discussion and Conclusion

The Feature Pyramid Network (FPN) [12] improves multi-level feature object detection architectures (first proposed by SSD [14]) by fusing information from feature maps of different scales. FPN compensates for the lack of semantic information in shallow feature maps by utilizing deeper feature maps.

DETR [1] first uses Transformer for object detection and as a single-level feature detection architecture. DETR achieves comparable results to the optimized Faster R-CNN [17] baseline on the challenging COCO [13] dataset by extracting only the final C5 feature from the backbone.

YOLOF [3] argues that the success of FPN is due to the divide and conquer optimization in object detection, rather than multi-scale feature fusion. Similar to DETR, YOLOF only uses C5 feature. To achieve the effect of the multi-level feature detection, YOLOF adopts an inflated convolution to process the C5 feature map and a new sample matching rule named Uniform Matching.

The feature pyramids in DBNet++ simply fuse several layers of features from the backbone network and do not use a divide and conquer approach. Therefore, we propose ACTM to replace it and use only C5 feature to achieve the performance of the original model. Our approach aims to improve the performance of scene text detection by recalibrating the channel and spatial information in the C5 feature map, rather than relying on multi-scale feature fusion. The proposed ACTM sequentially applies channel, channel, and spatial attention modules with transposed convolution insert them. By transferring more channel information to spatial dimension and capturing discriminative features of scene text, ACTM achieves improved efficiency and accuracy compared to the original model.

We propose two novel attention modules, Quantum-based Convolutional Attention Module (QCAM) and Adaptive Channel Information Transfer Module (ACTM), for enhancing the performance of scene text detection. QCAM is a simple plug-and-play module that can be directly applied to pre-trained models such as DBNet++. With just 0.5% extra parameters, QCAM significantly improves the performance of the DBNet++ model. ACTM replaces the feature pyramids and ASF module in DBNet++, effectively recalibrating the channel and spatial information in the C5 feature map. ACTM achieves a 35.9% reduction in FLOPs while still outperforming the baseline. These findings highlight the potential benefits of incorporating attention mechanisms in scene text detection.

References

1. Carion, N., Massa, F., Synnaeve, G., Usunier, N., Kirillov, A., Zagoruyko, S.: End-to-end object detection with transformers. In: Vedaldi, A., Bischof, H., Brox, T., Frahm, J.-M. (eds.) ECCV 2020. LNCS, vol. 12346, pp. 213–229. Springer, Cham (2020). https://doi.org/10.1007/978-3-030-58452-8_13
2. Chen, L., et al.: SCA-CNN: spatial and channel-wise attention in convolutional networks for image captioning. In: CVPR, pp. 6298–6306 (2017)
3. Chen, Q., Wang, Y., Yang, T., Zhang, X., Cheng, J., Sun, J.: You only look one-level feature. In: CVPR, pp. 13034–13043 (2021)
4. Ch'ng, C.K., Chan, C.S.: Total-Text: A comprehensive dataset for scene text detection and recognition. In: 2017 14th IAPR International Conference on Document Analysis and Recognition (ICDAR), vol. 01, pp. 935–942 (2017)
5. Du, B., Ye, J., Zhang, J., Liu, J., Tao, D.: I3CL: intra-and inter-instance collaborative learning for arbitrary-shaped scene text detection. IJCV **130**(8), 1961–1977 (2022)
6. Gupta, A., Vedaldi, A., Zisserman, A.: Synthetic data for text localisation in natural images. In: CVPR, pp. 2315–2324 (2016)
7. He, K., Zhang, X., Ren, S., Sun, J.: Deep residual learning for image recognition. In: CVPR, pp. 770–778 (2016)
8. Karatzas, D., et al.: ICDAR 2015 competition on robust reading. In: 2015 13th International Conference on Document Analysis and Recognition (ICDAR), pp. 1156–1160 (2015)
9. Kuang, Z., et al.: MMOCR: a comprehensive toolbox for text detection, recognition and understanding. In: Proceedings of the 29th ACM International Conference on Multimedia, MM '21, pp. 3791–3794. Association for Computing Machinery, New York (2021)
10. Liao, M., Wan, Z., Yao, C., Chen, K., Bai, X.: Real-time scene text detection with differentiable binarization. In: AAAI, vol. 34, pp. 11474–11481 (2020)
11. Liao, M., Zou, Z., Wan, Z., Yao, C., Bai, X.: Real-time scene text detection with differentiable binarization and adaptive scale fusion. TPAMI **45**(1), 919–931 (2023)
12. Lin, T.Y., Dollár P., Girshick, R., He, K., Hariharan, B., Belongie, S.: Feature pyramid networks for object detection. In: CVPR, pp. 936–944 (2017)
13. Lin, T.Y., et al.: Microsoft COCO: common objects in context. In: Fleet, D., Pajdla, T., Schiele, B., Tuytelaars, T. (eds.) ECCV 2014. LNCS, vol. 8693, pp. 740–755. Springer, Cham (2014). https://doi.org/10.1007/978-3-319-10602-1_48
14. Liu, W., et al.: SSD: single shot multibox detector. In: Leibe, B., Matas, J., Sebe, N., Welling, M. (eds.) ECCV 2016. LNCS, vol. 9905, pp. 21–37. Springer, Cham (2016). https://doi.org/10.1007/978-3-319-46448-0_2
15. Liu, Y., Chen, H., Shen, C., He, T., Jin, L., Wang, L.: ABCNet: real-time scene text spotting with adaptive bezier-curve network. In: CVPR, pp. 9806–9815 (2020)
16. Liu, Y., Jin, L., Zhang, S., Luo, C., Zhang, S.: Curved scene text detection via transverse and longitudinal sequence connection. Pattern Recogn. **90**, 337–345 (2019)
17. Ren, S., He, K., Girshick, R., Sun, J.: Faster R-CNN: towards real-time object detection with region proposal networks. TPAMI **39**(6), 1137–1149 (2017)
18. Tang, J., et al.: Few could be better than all: feature sampling and grouping for scene text detection. In: CVPR, pp. 4553–4562 (2022)
19. Tolstikhin, I.O., et al.: MLP-Mixer: an all-MLP architecture for vision. In: NeurIPS, vol. 34, pp. 24261–24272 (2021)

20. Vaswani, A., et al.: Attention is all you need. In: Proceedings of the 31st International Conference on Neural Information Processing Systems, NIPS'17, pp. 6000–6010. Curran Associates Inc., Red Hook (2017)
21. Wang, F., Chen, Y., Wu, F., Li, X.: TextRay: contour-based geometric modeling for arbitrary-shaped scene text detection. In: Proceedings of the 28th ACM International Conference on Multimedia, MM '20, pp. 111–119. Association for Computing Machinery, New York (2020)
22. Wang, H., et al.: All you need is boundary: toward arbitrary-shaped text spotting. In: AAAI, vol. 34, pp. 12160–12167 (2020)
23. Wang, W., et al.: Shape robust text detection with progressive scale expansion network. In: CVPR, pp. 9328–9337 (2019)
24. Wang, Y., Xie, H., Zha, Z.J., Xing, M., Fu, Z., Zhang, Y.: ContourNet: taking a further step toward accurate arbitrary-shaped scene text detection. In: CVPR, pp. 11750–11759 (2020)
25. Woo, S., Park, J., Lee, J.-Y., Kweon, I.S.: CBAM: convolutional block attention module. In: Ferrari, V., Hebert, M., Sminchisescu, C., Weiss, Y. (eds.) ECCV 2018. LNCS, vol. 11211, pp. 3–19. Springer, Cham (2018). https://doi.org/10.1007/978-3-030-01234-2_1
26. Zhang, J., et al.: An application of quantum mechanics to attention methods in computer vision. In: ICASSP, pp. 1–5 (2023)
27. Zhang, S.X., et al.: Deep relational reasoning graph network for arbitrary shape text detection. In: CVPR, pp. 9696–9705 (2020)
28. Zhu, Y., Chen, J., Liang, L., Kuang, Z., Jin, L., Zhang, W.: Fourier contour embedding for arbitrary-shaped text detection. In: CVPR, pp. 3122–3130 (2021)

NCMatch: Semi-supervised Learning with Noisy Labels via Noisy Sample Filter and Contrastive Learning

Yuanbo Sun and Can Gao[✉]

Shenzhen University, Shenzhen, China
sunyuanbo2021@email.szu.edu.cn, 2005gaocan@163.com

Abstract. Semi-supervised learning (SSL) has been widely studied in recent years, which aims to improve the performance of supervised learning by utilizing unlabeled data. However, the presence of noisy labels on labeled data is an inevitable consequence of either limited expertise in the labeling process or inadvertent labeling errors. While these noisy labels can have a detrimental impact on the performance of the model, leading to decreased accuracy and reliability. In this paper, we introduce the paradigm of Semi-Supervised Learning with Noisy Labels (SSLNL), which aims to address the challenges of semi-supervised classification when confronted with limited labeled data and the presence of label noise. To address the challenges of SSLNL, we propose the Noisy Samples Filter (NSF) module, which effectively filters out noisy samples by leveraging class agreement. Additionally, we propose the Semi-Supervised Contrastive Learning (SSCL) module, which harnesses the power of high-confidence pseudo-labels generated by a semi-supervised model to enhance the extraction of robust features through contrastive learning. Extensive experiments on CIFAR-10, CIFAR-100, and SVHN datasets demonstrate that our proposed method, outperforms state-of-the-art methods in addressing the problems of SSL and SSLNL, thus validating the effectiveness of our solution.

Keywords: Semi-supervised learning with Noisy labels ·
Semi-supervised learning · Self-supervised contrastive learning

1 Introduction

With the proliferation of extensive datasets [6] and computational resources, deep neural networks (DNNs) have demonstrated remarkable effectiveness in a wide range of computer vision (CV) applications, encompassing image classification, object detection, and segmentation [6,9,25]. However, collecting large amounts of labeled data can be costly and time-consuming, rendering the training of DNNs in practical settings a formidable task. As a solution to this chal-

© The Author(s), under exclusive license to Springer Nature Singapore Pte Ltd. 2024
Q. Liu et al. (Eds.): PRCV 2023, LNCS 14432, pp. 15–27, 2024.
https://doi.org/10.1007/978-981-99-8543-2_2

lenge, semi-supervised learning (SSL) has emerged by utilizing only a limited number of annotated samples, achieving comparable performance to supervised learning [2,21,24].

In practical scenarios, the existence of noisy labels in labeled data is an unavoidable outcome resulting from either insufficient expertise or inadvertent labeling. This phenomenon tends to considerably impact the performance of the model. However, with the development of Learning with Noisy Labels (LNL) [19], fully supervised models incorporating noisy labels have achieved impressive results, mitigating the negative effects to some extent.

Nevertheless, in numerous practical scenarios, achieving complete supervision may be excessively demanding or even infeasible. We are only able to obtain a limited amount of labeled data. Moreover, the labeling process may introduce some noise, which could affect the quality of the annotations. Therefore, in this study, we propose the paradigm of Semi-Supervised Learning with Noisy Labels (SSLNL), which aims to tackle the challenges of semi-supervised classification in the presence of limited labeled data with noisy labels. By combining the advantages of SSL and LNL, SSLNL provides a practical and effective solution for real-world scenarios. To address the challenges of SSLNL, it is necessary to filter out the noisy samples from the labeled data while enhancing the robustness of feature extraction.

To this end, we introduce NCMatch, a novel SSLNL framework that consists of two modules: the Noisy Samples Filter (NSF) module and the Semi-Supervised Contrastive Learning (SSCL) module. NCMatch can simultaneously address the SSL and SSLNL problems, achieving state-of-the-art performance on CIFAR-10, CIFAR-100, and SVHN datasets. Through extensive experiments, we demonstrate the effectiveness of these algorithms in various settings. The main contributions of our work can be summarized as follows:

- We address the problem of SSLNL by introducing the NSF module, which effectively filters out noisy samples and enhances the model's robustness against noisy labels.
- To enhance the robustness of the extracted features, we propose the SSCL module. By leveraging high-confidence pseudo-labels, this module guides the learning process and facilitates the extraction of discriminative and reliable features.
- We propose NCMatch, an end-to-end framework that simultaneously tackles the challenges of SSL and SSLNL. Through extensive experiments on CIFAR-10, CIFAR-100, and SVHN datasets, NCMatch achieves state-of-the-art results, surpassing baseline methods and demonstrating its effectiveness.

2 Related Work

2.1 Semi-supervised Learning

Semi-supervised learning (SSL) is a learning paradigm that utilizes both labeled and unlabeled data to train a model. Several methods have been proposed

to tackle this problem, including pseudo-labeling [13], consistency regularization [23], and hybrid methods [1,17,26]. Pseudo-labeling methods involve training a model using labeled data and subsequently utilizing the trained model to
generate pseudo-labels for unlabeled data. These pseudo-labels are then incorporated into the training process to retrain the model iteratively. Consistency
regularization methods enforce consistent predictions on augmented versions of
the same sample to enhance generalization performance. Hybrid methods combine both approaches and have achieved high accuracy in SSL, comparable to
fully-supervised learning. Notably, MixMatch [1] and FixMatch [17] are two popular methods that leverage pseudo-labeling and augmentation consistency. Flex-
Match [26] further improves on FixMatch [17] by using curriculum learning to
produce thresholds for each category dynamically, resulting in state-of-the-art
performance in SSL.

2.2   Self-supervised Contrastive Learning

Self-Supervised Contrastive Learning is a powerful method that leverages unlabeled data to learn robust representations. It operates by ensuring that different
augmentations of the same image yield similar embeddings, while embeddings of
different images remain distinct. In the field of self-supervised contrastive learning, prominent approaches such as SimCLR [3] and MoCo [8] have emerged.
SimCLR calculates pairwise similarity between augmented versions of the same
image and maximizes the similarity between them. MoCo [8] utilizes a momentum encoder to generate a queue of negative samples and a query encoder to
generate positive samples. The query encoder is updated by minimizing the
contrastive loss between the query and the queue. However, it is important to
note that self-supervised learning is task-agnostic and may lead to variations
in representations for images belonging to the same class. This variability is
undesirable for classification tasks. To address this challenge, two methods have
been proposed: SupCon [11] and UniMoCo [5]. These methods aim to enhance
self-supervised models with supervised information to improve their learning
capabilities and align the representations with supervised objectives.

2.3   Learning with Noisy Labels

Learning with Noisy Labels (LNL) aims to tackle the problem of noisy labels in
training datasets. Various methods have been proposed to mitigate the impact
of unreliable or noisy labels. One common approach is "sample selection", which
involves training the model using samples that exhibit low losses. Several methods have been proposed, such as MentorNet [10], Co-teaching [7], SELFIE [18],
DivideMix [14], and JoCoR [22], have been proposed to address this issue. However, these methods rely on fully labeled data, even though these labels may
contain noisy labels, making them insufficient in effectively resolving the SSLNL
problem with a limited amount of noisy labeled data.

3   Method

In this section, we will present our proposed method NCMatch. The framework of NCMatch is shown in Fig. 1. NCMatch consists of two modules: the Noisy Samples Filter (NSF) module and the Semi-Supervised Contrastive Learning (SSCL) module. The two modules are trained in an end-to-end manner. In the following subsections, we will describe the two modules in detail.

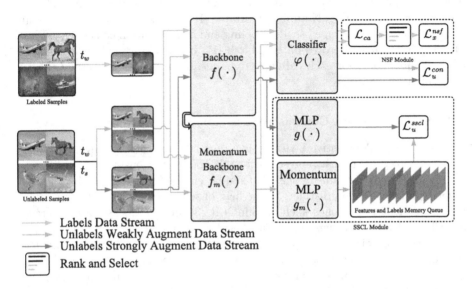

Fig. 1. The Framework of NCMatch. NCMatch consists of two modules: the Noisy Samples Filter (NSF) module and the Semi-Supervised Contrastive Learning (SSCL) module. The two modules are trained in an end-to-end manner.

3.1   Preliminaries

We define the problem of semi-supervised image classification with noisy labels as follows. Given a batch of B labeled samples $\mathcal{X} = \{(x_b : \widetilde{y_b})\}_{b=1}^{B}$, where $\widetilde{y_b}$ represents one-hot targets with noisy labels. We also define a batch of μB unlabeled samples $\mathcal{U} = \{u_b\}_{b=1}^{\mu B}$, where μ represents the ratio of unlabeled samples to labeled samples. The goal of SSLNL is to train a backbone $f(\cdot)$ and a classifier $\varphi(\cdot)$ to classify the samples in the test set $\mathcal{T} = \{x_t, y_t\}$, where x_t represents the test sample, and y_t represents the ground truth label.

3.2   Overall Framework

In our proposed method, NCMatch, we follow semi-supervised framework [17] to handle labeled and unlabeled samples. For each batch of labeled samples with noisy labels $\mathcal{X}$, weak augmentation function $t_w(\cdot)$, such as flipping and cropping, is applied to obtain weakly augmented samples. Similarly, for each batch of unlabeled samples $\mathcal{U}$, both weak and strong augmentations, $t_w(\cdot)$ and $t_s(\cdot)$ respectively, are applied to obtain weakly and strongly augmented samples. The feature information from these samples is extracted using a convolutional neural network-based encoder $f(\cdot)$, resulting in $\mathbf{h} = f(t(x))$. Additionally, a multi-layer perceptron projection head $g(\cdot)$ is utilized along with an encoder $f_m(\cdot)$ and a projection head $g_m(\cdot)$, which incorporate momentum updates derived from f and g respectively. The feature representation $\mathbf{h}$ is mapped into a latent space $\mathbf{z}$ using g, and into a momentum latent space $\mathbf{z_m}$ using g_m. These steps serve as a prerequisite for implementing the NSF and SSCL modules. Subsequently, a fully connected classifier $\varphi(\cdot)$ is employed to map $\mathbf{h}$ into class probabilities denoted as $p = \varphi(\mathbf{h})$. This enables us to obtain the weakly augmented probabilities p^w, and if $\max p^w > \tau$, p^w is considered as pseudo labels, as well as the strongly augmented probabilities p^s for the unlabeled samples. The consistency loss is then calculated using p^w and p^s, defined as the cross-entropy between the two probabilities.

$$\mathcal{L}_u^{con} = \frac{1}{\mu B} \sum \mathbb{1}(\max p^w > \tau) \mathrm{H}\left(p^w, p^s\right) \tag{1}$$

where τ is the threshold, $\mathbb{1}$ is the indicator function, and H is the cross-entropy loss. The consistency loss is only calculated for unlabeled samples with a high confidence score, which is determined by the threshold τ_c. Finally, we utilize the NSF module to compute the loss for labeled samples with noisy labels, and the SSCL module to calculate the contrastive loss for unlabeled samples.

3.3   Noisy Sample Filter (NSF)

In the proposed model, we introduce the NSF module, which aims to remove noisy samples from the labeled dataset. The NSF module ensures that these noisy samples are not utilized in the loss computation, thereby improving the quality of the training process.

In the first stage, the class-agreement loss for each sample is computed. In the second stage, a certain proportion of noisy samples is filtered out based on the magnitude of the class-agreement loss.

Class-Agreement Loss Calculation. In the first stage, the class-agreement loss for each sample is computed using two networks updated by gradients and momentum, respectively. The class-agreement loss is similar to the loss defined in [22] and is formulated as follows:

$$\mathcal{L}_{ca} = \lambda_{ca}(\mathrm{H}(p_x, \widetilde{y}) + \mathrm{H}(p_x^m, \widetilde{y})) + (1 - \lambda_{ca})(D_{KL}(p_x \| p_x^m) + D_{KL}(p_x^m \| p_x)), \tag{2}$$

where $\mathrm{H}(\cdot)$ is the cross-entropy loss, $D_{KL}(\cdot \| \cdot)$ is the Kullback-Leibler divergence, p_x and p_x^m are the predicted probability distributions of sample x from the networks updated by gradients and momentum, respectively, and $\widetilde{y}$ is the noisy label of sample x. The hyperparameter λ_{ca} balances the two terms.

Noisy Sample Removal. In the second stage, a proportional approach is employed to filter out a certain percentage of noisy samples. The samples are ranked based on their loss, and those with the highest losses are removed. This removal focuses on the samples with the highest losses, as they are likely to be the most noisy and have a significant negative impact on the model's performance [22]. By removing a specific percentage of these samples, we can enhance the quality of the training data and mitigate the adverse effects of noisy labels. The $\mathcal{L}_x^{nsf}$ loss is defined as follows:

$$\mathcal{L}_x^{nsf} = \frac{1}{N} \sum \mathbb{1}(\mathcal{B}(x, s))\mathcal{L}_{ca}(x), \tag{3}$$

where $\mathcal{B}$ is a binary variable that denotes the selection of the sample x for training. We select the top-s losses for removal and utilize the remaining samples for training. The value of hyperparameter s determines the proportion of samples to be eliminated. In our experiments, we consider s as a prior.

3.4 Semi-supervised Contrastive Learning (SSCL)

The SSCL module is designed to mitigate the impact of noisy labels on the model and enhance its robustness. This is achieved by leveraging the pseudo-labels provided by the classifier φ to perform contrastive learning, thereby obtaining more reliable and discriminative features. To enforce similarity between features of samples with strongly augmented views and their weakly augmented counterparts sharing the same labels, a feature memory queue is maintained using a momentum encoder f_m and a momentum projection head g_m, which are updated using momentum similar to MoCo [8].

Given a batch of unlabeled samples in $\mathcal{U}$, the weakly augmented features z^w are obtained by applying the models f_m and g_m to the weakly augmented samples, yielding $z^w = g_m(f_m(t_w(u)))$. Then, probabilities p^w are computed by applying the classifier φ to the weakly augmented features z^w, yielding $p^w = \varphi(z^w)$. Pseudo-labels with the largest class probability p^w above a threshold τ are retained. The z^w and p^w are then stored in the feature memory queue and pseudo-label memory queue, respectively. Strongly augmented features z are obtained by applying f and g to the strongly augmented samples, and probabilities p^s are computed using φ, like the weakly augmented features.

All features z^m are obtained from the feature memory queue. The features z^m and z are subsequently used to compute the contrastive loss. Positive examples are identified as samples with the same pseudo-label as the current sample, while negative examples correspond to samples with different pseudo-labels. In cases where no samples share the same pseudo-label, only the strongly augmented sample of the current sample is used for contrastive learning. Following [5], the contrast loss is defined as follows:

$$\mathcal{L}_u^{sscl} = \log \left(1 + \sum_{\{k^-\}} \exp\left(\text{sim}(z_k, z_{k^-}^m)/t\right) \sum_{\{k^+\}} \exp\left(-\text{sim}(z_k, z_{k^+}^m)/t\right) \right) \tag{4}$$

where t is the temperature parameter that controls the sharpness of the distribution, z_{k-}^m represents the negative sample of z in z^m, while z_{k+}^m denotes the positive sample. The similarity function $\text{sim}(\cdot)$ is defined as follows: $\text{sim}(\mathbf{u}, \mathbf{v}) = \frac{\mathbf{u} \cdot \mathbf{v}}{\|\mathbf{u}\|\|\mathbf{v}\|}$, where $\mathbf{u}$ and $\mathbf{v}$ are the features of the samples.

To enhance the SSCL module, we introduce a nearest neighbor (NN) approach that utilizes cosine similarity to identify positive samples. Given a sample, we calculate the cosine similarity between its features and the features stored in the features memory queue. The top k samples with the highest cosine similarity are selected as positive samples for contrastive learning.

The SSCL module aims to enhance the extraction of robust features, as the model learns to discriminate between similar and dissimilar samples. This approach also helps mitigate the influence of noisy labels, as the model focuses on the intrinsic characteristics of the samples rather than relying solely on the provided labels.

Finally, the overall loss function of NCMatch is defined as follows:

$$\mathcal{L} = \mathcal{L}_x^{nsf} + \lambda_u^{con} \mathcal{L}_u^{con} + \lambda_u^{sscl} \mathcal{L}_u^{sscl} \tag{5}$$

where $\mathcal{L}_x^{nsf}$ is the loss for labeled samples with noisy labels, $\mathcal{L}_u^{con}$ is the consistency loss, $\mathcal{L}_u^{sscl}$ is the contrastive loss, and λ_u^{con} and λ_u^{sscl} are the corresponding weights.

4 Experiments

In this section, we provide an experimental evaluation of NCMatch. Firstly, we describe the experimental setup of NCMatch under the SSL settings. Since the SSLNL problem is essentially an SSL problem. Next, we conduct experiments on the SSLNL problem to validate the effectiveness of NCMatch. Finally, we examine the impact of NCMatch's hyperparameters on its performance.

4.1 Datasets

We evaluated the performance of NCMatch and validated the effectiveness of their individual components on three commonly employed datasets: CIFAR-10 [12], CIFAR-100 [12], and SVHN [15]. CIFAR-10 consists of 60,000 32×32 color images distributed across 10 categories, with 50,000 training samples and 10,000 test samples. CIFAR-100 contains 100 classes, with each class having only 600 images, making a total of 50,000 training samples and 10,000 test samples. The SVHN dataset consists of real-world street view house numbers obtained from Google Street View images. The dataset contains 73,257 training images, 26,032 test images, and 531,131 additional extra training images.

4.2 Experimental for SSL

When using NCMatch to address the SSL problem, we set the hyperparameter s in the NSF module to 0. In this configuration, NCMatch essentially becomes a traditional SSL method. To assess the performance of NCMatch in the traditional SSL setting, we conducted experiments accordingly.

Experimental Setup. We use the experimental setup from [17, 26] to ensure a fair comparison in our popular semi-supervised classification experiments. For CIFAR-10 and CIFAR-100, we randomly select 4, 25, and 400 samples from the training set of each class as labeled data and use the remaining samples as unlabeled data. For SVHN, we use 25 and 100 samples from each class for labeling and the remaining data as unlabeled. We note that the SVHN dataset used in our experiment includes an additional set of 531,131 supplementary samples. We report the performance of the NCMatch's exponential moving average (EMA) model.

Baseline. For the SSL setting, we selected MixMatch [1], UDA [23], Fix-Match [17], and FlexMatch [26]. We evaluate the performance of NCMatch against these methods to validate its effectiveness.

Implementation Details. For a fair comparison in our widely-adopted semi-supervised classification experiments, we adopt the backbone and hyperparameters from prior research [17, 26]. Our strong augmentation choice is RandAugment [4]. The backbone network is the Wide ResNet (WRN) [25] and its variants: WRN28-2 for CIFAR-10, WRN28-8 for CIFAR-100, and WRN28-2 for SVHN. A 2-layer MLP projection head yields 128-dimensional embeddings for the SSCL module. Standard stochastic gradient descent (SGD) optimizes all experiments with a momentum of 0.9 [16, 20]. Batch size is fixed at 64 for all datasets, following the strategy of setting the batch size for unlabeled data μ times that of labeled data, where μ is 7, as established by UDA [23], FixMatch [17], and FlexMatch [26]. Parameters like τ are set to 0.95, and the K-Nearest-Neighbors count k is set to 5. We adhere to a dynamic threshold strategy akin to FlexMatch [26] within the SSL experimental framework.

Table 1. Semi-supervised classification results on CIFAR-10, CIFAR-100, and SVHN.

Method	CIFAR-10			CIFAR-100			SVHN	
	40 labels	250 labels	4000 labels	400 labels	2500 labels	10000 labels	250 labels	1000 labels
Supervised	95.38±0.05	95.39±0.04	95.38±0.05	80.70±0.09	80.7±0.09	80.73±0.05	97.87±0.01	97.86±0.01
MixMatch [1]	–	88.20±0.87	93.76±0.06	–	–	74.12±0.30	97.78±0.08	97.82±0.06
UDA [23]	–	94.57±0.96	95.68±0.08	–	–	–	–	97.77±0.07
FixMatch [17]	86.19±3.37	94.93±0.65	95.74±0.05	51.15±1.75	71.71±0.11	77.40±0.12	97.52±0.38	97.72±0.11
FlexMatch [26]	95.03±0.06	95.02±0.09	95.81±0.01	**60.06±1.62**	73.51±0.20	78.10±0.15	93.41±2.29	93.28±0.30
NCMatch	**95.65±0.06**	**95.54±0.09**	**96.08±0.03**	54.87±1.16	**73.79±0.56**	**78.66±0.13**	**97.85±0.16**	**98.01±0.06**

Results. Table 1 presents our findings. Our approach, NCMatch, excels across multiple experiments, particularly on CIFAR-10. However, on CIFAR-100, Flex-Match surpasses NCMatch with 40 labels, while NCMatch outperforms Flex-Match with 250 and 1000 labels. We attribute this difference to SSCL during warm-up, overly relying on uncertain pseudo-labels with 40 labels. NCMatch performs best on SVHN compared to other semi-supervised methods. Overall, our results affirm NCMatch's superiority in semi-supervised classification, surpassing current state-of-the-art methods.

4.3 Experimental for SSLNL

We present the experimental evaluation of NCMatch on the SSLNL problem. We begin by describing the experimental setup specifically designed for the SSLNL problem. Subsequently, we evaluate the performance of NCMatch in addressing the SSLNL problem.

Generation of Noisy Labels. First, we used different proportions of Symmetric and Pairflip noise to generate the noisy labels. Symmetric noise is generated by flipping a certain percentage of the training dataset's labels evenly to all possible labels, while Pairflip noise flips the labels to adjacent class labels.

Experimental Setup. For CIFAR-10, we randomly selected 100 samples from the training set of each class as labeled data and generated Symmetric and Pairflip noisy labels at 20% and 40%, respectively. The rest of the training set was used as unlabeled data. Similarly, for CIFAR-100, we used 50 and 100 labeled samples for each class and generated Symmetric and Pairflip noisy labels to analyze the impact of the number of labeled data with noisy labels on the model's performance. For the SVHN dataset, we used 100 labeled samples for each class and generated Symmetric and Pairflip noisy labels at 20%, with the remaining training set serving as the unlabeled data.

Baseline. We selected FixMatch [17] as the baseline method. FixMatch is a state-of-the-art semi-supervised learning method that utilizes a pseudo-labeling strategy to train a model. Our objective was to examine the performance of semi-supervised learning methods in the SSLNL setting.

Table 2. Semi-supervised with noisy labels classification results on CIFAR-10, CIFAR-100, and SVHN. S: Symmetric noise, P: Pairflip noise.

Dataset	Labeled Per Classes	Noise Type	Noise Rate	FixMatch [17]	NCMatch	Δ (%)
CIFAR-10	100	S	20%	92.23	**94.80**	+2.57
	100	P	20%	87.31	**93.41**	+6.10
	100	S	40%	83.06	**90.38**	+7.32
	100	P	40%	62.47	**92.33**	+29.86
CIFAR-100	50	S	20%	58.58	**65.31**	+6.73
	50	P	20%	57.74	**63.48**	+5.74
	100	S	20%	61.58	**70.06**	+8.48
	100	P	20%	61.58	**67.88**	+6.30
	100	S	40%	48.11	**64.30**	+16.19
	100	P	40%	43.52	**55.58**	+12.06
SVHN	100	S	20%	91.19	**96.73**	+5.54
	100	P	20%	87.02	**94.81**	+7.79

Implementation Details. For a fair comparison, we use similar experimental settings as SSL for SSLNL experiments. However, we set the total number of training steps to 2^{18} and use the WRN28-2 [25] as the backbone network for all experiments. The hyperparameter s is set as the proportion of noisy labels in the dataset.

Results. The experimental results are presented in Table 2. NCMatch achieves state-of-the-art performance in all experimental settings. When the number of labeled samples was consistent, the FixMatch and NCMatch methods showed similar accuracies at low noise rates. However, as the noise rate increased, Fix-Match's accuracy decreased significantly, whereas NCMatch demonstrated better robustness, maintaining higher accuracy even under high noise rates. As the number of labeled samples per class increased, the advantage of NCMatch over FixMatch became more pronounced, demonstrating NCMatch's more effective use of labeled data even with a limited number of labeled samples. These results demonstrate the effectiveness of our proposed method in the SSLNL setting.

4.4   Ablation Study

In this section, we perform ablation studies to assess the effectiveness of various components in NCMatch. In particular, we focus on evaluating the performance of NCMatch with different nearest-neighbor k values and different temperature values in the SSCL module. Furthermore, we evaluate the impact of noisy labels on the accuracy of pseudo-labels and analyze their correlation with the model's performance.

Nearest-Neighbor K. Firstly, we evaluate the performance of NCMatch with different nearest-neighbor k values in the SSCL module. The results in Fig. 2a demonstrate that the performance of NCMatch significantly improves with the nearest-neighbor k values in the SSCL module. The best performance in CIFAR-10 is achieved with a k value of 5.

Temperature. Furthermore, we investigate the effect of different temperature values on the performance of NCMatch. As shown in Fig. 2b, the best perfor-

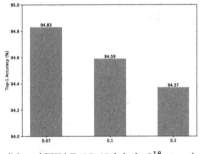

(a) k. (CIFAR-10 40 labels-2^{20} steps)   (b) t. (CIFAR-10 40 labels-2^{18} steps)

Fig. 2. Results of varying k and t values on CIFAR-10 for NCMatch.

mance is achieved with a temperature value of 0.07, which is consistent with previous research on contrastive learning [5].

Effectiveness of the NSF Module. We evaluate the effectiveness of the NSF module by comparing the performance of NCMatch with and without the NSF module. The results in Table 3 demonstrate that the NSF module significantly improves the performance of NCMatch. This demonstrates the effectiveness of the NSF module in addressing the SSLNL problem.

Table 3. Effectiveness of the NSF module on CIFAR-10 and CIFAR-100.

Dataset	Labeled Per Classes	Noise Type	Noise Rate	w/o. NSF	w. NSF	Δ (%)
CIFAR-10	100	S	40%	79.11	**90.38**	+11.21
	100	P	40%	61.27	**92.33**	+31.06
CIFAR-100	100	S	40%	47.36	**64.30**	+16.94
	100	P	40%	43.58	**55.58**	+12.00

Pseudo Label Accuracy. In Fig. 3, we present the impact of noisy labels on the pseudo-labels of NCMatch. The pseudo-label accuracy referred to here is the accuracy of the pseudo-labels for samples where $p^w > \tau$. At the beginning of Fig. 3, the pseudo-label accuracy of FixMatch is very high due to the high threshold of τ set at 0.95. However, as the influence of noisy labels in the labeled data increases, the model gradually overfits the noise, resulting in a sharp drop in pseudo-label accuracy. In contrast, our NCMatch model, with the NSF

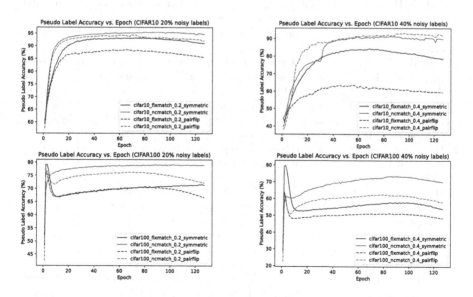

Fig. 3. Pseudo label accuracy of NCMatch on CIFAR-10 and CIFAR-100.

module, can remove the influence of some noisy labeled samples, resulting in a continuously increasing pseudo-label accuracy throughout training, and ultimately surpassing the pseudo-label accuracy of FixMatch. This demonstrates the effectiveness of our method.

5 Conclusion

In this paper, we provide an accurate description of the SSLNL problem and propose NCMatch, a novel semi-supervised learning framework that addresses both SSL and SSLNL problems. The NSF module effectively addresses the SSLNL problems, while the SSCL module enhances the performance of NCMatch through robust feature extraction from unlabeled data. Extensive experiments validate the effectiveness of each component in our framework, and the results on multiple datasets demonstrate the efficacy of NCMatch.

References

1. Berthelot, D., Carlini, N., Goodfellow, I., Papernot, N., Oliver, A., Raffel, C.A.: MixMatch: a holistic approach to semi-supervised learning. In: Advances in Neural Information Processing Systems, pp. 5050–5060 (2019)
2. Chapelle, O., Chi, M., Zien, A.: A continuation method for semi-supervised SVMs. In: Proceedings of International Conference on Machine Learning, pp. 185–192 (2006)
3. Chen, T., Kornblith, S., Norouzi, M., Hinton, G.: A simple framework for contrastive learning of visual representations. In: International Conference on Machine Learning, pp. 1597–1607 (2020)
4. Cubuk, E.D., Zoph, B., Shlens, J., Le, Q.V.: RandAugment: practical automated data augmentation with a reduced search space. In: Proceedings of the IEEE/CVF Conference on Computer Vision and Pattern Recognition Workshops, pp. 702–703 (2020)
5. Dai, Z., Cai, B., Chen, J.: UniMoCo: unsupervised, semi-supervised and fully-supervised visual representation learning. In: IEEE International Conference on Systems, pp. 3099–3106 (2022)
6. Deng, J., Dong, W., Socher, R., Li, L.J., Li, K., Li, F.F.: ImageNet: a large-scale hierarchical image database. In: Proceedings of the IEEE/CVF Conference on Computer Vision and Pattern Recognition, pp. 248–255 (2009)
7. Han, B., et al.: Co-teaching: robust training of deep neural networks with extremely noisy labels. In: Advances in Neural Information Processing Systems, pp. 8536–8546 (2018)
8. He, K., Fan, H., Wu, Y., Xie, S., Girshick, R.: Momentum contrast for unsupervised visual representation learning. In: Proceedings of the IEEE/CVF Conference on Computer Vision and Pattern Recognition, pp. 9729–9738 (2020)
9. He, K., Zhang, X., Ren, S., Sun, J.: Deep residual learning for image recognition. In: Proceedings of the IEEE/CVF Conference on Computer Vision and Pattern Recognition, pp. 770–778 (2016)
10. Jiang, L., Zhou, Z., Leung, T., Li, L.J., Li, F.F.: MentorNet: learning data-driven curriculum for very deep neural networks on corrupted labels. In: International Conference on Machine Learning, pp. 2304–2313 (2018)

11. Khosla, P., et al.: Supervised contrastive learning. In: Advances in Neural Information Processing Systems, pp. 18661–18673 (2020)
12. Krizhevsky, A.: Learning multiple layers of features from tiny images (2009)
13. Lee, D.H., et al.: Pseudo-label: the simple and efficient semi-supervised learning method for deep neural networks. In: International Conference on Machine Learning, p. 896 (2013)
14. Li, J., Socher, R., Hoi, S.C.: DivideMix: learning with noisy labels as semi-supervised learning. In: International Conference on Learning Representations, pp. 1–14 (2020)
15. Netzer, Y., Wang, T., Coates, A., Bissacco, A., Wu, B., Ng, A.Y.: Reading digits in natural images with unsupervised feature learning (2011)
16. Polyak, B.T.: Some methods of speeding up the convergence of iteration methods. USSR Comput. Math. Math. Phys. $\mathbf{4}$(5), 1–17 (1964)
17. Sohn, K., et al.: FixMatch: simplifying semi-supervised learning with consistency and confidence. In: Advances in Neural Information Processing Systems, pp. 596–608 (2020)
18. Song, H., Kim, M., Lee, J.G.: Selfie: refurbishing unclean samples for robust deep learning. In: International Conference on Machine Learning, pp. 5907–5915 (2019)
19. Song, H., Kim, M., Park, D., Shin, Y., Lee, J.G.: Learning from noisy labels with deep neural networks: a survey. arXiv preprint: arXiv:2007.08199 (2020)
20. Sutskever, I., Martens, J., Dahl, G., Hinton, G.: On the importance of initialization and momentum in deep learning. In: International Conference on Machine Learning, pp. 1139–1147 (2013)
21. Van Engelen, J.E., Hoos, H.H.: A survey on semi-supervised learning. Mach. Learn. $\mathbf{109}$(2), 373–440 (2020)
22. Wei, H., Feng, L., Chen, X., An, B.: Combating noisy labels by agreement: a joint training method with co-regularization. In: Proceedings of the IEEE/CVF Conference on Computer Vision and Pattern Recognition, pp. 13726–13735 (2020)
23. Xie, Q., Dai, Z., Hovy, E., Luong, T., Le, Q.: Unsupervised data augmentation for consistency training. In: Advances in Neural Information Processing Systems, pp. 6256–6268 (2020)
24. Yang, X., Song, Z., King, I., Xu, Z.: A survey on deep semi-supervised learning. IEEE Trans. Knowl. Data Eng., 1–20 (2022)
25. Zagoruyko, S., Komodakis, N.: Wide residual networks. arXiv preprint: arXiv:1605.07146 (2016)
26. Zhang, B., et al.: FlexMatch: boosting semi-supervised learning with curriculum pseudo labeling. In: Advances in Neural Information Processing Systems, pp. 18408–18419 (2021)

Data-Free Low-Bit Quantization via Dynamic Multi-teacher Knowledge Distillation

Chong Huang[1], Shaohui Lin[1,2(✉)], Yan Zhang[3], Ke Li[4], and Baochang Zhang[5,6]

[1] School of Computer Science and Technology,
East China Normal University, Shanghai, China
shlin@cs.ecnu.edu.cn
[2] KLATASDS-MOE, Shanghai, China
[3] Institute of Artificial Intelligence, Xiamen University, Shanghai, China
[4] Tencent Youtu Lab, Shanghai, China
[5] Beihang University, Beijing, China
[6] Nanchang Institute of Technology, Nanchang, China

Abstract. Data-free quantization is an effective way to compress deep neural networks under the situation where training data is unavailable, due to data privacy and security issues. Although Off-the-shelf data-free quantization methods achieve the relatively same accuracy as the fully-precision (FP) models for high-bit (*e.g.*, 8-bit) quantization, low-bit quantization performance drops significantly to restrict their extensive applications. In this paper, we propose a novel data-free low-bit quantization method via Dynamic Multi-teacher Knowledge Distillation (DMKD) to improve the performance of low-bit quantization models. In particular, we first introduce a generator to synthesize the training data based on the input of random noise. The low-bit quantization models are then trained on these synthetic images by the dynamic knowledge from the FP model and the high-bit quantization models, which are balanced by learnable loss weight factors. The factors are controlled by a tiny learnable FP network to adaptively allocate the balanced weights for the knowledge from the FP model and the high-bit quantization models during training. For inference, we only kept the low-bit quantization model by safely removing other additional networks, such as the generator and the tiny model. Extensive experiments demonstrate the effectiveness of DMKD for low-bit quantization of widely-used convolutional neural networks (CNNs) on different benchmark datasets. Our DMKD ooon methods.

Keywords: Data-Free Quantization · Knowledge Distillation · Dynamic knowledge · Loss Weight Factors

1 Introduction

Deep neural networks (DNNs) have achieved remarkable performance in computer vision, such as image classification [11] and object detection [24]. However,

Q. Liu et al. (Eds.): PRCV 2023, LNCS 14432, pp. 28–41, 2024.
https://doi.org/10.1007/978-981-99-8543-2_3

the success of DNNs is accompanied by significant computation and memory consumption, which restricts these models to be applied to resource-limited mobile phones and embedded devices. Network quantization [9,13,23] aims to reduce the model memory footprint and computation by converting the full-precision (FP) weights to low-bit ones (*e.g.*, 8-bit integers), so as to replace the original multiply-accumulate operations (MAC) with bitcounting and addition operations. It has received a great deal of research focus, due to its good compression performance.

Network quantization can be categorized into two classes either quantization-aware training or post-training quantization. For the former, quantization operations (*e.g.*, INT8) are replaced with the FP operation, and the quantized model is retrained using the whole training data [13,28]. For the latter, it still requires a small number of training data to fine-tune and learn quantization parameters such as scale and clipping interval [19,29,30]. Obviously, these methods need to access the training data to train the quantized models, which restricts their applications in the data-free situation where it is practical in many cases. For example, the data cannot be accessed due to data privacy and security issues in many companies. In addition, it requires significantly large memory storage to save the million-level or even billion-level training data.

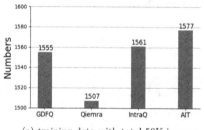
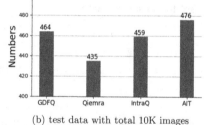

(a) training data with total 50K images   (b) test data with total 10K images

Fig. 1. The sample number on the original CIFAR-100, where 8-bit quantized models classify correctly while the FP ResNet-20 predicts wrongly during (a) training and (b) test.

To address the above problem, data-free quantization is proposed to first generate synthetic samples approximated to the original data by using the pre-trained FP models, which are then quantized using the synthetic data. For example, GDFQ [32], DSG [37], and Qimera [6] use a generator network to generate image samples, while ZeroQ [3] optimizes the initial Gaussian noise to be close to the training data. Although these methods can effectively quantize the FP models to be 8-bit ones, their performance drops significantly for lower-bit quantization, such as 4-bit. Two potential problems lead to performance degradation. On one hand, there is a large gap between the pre-trained FP model and the low-bit quantized network, such that the knowledge is insufficient and effective to be transferred from the FP model to the low-bit one. On the other hand, high-bit quantized models contain additional information, which is not necessarily held in the FP models. As shown in Fig. 1, we observe that 8-bit models trained by state-of-the-art data-free quantization methods correctly classify part of the original training and test data, while FP models misclassify instead. This begs our rethinking: *Why not adaptively merge both knowledge from the FP and high-bit models to fully balance their knowledge weights, which improves performance for data-free low-bit quantization?*

To answer the above question, we propose a novel data-free low-bit quantization via *Dynamic Multi-teacher Knowledge Distillation* (DMKD), which dynamically allocates the training weight factors to different teacher models during training for better quantization without accessing the original training data. Figure 2 depicts the DMKD framework. First, we construct a generator to generate the synthetic images using the Batch-Normalization (BN) statistic information from the pre-trained FP model. The low-bit quantized model (*e.g.*, 4-bit) is then trained by the dynamic knowledge from the FP model and the high-bit quantized model (*e.g.*, 8-bit), which is balanced by the learnable loss weight factors. To better learn these factors, we introduce a learnable tiny FP network to generate the factors, which are allocated to knowledge distillation losses from the FP model and the high-bit quantization models. During training, the generator and low-bit quantization network with a learnable tiny network can be alternatively optimized to obtain high-quality image outputs and achieve high-quantization performance, respectively. We only kept the low-bit quantized model for fast inference.

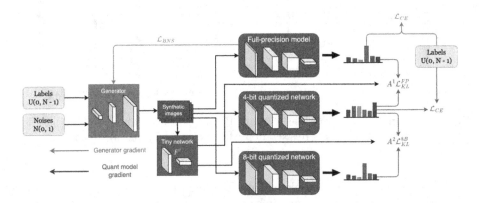

Fig. 2. An illustration of DMKD. We take 4-bit quantization for example. First, we construct a generator (blue box) to generate the synthetic images using the BN statistic information from the pre-trained FP model. 4-bit quantized model (green box) is then trained by the dynamic knowledge from the FP model (gray box) and the 8-bit quantization model (gray box), which is balanced by the learnable loss weight factors. These factors are controlled by a learnable tiny FP network (F^t) to dynamically allocate the knowledge from the FP model and the 8-bit quantized one to the 4-bit quantized one. The generator and low-bit quantization network with a learnable tiny network are alternatively optimized during training, while only the 4-bit quantized model is kept for fast inference. (Color figure online)

The main contributions of this paper are summarized as follows: (1) We propose a novel dynamic multi-teacher knowledge distillation (DMKD) for data-free low-bit quantization. To the best of our knowledge, DMKD is the first to use balanced knowledge from high-bit quantized models and a pre-trained FP model for data-free low-bit quantization during training. (2) The learnable loss weight factors are generated by a tiny FP network, which can be merged into the knowledge distillation losses from the FP model and the high-bit quantization models to dynamically control their loss balances. (3) Extensive experiments demonstrate the superior performance of our DMKD. On ImageNet ILSVRC 2012, our 4-bit quantized ResNet-18 achieves the highest Top-1 accuracy of 66.61%, compared to previous data-free 4-bit quantization methods.

2   Related Work

Model Quantization. Model quantization is an effective model compression method by converting the weight and activation of the network from full precision to low bit. One problem is that quantization can lead to a significant drop in model accuracy, especially low-bit quantization. To solve this problem, many methods have been paid attention to the quantization strategies. PACT [5] used learnable activation clipping parameters in the activation function. ACIQ [2] proposed a threshold selection method for activation quantization. AdaRound [19] abnegated the traditional rounding quantization method and proposed an optimization goal for rounding to find the best rounding strategy. There are also methods using specific quantization methods and strategies, such as randomly discarding some activation values [29,30]. Dorefa-net [40] not only quantized the weight and activations but also the gradient of the network is quantized. BNN [9] proposed to use binary representation of weights and activations to further compress the FP models, and XNOR-Net [23] introduced additional scaling factors to improve the quantization performance. However, these methods all rely on the original data set to train or fine-tune the quantized model, which restricts their applications in data-free scenarios.

Data-Free Quantization. It aims to quantize FP models without accessing training data [20,32]. In early time, DFQ [20] used weight equalization and bias correction techniques to improve quantization precision without accessing data. To effectively quantize the FP models, recent works have focused on generating the synthetic data from a generator or the image with Gaussian noise. For example, the works [21,33] reconstructed data from a pre-trained teacher model using prior information about the data distribution. ZeroQ [3] optimizes randomly generated noise pictures through the statistical data of the BN layer of the teacher network. GDFQ [32] uses the generator network and BN statistical information to generate images. DSG [37] introduced a slack on the mean and variance to solve the homogenization problem of generated images. Qimera [6] increases the quantization effect by stacking latent embeddings to generate boundary-supporting samples. AIT [7] found the contradiction between cross-entropy (CE) loss and Kullback-Leibler (KL) divergence loss and proposed the gradient inundation method to properly update the weights. These methods attempt to restore or reconstruct data for quantization training, which achieves high performance for high-bit quantization, such as 8-bit quantization. However, their performance drops significantly for low-bit quantization. Different from these methods, we proposed dynamic multi-teacher knowledge distillation to allocate the adaptive training weights factors to multiple teachers, which better quantizes the low-bit quantized model during training.

Knowledge Distillation. Knowledge distillation [12] aims to improve the performance of small student networks by transferring the knowledge from a large teacher network. The knowledge is formed from either class posterior probabilities [12] or intermediate feature [1,25,36]. Recently, multiple teachers have been utilized to construct more rich and instructive information to train the student model, where knowledge is from the ensemble logits [31] or features [10,34,39]. For example, You *et al* [34] use the incorporation of multiple teacher networks to extract the triplet ordering relationships in the intermediate layers of different examples, which is encouraged to be consistent with the student. Different from the above distillation methods, data-free knowledge distillation has been proposed to synthesize images without the requirement of real images. Generally, it can be roughly divided into two categories: (1) Directly learn images by using gradient descent on prior knowledge, such as activation statistics [18] and BN statistics [33]; (2) Adversarial training to learn a generator on a noise input [4,8,35]. DAFL [4] and DFQ [8] employ GAN to generate images in the first stage, which can be further used to learn the student model. Recently, ZAQ [17] proposed a two-level discrepancy modelling framework, which facilitates the discrepancy estimation on the intermediate features between a student and a teacher in an adversarial manner and knowledge transfer to learn the student. After training, the synthetic images and student model can be simultaneously obtained without retraining. Different from these data-free distillation methods, we introduce more balanced knowledge from multiple teachers controlled by the learnable loss weight factors to learn a low-bit quantized model during training.

3   Method

3.1   Preliminaries

For network quantization on the classification task, the quantized models are often optimized by minimizing the following loss:

$$\min_{\theta} \mathcal{L}(\theta) = \frac{1}{N} \sum_{i=1}^{N} \mathcal{L}_{CE}\big(\sigma(Q(x_i; \theta)), y_i\big) \tag{1}$$

where θ is the learnable parameters of the quantized model Q, including network weights and other additional scalars. (x_i, y_i) is the training data with total N samples, $\mathcal{L}_{CE}(\cdot, \cdot)$ denotes the cross-entropy loss function, and σ is the softmax operation. The key challenge for data-free quantization lies in the unavailable training data (x_i, y_i) for training. To effectively train the quantized models, various methods [3,32] use the feature distribution in the pre-trained full-precision model to construct synthetic training data $(\widehat{x}_i, \widehat{y}_i)$. As mentioned in Sect. 1, there are two ways to generate data. Here, we choose a generator to synthesize training samples. Therefore, for data-free quantization, we need to learn a generator G and quantized model Q to generate training data and also the parameters θ of Q, which can be formulated as:

$$\min_{\theta, G(\omega)} \mathcal{L}(\theta) = \frac{1}{N} \sum_{i=1}^{N} \mathcal{L}_{CE}\big(\sigma(Q(\widehat{x}_i; \theta)), \widehat{y}_i\big), \quad \widehat{x}_i = G(z, \widehat{y}_i; \omega) \qquad (2)$$

where z is a noise with the normal distribution, and ω is the parameters of generator G. By taking noise z and fake labels $\hat{y}$ as the input, the generator G will generate synthetic images $\widehat{x}$ and their corresponding one-hot labels.

To generate better synthetic images, the pre-trained FP model is used to update the parameters of the generator, as the pre-trained model contains the distribution information of the training data. Following [32], we use the mean and variance from the BN layer of the pre-trained FP model, which is aligned between the generated images and the real data. Thus, the aligning loss $\mathcal{L}_{BNS}$ can be formulated as:

$$\mathcal{L}_{BNS} = \sum_{i=1}^{L} \|\mu_l - \mu_l^r\|_2^2 + \|\sigma_l - \sigma_l^r\|_2^2 \qquad (3)$$

where μ_l and σ_l are the mean and variance extracted from the l-th BN layer of the pre-trained FP model by taking the generated image as input, respectively. Correspondingly, μ_l^r and σ_l^r are the mean and variance from the real data, which has been stored in the pre-trained FP model. To learn the generator G, $\mathcal{L}_{BNS}$ should be leveraged into the classification loss from the pre-trained FP model $\mathcal{L}_{CE}$, which can be formulated as:

$$\mathcal{L}_G(\omega) = \frac{1-\lambda}{N} \sum_{i=1}^{N} \mathcal{L}_{CE}\big(\sigma(F(\widehat{x}_i)), \widehat{y}_i\big) + \lambda \mathcal{L}_{BNS} \qquad (4)$$

where $F(\cdot)$ is the final output of the pre-trained FP model before the softmax operation, and λ is a balanced hyper-parameter. After accessing the synthetic images and their labels, the quantized model is further learned using knowledge distillation by transferring the softened logits from the pre-trained FP model to the quantized model, which can be formulated as:

$$\mathcal{L}_{KD}(\theta) = \mathcal{L}(\theta) + \mathcal{L}_{KL}\big(\sigma(F(\widehat{x}_i)/\tau), \sigma(Q(\widehat{x}_i; \theta)/\tau)\big) \qquad (5)$$

where τ is a temperature parameter and $\mathcal{L}_{KL}(\cdot, \cdot)$ is Kullback-Leibler (KL) divergence loss. Generally, Eq. 4 and Eq. 5 are alternatively optimized to learn the generator and the parameter θ of the quantized model. However, these methods have a large gap between the low-bit quantized model and the FP one. In the following sections, we will explore the important role of high-bit (e.g., 8-bit) quantized models, and propose dynamic multi-teacher knowledge distillation to improve low-bit quantized models.

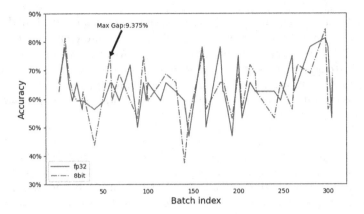

Fig. 3. Effect of 8-bit quantized model and FP model to classify the test images.

3.2   More Insight on 8-Bit Quantized Models

Direct optimization using the pre-trained FP model to minimize Eq. 5 for low-bit (*i.e.*, 4-bit) quantization generates a large gap between the pre-trained FP model and low-bit quantized model, which leads to unsuccessful knowledge transferring that reduces the accuracy of the low-bit quantized model. In addition, We assume that the 8-bit quantized model optimized by Eq. 4 and Eq. 5 generates some rich information, which is different from the pre-trained FP model. As shown in Fig 3, despite the overall accuracy of the FP model in all test data being higher than that of the 8-bit model, the 8-bit model achieves higher accuracy in some samples. Even, the maximized accuracy gap between the FP model and the 8-bit one attains 9.375%. In Fig. 1, the number is also large, where the 8-bit quantized model classifies correctly while the FP model predicts wrongly during both training and testing.

Based on the above finding, we try to use the 8-bit quantized model as an additional teacher together with the FP model to guide the lower-bit quantization. However, there is another question we should answer: *how to merge the FP model and high-bit quantized ones for effective knowledge distillation*. A simply equal balance for these models is not suitable to distinguish the importance of these models *w.r.t* different synthetic samples during training. To this end, we need to design a dynamic strategy to effectively allocate the knowledge from the FP model and high-bit quantized ones according to different samples. In Sect. 3.3, we propose Dynamic Multi-teacher Knowledge Distillation to solve the above questions.

3.3   Dynamic Multi-teacher Knowledge Distillation

Our innovation mainly lies in the training of the low-bit quantized model, so we keep the generator architecture and its optimization (*i.e.*, Eq. 4) be same as the previous works [32,37]. We take the 8-bit quantized model and FP model for example to dynamically transfer their knowledge to a lower-bit quantized model. Instead of the ensembled teacher in [34,39], we directly transfer the knowledge from the 8-bit quantized model and the FP model to low-bit quantized models, as shown in Fig. 2. Therefore, the loss for the low-bit quantized network can be formulated as:

$$\mathcal{L}_{KD}(\theta) = \frac{1}{B} \sum_{i=1}^{B} \alpha_1 \mathcal{L}_{KL}^{FP} \big(\sigma(F(\widehat{x}_i)/\tau), \sigma(Q(\widehat{x}_i;\theta)/\tau) \big) + \alpha_2 \mathcal{L}_{KL}^{8B} \big(\sigma(F^{8B}(\widehat{x}_i)/\tau), \sigma(Q(\widehat{x}_i;\theta)/\tau) \big)$$

$$+ \mathcal{L}_{CE} \big(\sigma(Q(\widehat{x}_i;\theta)), \widehat{y}_i \big),$$

$$(6)$$

where $\mathcal{L}_{KL}^{FP}$ and $\mathcal{L}_{KL}^{8B}$ are the same KL divergence loss. $F^{8B}(\cdot)$ and $Q(\cdot)$ denote the 8-bit and low-bit quantized models, respectively. α_1 and α_2 are lost weight factors to balance the knowledge distillation losses from the pre-trained FP model and the 8-bit quantized one, respectively. B means the mini-batchsize number. These factors play an important role in controlling the distillation process. Naturally, we can employ the cross-validation strategy to obtain them, which however has problems in the following two aspects: (1) *Heavy time-consuming search.* α_1 and α_2 can be regarded as hyper-parameters, which requires much time to be searched via the cross-validation strategy. (2) *The sub-optimal problem.* It is difficult or labor-intensive to determine the fixed α_1 and α_2 for different networks. Moreover, as discussed in Sect. 3.2, different synthetic images should be allocated different factors to implement more effective knowledge distillation during training. Setting them to be fixed during training leads to being sub-optimal to learn the quantized model.

To address the above issues, we introduce a learnable tiny network F^t to generate the loss weight factors α_1 and α_2. Considering the practical optimization in a batch-by-batch manner, each synthetic batch with the number of B images should generate B weight factors to the distillation losses from the FP model $\mathcal{L}_{KL}^{FP}$ and the 8-bit quantized one $\mathcal{L}_{KL}^{8B}$. Therefore, we formulate the following weight matrix $A \in \mathbf{R}^{B \times 2}$ as:

$$A = \sigma \big(F^t(\widehat{x}_{i=1}^B; \theta_t) \big).$$

$$(7)$$

Here, F^t is constructed by using the tiny FP network (*e.g.*, ResNet-18) with parameter θ_t and modifying the final fully-connected weights to generate two classes, rather than the original c classes. In addition, σ is operated at the row dimension of the output of F^t. Therefore, Eq. 6 can be reformulated as:

$$\mathcal{L}_{KD}(\theta, \theta_t) = \frac{1}{B} \sum_{i=1}^{B} A_i^1 \mathcal{L}_{KL}^{FP} \big(\sigma(F(\widehat{x}_i)/\tau), \sigma(Q(\widehat{x}_i;\theta)/\tau) \big) + A_i^2 \mathcal{L}_{KL}^{8B} \big(\sigma(F^{8B}(\widehat{x}_i)/\tau), \sigma(Q(\widehat{x}_i;\theta)/\tau) \big)$$

$$+ \mathcal{L}_{CE} \big(\sigma(Q(\widehat{x}_i;\theta)), \widehat{y}_i \big).$$

$$(8)$$

The Overall Loss. By combining Eq. 4 and Eq. 8, we construct the overall loss as:

$$\mathcal{L}_O(\theta, \theta_t, \omega) = \mathcal{L}_{KD}(\theta, \theta_t) + \mathcal{L}_G(\omega).$$

$$(9)$$

Equation 9 is minimized by using the Adam optimizer [14]. The generator and low-bit quantization network with a learnable tiny network can be alternatively optimized to obtain high-quality image outputs and achieve high-quantization performance. The detailed training process is presented in Algorithm 1.

Algorithm 1: The training process of DMKD.

Input: Full precision model F, 8-bit model F^{8B}, Total epoch T, Mini-batchsize
B, Tiny network F^t with parameter θ_t, generator G with ω.
Output: Low-bit quantized model Q with parameter θ.
for $i = 1, ..., T$ **do**
    for $j = 1, ..., B$ **do**
        Obtain random noise $z \sim \mathcal{N}(0, 1)$ and label $y \sim U(0, N-1)$;
        Taking z and y as input to generate synthetic sample $\widehat{x}$ via the
        generator G;
        for $l = 1, ...L$ **do**
           | Compute $\mathcal{L}_{BNS}$ using Eq. 3;
        end
        Update generator G by minimizing Eq. 4;
        Compute $F(\widehat{x})$ and $F^{8B}(\widehat{x})$;
        Obtain loss-weight factors $[A^1, A^2]$ using Eq. 7;
        Update the parameters of quantization model Q and tiny network F^t by
        minimizing Eq. 8.
    end
end

---

4   Experiments

4.1   Experimental Setups

Datasets and Models. We evaluate the proposed DMKD on three datasets, CIFAR-10 [15], CIFAR-100 [15], and ImageNet ILSVRC2012 [26]. CIFAR-10 contains 60K images of 10 categories with 50K training images and 10K test images. CIFAR-100 [15] has 100 categories, and each category contains 600 images, including 500 training images and 100 test images. ImageNet is a large-scale image classification dataset with 1000 categories, including about 1.2M training samples and 50k test samples. For CIFAR-10/100, we chose the widely-used ResNet-20 [11] model as the base model. For ImageNet, we tested three models, ResNet-18/50 and MobileNetV2 [27]. All pre-trained FP models are from PyTorch model zoo[1], and the 8-bit models are trained by AIT [7].

Implementation. We implement our method using Pytorch [22] with Adam optimizer [14]. The generator and quantized model with a learnable tiny network are alternatively optimized with a total epoch of 400, mini-batchsize of 200 and 16 on CIFAR-10/100 and ImageNet, respectively. Following [32], temperature τ and λ are fixed by 20 and 0.5, respectively. For the generator, we set an initial learning rate of 0.001 and a decay rate of 0.1 every 100 epochs. For the quantization stage, the initial learning rate and the momentum is set to 1e-4 and 0.9, respectively. The setting for training the tiny network F^t is the same as the generator. All experiments are trained using one NVIDIA TITAN RTX GPU and Intel(R) Core(TM) i9-10980XE CPU.

---

[1] https://pytorch.org/vision/stable/models.html.

Evaluation Metric. We select Top-1 accuracy to evaluate performance at the bit width of 4 and 5 for weight and activation quantization, whose models are trained without any training data.

Architecture of Tiny Network. In default, we use the pre-trained FP ResNet-18 model to initialize the tiny network F^t. As discussed in Sect. 3.3, we only modify the fully-connected layer while keeping other structures unchanged.

Table 1. Accuracy (%) comparison with previous data-free quantization methods on CIFAR-10, CIFAR-100 and ImageNet. $nwna$ represents the quantization bit for weights and activations.

Dataset	Model (Full precision)	Bit width	ZeroQ [3] (CVPR 2020)	GDFQ [32] (CVPR 2020)	GDFQ+AIT [7] (CVPR 2022)	IntraQ [38] (CVPR 2021)	DSG [37] (CVPR 2021)	Qimera [6] (NeurIPS 2021)	ARC [41] (IJCAI 2021)	DMKD (Ours)
CIFAR-10	ResNet-20	4w4a	79.30	90.25	91.23	91.49	88.74	91.26	88.55	**91.61**
	(93.89)	5w5a	91.34	93.38	93.41	-	-	93.46	92.88	**93.46**
CIFAR-100	ResNet-20	4w4a	47.45	63.58	65.80	64.98	62.36	65.10	62.76	**66.04**
	(70.33)	5w5a	65.61	66.12	69.26	-	-	69.02	68.40	**69.34**
ImageNet	ResNet-18	4w4a	22.58	60.60	65.51	66.47	60.12	63.84	61.32	**66.61**
	(71.47)	5w5a	59.26	68.40	70.01	69.94	69.53	69.29	68.88	**70.25**
	ResNet-50	4w4a	8.38	52.12	64.24	-	-	66.25	64.37	64.81
	(77.73)	5w5a	48.12	71.89	74.23	-	-	75.32	74.13	74.69
	MobileNetV2	4w4a	10.96	59.43	65.39	65.10	59.04	61.62	60.13	**65.63**
	(73.03)	5w5a	59.88	68.11	71.70	71.28	70.85	70.45	68.40	**71.81**

4.2 Comparison with Previous Data-Free Quantization Methods

Quantitative Comparison. In order to verify the effectiveness of the proposed DMKD, we select all data-free quantization methods, ZeroQ [3], GDFQ [32], AIT [7], IntraQ [38], DSG [37], Qimera [6] and ARC [41] to make a fair comparison. Table 1 summarizes their comparison results. 1) *ResNet-20 on CIFAR-10/100.* Our DMKD achieves the highest accuracy for 4-bit and 5-bit quantization, compared to all other methods. For 4-bit ResNet-20 quantization, DMKD outperforms the best previous method IntraQ by 0.12% on CIFAR-10, and achieves 0.24% accuracy gains over the best previous GDFQ+AIT on CIFAR-100. Similarly, Our DMKD also achieves the lowest accuracy gap between the 5-bit quantized models and the FP models, compared to other methods. For CIFAR-10, our gap is only 0.43%, which is better than GDFQ+AIT. 2) *ImageNet.* Our DMKD also achieves the highest accuracy for both 4-bit and 5-bit quantization on ResNet-18 and MobileNetV2. For example, our method outperforms IntraQ and GDFQ+AIT by 0.14% and 0.24% for 4-bit quantization and 5-bit quantization on ResNet-18, respectively. For a more compact MobileNetV2, our DMKD achieves the highest accuracies of 65.63% and 71.81%, compared to other methods. For ResNet-50, Qimera achieves the highest accuracy, which is due to the generation of synthetic boundary-supporting samples. It is orthogonal to our method, which can be leveraged into DMKD to achieve higher performance.

Visualization. We further visualize the synthetic images, as shown in Fig. 4. We found that our synthetic images have strong edge and pattern information, especially on 5-bit and 8-bit quantization. Although there is a gap between the original images and synthetic ones, the synthetic images are able to well support the training of quantized models.

4.3   Ablation Studies

To evaluate the effectiveness of DMKD, which lies in the usage of an 8-bit quantized model and dynamic loss weight factors, we select ResNet-20 on CIFAR-100 for an ablation study.

Effect of 8-Bit Model and Dynamic Factors. As shown in Fig. 5(a), our baseline is only using the pre-trained FP model without the 8-bit model and factors, which achieves Top-1 accuracy of 65.26% and 68.55% for 4-bit and 5-bit quantization, respectively. After adding the 8-bit quantized model for knowledge distillation and fixed factors ($A^1 = 0.8, A^2 = 0.2$ in Eq. 8), DMKD W. 8-bit + FF achieves 0.12% and 0.42% performance gains over baseline for 4-bit and 5-bit quantization, respectively. Our method employs multi-teacher knowledge distillation with dynamic loss weight allocation to significantly improve the baseline with 0.78% and 0.79% (*i.e.*, Top-1 accuracies of 66.04% and 69.34%) for 4-bit and 5-bit quantization. It indicates the effectiveness of the knowledge from the 8-bit quantized model and the dynamic loss-weight factors.

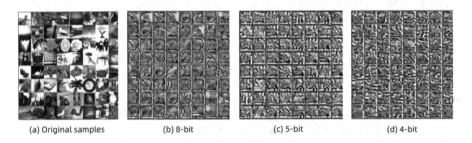

(a) Original samples   (b) 8-bit   (c) 5-bit   (d) 4-bit

Fig. 4. Visualization of the synthetic images from different-bit quantization.

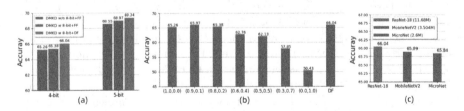

Fig. 5. Ablation study of DMKD. (a) Effect of 8-bit model and dynamic factors. W/O, W, FF and DF denote without, with, fixed factors and dynamic factors, respectively; (b) Fixed *vs.* dynamic factors. (c) Effect of tiny networks F^t.

Comparison of Fixed and Dynamic Factors. We further investigate the effect of loss-weight factors, as shown in Fig. 5(b). Although the 8-bit quantized model has rich information for low-bit data-free distillation, removing the FP model (*i.e.*, $A^1 = 0, A^2 = 1$) will significantly decrease the accuracy. With the fixed factors, the group of $(0.9, 0.1)$ achieves the highest accuracy of 65.97%, compared to other fixed combinations. It indicates that our method is more effective to adaptively allocate the different knowledge from the FP and 8-bit quantized model according to the synthetic images during training.

Effect of Tiny Network Structure. We also explore the effect of different tiny networks F^t, as shown in Fig. 5(c). We select ResNet-18, MobileNetV2, and MicroNet [16]. We report our results on the CIFAR-100 dataset with ResNet-20. We found that the impact of tiny FP network structure on performance is directly proportional to the amount of network parameters.

5 Conclusion

In this paper, we propose a novel dynamic multi-teacher knowledge distillation (DMKD) method to improve the performance for data-free low-bit quantization. Specifically, we analyze the 8-bit quantized model and find it has rich information different from the FP model. We design the learnable loss-weight factors by a tiny network to adaptively transfer their knowledge from the 8-bit quantized model and the FP model to a lower-bit quantized model according to the synthetic images. Extensive experiments demonstrate the effectiveness of DMKD for data-free low-bit quantization.

Acknowledgements. This work is supported by the National Natural Science Foundation of China (No. 62102151), Shanghai Sailing Program (21YF1411200), CCF-Tencent Open Research Fund, the Open Research Fund of Key Laboratory of Advanced Theory and Application in Statistics and Data Science, Ministry of Education (KLATASDS2305), the Fundamental Research Funds for the Central Universities.

References

1. Ba, J., Caruana, R.: Do deep nets really need to be deep? In: NeurIPS (2014)
2. Banner, R., Nahshan, Y., Soudry, D.: Post training 4-bit quantization of convolutional networks for rapid-deployment. In: NeurIPS, vol. 32 (2019)
3. Cai, Y., Yao, Z., Dong, Z., Gholami, A., Mahoney, M.W., Keutzer, K.: ZeroQ: a novel zero shot quantization framework. In: CVPR, pp. 13169–13178 (2020)
4. Chen, H., et al.: Data-free learning of student networks. In: ICCV (2019)
5. Choi, J., Wang, Z., Venkataramani, S., Chuang, P.I.J., Srinivasan, V., Gopalakrishnan, K.: Pact: parameterized clipping activation for quantized neural networks. arXiv preprint arXiv:1805.06085 (2018)
6. Choi, K., Hong, D., Park, N., Kim, Y., Lee, J.: Qimera: data-free quantization with synthetic boundary supporting samples. In: NeurIPS, vol. 34, pp. 14835–14847 (2021)

7. Choi, K., et al.: It's all in the teacher: zero-shot quantization brought closer to the teacher. In: CVPR, pp. 8311–8321 (2022)
8. Choi, Y., Choi, J., El-Khamy, M., Lee, J.: Data-free network quantization with adversarial knowledge distillation. In: CVPR Workshops (2020)
9. Courbariaux, M., Hubara, I., Soudry, D., El-Yaniv, R., Bengio, Y.: Binarized neural networks: training deep neural networks with weights and activations constrained to +1 or -1. arXiv preprint arXiv:1602.02830 (2016)
10. Gong, L., et al.: Adaptive hierarchy-branch fusion for online knowledge distillation. In: AAAI (2023)
11. He, K., Zhang, X., Ren, S., Sun, J.: Deep residual learning for image recognition. In: CVPR (2016)
12. Hinton, G., Vinyals, O., Dean, J.: Distilling the knowledge in a neural network. arXiv preprint arXiv:1503.02531 (2015)
13. Jacob, B., et al.: Quantization and training of neural networks for efficient integer-arithmetic-only inference. In: CVPR, pp. 2704–2713 (2018)
14. Kingma, D.P., Ba, J.: Adam: a method for stochastic optimization. arXiv preprint arXiv:1412.6980 (2014)
15. Krizhevsky, A., Hinton, G., et al.: Learning multiple layers of features from tiny images (2009)
16. Li, Y., et al.: Micronet: improving image recognition with extremely low flops. In: ICCV, pp. 468–477 (2021)
17. Liu, Y., Zhang, W., Wang, J.: Zero-shot adversarial quantization. In: CVPR (2021)
18. Lopes, R.G., Fenu, S., Starner, T.: Data-free knowledge distillation for deep neural networks. arXiv preprint arXiv:1710.07535 (2017)
19. Nagel, M., Amjad, R.A., Van Baalen, M., Louizos, C., Blankevoort, T.: Up or down? Adaptive rounding for post-training quantization. In: ICML, pp. 7197–7206 (2020)
20. Nagel, M., Baalen, M.V., Blankevoort, T., Welling, M.: Data-free quantization through weight equalization and bias correction. In: ICCV, pp. 1325–1334 (2019)
21. Nayak, G.K., Mopuri, K.R., Shaj, V., Radhakrishnan, V.B., Chakraborty, A.: Zero-shot knowledge distillation in deep networks. In: ICML, pp. 4743–4751 (2019)
22. Paszke, A., Gross, S., Chintala, S., Chanan, G.: Pytorch: tensors and dynamic neural networks in python with strong GPU acceleration. PyTorch **6**(3), 67 (2017)
23. Rastegari, M., Ordonez, V., Redmon, J., Farhadi, A.: XNOR-Net: ImageNet classification using binary convolutional neural networks. In: Leibe, B., Matas, J., Sebe, N., Welling, M. (eds.) ECCV 2016. LNCS, vol. 9908, pp. 525–542. Springer, Cham (2016). https://doi.org/10.1007/978-3-319-46493-0_32
24. Ren, S., He, K., Girshick, R., Sun, J.: Faster R-CNN: towards real-time object detection with region proposal networks. In: NeurIPS, pp. 91–99 (2015)
25. Romero, A., Ballas, N., Kahou, S.E., Chassang, A., Gatta, C., Bengio, Y.: Fitnets: hints for thin deep nets. In: ICLR (2015)
26. Russakovsky, O., et al.: Imagenet large scale visual recognition challenge. IJCV **115**, 211–252 (2015)
27. Sandler, M., Howard, A., Zhu, M., Zhmoginov, A., Chen, L.C.: Mobilenetv2: inverted residuals and linear bottlenecks. In: CVPR, pp. 4510–4520 (2018)
28. Tailor, S.A., Fernandez-Marques, J., Lane, N.D.: Degree-quant: quantization-aware training for graph neural networks. arXiv preprint arXiv:2008.05000 (2020)
29. Wang, P., Chen, Q., He, X., Cheng, J.: Towards accurate post-training network quantization via bit-split and stitching. In: ICML, pp. 9847–9856 (2020)

30. Wei, X., Gong, R., Li, Y., Liu, X., Yu, F.: QDrop: randomly dropping quantization for extremely low-bit post-training quantization. arXiv preprint arXiv:2203.05740 (2022)
31. Xiang, L., Ding, G., Han, J.: Learning from multiple experts: self-paced knowledge distillation for long-tailed classification. In: Vedaldi, A., Bischof, H., Brox, T., Frahm, J.-M. (eds.) ECCV 2020. LNCS, vol. 12350, pp. 247–263. Springer, Cham (2020). https://doi.org/10.1007/978-3-030-58558-7_15
32. Xu, S., et al.: Generative low-bitwidth data free quantization. In: Vedaldi, A., Bischof, H., Brox, T., Frahm, J.-M. (eds.) ECCV 2020. LNCS, vol. 12357, pp. 1–17. Springer, Cham (2020). https://doi.org/10.1007/978-3-030-58610-2_1
33. Yin, H., et al.: Dreaming to distill: data-free knowledge transfer via deepinversion. In: CVPR, pp. 8715–8724 (2020)
34. You, S., Xu, C., Xu, C., Tao, D.: Learning from multiple teacher networks. In: KDD (2017)
35. Yu, S., Chen, J., Han, H., Jiang, S.: Data-free knowledge distillation via feature exchange and activation region constraint. In: CVPR, pp. 24266–24275 (2023)
36. Zagoruyko, S., Komodakis, N.: Paying more attention to attention: improving the performance of convolutional neural networks via attention transfer. In: ICLR (2017)
37. Zhang, X., et al.: Diversifying sample generation for accurate data-free quantization. In: CVPR, pp. 15658–15667 (2021)
38. Zhong, Y., et al.: IntraQ: learning synthetic images with intra-class heterogeneity for zero-shot network quantization. In: CVPR, pp. 12339–12348 (2022)
39. Zhou, P., Mai, L., Zhang, J., Xu, N., Wu, Z., Davis, L.S.: M2KD: multi-model and multi-level knowledge distillation for incremental learning. arXiv preprint arXiv:1904.01769 (2019)
40. Zhou, S., Wu, Y., Ni, Z., Zhou, X., Wen, H., Zou, Y.: DoReFa-Net: training low bitwidth convolutional neural networks with low bitwidth gradients. arXiv preprint arXiv:1606.06160 (2016)
41. Zhu, B., Hofstee, P., Peltenburg, J., Lee, J., Alars, Z.: Autorecon: neural architecture search-based reconstruction for data-free compression. arXiv preprint arXiv:2105.12151 (2021)

LeViT-UNet: Make Faster Encoders with Transformer for Medical Image Segmentation

Guoping Xu[1], Xuan Zhang[1], Xinwei He[2], and Xinglong Wu[1](✉)

[1] School of Computer Sciences and Engineering, Hubei Key Laboratory of Intelligent Robot, Wuhan Institute of Technology, Wuhan 430205, Hubei, China
xwu@wit.edu.cn
[2] College of Informatics, Huazhong Agricultural University, Wuhan 430070, Hubei, China

Abstract. Medical image segmentation plays an essential role in developing computer-assisted diagnosis and treatment systems, yet it still faces numerous challenges. In the past few years, Convolutional Neural Networks (CNNs) have been successfully applied to the task of medical image segmentation. Regrettably, due to the locality of convolution operations, these CNN-based architectures have their limitations in learning global context information in images, which might be crucial to the success of medical image segmentation. Meanwhile, the vision Transformer (ViT) architectures own the remarkable ability to extract long-range semantic features with the shortcoming of their computation complexity. To make medical image segmentation more efficient and accurate, we present a novel light-weight architecture named LeViT-UNet, which integrates multi-stage Transformer blocks in the encoder via LeViT, aiming to explore the effectiveness of fusion between local and global features together. Our experiments on two challenging segmentation benchmarks indicate that the proposed LeViT-UNet achieved competitive performance compared with various state-of-the-art methods in terms of efficiency and accuracy, suggesting that LeViT can be a faster feature encoder for medical images segmentation. LeViT-UNet-384, for instance, achieves Dice similarity coefficient (DSC) of 78.53% and 90.32% with a segmentation speed of 85 frames per second (FPS) in the Synapse and ACDC datasets, respectively. Therefore, the proposed architecture could be beneficial for prospective clinic trials conducted by the radiologists. Our source codes are publicly available at https://github.com/apple1986/LeViT_UNet.

Keywords: Medical Image Segmentation · Transformer · Convolutional Neural Network

1 Introduction

Automated medical image segmentation has been widely studied in the research community since it would significantly reduce the tedious workload of radiolo-

Q. Liu et al. (Eds.): PRCV 2023, LNCS 14432, pp. 42–53, 2024.
https://doi.org/10.1007/978-981-99-8543-2_4

gists. In the past few years, Convolutional Neural Networks (CNNs) have made substantial progress in medical image segmentation. Fully convolutional networks (FCNs) [1] and their variants, e.g., U-Net [2], DeepLab [3] etc., are extensively used in cardiac segmentation from MRI [4], liver and tumor segmentation from CT [5] and etc.

Although powerful representation learning capabilities have made CNN-based approaches the de facto selection for the task of image segmentation, these approaches still have their own limitations. For instance, the insufficient capability to capture a larger context owing to the intrinsic locality of convolution operations. Though various methods, e.g., dilated convolution [3], image pyramids [6], prior-guided [4], multi-scale fusion [7], and attention mechanisms [8], are introduced to address these limitations, they still have deficiencies in extracting global context features in the task of medical image segmentation, especially for those objects that have large inter-patient variations in terms of shape, scale, and texture.

Transformer was initially proposed for sequence-to-sequence modeling in natural language processing (NLP) tasks, such as machine translation, sentiment analysis and information extraction. Recently, Transformed-based architectures (known as ViT [9]) have been applied to vision-related tasks and achieved state-of-the-art (SOTA) results for the task of image classification via pre-training on the large-scale dataset [10] [11]. They have also been studied for semantic segmentation, such as Swin Transformer [12], Swin-UNet [13], TransUNet [14]. However, the main limitation of these Transformer-based methods lies in the high requirement of computation power, which impedes them to be utilized in real-time applications, e.g., radiotherapy.

LeViT [11] is initially proposed for fast inference image classification with hybrid Transformer and convolution blocks, which optimize the trade-off between accuracy and efficiency. However, this architecture has not fully leveraged various scales of feature maps from Transformer and convolution blocks, which might be crucial for image segmentation. Inspired by LeViT, we propose LeViT-UNet for 2D medical image segmentation in this paper, aiming to make a faster encoder with Transformer and improve the segmentation performance by integrating the long-range spatial relations from Transformers into the features extracted from convolution layers. To the best of our knowledge, LeViT-UNet is one of the first few networks that focus on both the efficiency and accuracy of Transformer-based architecture for medical image segmentation.

The proposed LeViT-UNet mainly consists of an encoder, a decoder and several skip connections. Here, the encoder is built based on LeViT Transformer blocks, and the decoder based on convolution blocks. Motivated by the U-shape architecture design, we extracted multi-scale feature maps from Transformer blocks of LeViT and passed them into decode blocks via skip connections. We expect that such a design could integrate the merits of the Transformer for global features extraction and of the CNNs for local feature representation. Our experiments demonstrate that LeViT-UNet could exploit the merits of both Transformer and CNNs, and improve both accuracy and efficiency for the task of medical image segmentation. The main contributions of our work can be summarized as follows:

1. We propose a novel light-weight, fast and high accurate hybrid convolution and Transformer segmentation architecture, named LeViT-UNet, which use multi-stage Transformer blocks to extract global context features and convolution blocks to learn local high-resolution spatial information;
2. We explore the effect of skip connection and Transformer blocks in the encoder and decoder LeViT-UNet architecture, and find that it is helpful by integrating more low-level features from skip connection and provide global context information from Transformer for segmentation task;
3. Comprehensive experiments are conducted on two public datasets , and the results demonstrate that the proposed LeViT-UNet is competitive with other SOTA methods in terms of accuracy and efficiency. Our work will provide a benchmark comparison for the fast segmentation with Transformer in the field of medical image analysis.

2   Related Works

CNN-Based Methods: CNNs have been extensively studied in biomedical image segmentation. The typical U-shaped network, U-Net [2], which consists of a symmetric encoder and decoder network with skip connections, has become the de-facto choice for biomedical image analysis. Afterwards, various U-Net like architectures is proposed, such as Res-UNet [15], UNet++ [16], V-Net [17] and nnU-Net [18]. While CNN-based methods have achieved much progress in biomedical image segmentation, they still cannot fully meet the clinical application requirements for segmentation accuracy and efficiency owing to its intrinsic locality of convolution operations and its complex data access patterns.

Self-attention Mechanisms to Complement CNNs: Several works have attempted to integrate the self-attention mechanism into CNNs for segmentation. The main purpose is to catch the attention weight in terms of channel-wise or spatial shape. For instance, the squeeze-and-excite network built an attention-like module to extract the relationship between each feature map of CNN layers [19]. The dual attention network appended two types of attention modules to model the semantic interdependencies in spatial and channel dimensions, respectively [20]. The Attention U-Net proposed an attention gate to suppress irrelevant regions of a feature map while highlighting salient features for the segmentation task [8]. Although these strategies could improve the performance of segmentation, the ability to extract long-range semantic information still needs to be addressed.

Transformers: Recently, Vision Transformer (ViT) has achieved SOTA on ImageNet classification by using the transformer with pure self-attention to input images [9]. Afterwards, different ViT variants have been proposed, such as Swin [12], and LeViT [11] on natural images. Transformer structure have also been utilized to medical image segmentation. For example, Swin-UNet [13] employed pure transformer into the U-shaped encoder-decoder architecture for global semantic feature learning. In this paper, we attempt to exploit LeViT transformer block as a basic unit in the encoder of a U-shaped architecture, which aims to keep trade-off between accuracy and efficiency for biomedical image segmentation. Our work will provide a benchmark comparison for the fast segmentation with Transformer in the field of biomedical image analysis.

3   Method

Unlike the conventional U-Net which employs convolutional operations to encode and decode features, we apply LeViT module in the encoder part to build global context information and integrate them into the proposed LeViT-UNet architecture. In the following sections, the overall LeViT-UNet architecture is introduced in Sect. 3.1. Then, the components of the encoder and decoder in the LeViT-UNet will be detailed in Sect. 3.2 and 3.3, respectively.

3.1   Architecture of LeViT-UNet

The architecture of LeViT-UNet (Fig. 1) is composed of an encoder, a decoder and several skip-connections. Here, we apply a LeViT module in the encoder part to extract long-range structural information from the feature maps. The LeViT (Fig. 1) is a hybrid neural network with convolutional blocks and vision transformers.

3.2   LeViT as Encoder

Following [11], we apply LeViT architecture as the encoder, which consists of two main components: convolutional blocks and transformer blocks. Specifically, there are 4 layers of convolutions with stride 2 in the convolutional blocks, which could do the resolution reduction. These feature maps will be fed into the transformer block, in which long-range relations between each pixel are built. There are three types of LeViT encoders, which are named LeViT-128 s, LeViT-192 and LeViT-384, based on the number of channels fed into the first Transformer block. The block diagram of the LeViT-192 architecture is shown in Fig. 2. Note that we concatenate the features from convolution layers and transformer blocks in the last stage of the encoder, which could fully leverage the local and global features in various scales.

The transformer block can be formulated as:

$$\hat{z}^n = MLP(BN(z^{n-1})) + z^{n-1}, \tag{1}$$

$$z^n = MSA(BN(\hat{z}^n)) + \hat{z}^n, \tag{2}$$

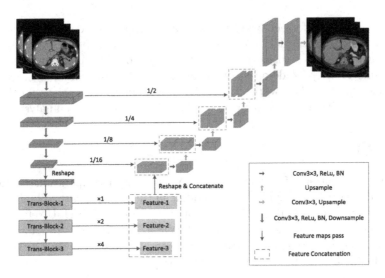

Fig. 1. The architecture of LeViT-UNet, which is composed of an encoder (purple boxes), a decoder (blue and green boxes) and several skip connections. Here, the encoder is constructed based on LeViT module. (Color figure online)

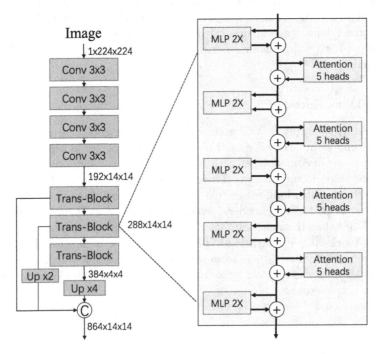

Fig. 2. Block diagram of LeViT-192 architecture, in which convolution and Transformer are integrated.

where $\hat{z}^n$ and z^n represent the outputs of MLP (Multiple Layer Perceptron) module and the MSA (Multi-head Attention) module of the nth block, respectively. BN means the batch normalization. Similar to the previous work [11], self-attention is computed as follows:

$$Attention(Q, K, V) = Softmax(\frac{QK^T}{\sqrt{d}} + b)V, \tag{3}$$

where Q, K, V are the query, key and value matrices, whose sizes are $M^2 \times d$. M^2 and d denote the number of patches and the dimension of the query or key. B represents attention bias, which takes place of positional embedding and could provide positional information within each attention block.

3.3   CNNs as Decoder

Similar to U-Net, we concatenate the features from the decoder with skip connection. The cascaded upsampling strategy is used to recover the resolution from the previous layer using CNNs. For example, there are feature maps with the shape of H/16 × W/16 × D from the encoder. Then, we use cascaded multiple upsampling blocks for reach the full resolution of H × W, where each block consists of two 3×3 convolution layers, batch normalization layer, ReLU layer, and an upsampling layer.

4   Experiments and Results

4.1   Dataset

Synapse Multi-organ Segmentation Dataset (Synapse): This dataset[1] includes 30 abdominal CT scans with 3779 axial contrast-enhanced abdominal clinical CT images in total. Following the same splitting in [14], we use 18 cases for training and the other 12 cases for testing. We evaluated the performance with the average Dice Similarity Coefficient (DSC) and the average Hausdorff Distance (HD) on 8 abdominal organs, which are Aorta, Gallbladder (Gall), Left kidney (KidL), Right kidney (KidR), Liver, Pancreas (Panc), Spleen, and Stomach (Stom) respectively.

Automated Cardiac Diagnosis Challenge Dataset (ACDC): This dataset[2] is collected from 150 patients using cine-MR scanners, splitting into 100 volumes with human annotations and the other 50 volumes which are private for the evaluation purpose. Here, we split the 100 annotated volumes into 80 training samples and 20 testing samples.

---

[1] https://www.synapse.org/#!Synapse:syn3193805/wiki/217789.
[2] https://www.creatis.insa-lyon.fr/Challenge/acdc/.

Table 1. Segmentation accuracy (average DSC% and average HD in mm, and DSC for each organ) of different methods on the Synapse multi-organ CT dataset.

Methods	DSC↑	HD↓	Aorta	Gallbladder	Kidney(L)	Kidney(R)	Liver	Pancreas	Spleen	Stomach
V-Net [16]	68.81	–	75.34	51.87	77.10	**80.75**	87.84	40.05	80.56	56.98
DARR [19]	69.77	–	74.74	53.77	72.31	73.24	94.08	54.18	89.90	45.96
U-Net [2]	76.85	39.70	89.07	**69.72**	77.77	68.60	93.43	53.98	86.67	75.58
R50 U-Net [14]	74.68	36.87	87.74	63.66	80.60	78.19	93.74	56.90	85.87	74.16
R50 Att-UNet [14]	75.57	36.97	55.92	63.91	79.20	72.71	93.56	49.37	87.19	74.95
Att-UNet [8]	77.77	36.02	**89.55**	68.88	77.98	71.11	93.57	58.04	87.30	75.75
R50-Deeplabv3+ [3]	75.73	26.93	86.18	60.42	81.18	75.27	92.86	51.06	88.69	70.19
TransUNet [14]	77.48	31.69	87.23	63.13	81.87	77.02	94.08	55.86	85.08	75.62
Swin-UNet [12]	**79.13**	21.55	85.47	66.53	83.28	79.61	**94.29**	56.58	**90.66**	**76.60**
LeViT-UNet-128 s	73.69	23.92	86.45	66.13	79.32	73.56	91.85	49.25	79.29	63.70
LeViT-UNet-192	74.67	18.86	85.69	57.37	79.08	75.90	92.05	53.53	83.11	70.61
LeViT-UNet-384	78.53	**16.84**	87.33	62.23	**84.61**	80.25	93.11	**59.07**	88.86	72.76

4.2 Implementation Details

All experiments are implemented based on a single Nvidia 3090, Python 3.8, PyTorch 1.8.0 in a workstation with Ubuntu 18.04 LTS. All training samples are resized to 224×224. The Adam optimizer is applied with a learning rate of 1e-5 and weight decay of 1e-4. The cross entropy and Generalized Dice loss as objective function during the training. We set batch size of 8 for each iteration. The training epochs are 350 and 400 for Synapse and ACDC datasets, respectively. Transformer backbones in the LeViT are pretrained on ImageNet-1k to initialize the network parameters. Following [14], all 3D volume datasets are trained by slice and the predicted 2D slices are stacked together to build 3D prediction for evaluation.

4.3 Experiment Results on Synapse Dataset

We perform experiments with other SOTA methods in terms of accuracy and efficiency as the benchmark for comparison with LeViT-UNet. Three variants of LeViT-UNet are designed. We identify them by the number of channels input to the first transformer block: LeViT-UNet-128 s, LeViT-UNet-192, and LeViT-UNet-384, respectively.

(1) Compare with Other SOTA Methods

The comparison of the proposed LeViT-UNet with other SOTA methods on the Synapse dataset (Table 1) demonstrates that LeViT-UNet-384 achieves the best performance in terms of average Hausdorff Distance (HD) with 16.84 mm, which is improved by about 14.85 mm and 4.71 mm comparing to Transformer-based methods TransUNet and Swin-UNet. In addition, our approach could achieve the competitive result in terms of DSC, such as left and right kidney, pancreas and spleen.

The segmentation results of different methods on the Synapse dataset are further visualized in Fig. 3, which shows that the other three methods (TransUNet, U-Net and DeepLabv3+) are more likely to under-segment or over-segment the organs. For example, the stomach is under-segmented by TransUNet and DeepLabv3+ (as indicated by the red arrow in the third panel of the upper row), and over-segmented by U-Net (as indicated by the red arrow in the fourth panel of the second row).

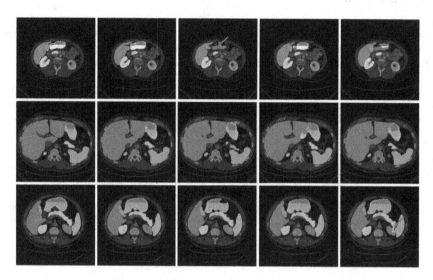

Fig. 3. Qualitative comparison of various methods by visualization From Left to right: Ground Truth, LeViT-UNet-384, TransUNet, U-Net, and DeepLabv3+.

(2) Comparison with Efficient Segmentation Methods

Firstly, it can be seen that LeViT-UNet-384 achieves 78.53% mean DSC and 16.84 mm mean HD, which is the best among all methods in Table 2. Compared to the number of parameters, our method has much fewer parameters than TransUNet, however, it still has space to other fast segmentation methods, like CGNet, ContextNet and ENet. Meanwhile, we evaluate the runtime at different methods. Here, ENet (114 fps) and FPENet (160 fps) are slightly faster than LeViT-UNet-128 s (114 fps), yet the HD is still needed to improve. Therefore, we conclude that LeViT-UNet is competitive with the current pure CNN efficient segmentation method with better performance.

(3) Ablation Study

We conduct a variety of ablation studies to thoroughly evaluate the proposed LeViT-UNet architecture and validate the performance under different settings, including: 1) without and with transformer blocks; 2) the number of skip-connections.

Effect of the Number of Transformer Blocks: Here, we compare the performance when Transformer blocks are utilized or not. We find that adding transformer blocks leads to a better segmentation performance in terms of DSC and HD in the Table 3. Moreover, the channel number of feature maps that input to the transformer block (here we set 128, 192 and 384, respectively) could improve the HD performance significantly. Moreover, the DSC is boosted to 1.25%, 0.2%, and 3.94% with transformer blocks comparing to only use convolution operations.

Table 2. Mean DSC and HD of the proposed LeViT-UNet compared to other efficient semantic segmentation methods on the Synapse dataset in terms of parameters (Para), floating-point operations (Flops) and inference speed by FPS (frame per second).

Methods	DSC↑	HD↓	Aorta	Gall	KidL	KidR	Liver	Panc	Spleen	Stom	Para(M)	Flops(G)	FPS
CGNet [21]	75.08	24.99	83.48	65.32	77.91	72.04	91.92	57.37	85.47	67.15	0.49	0.66	124
ContextNet [22]	71.17	36.41	79.92	51.17	77.58	72.04	91.74	43.78	86.65	66.51	0.87	0.16	280
DABNet [23]	74.91	26.39	85.01	56.89	77.84	72.45	93.05	54.39	88.23	71.45	0.75	0.99	221
EDANet [24]	75.43	29.31	84.35	62.31	76.16	71.65	93.20	53.19	85.47	77.12	0.69	0.85	213
ENet [25]	77.63	31.83	85.13	64.91	81.10	77.26	93.37	57.83	87.03	74.41	0.36	0.50	141
FPENet [26]	68.67	42.39	78.98	56.35	74.54	64.36	90.86	40.60	78.30	65.35	0.11	0.14	160
FSSNet [27]	74.59	35.16	82.87	64.06	78.03	69.63	92.52	53.10	85.65	70.86	0.17	0.33	213
SQNet [28]	73.76	40.29	83.55	61.17	76.87	69.40	91.53	56.55	85.82	65.24	16.25	18.47	241
FastSCNN [29]	70.53	32.79	77.79	55.96	73.61	67.38	91.68	44.54	84.51	68.76	1.14	0.16	**292**
TransUNet [14]	77.48	31.69	87.23	63.13	81.87	77.02	94.08	55.86	85.08	75.62	105.28	24.64	50
LeViT-UNet-128 s	73.69	23.92	86.45	66.13	79.32	73.56	91.85	49.25	79.29	63.70	15.91	17.55	114
LeViT-UNet-192	74.67	18.86	85.69	57.37	79.08	75.90	92.05	53.53	83.11	70.61	19.90	18.92	95
LeViT-UNet-384	**78.53**	**16.84**	87.33	62.23	84.61	80.25	93.11	59.07	88.86	72.76	52.17	25.55	85

Table 3. Ablation study w/o Transformer blocks. Note that '-Conv' means that no transformer blocks are used, and '-128', '-192', '-384' denote the channel number of feature maps, which input to the transformer block.

Methods	DSC↑	HD↓	Aorta	Gall	KidL	KidR	Liver	Panc	Spleen	Stom	Para(M)	Flops(G)
LeViT-UNet-128 s-Conv	72.44	41.63	84.29	59.11	77.70	69.20	91.93	44.18	87.60	65.52	5.46	14.54
LeViT-UNet-192-Conv	74.42	35.41	85.34	62.90	81.39	72.80	91.76	44.95	88.84	67.36	5.97	15.30
LeViT-UNet-384-Conv	74.59	30.19	85.49	62.52	83.00	73.87	91.91	43.47	88.75	67.69	11.94	17.70
LeViT-UNet-128 s	73.69	23.92	86.45	**66.13**	79.32	73.56	91.85	49.25	79.29	63.70	15.91	17.55
LeViT-UNet-192	74.67	18.86	85.69	57.37	79.08	75.90	92.05	53.53	83.11	70.61	19.90	18.92
LeViT-UNet-384	**78.53**	**16.84**	**87.33**	62.23	**84.61**	**80.25**	**93.11**	**59.07**	**88.86**	**72.76**	52.17	25.55

Effect of the Number of Skip Connections: We investigate the influence of skip-connections on LeViT-UNet. The results can be seen in Table 4. Note that "1-skip" setting means that we only apply one time of skip-connection at the 1/2 resolution scale, and "2-skip", "3-skip" and "4-skip" are inserting skip-connections at 1/4, 1/8 and 1/16, respectively. We can find that adding more skip-connections could result in better performance. Moreover, the performance gain of smaller organs, like the aorta, gallbladder, kidneys, is more obvious than that of larger organs, like the liver, spleen and stomach.

4.4   Experiment Results on ACDC Dataset

To demonstrate the generalization ability of LeViT-UNet, we train our model on ACDC MR dataset for automated cardiac segmentation. We can observe that our proposed LeViT-UNet could achieve better results in terms of DSC in the Table 5. Compared with Swin-UNet [13] and TransUNet [14], LeViT-UNet achieves comparable DSC, for instance, the LeViT-UNet-192 and LeViT-UNet-384 achieve 90.08% and 90.32% DSC, respectively.

Table 4. Ablation study on the number of skip-connection in LeViT-UNet. ('_N' means the number of skip connections)

Number of skip-connection	DSC↑	HD↓	Aorta	Gall	KidL	KidR	Liver	Panc	Spleen	Stom
LeViT-UNet-384_N0	67.19	27.89	73.70	47.08	69.85	65.03	89.92	45.53	82.22	64.18
LeViT-UNet-384_N1	68.72	27.97	73.59	48.73	75.05	67.96	91.15	45.03	84.13	64.09
LeViT-UNet-384_N2	74.54	25.85	84.98	59.27	75.43	69.16	92.53	57.20	87.18	70.58
LeViT-UNet-384_N3	76.91	20.87	86.89	61.01	81.57	76.18	92.86	56.00	87.62	73.19
LeViT-UNet-384_N4	78.53	16.84	87.33	62.23	84.61	80.25	93.11	59.07	88.86	72.76

Table 5. Segmentation performance of different methods on the ACDC dataset.

Methods	DSC↑	RV	Myo	LV
R50 U–Net	87.55	87.10	80.63	94.92
R50 Att-UNet	86.75	87.58	79.20	93.47
R50 ViT	87.57	86.07	81.88	94.75
TransUNet	89.71	88.86	84.53	95.73
Swin-UNet	90.00	88.55	85.62	95.83
LeViT-UNet-128 s	89.39	88.16	**86.97**	93.05
LeViT-UNet-192	90.08	88.86	87.50	**93.87**
LeViT-UNet-384	**90.32**	**89.55**	87.64	93.76

5   Conclusion

In this work, we introduced a Transformer and convolution-based framework, named LeViT-UNet, for fast medical image segmentation. The performance of LeViT-UNet was assessed on two different modality datasets. The results demonstrated that the proposed method has achieved competitive SOTA performance with better trade-off between accuracy and efficiency in medical image segmentation. We would like to explore the applications of LeViT-UNet for 3D medical image segmentation in the future.

Acknowledgments. This work is supported by the Guangdong Provincial Key Laboratory of Human Digital Twin (No. 2022B1212010004), and the Hubei Key Laboratory of Intelligent Robot in Wuhan Institute of Technology (No. HBIRL202202 and No. HBIRL202206).

References

1. Shelhamer, E., Long, J., Darrell, T.: Fully convolutional networks for semantic segmentation. IEEE Trans. Pattern Anal. Mach. Intell. **39**(4), 640–651 (2017)
2. Ronneberger, O., Fischer, P., Brox, T.: U-Net: convolutional networks for biomedical image segmentation. In: Navab, N., Hornegger, J., Wells, W.M., Frangi, A.F. (eds.) MICCAI 2015. LNCS, vol. 9351, pp. 234–241. Springer, Cham (2015). https://doi.org/10.1007/978-3-319-24574-4_28
3. Chen, L.C., Papandreou, G., Kokkinos, I., Murphy, K., Yuille, A.L.: DeepLab: semantic image segmentation with deep convolutional nets, Atrous convolution, and fully connected CRFs. IEEE Trans. Pattern Anal. Mach. Intell. **40**(4), 834–848 (2018)
4. Cheng, F., et al.: Learning directional feature maps for cardiac MRI segmentation. In: Martel, A.L., et al. (eds.) MICCAI 2020. LNCS, vol. 12264, pp. 108–117. Springer, Cham (2020). https://doi.org/10.1007/978-3-030-59719-1_11
5. Jin, Q., Meng, Z., Sun, C., Cui, H., Ran, S.: RA-UNet: a hybrid deep attention-aware network to extract liver and tumor in CT scans. Front. Bioeng. Biotechnol. **8**, 605132 (2020)
6. Zhao, H., Shi, J., Qi, X., Wang, X., Jia, J.: Pyramid scene parsing network. In: 2017 IEEE Conference on Computer Vision and Pattern Recognition (CVPR). IEEE (2017)
7. Sun, K., Xiao, B., Liu, D., Wang, J.: Deep high-resolution representation learning for human pose estimation. In: 2019 IEEE/CVF Conference on Computer Vision and Pattern Recognition (CVPR). IEEE (2019)
8. Oktay, O., et al.: Attention U-Net: learning where to look for the pancreas (2018)
9. Dosovitskiy, A., et al.: An image is worth 16x16 words: transformers for image recognition at scale (2020)
10. Touvron, H., Cord, M., Douze, M., Massa, F., Sablayrolles, A., Jegou, H.: Training data-efficient image transformers & distillation through attention. In: Meila, M., Zhang, T. (eds.) Proceedings of the 38th International Conference on Machine Learning, volume 139 of Proceedings of Machine Learning Research, pp. 10347–10357. PMLR (2021)
11. Graham, B., et al.: LeViT: a vision transformer in ConvNet's clothing for faster inference. In: 2021 IEEE/CVF International Conference on Computer Vision (ICCV). IEEE (2021)
12. Liu, Z., et al.: Swin transformer: hierarchical vision transformer using shifted windows. In: 2021 IEEE/CVF International Conference on Computer Vision (ICCV). IEEE (2021)
13. Cao, H., et al.: Swin-Unet: Unet-like pure transformer for medical image segmentation (2021)
14. Chen, J., et al.: Transformers make strong encoders for medical image segmentation. TransUNet (2021)
15. Xiao, X., Lian, S., Luo, Z., Li, S.: Weighted res-UNet for high-quality retina vessel segmentation. In: 2018 9th International Conference on Information Technology in Medicine and Education (ITME). IEEE (2018)

16. Zongwei Zhou, Md., Siddiquee, M.R., Tajbakhsh, N., Liang, J.: UNet++: redesigning skip connections to exploit multiscale features in image segmentation. IEEE Trans. Med. Imaging **39**(6), 1856–1867 (2020)

17. Milletari, F., Navab, N., Ahmadi, S.A.: V-Net: fully convolutional neural networks for volumetric medical image segmentation. In: 2016 Fourth International Conference on 3D Vision (3DV). IEEE (2016)

18. Isensee, F., Jaeger, P.F., Kohl, S.A., Petersen, J., Maier-Hein, K.H.: nnU-Net: a self-configuring method for deep learning-based biomedical image segmentation. Nat. Methods **18**(2), 203–211 (2021)

19. Jie, H., Shen, L., Albanie, S., Sun, G., Enhua, W.: Squeeze-and-Excitation networks. IEEE Trans. Pattern Anal. Mach. Intell. **42**(8), 2011–2023 (2020)

20. Fu, et al.: Dual attention network for scene segmentation. In: 2019 IEEE/CVF Conference on Computer Vision and Pattern Recognition (CVPR). IEEE (2019)

21. Wu, T., Tang, S., Zhang, R., Cao, J., Zhang, Y.: CGNet: a light-weight context guided network for semantic segmentation. IEEE Trans. Image Process. **30**, 1169–1179 (2021)

22. Poudel, R.P., Bonde, U., Liwicki, S., Zach, C.: ContextNet: exploring context and detail for semantic segmentation in real-time. In: British Machine Vision Conference 2018, BMVC 2018, Newcastle, UK, September 3–6 2018, p. 146. BMVA Press (2018)

23. Li, G., Yun, I., Kim, J., Kim, J.: DabNet: depth-wise asymmetric bottleneck for real-time semantic segmentation. In: 30th British Machine Vision Conference 2019, BMVC 2019, Cardiff, UK, September 9–12 2019, pp. 259. BMVA Press (2019)

24. Lo, S.Y., Hang, H.M., Chan, S.W., Lin, J.J.: Efficient dense modules of asymmetric convolution for real-time semantic segmentation. In: Xu, C., Kankanhalli, M.S., Aizawa, K., Jiang, S., Zimmermann, R., Cheng, W.-H. (eds.) MMAsia '19, ACM Multimedia Asia, Beijing, China, December 16–18 2019, pp. 1– 6. ACM (2019)

25. Paszke, A., Chaurasia, A., Kim, S., Culurciello, E.: ENet: a deep neural network architecture for real-time semantic segmentation. CoRR, abs/1606.02147 (2016)

26. Liu, M., Yin, H.: Feature pyramid encoding network for real-time semantic segmentation. In: 30th British Machine Vision Conference 2019, BMVC 2019, Cardiff, UK, September 9–12 2019, pp. 260. BMVA Press (2019)

27. Zhang, X., Chen, Z., Wu, Q.J., Cai, L., Lu, D., Li, X.: Fast semantic segmentation for scene perception. IEEE Trans. Industr. Inf. **15**(2), 1183–1192 (2019)

28. Treml, M., Arjona-Medina, J., Unterthiner, T., Durgesh, R., Hochreiter, S.: Speeding up semantic segmentation for autonomous driving. In: NIPS 2016 Workshop - MLITS (2016)

29. Rudra P. K. Poudel, Stephan Liwicki, and Roberto Cipolla. Fast-SCNN: fast semantic segmentation network. In: 30th British Machine Vision Conference 2019, BMVC 2019, Cardiff, UK, September 9–12 2019, p. 289. BMVA Press (2019)

DUFormer: Solving Power Line Detection Task in Aerial Images Using Semantic Segmentation

Deyu An[1,2,3,4], Qiang Zhang[1(✉)], Jianshu Chao[2,3,4], Ting Li[1,2,3], Feng Qiao[5], Zhenpeng Bian[1], and Yong Deng[1]

[1] Autel Robotics, Shenzhen 518000, China
zq18487102396@gmail.com
[2] Fujian College, University of Chinese Academy of Sciences, Fuzhou 35000, China
[3] Quanzhou Institute of Equipment Manufacturing Fujian Institute of Research on the Structure of Matter, Chinese Academy of Sciences, Quanzhou 36200, China
[4] Fujian Agriculture and Forestry University, Fuzhou 35000, China
[5] RWTH Aachen University, 52062 Aachen, Germany

Abstract. Unmanned aerial vehicles (UAVs) are frequently used for inspecting power lines and capturing high-resolution aerial images. However, detecting power lines in aerial images is difficult, as the foreground data (i.e., power lines) is small and the background information is abundant. To tackle this problem, we introduce DUFormer, a semantic segmentation algorithm explicitly designed to detect power lines in aerial images. We presuppose that it is advantageous to train an efficient Transformer model with sufficient feature extraction using a convolutional neural network (CNN) with a strong inductive bias. With this goal in mind, we introduce a heavy token encoder that performs overlapping feature remodeling and tokenization. The encoder comprises a pyramid CNN feature extraction module and a power line feature enhancement module. After successful local feature extraction for power lines, feature fusion is conducted. Then, the Transformer block is used for global modeling. The final segmentation result is achieved by amalgamating local and global features in the decode head. Moreover, we demonstrate the importance of the joint multi-weight loss function in power line segmentation. Our experimental results show that our proposed method outperforms all state-of-the-art methods in power line segmentation on the publicly accessible TTPLA dataset.

Keywords: Semantic Segmentation · Power Line Detection · Aerial Image

1 Introduction

Unmanned aerial vehicles (UAVs) are more popular in applications such as geographic information systems (GIS), power line inspection, security surveillance, and agricultural and forestry protection, surpassing traditional tools and human labour. However, detecting power lines is challenging because of the slender characteristics of the

---

D. An and T. Li—Interns at Autel Robotics. Deyu An and Qiang Zhang contribute equally. This work was partially supported by Guiding Project of Fujian Science and Technology Program (No. 2022H0042).

Q. Liu et al. (Eds.): PRCV 2023, LNCS 14432, pp. 54–66, 2024.
https://doi.org/10.1007/978-981-99-8543-2_5

power line and complex backgrounds. To address this problem, DUFormer, a CNN-Transformer hybrid algorithm, is specifically designed to detect power lines in aerial images.

Recently, the Vision Transformer [8] and its variants have been used for challenging prediction tasks by performing global self-attention on high-resolution tokens. However, this approach is computationally and memory-wise inefficient as it results in quadratic complexity. Inspired by the work of Zhang et al. on TopFormer [23], we argue that using a convolutional neural network (CNN) with a strong inductive bias to perform ample feature extraction could facilitate the training of an efficient Transformer model. Therefore, we first introduce the concept of a heavy token encoder. In the previous Transformer algorithms, as shown in Fig. 1, we observe that the parameter ratios of the token encoder (i.e., patch embedding) to the Transformer blocks are small, even less than 0.05. In contrast, the proposed DUFormer improves this parameter ratio to 0.7 with a heavy token encoder. The explanation of the heavy token encoder is elaborated in the upcoming sections.

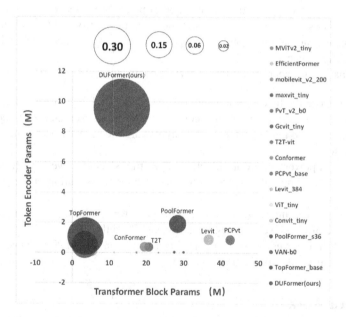

Fig. 1. Parameter ratios. The parameter ratio refers to the ratio of the token encoder's parameters to the Transformer blocks' parameters. The circle's size represents the parameter ratio.

We improve the efficiency of the Transformer model by leveraging the advantages of CNNs for feature extraction while maintaining the effectiveness of the Transformer's self-attention mechanism. First, the input feature maps are projected to four scales via a pyramidal CNN feature extraction module (DUB in Sect. 3.2). Subsequently, the feature maps go through a transition layer to enhance the receptive field. Altogether, five feature maps with different scales are generated separately. Then they are tokenized

by multi-scale average pooling and fed into the Transformer blocks for global attention calculation. Our method can produce tokens with relatively low resolutions and
enables the Vision Transformer to perform computations with acceptable throughput,
even when dealing with numerous feature map channels.

In the local feature extraction stage of the power line detection task, we propose a
power line feature enhance module, consisting of an asymmetric dilated convolution-
based Power Line Aware Block (PLAB in Sect. 3.3) and a BiscSE module (in Sect. 3.4).
These modules extract slender power line features at shallow layers and enhance semantic information at deeper layers. The network follows the U-Net [17] architecture, with
the output of the Transformer block serving as the upsampling source. This output is
upsampled four times in separate channels and then multiplied with our proposed power
line aware block element-wisely before being concatenated with the DUB output in
the corresponding decoding stage. Five stages generate five segmentation results, with
losses calculated separately. The final segmentation result is obtained by fusing the five
results.

In aerial image power line detection tasks, imbalanced data samples pose a significant problem, as foreground (i.e. power line) pixels are considerably smaller than
background pixels. To solve this, we introduce a joint multi-weight loss function. In
our experiments, the proposed method outperforms existing methods, achieving state-
of-the-art performance.

In summary, our paper makes the following contributions:

- We first propose the theory of a heavy token encoder and demonstrate that a Transformer model with sufficient token encoding is easier to train efficiently. Accordingly,
we further propose a CNN-Transformer hybrid architecture for aerial image power
line detection.
- We propose a power line aware block for detecting slender power line features and
an improved scSE module for enhancing the semantic information of the network at
deeper layers.
- We investigate the importance of the joint multi-weight loss, which improves the
performance of the power line segmentation significantly.
- Our approach achieves state-of-the-art performance on the TTPLA dataset by conducting a number of experiments.

2   Related Work

2.1   Vision Transformer

The original Vision Transformer [8] slices the image into multiple non-overlapping
patches, which is good at capturing long-distance dependencies between patches but
ignores local feature extraction. TNT [11] further divides patches into multiple sub-
patches and introduces a new structure, Transformer-iN-Transformer, which uses internal Transformer blocks to model the relationship between patches and external Transformer blocks to achieve patch-level information exchange. Twins [6] and CAT [14]
alternate local and global attention layer by layer. Swin Transformer [15] performs local

attention in the window and introduces a shift window partitioning method for cross-window connections. In addition, some works combine CNN with the Transformer. CPVT [7] proposes a conditional position encoding (CPE) method, which is conditional on the local neighborhood of the input tokens. It applies to arbitrary input sizes for fine feature encoding using convolution. CVT [19], CeiT [20], LocalViT [13], and CMT [9] analyze the potential pitfalls of directly applying Transformer architecture to images. The mitigation method is proposed in their papers, i.e., combining convolution with Transformer. Specifically, a feedforward network (FFN) in each converter block is combined with a convolutional layer to facilitate the association between adjacent tokens.

2.2 Semantic Segmentation

The FCN [16] proposed by Long et al. in 2015 pioneered semantic segmentation in deep learning. It replaces the fully-connected layer in traditional CNN models with a convolutional layer, gradually deconvoluting to restore the original image size and obtain the final semantic segmentation result. In the same year, Ronneberger et al. proposed U-Net [17], also based on the FCN. The U-Net structure resembles the letter U with a encoding and decoding structure. Initially used for medical images, it is suitable for small datasets. In 2017, SegNet [2], proposed by Badrinarayanan et al., also has an encoder-decoder structure but with the difference that max-pooling with returned coordinates is used. Then, the returned coordinates are used for feature recovery during upsampling. As a result, it significantly reduces the model's parameters. PSPNet [24], also proposed in 2017, introduced the pyramid pooling module (PPM). It concatenates four global pooling layers of different sizes to generate feature maps at different levels, aggregating multi-scale image features. The DeepLab series [3–5] use dilated convolution to propose Atrous Spatial Pyramid Pooling (ASPP) and incorporate conditional random fields (CRF) in the final structured prediction to improve model accuracy. Guo et al. proposed SegNeXt [10], which introduces a new multi-scale convolutional attention (MSCA) module using a larger kernel size to capture global features. They designed the parallelization of multiple kernels to increase information combination of dense contexts, which is essential for semantic segmentation.

3   Proposed Architecture

3.1   Overview

Our network for the high-resolution (1K) aerial image power line detection is presented in Fig. 2. The overall architecture design follows the U-Net structure. The Stem block is designed to handle high-resolution data without consuming too much GPU memory or significantly increasing FLOPs. It consists of a parallel max-pooling layer and average-pooling layer, followed by channel feature fusion using a 1×1 convolution. The heavy token encoder implementation is divided into two parts: the pyramidal Double U Blocks (DUB in Sect. 3.2) for obtaining feature maps with different resolutions and the Power Line Optimization Module, including PLAB (Sect. 3.3) and BiscSE (Sect. 3.4)

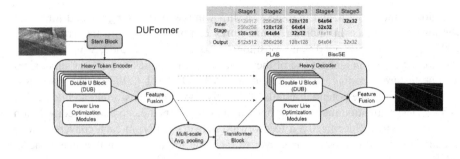

Fig. 2. Overall architecture of **DUFormer**. The table shows the feature map resolutions in different stages. The overlapping features in various stages represent our novel idea of feature re-mining.

for enhancing power line features. Then, the output results of the DUB and Power Line Optimization Modules are fused. The fused result is used as the output of the heavy token encoder, and the feature fusion method is shown in Eq. 1 and Eq. 2:

$$output_{stage_{1,2,3}} = out_{DUB_{1,2,3}} \odot out_{PLAB_{1,2,3}} \tag{1}$$

$$output_{stage_{4,5}} = out_{DUB_{4,5}} \odot out_{BiscSE_{4,5}} \tag{2}$$

The Transformer block can effectively capture information about the thin power lines in the entire image. Its input needs to be tokenized, and the operation of tokenization is shown in Eq. 3:

$$x = concat(AdaptiveAvgPool(output_{stage_1}),$$
$$..., AdaptiveAvgPool(output_{stage_5}) \tag{3}$$

In the Transformer block, a multi-headed attention mechanism is used for global modeling. The multi-head attention is formulated as follows:

$$MultiHead(Q, K, V) = Concat(head_1, \ldots, head_h) \cdot W^O \tag{4}$$

where W^O is the learnable weight that maps the concatenate result back to the input dimension, $head_i$ represents the calculation of the ith attention head, as shown in Eq. 5:

$$Head\_Attention_i(Q_i, K_i, V_i) = softmax\left(\frac{Q_i K_i^T}{\sqrt{d_{k_i}}}\right) V_i \tag{5}$$

where the $softmax$ function is applied to the rows of the similarity matrix and d_{k_i} provides a normalization. Q_i, K_i, V_i represent the query matrix, key matrix, and value matrix, respectively.

Finally, The decoder part is symmetric to the encoder in its structure. Following sections provide detailed descriptions of our proposed components.

3.2   Double U Block (DUB)

Double U Block (DUB) is an integral part of the heavy token encoder and comprises two U-shaped networks named U1 and U2, as illustrated in Fig. 3. Each U-shaped network consists of two downsampling and upsampling layers, resulting in a larger receptive field. Furthermore, the feature maps of each resolution in U1 and U2 are connected through a shortcut, which allows information not mined in U1 to be mined again in U2. This method enhances the network's information mining capability. The joint output of residuals from both shallow and deep features is beneficial to constructing deep networks.

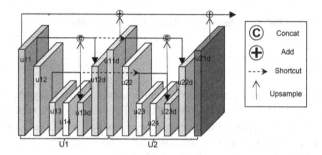

Fig. 3. Architecture of Double U-Block.

The feature extraction component of DUFormer consists of four DUBs and one transition layer. Each DUB operates at a specific multi-level resolution as shown in Fig. 3 and employs overlapping feature mining, as shown in the table located in the upper-right corner of Fig. 2. The degree of repetition in feature map mining is indicated by the darkness of its corresponding color. The transition layer employs dilated convolution, capturing features at different scales without downsampling or upsampling the feature maps. In other words, big receptive field ensures the capture of fine-grained detail while maintaining a high-level understanding of the input image. This method facilites effective global modeling by the subsequent Transformer block.

3.3   Power Line Aware Block (PLAB)

We present the PLAB, a module designed to effectively extract slender power line features from high-resolution aerial images. The module is designed to leverage the rich, detailed information available in the shallow network layers to extract power line features precisely. The PLAB structure, illustrated in Fig. 4, comprises two parallel asymmetric dilated convolutions that efficiently extract features in both vertical and horizontal directions, while also complementing the features with paralleled original convolutions. The feature fusion enables the subsequent network to exhibit an enhanced ability to perceive power lines.

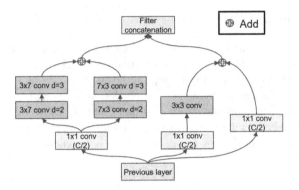

Fig. 4. Architecture of **P**ower**L**ine **A**ware **B**lock.

3.4   BiscSE Block

Although the feature map resolution is lower in the network's deep layers, the extracted semantic features are more robust. To enhance the effectiveness of the scSE [18] module as depicted in Fig. 5, we replace the average-pooling with max-pooling in the channel SE branch. Max pooling retains the maximum activation within each pooling region, which helps capture the most discriminative features (i.e., power line feature) present in the input data. This is particularly beneficial for power line detection tasks as it enhances the localization of power line feture in channel dimension. In addition, we extend the original module to improve the spacial SE branch by using various convolutional kernels for feature extraction under different receptive field. This idea achievs proper spatial squeeze and expansion, and further enhances the performance of the spatial SE branch.

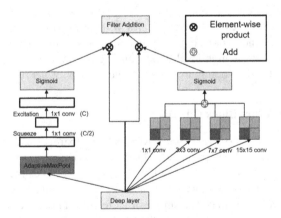

Fig. 5. Architecture of **BiscSE**

3.5 Loss Function

In conventional semantic segmentation algorithms, the input images typically includes valid semantic information. However, in the power line detection task, the background can introduce much irrelevant and redundant information. Using the cross-entropy loss function, which is a standard loss function for semantic segmentation, may not adequately suppress this redundant information. To address this issue, we investigate multiple loss functions, i.e., FocalLoss, PhiLoss, and DiceLoss, and devise an approach to combine them to improve the model's focus on detecting power lines. The combination is formulated as follows:

$$FocalLoss(p, \hat{p}) = -(\alpha(1 - \hat{p})^{\gamma}plog(\hat{p}) \\ + (1 - \alpha)\hat{p}^{\gamma}(1 - p)log(1 - \hat{p})) \tag{6}$$

$$PhiLoss = (1 - MCC)^{\theta} \tag{7}$$

$$DiceLoss = 1 - \frac{2 * TP}{2 * TP + FP + FN} \tag{8}$$

where α in Eq. 6 is used to adjust the ratio between positive and negative sample losses, γ is used to reduce the loss contribution of the easy samples. MCC in Eq. 7 refers to the Matthews correlation coefficient. Equation 9 shows the final loss used in our experiments. By conducting the experiment, we get the values of ρ, τ, ϕ as 3.0, 1.5, 3.0, respectively.

$$Loss = \rho * FocalLoss_{weight=1:5} + \tau * PhiLoss \\ + \phi * DiceLoss_{weight=1:5} \tag{9}$$

4 Experiments

In this section, we evaluate DUFormer's ability to detect power lines by conducting many comparative and ablation experiments and demonstrate the reliability and feasibility of our proposed model. We compare the proposed method with classical and state-of-the-art algorithms.

4.1 Experimental Settings

Datasets. We conduct experiments on the challenging power line dataset TTPLA [1], which consists of 1124 training images and 107 validation images with a resolution of 3840×2160.

Training. Our proposed approach is implemented based on the MMSegmentation framework. The network model is trained from scratch, specifically for power line data, without using any pre-trained weights. We set the training maximum iteration to 80k, the initial learning rate to 9e-4, and the weight decay to 0.01 for all experiments. We use a 'poly' learning rate strategy with a factor of 1.0. To ensure fair comparisons, we fix the random seeds in all experiments. A batch size of 8 is used, and all experiments are conducted on two NVIDIA GeForce RTX 3090 GPUs.

Testing. We test with an inference resolution of 1080×1920 for all the methods during the test procedure.

Evaluation Criteria. When detecting power lines in the UAV aerial data, we aim to achieve a model with high sensitivity, which reflects in a high Recall value, i.e., a low power line miss-detect rate. However, we must also consider Precision and ensure it falls within an acceptable range while maintaining a high Recall score. Therefore, we adjust the calculation of the F-score accordingly.

$$\text{F-score} = (1 + \beta^2) * \frac{Precision * Recall}{\beta^2 * Precision + Recall} \tag{10}$$

where β is set to 2, giving more weight to Recall than Precision and making the F-score consistent with our desired low power line miss-detect rate.

4.2 Comparative Experiments

DUFormer vs. Other Methods. Table 1 presents the performance of our DUFormer method for power line detection, demonstrating superior results compared to other classical methods. Our method achieves an F-score of 85.96% and an IoU of 74.4% with only 28.51M parameters, outperforming HRNet-OCR [21], which has 70.37M parameters. DUFormer only accounts for 1/5 of HRNet-OCR's FLOPs, which is a substantial improvement in limited computing resources. Furthermore, we test the inference speed of each model, The proposed method outperforms other models with $200.40ms/img$, which can effectively improve the efficiency of power line detection tasks.

Table 1. Comparison results of various models.

Method	#Params (M)	FLOPsc (G)	F-score (%)	Precision (%)	Recall (%)	IoU (%)	Latency (ms)
Deeplab	29.06	813.71	81.03	85.61	79.96	70.48	763.35
PSPNet	29.05	791.02	81.16	85.57	80.13	70.59	833.33
Sem-FPN-r50	28.51	182.48	80.03	84.17	79.06	68.82	366.30
U-Net	29.06	810.23	82.11	**86.95**	80.99	72.21	401.61
SegNeXt-Base	28.0	128.1	79.07	82.26	78.31	67.17	361.01
SegFormer-b2	24.76	74.21	82.51	85.05	81.9	71.59	543.47
EncNet	35.89	563.28	76.14	82.08	74.78	64.28	520.8
CCNet	49.81	801.52	78.04	83.48	76.79	66.66	595.23
HRNet-OCR	70.37	648.39	83.91	86.0	83.41	73.43	800.00
DUFormer(ours)	28.51	123.41	**85.96**	84.35	**86.37**	**74.44**	**200.40**

Joint Multi-weighted Loss Function. In the following experiments, we investigate the effectiveness of the joint multi-weight loss function in addressing category imbalance issues in power line detection due to the slender characteristics of the power line. We employ the traditional Cross-Entropy loss function as a baseline and demonstrate in

Table 2. Results of different loss functions.

Method	Loss	Fscore(%)	Precision(%)	Recall(%)	IoU(%)
DUFormer (ours)	CE Loss	82.38	86.44	81.42	72.2
DUFormer (ours)	Multi Loss	**85.96**	84.35	**86.37**	**74.44**
SegNeXt	CE Loss	79.07	82.26	78.31	67.17
SegNeXt	Multi Loss	**81.91**	78.35	**82.85**	**67.49**
EncNet	CE Loss	76.14	82.08	74.78	64.28
EncNet	Multi Loss	**82.67**	79.48	**83.51**	**68.7**
CCNet	CE Loss	78.04	83.48	76.79	66.66
CCNet	Multi Loss	**82.73**	78.6	**82.83**	**68.26**

Table 2 that using the joint multi-weight loss function improves the network's ability to detect power lines and reduces cluttered background information. Moreover, we apply the joint multi-weight loss function to other existing algorithms, e.g., SegNeXt [10], EncNet [22], and CCNet [12]. We observe a significant improvement in the Recall metrics, indicating a lower power line miss-detect rate. These similar observations further validate the effectiveness of our proposed method.

4.3 Ablation Experiments

Effect of Heavy Token Encoder. As noted in the previous section, the heavy token encoder can significantly enhance the Transformer models' performance on small data sets. Table 3 presents the experimental results. The lightweight token encoder serves as the baseline, and the repetitions of the Transformer block produce minimal improvement. In contrast, when the proposed heavy token encoder is applied, the performance improves significantly, resulting in a 3.2% increase in both IoU and F-score compared to the baseline.

Table 3. Results of Heavy Token Encoder.

Method	Tok. Encoder Parameters	Trans. Block Parameters	Param. Ratios	F-score (%)	Precision (%)	Recall (%)	IoU (%)
Baseline	0.881	13.28	0.066	82.76	83.96	82.46	71.24
Repetitive Transformer block	0.881	26.56	0.033	84.09	82.25	84.56	71.51
Heavy Token Encoder (ours)	9.594	13.28	0.722	**85.96**	**84.35**	**86.37**	**74.44**

The Impact of Each Module. Table 4 shows the impact of each module on the model. First, we establish a baseline model by removing the proposed modules from the network.

Table 4. The impact of each module.

+DUB	+PLAB&BiscSE	+Multi Loss	Fscore(%)	Precision(%)	Recall(%)	IoU(%)
✗	✗	✗	81.01	81.11	80.98	68.14
✔	✗	✗	82.13	84.29	81.6	70.83
✔	✔	✗	82.38	86.44	81.42	72.2
✔	✔	✔	**85.96**	84.35	**86.37**	**74.44**

Then, we gradually add each proposed module to show their contributions. The results show that Double U Block, Power Line Aware Block and BiscSE can significantly improve the Precision. It should be noted that the high Recall is more in line with the industrial requirements, as the power lines in the aerial images should be detected in their entirety as much as possible. The Multi-Loss function is able to regulate Recall and Precision better, making the model more sensitive (i.e., higher Recall) to power lines. In summary, our proposed methods can enhance the power line detection capability of the model in various aspects. We further evaluate the performance of the PLAB at different positions in the network and compare the effectiveness of the improved scSE module. The experimental results can be found in Sect. 2 in the supplementary material. The supplementary material can be found at https://drive.google.com/file/d/1AApoQRMlJAkVhzTAv3jtxnkeQs1tCl84/view?usp=sharing.

5 Conclusion

In this work, we present a comprehensive methodology for tackling the problem of detecting power lines in aerial images captured by UAVs. The proposed solutions include a heavy token encoder, which captures fine-grained features by performing feature re-mining on feature maps of different resolutions at different stages. We also introduce a Power Line Aware Block (PLAB), composed of asymmetric dilated convolutions, which particularly enhances power line features while suppressing background information. Moreover, we propose an improved scSE module, i.e., BiscSE, optimized for dichotomous segmentation for power lines, effectively enhancing the Precision and Recall metrics. Through extensive experiments, we demonstrate that our proposed method significantly improves the performance in terms of the number of parameters, computational complexity, and accuracy. The proposed DUFormer sets a new state-of-the-art record on the public dataset TTPLA.

References

1. Abdelfattah, R., Wang, X., Wang, S.: TTPLA: an aerial-image dataset for detection and segmentation of transmission towers and power lines. In: Proceedings of the Asian Conference on Computer Vision (2020)
2. Badrinarayanan, V., Kendall, A., Cipolla, R.: SegNet: a deep convolutional encoder-decoder architecture for image segmentation. IEEE Trans. Pattern Anal. Mach. Intell. **39**(12), 2481–2495 (2017)

3. Chen, L.C., Papandreou, G., Kokkinos, I., Murphy, K., Yuille, A.L.: Semantic image segmentation with deep convolutional nets and fully connected CRFs. In: International Conference on Learning Representations (2015). http://arxiv.org/abs/1412.7062
4. Chen, L.C., Papandreou, G., Kokkinos, I., Murphy, K., Yuille, A.L.: DeepLab: semantic image segmentation with deep convolutional nets, Atrous convolution, and fully connected CRFs. IEEE Trans. Pattern Anal. Mach. Intell. **40**(4), 834–848 (2017)
5. Chen, L.C., Papandreou, G., Schroff, F., Adam, H.: Rethinking Atrous convolution for semantic image segmentation. arXiv preprint: arXiv:1706.05587 (2017)
6. Chu, X., et al.: Twins: revisiting the design of spatial attention in vision transformers. In: Advances in Neural Information Processing Systems, vol. 34, pp. 9355–9366 (2021)
7. Chu, X., Tian, Z., Zhang, B., Wang, X., Shen, C.: Conditional positional encodings for vision transformers. In: International Conference on Learning Representations (2023)
8. Dosovitskiy, A., et al.: An image is worth 16x16 words: transformers for image recognition at scale. In: International Conference on Learning Representations (2021)
9. Guo, J., et al.: CMT: convolutional neural networks meet vision transformers. In: Proceedings of the IEEE/CVF Conference on Computer Vision and Pattern Recognition, pp. 12175–12185 (2022)
10. Guo, M.H., Lu, C.Z., Hou, Q., Liu, Z.N., Cheng, M.M., Hu, S.M.: SegNeXt: rethinking convolutional attention design for semantic segmentation. In: Oh, A.H., Agarwal, A., Belgrave, D., Cho, K. (eds.) Advances in Neural Information Processing Systems (2022)
11. Han, K., Xiao, A., Wu, E., Guo, J., Xu, C., Wang, Y.: Transformer in transformer. In: Advances in Neural Information Processing Systems, vol. 34, pp. 15908–15919 (2021)
12. Huang, Z., Wang, X., Huang, L., Huang, C., Wei, Y., Liu, W.: CCNet: criss-cross attention for semantic segmentation. In: Proceedings of the IEEE/CVF International Conference on Computer Vision (2019)
13. Li, Y., Zhang, K., Cao, J., Timofte, R., Van Gool, L.: LocalViT: bringing locality to vision transformers. arXiv preprint: arXiv:2104.05707 (2021)
14. Lin, H., Cheng, X., Wu, X., Shen, D.: CAT: cross attention in vision transformer. In: 2022 IEEE International Conference on Multimedia and Expo (ICME), pp. 1–6. IEEE (2022)
15. Liu, Z., et al.: Swin transformer: hierarchical vision transformer using shifted windows. In: Proceedings of the IEEE/CVF International Conference on Computer Vision, pp. 10012–10022 (2021)
16. Long, J., Shelhamer, E., Darrell, T.: Fully convolutional networks for semantic segmentation. In: Proceedings of the IEEE/CVF Conference on Computer Vision and Pattern Recognition, pp. 3431–3440 (2015)
17. Ronneberger, O., Fischer, P., Brox, T.: U-Net: convolutional networks for biomedical image segmentation. In: Navab, N., Hornegger, J., Wells, W.M., Frangi, A.F. (eds.) MICCAI 2015. LNCS, vol. 9351, pp. 234–241. Springer, Cham (2015). https://doi.org/10.1007/978-3-319-24574-4_28
18. Roy, A.G., Navab, N., Wachinger, C.: Recalibrating fully convolutional networks with spatial and channel "squeeze and excitation" blocks. IEEE Trans. Med. Imaging **38**(2), 540–549 (2018)
19. Wu, H., et al.: CVT: introducing convolutions to vision transformers. In: Proceedings of the IEEE/CVF International Conference on Computer Vision, pp. 22–31 (2021)
20. Yuan, K., Guo, S., Liu, Z., Zhou, A., Yu, F., Wu, W.: Incorporating convolution designs into visual transformers. In: Proceedings of the IEEE/CVF International Conference on Computer Vision, pp. 579–588 (2021)
21. Yuan, Y., Chen, X., Wang, J.: Object-contextual representations for semantic segmentation. In: Proceedings of the European Conference on Computer Vision (2020)
22. Zhang, H., et al.: Context encoding for semantic segmentation. In: Proceedings of the IEEE/CVF Conference on Computer Vision and Pattern Recognition (2018)

23. Zhang, W., et al.: TopFormer: token pyramid transformer for mobile semantic segmentation. In: Proceedings of the IEEE/CVF Conference on Computer Vision and Pattern Recognition, pp. 12083–12093 (2022)
24. Zhao, H., Shi, J., Qi, X., Wang, X., Jia, J.: Pyramid scene parsing network. In: Proceedings of the IEEE/CVF International Conference on Computer Vision, pp. 2881–2890 (2017)

Space-Transform Margin Loss with Mixup for Long-Tailed Visual Recognition

Fangyu Zhou[1], Xicheng Chen[1], and Haibo Ye[1,2,3]($\boxtimes$)

[1] College of Computer Science and Technology, Nanjing University of Aeronautics and Astronautics, Nanjing, China
{saya,ccmo_cxc,yhb}@nuaa.edu.cn
[2] MIIT Key Laboratory of Pattern Analysis and Machine Intelligence, Nanjing, China
[3] Collaborative Innovation Center of Novel Software Technology and Industrialization, Nanjing, China

Abstract. In the real world, naturally collected data often exhibits a long-tailed distribution, where the head classes have a larger number of samples compared to the tail classes. This long-tailed data distribution often introduces a bias in classification results, leading to incorrect classifications that harm the tail classes. Mixup is a simple but effective data augmentation method that transforms data into a new shrinking space, resulting in a regularization effect that is beneficial for classification. Therefore, many researchers consider using Mixup to promote the performance of long-tailed learning. However, these methods do not consider the special space transformation of data caused by Mixup in long-tail learning. In this paper, we present the Space-Transform Margin (STM) loss function, which offers a novel approach to dynamically adjusting the margin between classes by leveraging the shrinking strength introduced by Mixup. In this way, the margin of data can adapt to the special space transformation of Mixup. In the experiments, our solution achieves state-of-the-art performance on benchmark datasets, including CIFAR10-LT, CIFAR100-LT, ImageNet-LT, and iNaturalist 2018.

Keywords: Long-tail learning · Mixup · Visual recognition

1 Introduction

Deep neural networks heavily depend on large-scale datasets [1,2] which are well-balanced in terms of class distribution. Unfortunately, real-world datasets tend to be the long-tailed distribution, where most of the data is concentrated in the head classes and fewer data belong to the tail classes [3,4]. Training deep learning models directly on long-tailed datasets often leads to biased prediction outcomes that favor the head class and detrimentally impact the tail classes [5,6]. In recent years, addressing the long-tailed problem and improving the accuracy of tail classes has been a key focus of research [7–12,33,34].

© The Author(s), under exclusive license to Springer Nature Singapore Pte Ltd. 2024
Q. Liu et al. (Eds.): PRCV 2023, LNCS 14432, pp. 67–79, 2024.
https://doi.org/10.1007/978-981-99-8543-2_6

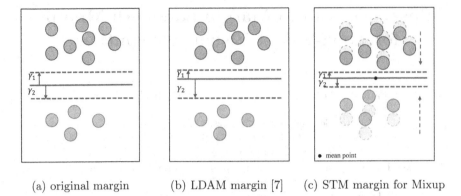

(a) original margin   (b) LDAM margin [7]   (c) STM margin for Mixup

Fig. 1. Suppose there is a binary problem, the points above and below represent the head class and the tail class respectively. Our method could effectively suit the transformation of Mixup.

Some previous methods, such as deferred re-weighting (DRW) [7], deferred re-sampling (DRS) [7], and two-stage strategies [8,13], can alleviate the long-tailed problem to a certain extent. Considering using re-weighting, re-sampling, and other strategies in the later stage, these methods are believed to model a more conductive feature space. Some other methods [7,11,14,15] focus on adjusting the margin between classes by modifying the logits of the loss function, aiming to create a larger margin specifically for tail classes, thus enhancing their overall robustness. Additionally, Mixup [16] is a simple but effective data augmentation method to improve the robustness of the model in long-tail learning. It mixes two samples randomly. Zhong et al. [12] have shown the effectiveness of Mixup for the tail classes. However, directly applying Mixup is hard to achieve an improvement effect because of the long-tailed data distribution in our experiments. And we consider readjusting the margin of classes through logits adjust methods [7,11, 14,15]. The problem is that the changed data distribution caused by Mixup influences the margin between classes.

In a binary classification problem, the original classification boundary typically gives the same-sized margin between the two classes, as shown in Fig. 1(a). But this method does not significantly improve the tail class. The margin shown in Fig. 1(b) represents a common approach in long-tail learning, where the decision boundary is shifted towards the head class. This adjustment aims to provide a larger margin for the tail classes, allowing for improved discrimination and increased robustness for the tail classes. However, when applying Mixup, the original data undergoes a transformation that causes it to move towards the mean point [18], resulting in a smaller distance between them, as observed in Fig. 1(c). As a consequence, the optimal margin between the head class and the tail class, should also be adjusted accordingly to account for this shrinking effect. In this paper, we introduce the strength of shrinking to design a Space-Transform Margin (STM) loss function, allowing the margin between classes to dynamically adapt and transform along with the data shrinking.

In summary, the main contributions of this paper are: 1) we analyze the data shrinking process in Mixup and find its mismatch with traditional margins, 2) we rethink the margin loss function in Mixup and propose a novel Space-Transform Margin loss function to suit the new shrinking margin of Mixup, 3) experiment results on benchmark datasets illustrate the effectiveness of our method.

2 Related Work

2.1 Mixup and Its Space Transformation

Different from Empirical Risk Minimization (ERM), Mixup [16] and its derived variants, such as Manifold Mixup [19] and CutMix [20], follow the principle of Vicinal Risk Minimization (VRM) [21], improve the robustness of neural networks by leveraging augmented training examples. The core of Mixup is to randomly mix two samples (x_i, y_i), (x_j, y_j) through a coefficient θ:

$$\begin{cases} \widetilde{x}_{ij} = \theta x_i + (1 - \theta)x_j, \\ \widetilde{y}_{ij} = \theta y_i + (1 - \theta)y_j, \end{cases} \tag{1}$$

where $\theta \sim Beta(\alpha, \alpha)$, $\theta \in [0, 1]$, $\alpha \in (0, +\infty)$. The resulting mixed sample is denoted as $(\widetilde{x}_{ij}, \widetilde{y}_{ij})$.

Some studies have proved the effectiveness of Mixup from the perspective of regularization [18,20,22]. Especially Carratino et al. [18] explain that Mixup shrinks inputs and outputs towards their mean and adds a zero-mean perturbation.

2.2 Long-Tailed Learning with Mixup

Mixup [16] has been shown to significantly improve the performance of the tail classes in long-tail learning scenarios [12]. However, simply applying Mixup to long-tailed datasets does not yield satisfactory results because the newly generated samples through Mixup still inherit the imbalanced nature of the original dataset. Some methods [8,9,11,12] primarily focus on increasing the samples of the tail class to solve this problem from the input perspective, while this kind of method may lead to overfitting to the tail classes. In this paper, based on the phenomenon of shrinking samples generated by Mixup, we propose to readjust the margin between classes through a novel Space-Transform Margin (STM) loss.

2.3 Re-balanced Loss Function Modification Methods

Inspired by the concept of margin loss [23–25], some methods [7,11,15] introduce a margin term in logits to create a larger margin for the tail classes. It profits the tail classes as a larger margin is beneficial for robustness. In addition, it has been demonstrated by certain methods [7,8,13] that applying re-balanced strategies

at a later stage of the training process often leads to improved feature representation. Among these methods, deferred re-weighting (DRW) [7] has emerged as
a prevalent strategy. In this paper, we also adopt the concept of a larger margin
for the tail classes, as observed in previous works [7]. Also, we leverage the DRW
technique to enhance the effectiveness of our proposed STM loss function.

3 Method

3.1 Space Transformation in Mixup

Assume an imbalanced dataset denoted as $S = \{(x_i, y_i)\}, i \in \{1, \cdots, N\}$, where
x_i represents the input data and y_i represents the corresponding one-hot label.
Additionally, the transformed dataset obtained through Mixup is represented as
$\widetilde{S} = \{(\widetilde{x}_i, \widetilde{y}_i)\}, i \in \{1, \cdots, N\}$. Here, $\widetilde{x}_i$ refers to the transformed input data,
and $\widetilde{y}_i$ corresponds to the transformed label for sample $\widetilde{x}_i$. The model function
is denoted as $f(x)$.

Carratino et al. [18] have demonstrated that Mixup causes the original data
(x_i, y_i) to shrink towards their mean $(\overline{x}, \overline{y})$ and introduces a zero-mean perturbation $(\delta_i, \varepsilon_i)$ to the data. Specifically, if θ in Eq. (1) is greater than 0.5, the
sample (x_i, y_i) is considered the major term, while the other sample (x_j, y_j)
is considered the perturbation term. The following theorem provides a way to
conceal the presence of (x_j, y_j) in this case:

Theorem 1. *Let $\theta \sim Beta_{[\frac{1}{2}, 1]}(\alpha, \alpha)$, $\alpha \in (0, +\infty)$, $j \sim Unif(1, N)$, N denotes
the number of samples, the modified (x_i, y_i) obtained through Mixup can be rep-
resented as $(\widetilde{x}_i + \delta_i, \widetilde{y}_i + \varepsilon_i)$, and $(\widetilde{x}_i, \widetilde{y}_i)$ is given by*

$$\begin{cases} \widetilde{x}_i = \overline{x} + \overline{\theta}(x_i - \overline{x}), \\ \widetilde{y}_i = \overline{y} + \overline{\theta}(y_i - \overline{y}), \end{cases} \tag{2}$$

*where $\overline{\theta}$ represents the mean of θ, $\overline{x} = \frac{1}{N}\sum_{i=1}^{N} x_i$, $\overline{y} = \frac{1}{N}\sum_{i=1}^{N} y_i$. $(\delta_i, \varepsilon_i)$ is a
zero-mean random perturbation, which can be expressed as:*

$$\begin{cases} \delta_i = (\theta - \overline{\theta})x_i + (1 - \theta)x_j - (1 - \overline{\theta})\overline{x}, \\ \varepsilon_i = (\theta - \overline{\theta})y_i + (1 - \theta)y_j - (1 - \overline{\theta})\overline{y}. \end{cases} \tag{3}$$

The proof of Theorem 1 can be found in [18]. Assume that $\theta \sim Beta_{[0, \frac{1}{2}]}(\alpha, \alpha)$,
it has been shown in the proof provided in [18] that Eq. (2) and Eq. (3) still
hold. Eq. (2) demonstrates that the sample (x_i, y_i) undergoes a shrinking process
towards their mean values $(\overline{x}, \overline{y})$. The extent of shrinking is determined by the
distribution of θ. This implies that the samples are pulled towards the central
point defined by their mean values, resulting in a contraction of the data points.

Now, let's extend the distribution of θ to $Beta_{[0, 1]}(\alpha, \alpha)$. For the sample
(x_i, y_i), the sample (x_j, y_j) is a perturbation term. Also, for sample (x_j, y_j),

(x_i, y_i) is pertubation. In a specific training batch, we ignore the slight perturbation $(\delta_i, \varepsilon_i)$, so $\widetilde{x}_i, \widetilde{y}_i$ can be expressed as

$$\begin{cases} \widetilde{x}_i = \overline{x} + \theta(x_i - \overline{x}), \\ \widetilde{y}_i = \overline{y} + \theta(y_i - \overline{y}), \end{cases} \tag{4}$$

$(\widetilde{x}_i, \widetilde{y}_i)$ shrink based on θ.

3.2   Space-Transform Margin Loss Function

In long-tail learning, it has been shown that a larger margin for the tail classes
is beneficial for classification [7]. However, in the context of Mixup, where data
is transformed by shrinking towards the mean, the margin between classes also
tends to shrink.

Here, for the sample (x, y), y denotes the one-hot label of input x. n_k represents the number of samples belonging class k, where $k \in \{1, \cdots, K\}$. K
represents the total number of classes. $f(x)_k$ denotes the k-th output logit of
sample (x, y). The set $S_k = \{i : y_i = k\}$ is defined as the subset of indices
that correspond to samples with label k. We can define $\gamma(x, y)$ as the margin of
sample (x, y):

$$\gamma(x, y) = f(x)_y - \max_{k \neq y} f(x)_k, \tag{5}$$

which represents the difference between the predicted logit of the true label and
the maximum value of the predicted logit among the other labels. Next, the
margin of class k can be defined as

$$\gamma_k = \min_{i \in S_k} \gamma(x_i, y_i), \tag{6}$$

Let $C(\mathcal{F})$ be some complexity measure of hypothesis class $\mathcal{F}$. In a balanced
dataset, the total test error is the sum of class error [7]. From [7], we can directly
get

$$err_{bal} \lesssim \frac{1}{K} \sum_{k=1}^{K} \left(\frac{1}{\gamma_k} \sqrt{\frac{C(\mathcal{F})}{n_k}} + \frac{\log N}{\sqrt{n_k}} \right), \tag{7}$$

Equation (7) denotes that the total error is associated with γ_k of different class.
To minimize the generalization error bound, the γ_k should be larger. However the
margin of different classes is not independent, there exists a trade-off between
them. In long-tail learning, it should be larger for the tail classes, while it will
be smaller for the head classes. Cao et al. [7] have proved this optimal trade-off
margin is

$$\gamma_k = \frac{C}{n_k^{1/4}}, \tag{8}$$

where C is a constant, the detailed provement of Eq. (8) is in [7].

In an ideal scenario, the margin of sample (x, y) is $\gamma(x, y) = 1$, indicating that the logit of the true label is 1, while the logit values for other classes are 0. However, in Mixup, we get the transformed data $(\tilde{x}, \tilde{y})$, and $\tilde{y}$ is equal to

$$\tilde{y} = \overline{y} + \theta(y - \overline{y})$$
$$= (1 - \theta)\frac{1}{N}\sum_{k=1}^{K} n_k \cdot y_k + \theta y , \tag{9}$$

from Eq. (9), the ideal output logits is $[\frac{(1-\theta)}{N}n_1, \frac{(1-\theta)}{N}n_2, \cdots, \frac{(1-\theta)}{N}n_y + \theta, \cdots, \frac{(1-\theta)}{N}n_K]$. The margin of $\gamma(\tilde{x}, y)$ is

$$\gamma(\tilde{x}, y) = f(\tilde{x})_y - \max_{k \neq y} f(\tilde{x})_k \tag{10}$$

To ensure the nonnegativity of the margin in the experiments, we replace the maximum value with the minimum value:

$$\gamma(\tilde{x}, y) \leq f(\tilde{x})_y - \min_{k \neq y} f(\tilde{x})_k$$
$$= \frac{(1 - \theta)}{N}n_y + \theta - \left(\frac{(1 - \theta)}{N}\min_{k \neq y} n_k\right) \tag{11}$$
$$= \frac{(1 - \theta)}{N}(n_y - \min_{k \neq y} n_k) + \theta.$$

It represents the $\gamma(\tilde{x}, y)$ shrinks to $\frac{(1-\theta)}{N}(n_y - \min_{k \neq y} n_k) + \theta$ time of $\gamma(x, y)$ approximately. And the fewer samples in the class, the greater the degree of shrinking. We set $\frac{(1-\theta)}{N}(n_y - \min_{k \neq y} n_k) + \theta$ as the shrinkage strength s. The margin in Eq. (8) should be readjusted to its s-fold:

$$\tilde{\gamma}_{y,\theta} \approx s \cdot \gamma_y$$
$$= \left(\frac{(1 - \theta)}{N}(n_y - \min_{k \neq y} n_k) + \theta\right) \cdot \frac{C}{n_y^{1/4}}. \tag{12}$$

The final form of Space-Transform Margin loss is

$$\ell(z, y, \theta) = -\log \frac{e^{z_y - \tilde{\gamma}_{y,\theta}}}{e^{z_y - \tilde{\gamma}_{y,\theta}} + \sum_{k \neq y}^{K} e^{z_k}}, \tag{13}$$

where z_y is the logit of label y, in Mixup, for the random two sampled (x_i, y_i), (x_j, y_j), the total loss is

$$\ell_{total} = \theta\ell(z, y_i, \theta) + (1 - \theta)\ell(z, y_j, 1 - \theta). \tag{14}$$

4   Experiments

4.1   Datasets

CIFAR10-LT and CIFAR100-LT. CIFAR10 and CIFAR100 are small-scale datasets used for image recognition. They contain 60,000 images of size 32×32, the number of training images is 50,000, and the number of testing images is 10,000. For CIFAR10-LT and CIFAR100-LT, we refer to [7] to set the imbalance ratio ρ of datasets. The sample size of each class is exponential decay with the class index increasing and the imbalance ratio is set to $\rho = \max_i\{n_i\}/min_i\{n_i\}$. In experiments, we set the imbalance ratio to 200, 100, 50, and 10.

ImageNet-LT and iNaturalist 2018. ImageNet-LT is a large-scale image database that comes from ImageNet [2] and follows the Pareto distribution with $\alpha = 6$. iNaturalist [3] is a large and extremely imbalanced dataset with 8,142 classes, 437,513 training images, 24,426 validation images, and 149,394 testing images. Since the number of species in nature is not the same, there is a long-tailed problem in this kind of dataset. In experiments, we use iNaturalist 2018 to train and test.

4.2   Implementations Details

Implementation Details on CIFAR. In our experiments on CIFAR10-LT and CIFAR100-LT, we adopt the ResNet-32 backbone as our base model. We conduct our model training on an NVIDIA 3080Ti GPU with 400 epochs and 128 batch size. The DRW strategy [7] is implemented at the 320th epoch. The α of Mixup is set to 1.0, and C is the same as [7] which enforces the max-margin to 0.5. To further refine the model's performance, we adopt a fine-tuning strategy during the last 40 epochs of the training process without Mixup.

Implementation Details on ImageNet-LT and iNaturalist 2018. We conduct experiments on ImageNet-LT using two different backbone architectures: ResNet-10 and ResNet-50. For the training of the ResNet-10 backbone model on ImageNet-LT, we utilize a single NVIDIA 3080Ti GPU and set the batch size to 256. We employ two NVIDIA 3080Ti GPUs to train ResNet-50 and the batch size is set to 128. The total epoch is set to 180. The α is set to 1.0, and C is set to be the same as [7] which enforces the max-margin to 0.5. DRW [7] is used at the 150th, and fine-tuning is used at the 160th. On iNaturalist 2018 [3], the backbone of the model is ResNet-50. It is trained on a single NVIDIA 3090Ti GPU with 360 epochs and 128 batch size. α of Mixup is 0.4, C is same as [7] which enforces the max margin to 0.3. DRW [7] is used at the 240th. At the 320th epoch, the Mixup is moved to fine-tune the model.

4.3   Main Results

Results on CIFAR10-LT and CIFAR100-LT. In our experiments, we evaluate the performance of our method on long-tailed datasets. Table 1 presents

Table 1. Top-1 accuracy (%) of ResNet-32 on CIFAR10-LT and CIFAR100-LT with 200, 100, 50, and 10 imbalance ratio. * denotes the reproduced results from our own experiments.

Dataset	CIFAR10-LT				CIFAR100-LT			
Imbalance ratio	200	100	50	10	200	100	50	10
Mixup* [16]	67.00	73.45	79.24	87.02	37.22	40.21	45.24	58.56
BBN[8]	–	79.82	82.18	88.32	–	42.56	47.02	59.12
LDAM-DRW [7]	–	77.03	81.03	88.16	–	42.54	46.62	58.71
LDAM-DRW+Mixup*	78.84	82.23	84.67	88.42	42.20	46.09	49.97	58.60
remix-DRW [9]	–	79.82	–	89.02	–	46.77	–	61.23
UniMix [11]	78.48	82.75	84.32	89.66	42.07	45.45	51.11	61.25
CRT+Mixup [12]	–	79.10	84.20	89.80	–	45.10	50.90	62.10
LWS+Mixup [12]	–	76.30	82.60	89.60	–	44.20	50.70	62.30
MiSLAS [12]	–	82.10	85.70	90.00	–	47.00	52.30	63.20
Ours	**79.38**	**82.98**	**85.88**	**90.45**	**44.62**	**48.82**	**53.68**	**63.24**

the Top-1 accuracy of the ResNet-32 model compared to several other methods related to Mixup [16] and long-tail learning. Our method demonstrates clear advantages over other existing methods. Specifically, on CIFAR100-LT with imbalance ratios of 200, 100, and 50, our method outperforms the state-of-the-art method by 1.2% to 2.5% in terms of accuracy.

Results on Large-Scale Datasets. The results of large-scale datasets ImageNet-LT [2] and iNaturalist 2018 [3] are presented in Table 2. Our experiment results are the best among these methods. Especially, on ImageNet-LT with ResNet-10 and ResNet-50, our method is higher than the state-of-the-art method by 0.57% and 2.2% respectively.

4.4   Feature Visualization and Analysis of STM Loss

In this section, we aim to gain a better understanding of the margin by visualizing the features. In our experiments, we employ three different loss functions during the training process: Cross Entropy (CE) loss, LDAM-DRW [7] loss with Mixup, and STM-DRW loss with Mixup. The differences between the features learned using CE, LDAM-DRW, and STM-DRW can be effectively demonstrated through Fig. 2. As shown in Fig. 2(a), CE does not consider the margin in long-tail learning, thus the margin between classes appears to be relatively close. LDAM-DRW incorporates the consideration of margin between classes in long-tail learning. However, when LDAM-DRW is applied in combination with Mixup, the margin becomes less suitable. The larger margin results in the features of some classes being cut off by other classes. While our STM-DRW loss function creates a transformed margin, successfully adapting the transformation of Mixup as shown in Fig. 2(c).

Table 2. Top-1 accuracy (%) of ResNet-10 and ResNet-50 on ImageNet-LT and top-1 accuracy (%) of ResNet-50 on iNaturalist 2018. † denotes the reproduced results from [11].

Dataset	ImageNet-Lt		iNaturalist 2018
Method	ResNet-10	ResNet-50	ResNet-50
LDAM-DRW [7]	38.22†	45.75†	68.00
BBN [8]	–	48.30	69.62
Remix-DRW [9]	–	–	70.49
UniMix [11]	42.90	48.41	69.15
CRT+Mixup [12]	–	51.70	70.20
LWS+Mixup [12]	–	52.00	70.90
MiSLAS [12]	–	52.70	71.60
Ours	**43.47**	**54.99**	**71.64**

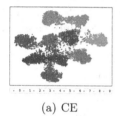      

(a) CE   (b) LDAM-DRW [7]   (c) STM-DRW

Fig. 2. The t-SNE feature visualiazation of CE, LDAM-DRW [7] with Mixup, and STM-DRW with Mixup on CIFAR10-LT. The imbalance ratio is 10. 10 colors represent 10 classes.

Figure 3 displays the log-confusion matrices of the three loss functions: Cross Entropy (CE), LDAM-DRW [7] with Mixup, and STM-DRW with Mixup. Apparently, the prediction of the CE loss function is biased toward the head classes as shown in Fig. 3(a). LDAM-DRW alleviates the long-tailed problem as shown in Fig. 3(b). Figure 3 indicates reduced misclassifications between head and tail classes. This visual analysis of the confusion matrix strongly supports STM-DRW's superiority in addressing long-tailed learning challenges with Mixup, outperforming LDAM-DRW.

4.5   Ablation Study

In our ablation experimental setup, we train the ResNet-32 model on the CIFAR100-LT with the imbalance ratio of 200, 100, 50, and 10 for 400 epochs as shown in Table 3.

The baseline in our experiments is the Empirical Risk Minimization (ERM) approach. This baseline does not utilize Mixup and does not incorporate any rebalanced loss function. The experimental results indicate that using the ERM

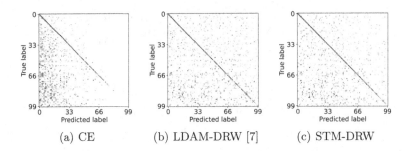

(a) CE   (b) LDAM-DRW [7]   (c) STM-DRW

Fig. 3. The log-confusion matrix on CIFAR100-LT validate dataset with Mixup. The imbalance ratio is 200. The x-axis is the predicted label and the y-axis is the true label. The value is larger, the color is deeper. (Color figure online)

approach with Mixup does not lead to significant improvements in accuracy. When the STM-DRW loss is applied without Mixup, it is essentially equivalent to the LDAM-DRW loss in the original long-tailed datasets. The results show that solely using STM-DRW yields slightly better performance than using the ERM baseline. Finally, our proposed STM-DRW loss combined with Mixup effectively adjusts the margin according to the Mixup space transformation. This combination outperforms the other methods mentioned above.

Table 3. Ablation study on CIFAR100-LT with 200, 100, 50, and 10 imbalance ratio.

Dataset	CIFAR100-LT			
Imbalance ratio	200	100	50	10
ERM	36.93	40.36	45.97	58.94
w Mixup	37.22	40.21	45.24	58.56
w STM-DRW Loss	38.16	41.94	47.28	58.35
w Mixup and STM-DRW Loss	**44.22**	**48.02**	**52.37**	**62.43**

5 Conclusion

In this paper, we analyze the limitations of existing Mixup methods for long-tail learning and propose a novel loss function called the Space-Transform Margin (STM) loss, which takes into account the data transformation process of Mixup. This approach allows us to effectively adjust the margin for different samples, resulting in improved performance compared to existing methods. In our experiments, we conduct a comprehensive comparison of our STM loss function with the commonly used CE and LDAM loss functions and prove the adaptability of our STM loss to Mixup. Our STM loss function outperforms other existing methods, demonstrating state-of-the-art performance.

Acknowledgement. We thank the anonymous reviewers for their helpful comments. This work is supported by the National Key R&D Program of China (2020AAA0107000) and Postgraduate Research & Practice Innovation Program of NUAA (xcxjh20221611).

References

1. Lin, T.-Y., et al.: Microsoft COCO: common objects in context. In: Fleet, D., Pajdla, T., Schiele, B., Tuytelaars, T. (eds.) ECCV 2014. LNCS, vol. 8693, pp. 740–755. Springer, Cham (2014). https://doi.org/10.1007/978-3-319-10602-1_48

2. Deng, J., Dong, W., Socher, R., Li, L.J., Li, K., Fei-Fei, L.: Imagenet: a large-scale hierarchical image database. In: IEEE Conference on Computer Vision and Pattern Recognition, pp. 248–255. IEEE (2009)

3. Van Horn, G., et al.: The inaturalist species classification and detection dataset. In: Proceedings of the IEEE Conference on Computer Vision and Pattern Recognition, pp. 8769–8778 (2018)

4. Van Horn, G., Perona, P.: The devil is in the tails: fine-grained classification in the wild. arXiv preprint arXiv:1709.01450 (2017)

5. Tan, J., et al.: Equalization loss for long-tailed object recognition. In: Proceedings of the IEEE/CVF Conference on Computer Vision and Pattern Recognition, pp. 11662–11671 (2020)

6. Tang, K., Huang, J., Zhang, H.: Long-tailed classification by keeping the good and removing the bad momentum causal effect. In: Advances in Neural Information Processing Systems, vol. 33, pp. 1513–1524 (2020)

7. Cao, K., Wei, C., Gaidon, A., Arechiga, N., Ma, T.: Learning imbalanced datasets with label-distribution-aware margin loss. In: Advances in Neural Information Processing Systems, vol. 32 (2019)

8. Zhou, B., Cui, Q., Wei, X.S., Chen, Z.M.: BBN: bilateral-branch network with cumulative learning for long-tailed visual recognition. In: Proceedings of the IEEE/CVF Conference on Computer Vision and Pattern Recognition, pp. 9719–9728 (2020)

9. Chou, H.-P., Chang, S.-C., Pan, J.-Y., Wei, W., Juan, D.-C.: Remix: rebalanced mixup. In: Bartoli, A., Fusiello, A. (eds.) ECCV 2020. LNCS, vol. 12540, pp. 95–110. Springer, Cham (2020). https://doi.org/10.1007/978-3-030-65414-6_9

10. Lin, T.Y., Goyal, P., Girshick, R., He, K., Dollár, P.: Focal loss for dense object detection. In: Proceedings of the IEEE International Conference on Computer Vision, pp. 2980–2988 (2017)

11. Xu, Z., Chai, Z., Yuan, C.: Towards calibrated model for long-tailed visual recognition from prior perspective. In: Advances in Neural Information Processing Systems, vol. 34 (2021)

12. Zhong, Z., Cui, J., Liu, S., Jia, J.: Improving calibration for long-tailed recognition. In: Proceedings of the IEEE/CVF Conference on Computer Vision and Pattern Recognition, pp. 16489–16498 (2021)

13. Kang, B., et al.: Decoupling representation and classifier for long-tailed recognition. arXiv preprint arXiv:1910.09217 (2019)

14. Ren, J., Yu, C., Ma, X., Zhao, H., Yi, S., et al.: Balanced meta-softmax for long-tailed visual recognition. In: Advances in Neural Information Processing Systems, vol. 33, pp. 4175–4186 (2020)

15. Menon, A.K., Jayasumana, S., Rawat, A.S., Jain, H., Veit, A., Kumar, S.: Long-tail learning via logit adjustment. arXiv preprint arXiv:2007.07314 (2020)

16. Zhang, H., Cisse, M., Dauphin, Y.N., Lopez-Paz, D.: Mixup: beyond empirical risk minimization. arXiv preprint arXiv:1710.09412 (2017)
17. Galdran, A., Carneiro, G., González Ballester, M.A.: Balanced-mixup for highly imbalanced medical image classification. In: de Bruijne, M., et al. (eds.) MICCAI 2021. LNCS, vol. 12905, pp. 323–333. Springer, Cham (2021). https://doi.org/10. 1007/978-3-030-87240-3_31
18. Carratino, L., Cissé, M., Jenatton, R., Vert, J.P.: On mixup regularization. arXiv preprint arXiv:2006.06049 (2020)
19. Verma, V., et al.: Manifold mixup: better representations by interpolating hidden states. In: International Conference on Machine Learning, pp. 6438–6447. PMLR (2019)
20. Yun, S., Han, D., Oh, S.J., Chun, S., Choe, J., Yoo, Y.: Cutmix: regularization strategy to train strong classifiers with localizable features. In: Proceedings of the IEEE/CVF International Conference on Computer Vision, pp. 6023–6032 (2019)
21. Chapelle, O., Weston, J., Bottou, L., Vapnik, V.: Vicinal risk minimization. In: Advances in Neural Information Processing Systems, vol. 13 (2000)
22. Guo, H., Mao, Y., Zhang, R.: Mixup as locally linear out-of-manifold regularization. In: Proceedings of the AAAI Conference on Artificial Intelligence, vol. 33, pp. 3714–3722 (2019)
23. Elsayed, G., Krishnan, D., Mobahi, H., Regan, K., Bengio, S.: Large margin deep networks for classification. In: Advances in Neural Information Processing Systems, vol. 31 (2018)
24. Wang, F., Cheng, J., Liu, W., Liu, H.: Additive margin softmax for face verification. IEEE Sig. Process. Lett. **25**, 926–930 (2018)
25. Liu, W., Wen, Y., Yu, Z., Yang, M.: Large-margin softmax loss for convolutional neural networks. arXiv preprint arXiv:1612.02295 (2016)
26. Cui, Y., Jia, M., Lin, T.Y., Song, Y., Belongie, S.: Class-balanced loss based on effective number of samples. In: Proceedings of the IEEE/CVF Conference on Computer Vision and Pattern Recognition, pp. 9268–9277 (2019)
27. Wang, J., et al.: Seesaw loss for long-tailed instance segmentation. In: Proceedings of the IEEE/CVF Conference on Computer Vision and Pattern Recognition, pp. 9695–9704 (2021)
28. Ren, M., Zeng, W., Yang, B., Urtasun, R.: Learning to reweight examples for robust deep learning. In: International Conference on Machine Learning, pp. 4334–4343. PMLR (2018)
29. He, Z., Xie, L., Chen, X., Zhang, Y., Wang, Y., Tian, Q.: Data augmentation revisited: rethinking the distribution gap between clean and augmented data. arXiv preprint arXiv:1909.09148 (2019)
30. Zhang, Y., Wei, X.S., Zhou, B., Wu, J.: Bag of tricks for long-tailed visual recognition with deep convolutional neural networks. In: Proceedings of the AAAI Conference on Artificial Intelligence, vol. 35, pp. 3447–3455 (2021)
31. Goyal, P., et al.: Accurate, large minibatch SGD: training imagenet in 1 hour. arXiv preprint arXiv:1706.02677 (2017)
32. Cubuk, E.D., Zoph, B., Mane, D., Vasudevan, V., Le, Q.V.: Autoaugment: learning augmentation strategies from data. In: Proceedings of the IEEE/CVF Conference on Computer Vision and Pattern Recognition, pp. 113–123 (2019)

33. Chen, X., et al.: Imagine by reasoning: a reasoning-based implicit semantic data augmentation for long-tailed classification. In: Proceedings of the AAAI Conference on Artificial Intelligence, vol. 36, pp. 356–364 (2022)
34. Du, F., Yang, P., Jia, Q., Nan, F., Chen, X., Yang, Y.: Global and local mixture consistency cumulative learning for long-tailed visual recognitions. In: Proceedings of the IEEE/CVF Conference on Computer Vision and Pattern Recognition, pp. 15814–15823 (2023)

A Multi-perspective Squeeze Excitation Classifier Based on Vision Transformer for Few Shot Image Classification

Zebao Zhang[ID], Yuzhao Li[ID], and Ming He[✉][ID]

Harbin Engineering University, Harbin 150000, China
heming@hrbeu.edu.cn

Abstract. Few-shot image classification is a task that uses a small number of labeled samples to train a model to complete the classification task. Most few-shot image classification methods use small CNN-based models due to its good performance under supervised learning. However, small CNN-based models have performance bottlenecks under self-supervised learning with a large amount of unlabeled data. So we propose a model based on ViT for few-shot image classification. We propose a method combining Mask Image Modeling self-supervised learning and cross-architecture knowledge distillation to improve ViT. For few-shot image classification task, we propose a multi-perspective squeeze-excitation projector that is able to exploits the mutual information between samples in different perspectives, and aggregate in-class samples and discretize out-of-class samples. Finally, we construct a classifier based on it. Experimental results on Mini-ImageNet and Tiered-ImageNet show that our model achieves an average of 2% improvement over the previous state-of-the-art.

Keywords: Few-shot learning · Vision transformer · Image classification

1 Introduction

Few-shot image classification is a sub-research topic in computer vision, so it is inevitably affected by the current hot work in computer vision. With the rise of self-supervised learning, some works [4,12,15] propose self-supervised siamese networks for few-shot image classification. Due to self-supervised learning using large amounts of unlabeled data, the performance bottleneck of small CNN becomes more and more obvious. More and more researchers [4,12,14–16,19] try to use large backbone networks to complete few-shot tasks for better performance. ViT [9] has been shown to perform well in many vision tasks, but it has not attracted enough attention in few-shot learning due to its requirement

This work is supported by the National Key R&D Program of China (2022YFC3 301800) and the National Natural Science Foundation of China (Grant No. 62072135).

Q. Liu et al. (Eds.): PRCV 2023, LNCS 14432, pp. 80–92, 2024.
https://doi.org/10.1007/978-981-99-8543-2_7

for large amounts of training data. With more and more efforts to improve the training efficiency of ViT, this problem has been solved to some extent. [20,21] improve the training efficiency of ViT by knowledge distillation, [2,10,13,23] use Masked Image Modeling (MIM) to improve the performance of ViT. We are inspired by these works and construct a few-shot image classification model using ViT as the backbone network. We propose a MIM self-supervised cross-architecture distillation method to make ViT more suitable for few-shot learning tasks.

For few-shot image classification, we find that the different perspectives of the sample have a significant impact on its distribution in the latent space. We prove this in 4.3. So we propose a multi-perspective squeeze-excitation projector, which obtains different perspectives of the image, processed by its squeeze-excitation method, to aggregate the in-class samples and discretize the out-of-class samples. Finally, it improves distribution of feature maps in the latent space. We construct a classifier based on it. In the experiment, our model achieves an average of 2% improvement($92.01 \pm 0.16\%$) over the previous state-of-the-art on the Mini-ImageNet dataset for 5-way 5-shot classification task. Our contributions are as follows:

1. We propose a cross-architecture knowledge distillation method for MIM self-supervised learning on ViT, which improves the adaptability of ViT for few-shot tasks.
2. We propose a perspective adaptive fine-tune method. Based on it, we propose a multi-perspective squeeze-excitation projector, which improves the distribution of samples in the latent space.
3. Our model achieves an average of 2% improvement over the previous state-of-the-art on the Mini-ImageNet dataset.

2   Related Work

The research on few-shot image classification has developed rapidly in recent years. Here we only review recent studies that are most relevant to our work. [18] classifies the samples according to their position in the metric space. However, it is easily disturbed by wrong samples. We improve this calculation process to reduce the error. [7] proposes a few-shot learning model based on fine-tuning. This work brings us to the focus of our work: how to train a good backbone network for few-shot image classification? [8] introduce ViT in few-shot learning and get good results. These works inspire us to improve ViT as backbone network for few-shot image classification. [20,21] introduce distillation token into ViT. [19] introduces knowledge distillation into fine-tuning based few-shot learning model. This work inspires us to build an improved ViT distillation model for few-shot learning. [4,12,15] introduce self-supervised learning to improve the training process. These works inspire us to use self-supervised learning method to improve our backbone network. [14,16] improve the feature maps of the support samples based on the query samples. This idea of exploiting mutual information between samples inspires us. We want to build a classifier that fully exploits the mutual information.

Self-supervised learning is a method of training model using unlabeled datasets. Unlike the works [3,5,6] that uses contrastive self-supervised learning, our self-supervised method is more like Masked Image Modeling [2,10,13,22,23]. With the development of Masked Image Modeling(MIM), it shows good cross-domain adaptability than contrastive learning.

3   Method

In this section, we introduce the problem definition and our model architecture. We divide our work into two phases: meta-training and meta-testing. We propose a MIM self-supervised cross-architecture distillation model in the meta-training phase, and propose a multi-perspective squeeze-excitation classifier in the meta-testing phase.

3.1   Problem Definition

We divide the dataset into training set D_{train} and test set D_{test}. In few-shot image classification, D_{train} is used in the meta-training phase, and the D_{test} is used in the meta-testing phase. In the meta-testing phase, D_{test} is divided into many few-shot tasks. There is a support set $D_{test}^{support}$ for training and a query set D_{test}^{query} for testing in each task. Take our few-shot image classification work as an example. In the meta-training phase, we use D_{train} to train an encoder(backbone network) $y = f_\phi(x)$. In the meta-testing phase, we feed the output of the encoder to the classifier. We fine-tune the whole model with $D_{test}^{support}$ for each task as follows:

$$\theta^* = \underset{\theta^*}{argmin}\, Loss(Cls(f_\phi(D_{test}^{support})), \theta^*) \tag{1}$$

Where θ^* means the weights of whole model, $Loss$ means loss and Cls means our classifier. Then we test model with D_{test}^{query}. In the work of few-shot image classification, N-way K-shot indicates that the task is an N-classification task. Each class has K samples.

3.2   Meta-Training Phase

In the meta-training phase, we train an encoder in a MIM self-supervised way, and use a cross-architecture method to distill the encoder.

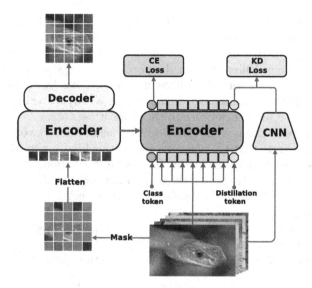

Fig. 1. Meta-training phase: The D_{train} are fed into Encoder-Decoder for MIM Self-supervised learning. CNN is then used as a teacher network for cross-architecture knowledge distillation. The Encoder is fine-tuned by reducing the cross-entropy loss and knowledge distillation loss.

MIM Self-supervised Learning. Masked Image Modeling is a kind of self-supervised learning that train model by predicting the masked image. Compared with contrastive learning, it has higher training efficiency [10,13]. Usually, few-shot learning uses a small CNN as the backbone network [7,19]. But there is a performance bottleneck for small CNN that are not suitable for self-supervised pre-training on big datasets. So we use the ViT-Base [9] as backbone network (enocder). ViT is a data-inclusive model that fits large datasets well. We want to get a pre-trained encoder with good scaling performance on large datasets.

As shown in Fig. 1, we first segment the input image into patches and mask them randomly. We put the uncovered patches into encoder to obtain feature maps. Then, we combine these feature maps with the covered patch, put them into decoder for predicting. With this "cloze" task, we can get a well-trained encoder.

Unlike [2,15,23] that use a simple MLP as decoder, our work use a ViT as decoder to get higher linear probing accuracy. Linear probing accuracy is important for few-shot image classification because there is only a few training set to fine-tune model. In addition, we use a higher mask rate(75%). Finally, mean squared error (MSE) is used, which is the difference between the original image and the predicted image on each pixel as follows:

$$Loss = \frac{\sum_{i=0}^{n} MSE(D_{train}^{predicted(i)} - D_{train}^{(i)})}{n} \tag{2}$$

In order to get better encoder, pixel differences on masked patches are calculated as loss.

Cross-Architecture Knowledge Distillation. ViT is different from CNN that ViT lacks inductive bias [9] such as translation invariance. Due to the absence of this inductive bias, the performance of ViT in some specific cases needs to be improved. Different from the previous work [8] modifying the ViT, we choose to compensate for the absence of inductive bias by cross-architecture knowledge distillation. Knowledge distillation is a way to transfer the prior knowledge from the teacher network to the student network.

There are great differences in the effect of teacher networks on student networks. [1] demonstrate that the absence of inductive bias due to different structural networks can be compensated by knowledge distillation. The student network will learn the inductive bias in the teacher network. Inspired by these works [1,19–21], we choose CNN as the teacher network to make ViT learn inductive bias by knowledge distillation.

As shown in Fig. 1, we use D_{train} to train the teacher model $y = f_\sigma(x)$. To be able to distill ViT, we modify the patch embedding layer and we incorporate a distillation token. We use a similar distillation method from [20, 21] as follows:

$$Loss_{CE} = CE(f_\phi(D_{train}), label) \tag{3}$$

$$Loss_{KL} = \tau^2 * KL(f_\phi(D_{train})/\tau, f_\sigma(D_{train})/\tau) \tag{4}$$

$$Loss_{KD} = \lambda * Loss_{CE} + (1 - \lambda) * Loss_{KL} \tag{5}$$

As shown in Eqs. 3, 4, 5, $f_\phi(D_{train})$ is the probability distribution obtained by feeding D_{train} into the encoder, and $f_\sigma(D_{train})$ is the probability distribution obtained by feeding D_{train} into the CNN. KL means Kullback-Leibler divergence. τ represents the distillation temperature. λ is a hyperparameter, which can be set slightly differently for different datasets.

After cross-architecture knowledge distillation, the encoder is improved. As shown in Sect. 4.3, ViT get a better result in few-shot learning. In this phase, we obtain a good encoder that scales well on large datasets.

3.3   Meta-test Phase

In the meta-test phase, we propose a multi-perspective squeeze-excitation projector. We propose a perspective adaptive fine-tune (PAF) method to train the projector. Finally, we build a classifier based on the projector for few-shot image classification.

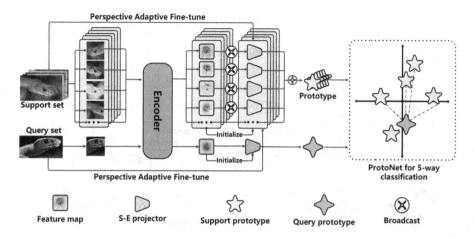

Fig. 2. Meta-test Architecture: The support are fed into the encoder after data augment. The obtained multi-perspective feature maps are used to initialize projectors. Then, PAF method are used to fine-tune the projectors. Finally, the prototypes are obtained by averaging the output from projectors.

Multi-perspective Squeeze-Excitation Projector. The samples contain different latent layer information in different perspectives. Different kinds of latent information have different importance. In order to dynamically match various latent layer information with samples, we propose the multi-perspective squeeze-excitation projector. As shown in the Fig. 2, we can obtain feature maps of $D_{test}^{support}$ in different perspectives from encoder after data augmentation. To improve the feature maps in each perspective, we propose a squeeze-excitation projector(S-E projector), which is a linear layer. We initialize it with support-base initialization. This initialization method takes the feature map as the initial weight, so the projector can quickly get the key information in different perspectives. As shown in Eq. 6 and Eq. 7, x^i is the ith sample in a class of $D_{test}^{support}$. We feed n samples into the encoder f_ϕ to get n feature maps. We average the maps to get the v. We use v divided by $||v||$ to get attention distribution w, which is the initial weights of projector. Then, we propose a Perspective Adaptive Fine-tune(PAF) method to fine-tune the S-E projector.

$$v = \frac{\sum_{i=1}^{n} f_\phi(x^i)}{n} \tag{6}$$

$$w = \frac{v}{||v||} \tag{7}$$

Perspective Adaptive Fine-Tune. After weights initialization, we propose a Perspective Adaptive Fine-tune(PAF) method to fine-tune the weights. The PAF method calculate the mutual information between support sample and query sample in each perspective. The interaction between samples enables projector to get inter-perspective knowledge. The knowledge enables projector to increase the key attributes of feature map for similar samples. For different classes of samples, projector will increase different attributes to expand diversity.

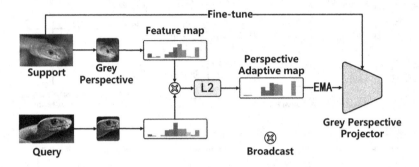

Fig. 3. Perspective Adaptive Fine-tune

As shown in Fig. 3, We first get the feature maps of the support sample and the query sample in the grayscale perspective(other perspectives will also be calculated). Then we flatten the feature map and do a broadcast operation to multiply the values at the corresponding position. Finally, after L2 normalization, we get a perspective adaptive map to update the parameters of projector. To avoid drastic parameter change, we use the EMA to update it smoothly.

After PAF, we get a trained projector. We use the projector to aggregate samples in the same class and discretize samples in different classes. This is done with broadcast operation. As shown in Fig. 2, we get the feature map before broadcast operation, and get the prototype after averaging the multi-perspective feature maps. The whole process is as follows:

$$P = \frac{\sum_{i=1}^{n} \sum_{j=1}^{m} \rho_j(f_\phi(x_i))}{n * m} \tag{8}$$

As shown in Eq. 8, we get m projectors $\rho_j(1 \leq j \leq m)$ after PAF. We feed $m * n$ perspectives of n samples into encoder f_ϕ, then feed the obtained feature maps into ρ, and average the outputs $\rho_j(f_\phi(x_i))$ to get the prototype center P of the class.

The prototype calculated by our projector has a noticeable effect. As shown in Fig. 5, the distance of similar prototype is pulled closer, and the distance of different prototype is pulled further. As shown in Fig. 2, prototype center represents a class, we measure the distance between the query set D_{test}^{query} and the prototype to classify them.

4   Experimental Results

4.1   Datasets and Training Details

Mini-ImageNet is a subset of ILSVRC-12, which consists of 100 classes, each containing 600 images. Tiered-ImageNet is also a subset of ILSVRC-12, it is divided into 34 broad categories, each containing 10 to 30 classes. We use ViT-Base as our encoder. In the knowledge distillation phase, we use RegNetY-12GF as the teacher network. In the MIM self-supervised learning phase, we adopt random-size cropping and horizontal flipping to augment datasets. We adopt AdamW as the optimizer. lr is set to $1e-3$ and weight decay is set to $1e-2$. The encoder is trained for 300 epochs. 20 warmup epochs are used. In the cross-architecture knowledge distillation phase, lr is set to $1e-4$ and weight decay is set to $5e-2$.

4.2 Evaluation Results

For experimental comparison, we use the same task as in the previous work [4,8,15,19]. We randomly sample 1000 few-shot tasks from the test set and take the average accuracy (95% confidence interval) as the experimental result. As shown in Table 1, our model achieves an accuracy of $82.67 \pm 0.46\%$ in the 5-way 5-shot scenarios on Mini-ImageNet. [4,7,19] use CNN-based network as backbone network. Their performance is limited by CNN. [19] achieves an accuracy of $82.14 \pm 0.43\%$ by knowledge distillation. [4] uses self-supervised learning to improve CNN. Instead of backbone selection, [24] focus on extracting mutual information between support samples and query samples to help classification. Our model takes the advantages of the above work [4,7,18,19,24] to achieve higher accuracy.

[8] use ViT as backbone instead of CNN-based network to get better result. [8] modify loss to make ViT more suitable for few-shot task. [8] get the best result of $91.72 \pm 0.11\%$ on Tiered-ImageNet because of its improvement to architecture of ViT. [11,16] take extracting mutual information between samples a step further, and [11] get a better result of $89.96 \pm 0.55\%$ on Tiered-ImageNet. For better

Table 1. Classification Accuracies on Mini-Imagenet and Tiered-Imagenet

Model	Mini-ImageNet		Tiered-ImageNet	
	5-way 1-shot	5-way 5-shot	5-way 1-shot	5-way 5-shot
ProtoNet [18]	53.31 ± 0.89	72.69 ± 0.74		
Baseline [7]	57.73 ± 0.62	78.17 ± 0.49	66.58 ± 0.70	85.55 ± 0.48
R-distill [19]	64.82 ± 0.60	82.14 ± 0.43	71.52 ± 0.69	86.03 ± 0.49
DeepEMD [24]	65.91 ± 0.82	82.41 ± 0.56	71.16 ± 0.87	86.03 ± 0.58
AmdimNet [4]	64.03 ± 0.20	81.15 ± 0.14		
Ours	$\mathbf{64.96 \pm 0.55}$	$\mathbf{82.67 \pm 0.46}$	$\mathbf{72.56 \pm 0.55}$	$\mathbf{86.78 \pm 0.62}$
SUN [8]	67.80 ± 0.45	83.25 ± 0.30	72.99 ± 0.50	86.74 ± 0.33
INSTA-DeepEMD [16]	68.46 ± 0.48	84.21 ± 0.82	73.87 ± 0.31	88.02 ± 0.61
RGFS [11]	72.40 ± 0.78	86.38 ± 0.49	76.32 ± 0.87	89.96 ± 0.55
HCTransformers [8]	74.74 ± 0.17	89.19 ± 0.13	79.67 ± 0.20	91.72 ± 0.11
Ours++	$\mathbf{77.73 \pm 0.21}$	$\mathbf{92.01 \pm 0.16}$	$\mathbf{78.91 \pm 0.11}$	$\mathbf{91.50 \pm 0.23}$

Ours++ means using more unlabeled data in MIM self-supervised learning phase.

performance, our projector take the advantages of [11,16] to extract more mutual information between samples. And we use more unlabeled data(ImageNet1K without validation set) in meta-training phase. Our model achieves the best result of 92.01 ± 0.16% for 5-way 5-shot on Mini-ImageNet. This proves that using more unlabeled data in training can improve the adaptation of the ViT for few-shot learning, and also proves that our model has good scalability.

4.3   Ablation Study

Table 2. Few-Shot Image Classification Accuracy

5-shot	CUB200	FC-100	M-imageNet	T-ImageNet
CNN	54.93	48.67	68.20	74.09
ResNet	60.61	69.10	79.93	84.41
ViT+CL	59.89	67.86	82.11	86.40
ViT+MIM	63.41	68.44	82.67	86.78
	Cross-Domain			
5-shot	CUB200	StanfordCars	Omniglot	CIFAR-FS
ViT+CL	56.68	46.78	90.51	77.37
ViT+MIM	56.23	46.90	90.92	78.05

MIM Self-supervised Learning for Few-Shot Image Classification. In order to verify the MIM self-supervised learning phase, we remove the MIM phase and retain supervised training. As shown in Table 2, CL means contrastive learning is used after the removal. CNN and ResNet mean methods use CNN-12 or ResNet-18 as the backbone network. We validate our model on four few-shot image classification datasets. Among them, CUB-200 is a 200-class bird dataset and FC-100 is a 100-class flower dataset. As shown in Table 2, there is obviously a performance bottleneck for small CNN-based networks that are not suitable for self-supervised learning on big data. On small datasets, ResNet get a better

Table 3. Few-Shot Image Classification Results Using Different Teacher Networks

Teacher Network	1-shot	5-shot	S.D.
No distillation	62.41	79.27	14.43
ViT-Base	59.82	75.14	14.21
Self-distillation	62.93	80.81	12.76
RegNetY-8GF	64.78	82.49	11.12
RegNetY-12GF(ours)	64.96	82.67	10.90
RegNetY-16GF	64.38	82.21	11.04

result. On large datasets, ViT get better results. In addition, When the MIM phase is not used, the classification accuracy on CUB-200 is 59.89%, and after using it, the accuracy is improved to 63.41%. There is a similar result for the cross-domain dataset. The results demonstrate the effect of MIM phase. MIM self-supervised learning allows backbone to learn more information about the samples. To some extent, it solves the drawbacks of "data hungry" [9].

Cross-Architecture Knowledge Distillation for Few-Shot Image Classification. As shown in Table 3, we adopt different architectures as teacher networks to verify the effect of cross-architecture knowledge distillation. We use the average accuracy over 1000 runs on Mini-ImageNet as the result. The S.D. represent the standard deviation of it. According to Table 3, we find that using knowledge distillation is generally better than not using it. Without knowledge distillation, the classification accuracy is $79.27 \pm 0.64\%$. With knowledge distillation, the accuracy is improved. RegNetY-12GF (ours) as the teacher network has the best effect, with an accuracy of $82.67 \pm 0.39\%$. We notice that the performance of using ViT-Base as the teacher model instead drops. It makes sense that ViT does not perform well on small datasets as a "data hungry" [9] model, which fails to guide the loss of the student network close to the global minimum. In addition, We find that the model using knowledge distillation has a smaller standard deviation, which means that the distillation model is more stable for different few-shot classification tasks. We argue that cross-architecture knowledge distillation enables ViT to learn the inductive bias in RegNetY-12GF [17], which improves the performance of ViT and thus improves few-shot image classification accuracy.

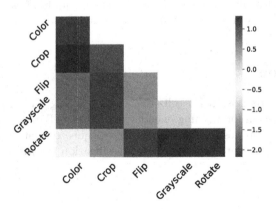

Fig. 4. Multi-perspective Squeeze-excitation Projector Heatmap

Multi-perspective Squeeze-Excitation Projector for Few-Shot Image Classification. As shown in Fig. 4, we use a heatmap to represent the effect of different combinations of multi-perspective squeeze-excitation projectors. Each

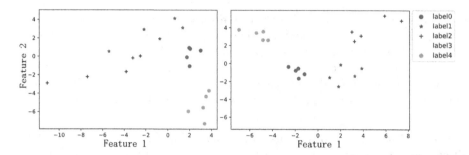

Fig. 5. 2D Scatter Plot Of Image Features: the left is the scatter plot of the original 5-way 5-shot sample features, and the right is the the scatter plot after processing by our projector.

value represents the increment of classification accuracy on Mini-ImageNet. We take the original ProtoNet [18] as the baseline. We see that the use of Rotate alone decreases the classification accuracy. We conjecture that Rotate interferes with the translation invariance that encoder learns from CNN-based networks. Besides, the use of Horizontal Filp, Grayscale, Crop and Color improves the classification accuracy. Among them, the effect of Color and Crop is more significant, which improves the original classification accuracy by 0.92% and 0.75% respectively. For better results, we combine multiple projectors to build our classifier.

To visualize the effect of our multi-perspective squeeze-excitation projectors, we use PCA method to reduce the dimension of feature maps, and draw it by two-dimensional scatter plot. As shown in Fig. 5, our projectors aggregate the in-class samples and discretize the out-of-class samples. Compared with the original scatter plot, our projectors make label1 and label3 more separated, and the label1 more clustered.

5   Conclusion

In this paper, we propose MIM self-supervised learning and cross-architecture knowledge distillation for training encoder. It makes ViT more suitable for few-shot learning. We propose a multi-perspective squeeze-excitation classifier for few-shot image classification. It improves classification accuracy.

References

1. Abnar, S., Dehghani, M., Zuidema, W.: Transferring inductive biases through knowledge distillation (2020)
2. Bao, H., Dong, L., Piao, S., Wei, F.: Beit: BERT pre-training of image transformers. In: International Conference on Learning Representations (2021)
3. Caron, M., et al.: Emerging properties in self-supervised vision transformers. In: Proceedings of the IEEE/CVF International Conference on Computer Vision, pp. 9650–9660 (2021)

4. Chen, D., Chen, Y., Li, Y., Mao, F., He, Y., Xue, H.: Self-supervised learning for few-shot image classification. In: ICASSP 2021–2021 IEEE International Conference on Acoustics, Speech and Signal Processing (ICASSP), pp. 1745–1749 (2021). https://doi.org/10.1109/ICASSP39728.2021.9413783

5. Chen, X., He, K.: Exploring simple Siamese representation learning. In: Proceedings of the IEEE/CVF Conference on Computer Vision and Pattern Recognition, pp. 15750–15758 (2021)

6. Chen, X., Xie, S., He, K.: An empirical study of training self-supervised vision transformers. In: Proceedings of the IEEE/CVF International Conference on Computer Vision, pp. 9640–9649 (2021)

7. Dhillon, G.S., Chaudhari, P., Ravichandran, A., Soatto, S.: A baseline for few-shot image classification. In: International Conference on Learning Representations (2019)

8. Dong, B., Zhou, P., Yan, S., Zuo, W.: Self-promoted supervision for few-shot transformer. In: Avidan, S., Brostow, G., Cissé, M., Farinella, G.M., Hassner, T. (eds.) ECCV 2022, Part XX. LNCS, vol. 13680, pp. 329–347. Springer, Cham (2022). https://doi.org/10.1007/978-3-031-20044-1_19

9. Dosovitskiy, A., et al.: An image is worth 16x16 words: transformers for image recognition at scale. In: International Conference on Learning Representations (2020)

10. He, K., Chen, X., Xie, S., Li, Y., Dollár, P., Girshick, R.: Masked autoencoders are scalable vision learners. In: Proceedings of the IEEE/CVF Conference on Computer Vision and Pattern Recognition, pp. 16000–16009 (2022)

11. Hiller, M., Ma, R., Harandi, M., Drummond, T.: Rethinking generalization in few-shot classification. In: Advances in Neural Information Processing Systems (2022)

12. Hu, S.X., Li, D., Stühmer, J., Kim, M., Hospedales, T.M.: Pushing the limits of simple pipelines for few-shot learning: external data and fine-tuning make a difference. In: 2022 IEEE/CVF Conference on Computer Vision and Pattern Recognition (CVPR), pp. 9058–9067 (2022). https://doi.org/10.1109/CVPR52688.2022.00886

13. Huang, L., You, S., Zheng, M., Wang, F., Qian, C., Yamasaki, T.: Green hierarchical vision transformer for masked image modeling. In: Advances in Neural Information Processing Systems (2022)

14. Li, P., Gong, S., Wang, C., Fu, Y.: Ranking distance calibration for cross-domain few-shot learning. In: 2022 IEEE/CVF Conference on Computer Vision and Pattern Recognition (CVPR), pp. 9089–9098 (2022). https://doi.org/10.1109/CVPR52688.2022.00889

15. Lin, H., Han, G., Ma, J., Huang, S., Lin, X., Chang, S.F.: Supervised masked knowledge distillation for few-shot transformers. arXiv preprint arXiv:2303.15466 (2023)

16. Ma, R., et al.: Learning instance and task-aware dynamic kernels for few-shot learning. In: Avidan, S., Brostow, G., Cissé, M., Farinella, G.M., Hassner, T. (eds.) ECCV 2022. LNCS, vol. 13680, pp. 257–274. Springer, Cham (2022). https://doi.org/10.1007/978-3-031-20044-1_15

17. Radosavovic, I., Kosaraju, R.P., Girshick, R., He, K., Dollár, P.: Designing network design spaces. In: Proceedings of the IEEE/CVF Conference on Computer Vision and Pattern Recognition, pp. 10428–10436 (2020)

18. Snell, J., Swersky, K., Zemel, R.: Prototypical networks for few-shot learning. In: Advances in Neural Information Processing Systems, vol. 30 (2017)

19. Tian, Y., Wang, Y., Krishnan, D., Tenenbaum, J.B., Isola, P.: Rethinking few-shot image classification: a good embedding is all you need? In: Vedaldi, A., Bischof, H., Brox, T., Frahm, J.-M. (eds.) ECCV 2020. LNCS, vol. 12359, pp. 266–282. Springer, Cham (2020). https://doi.org/10.1007/978-3-030-58568-6_16
20. Touvron, H., Cord, M., Douze, M., Massa, F., Sablayrolles, A., Jégou, H.: Training data-efficient image transformers & distillation through attention. In: International Conference on Machine Learning, pp. 10347–10357. PMLR (2021)
21. Touvron, H., Cord, M., Jégou, H.: DeiT III: revenge of the ViT. In: Avidan, S., Brostow, G., Cissé, M., Farinella, G.M., Hassner, T. (eds.) ECCV 2022, Part XXIV. LNCS, vol. 13684, pp. 516–533. Springer, Cham (2022). https://doi.org/10.1007/978-3-031-20053-3_30
22. Wang, H., Tang, Y., Wang, Y., Guo, J., Deng, Z.H., Han, K.: Masked image modeling with local multi-scale reconstruction. In: Proceedings of the IEEE/CVF Conference on Computer Vision and Pattern Recognition (CVPR), pp. 2122–2131 (2023)
23. Xie, Z., et al.: Simmim: a simple framework for masked image modeling. In: 2022 IEEE/CVF Conference on Computer Vision and Pattern Recognition (CVPR), pp. 9643–9653 (2022). https://doi.org/10.1109/CVPR52688.2022.00943
24. Zhang, C., Cai, Y., Lin, G., Shen, C.: DeepEMD: few-shot image classification with differentiable earth mover's distance and structured classifiers. In: Proceedings of the IEEE/CVF Conference on Computer Vision and Pattern Recognition, pp. 12203–12213 (2020)

ITCNN: Incremental Learning Network Based on ITDA and Tree Hierarchical CNN

Pengyu Wang[1,3], Tao Ren[1(✉)], Jiaxin Liu[2,3], Wei Liu[2,3], Jun Hu[3], Shuai Cheng[3], and Dazong Zhang[4]

[1] Institute of Network Science and Big Data Technology,
Software College, Northeastern University, Shenyang 110169, China
chinarentao@163.com
[2] School of Computer Science and Engineering,
Northeastern University, Shenyang 110169, China
[3] Neusoft Reach Automotive Technology Company, Shenyang 110179, China
[4] BYD Company Limited, Shenzhen 518118, China

Abstract. In class incremental learning, sensitive changes in network parameter tuning pose a significant problem. To address this challenge, we propose a new model called ITCNN, which combines a tree-like hierarchical network structure with incremental tensor discriminant analysis (ITDA). Unlike previous deep convolutional neural networks (DCNNs) that rely on data organization, ITCNN is feature-driven, using a tree hierarchy that can learn and grow with new data while retaining the classification functions of previous categories. The use of ITDA obviates the need to represent high-dimensional training samples in vector form, and the incremental computation has less space and time complexity. Compared with other incremental learning methods, our proposed network achieves competitive accuracy on multiple datasets. Our results suggest that ITCNN is an effective model for addressing the challenges of class incremental learning, providing increased accuracy while retaining efficiency and scalability.

Keywords: Class incremental learning · Incremental tensor discriminant analysis · Tree hierarchical CNN · Tensor analysis

1 Introduction

Deep Convolutional Neural Networks (DCNNs) are widely used for computer vision tasks. They use nonlinear transformations to learn hierarchical representations of data through feature extraction and classification. Shallow layers

This work is partially supported by National Natural Science Foundation of China (62276058, 61902057), Fundamental Research Funds for the Central Universities (N2217003), Joint Fund of Science and Technology Department of Liaoning Province and State Key Laboratory of Robotics, China(2020-KF-12-11).

retrieve coarse-grained features, while deep layers retrieve fine-grained features [17]. These features can be semantic groupings or feature-driven groupings [18]. However, adjusting certain parameters can negatively impact accuracy and generalization, and it is difficult to determine where errors occur in the network. In addition, small features can significantly affect classification results [12].

To address these challenges, researchers have developed methods such as Learning without Forgetting (LwF) [15] and incremental learning (IL) methods such as iCaRL [11], which refine and improve DCNN performance through fine-tuning, joint training, and feature extraction. The latter is commonly known as class incremental learning (CIL). CIL gradually learns a unified classifier to recognize all new classes observed so far. However, these approaches are often highly dependent on the specific task at hand.

To overcome the limitations of DCNNs, researchers have also combined them with decision trees (DTs) to create hybrid models [6,16,20]. DTs can optimize their architecture based on training data, which is especially helpful when data is scarce [2]. These hybrid models also require prior knowledge of all categories and their attributes to build a hierarchical model [16]. Neural-Backed Decision Trees (NBDT) are a breakthrough in this field [21]. Although NBDT is not as accurate as traditional neural networks, it is better than classical decision trees. However, combining the two approaches can provide complementary benefits and allow for a better understanding of issues at different levels of the hierarchy. It is important to note that building and implementing these hybrid models requires expertise and significant training data [4,8,10].

However, some of the current disadvantages of incremental learning include:

1. Knowledge drift: IL typically requires incremental adjustments to existing models, which can lead to previous knowledge being "forgotten" or "drifting," affecting model performance [24].
2. Training time: Since CIL requires continuous IL, it usually requires more training time and computational resources, especially when dealing with large datasets.
3. Sample imbalance: CIL requires the continuous addition of new classes, which may lead to sample imbalance between classes and thus affect model performance.
4. Hyperparameter tuning: CIL requires more hyperparameter tuning, such as learning rate, regularization parameters, etc., which requires more human and computational resources.

Advances in technology lead to the gradual resolution of certain problems. Knowledge drift, for example, can be addressed through methods such as knowledge distillation [3]. In addition, IL networks can help improve training efficiency and sample balance [5,7,13,14,19]. These developments contribute significantly to the advancement of the academic field.

ITCNN is a hierarchical model that addresses CIL limitations by incorporating a tree structure. ITCNN uses shallow layers to extract coarse-grained features and employs incremental tensor discriminant analysis (ITDA) [23] and branch routing strategy to control node splitting. When new data is added to the dataset, ITCNN infers and analyzes new samples to locate nodes with inaccurate inferences. ITCNN effectively mitigates these issues while maintaining competitive accuracy across multiple datasets.

2 Proposed Network

2.1 Network Structure

ITCNN uses Leaf Cells and Normal Cells to achieve its structure, as shown in Fig. 1.

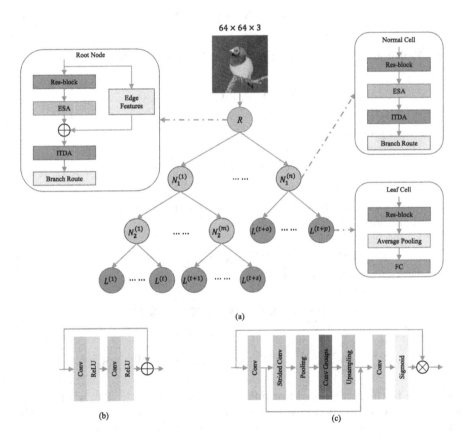

Fig. 1. The structure of ITCNN. (a) The tree-like branching structure of ITCNN and the modules it employs. (b) residual block. (c) Enhanced Spatial Attention.

The architecture of ITCNN is comparable to Tree-CNN [16], but with the integration of residual blocks and Enhanced Spatial Attention (ESA) blocks to enhance the generalization performance of extracted features using fewer parameters. Since edge features extracted by conventional operators such as Sobel and

Canny can provide a rough classification of samples, ITCNN integrates these features at the root node to reduce the number of parameters used in the residual module and ESA. In order for ITCNNs to function properly, it is essential to have a pre-designed classification of artificial datasets into superclasses or a predetermined number of node splits.

2.2   ITDA

For multi-dimensional image data $\mathcal{X} = \{\mathcal{X}_1, \cdots, \mathcal{X}_K\}$, $\mathcal{X}_i \in \mathbb{R}^{I_1 \times \cdots \times I_N}$, its corresponding category is $l(i) \in [1, C]$ where C denotes the number of categories, and let the sample of category c be n_c, then the within-class scatter tensor is

$$\mathcal{S}_w = \sum_{c=1}^{C} \sum_{i=1}^{n_c} \left\| \mathcal{X}_i - \bar{\mathcal{X}}_c \right\|^2 \tag{1}$$

where $\bar{\mathcal{X}}_c$ denotes the mean tensor of class c. The between-class scatter tensor is

$$\mathcal{S}_b = \sum_{c=1}^{C} n_c \left\| \bar{\mathcal{X}}_c - \bar{\mathcal{X}} \right\|^2 \tag{2}$$

and the total scatter tensor is

$$\mathcal{S}_t = \sum_{i=1}^{K} \left\| \bar{\mathcal{X}}_i - \bar{\mathcal{X}} \right\|^2 = \mathcal{S}_w + \mathcal{S}_b \tag{3}$$

Thus, mode-n within-class scatter matrix, mode-n between-class scatter matrix, and mode-n total scatter matrix are

$$\mathbf{S}_w^{(n)} = \sum_{c=1}^{C} \sum_{i=1}^{n_c} (\mathbf{X}_i^{(n)} - \bar{\mathbf{X}}_c^{(n)})(\mathbf{X}_i^{(n)} - \bar{\mathbf{X}}_c^{(n)})^T \tag{4}$$

$$\mathbf{S}_b^{(n)} = \sum_{c=1}^{C} n_c (\bar{\mathbf{X}}_c^{(n)} - \bar{\mathbf{X}}^{(n)})(\bar{\mathbf{X}}_c^{(n)} - \bar{\mathbf{X}}^{(n)})^T \tag{5}$$

$$\mathbf{S}_t^{(n)} = \sum_{i=1}^{K} (\bar{\mathbf{X}}_i^{(n)} - \bar{\mathbf{X}}^{(n)})(\bar{\mathbf{X}}_i^{(n)} - \bar{\mathbf{X}}^{(n)})^T = \mathbf{S}_w^{(n)} + \mathbf{S}_b^{(n)} \tag{6}$$

According to the incremental learning of multiple samples in [23], when several samples are added, new added samples $\mathcal{X}_{new} = \{\mathcal{X}_K, \cdots, \mathcal{X}_{K+T}\}$, $T \geq 1$, the corresponding category are $l_{new} = l_1, \cdots, l_T$ it is assumed that n_{new_r} samples belong to the r-th class, then the mean tensor of the r-th class is

$$\bar{\mathcal{X}}_r = \frac{1}{n_{old\_r} + n_{new\_r}} \left(n_{old\_r} \bar{\mathcal{X}}_{old\_r} + \sum_{l(i)=r, i=1}^{n_{new\_r}} \mathcal{X}_i \right)$$
$$= \frac{(n_{old\_r} \bar{\mathcal{X}}_{old\_r} + n_{new\_r} \bar{\mathcal{X}}_{new\_r})}{n_{old\_r} + n_{new\_r}} \tag{7}$$

Thus the total mean tensor is updated as $\bar{\mathcal{X}} = (K\bar{\mathcal{X}}_{old} + T\bar{\mathcal{X}}_{new})/(K + T)$ and the inter-class scatter mean tensor is updated as $\mathcal{S}_b = \sum_{c=1}^{C} n_c \left\| \bar{\mathcal{X}}_c - \bar{\mathcal{X}} \right\|^2$. For T new samples, there are n_{C+1} samples belonging to new class label $C+1$, then updated mode-n inter-class scatter matrix is

$$\mathbf{S}_b^{(n)} = \sum_{c=1}^{C+1} n_c (\bar{\mathbf{X}}_c^{(n)} - \bar{\mathbf{X}}^{(n)})(\bar{\mathbf{X}}_c^{(n)} - \bar{\mathbf{X}}^{(n)})^T \tag{8}$$

and the corresponding mode- n intra-class scatter matrix is

$$\mathbf{S}_w^{(n)} = \sum_{c=1}^{C}\sum_{i=1}^{n_c}(\mathbf{X}_i^{(n)} - \bar{\mathbf{X}}_c^{(n)})(\mathbf{X}_i^{(n)} - \bar{\mathbf{X}}_c^{(n)})^T + \sum_{i=1}^{n_{c+1}}(\mathbf{X}_i^{(n)} - \bar{\mathbf{X}}_{C+1}^{(n)})(\mathbf{X}_i^{(n)} - \bar{\mathbf{X}}_{C+1}^{(n)})^T \tag{9}$$

To make within-class scatter tensor smaller and between-class scatter tensor larger, the objective function of ITDA is

$$J(\mathbf{U}^{(n)}) = \arg\max \frac{\mathbf{U}^{(n)^T}\mathbf{S}_b^{(n)}\mathbf{U}^{(n)}}{\mathbf{U}^{(n)^T}\mathbf{S}_w^{(n)}\mathbf{U}^{(n)}} \tag{10}$$

2.3 Branch Route

Equation (10) can be converted into trace difference form and solved using the iterative method

$$J(\mathbf{U}_t^{(n)}) = \arg\max tr(\mathbf{U}_t^{(n)^T}(\mathbf{S}_b^{(n)} - \gamma\mathbf{S}_w^{(n)})\mathbf{U}_t^{(n)}) \tag{11}$$

where t denotes the number of iterations and $\gamma = \frac{\mathbf{U}_{t-1}^{(n)^T}\mathbf{S}_b^{(n)}\mathbf{U}_{t-1}^{(n)}}{\mathbf{U}_{t-1}^{(n)^T}\mathbf{S}_w^{(n)}\mathbf{U}_{t-1}^{(n)}}$ denotes the trace ratio for the $t-1$ iteration, $\mathbf{U}^{(n)^T}\mathbf{U}^{(n)} = \mathbf{I}$, and the iterative process converges to a local optimum. In CNN, the feature maps obtained by some convolution kernels are sparse and therefore the feature tensor obtained is also sparse. Therefore, the sparse method [9] can be used to accelerate the solution. The nearest neighbor algorithm based on Euclidean distance is used for classification.

Figure 2 illustrates the IL process for ITCNN, where a small number of samples are added to enable the identification of new classes. ITCNN uses the ITDA to determine the branch that corresponds to the newly added data, and then infer the image.

After determining the number of node splits, the ITDA and Branch Route in Root Node and Normal Cell are replaced with fully connected layers. ITCNN then trains its parameters solely on the path from the root to the new leaf. The branch is further tuned using the corresponding sub-structure data, whereby parameter updates are confined to the Leaf Cell if the best case occurs, and on the path from the root node to the new class or leaf nodes if the worst case occurs. Through this methodology, ITCNN achieves efficient learning without the need for excessive computational resources.

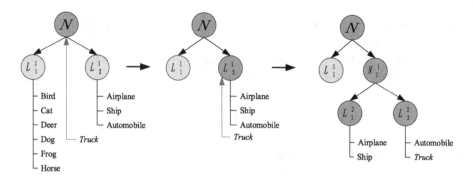

Fig. 2. The model update using the Cifar-10 data set. The structure on the right side of the model (blue node) is the updated and generated new structure. (Color figure online)

2.4   Training Strategies

The ITCNN adopts a hierarchical training strategy. Initially, the model is pretrained on a limited number of samples (specifically 20% of the data) during training. The procedure can be summarized as follows:

1. We train the root node to partition the dataset into exactly N superclasses, denoted as $S_1, S_2, \ldots, S_N$. The root node parameters are fixed.
2. The samples assigned to the superclass S_i are then fed through the root node to generate significant feature maps for node N_i^1. The Normal Cell is used to train the nodes N_i^1.
3. If node N_i^1 can classify its assigned superclass, the training process continues for subsequent nodes N_u^j in the next layer using the same approach. However, if the node N_i^1 fails to divide the assigned superclass, it is instead trained using the Leaf Cell technique.

2.5   Optimization Strategies

The cellular structure depicted in Fig. 1 presents a problem of parameter redundancy or insufficiency in nodes. To optimize each node, we implemented the Efficient Neural Architecture Search (ENAS) method. This process involves predefining search space boundaries, considering the hierarchical relationship of classifications in the dataset:

1. The number of residual blocks in Normal Cell and Leaf Cell does not exceed 5.
2. The maximum number of splits between the root node and the Normal Cell is not greater than 5.
3. The leaf nodes only use 1 layer of FC.

ENAS searches the number of residual blocks for each node. The model reuses the structure and parameters of the previous node in searching for child nodes.

3   Experiments and Results

3.1   Experiment on Classification

We test the ITCNN on Cifar-100 and ImageNet. We use three different structures of ITCNN:

1. ITCNN-20 uses the Cell structure shown in Fig. 1, which connects 20 leaf nodes directly under the root node according to the super-class division rule of the Cifar-100 dataset.
2. ITCNN-Cell is a structure generated from a data set.
3. ITCNN-NAS is a structure optimized by ENAS.

Cifar-100 samples are resized to $64 \times 64 \times 3$, and ImageNet images are resized to $224 \times 224 \times 3$. Despite this difference, ITCNN achieves competitive accuracy, as shown in Table 1, while having a faster computational speed than most competing models. Although ITCNN has more parameters, only a portion of them are used in inference. The number of parameters of ITCNN-NAS is close to that of lightweight networks, but the accuracy is significantly better than them.

Table 1. Model performance and results on Cifar-100 and ImageNet

Method	MAdds	Params.	Cifar-100		ImageNet	
			Top-1 ACC (%)	Top-5 ACC (%)	Top-1 ACC (%)	Top-5 ACC (%)
MobileNet	0.569B	4.2M	65.98	89.44	70.60	89.22
MobileNet v2	0.3B	3.5M	68.08	90.98	74.70	90.18
SqueezeNet	0.35B	**1.24M**	69.41	91.64	60.40	82.50
Shufflenet	**0.292B**	1.4M	70.06	91.65	69.40	/
VGG-16	15.3B	138M	72.93	91.16	76.30	93.20
Resnet-18	1.82B	11.7M	75.61	93.05	72.70	89.09
Resnet-50	4.14B	25.6M	77.39	93.96	75.30	93.29
Densenet-201	4.39B	20.0M	**78.54**	**94.1**	76.90	93.56
GoogLeNet	1.55B	7M	78.03	94.06	68.93	89.14
Inception v3	5.75B	23.9M	77.19	93.61	77.90	93.81
InceptionResnetv2	/	55.9M	72.49	90.89	80.14	95.10
Xception	/	22.9M	74.93	92.68	79.00	94.50
NasNet	5.64B	88.9M	77.29	94.09	**82.70**	**96.20**
NBDT [21]	/	/	77.09	/	76.60	/
Tree-CNN-10 [16]	/	/	69.53	91.85	/	/
Tree-CNN-20 [16]	/	/	68.49	90.76	/	/
ITCNN-20	2.01B	25.36M	75.52	93.04	78.11	93.37
ITCNN-Cell	1.54B	11.68M	77.34	93.13	79.14	94.40
ITCNN-NAS	0.51B	5.03M	77.47	94.04	81.67	95.41

ITCNN-NAS has a deeper tree structure for more detailed data classification, as shown in Fig. 3. Classification criteria are closer to manual methods and do not rely on weak scalability features such as "Red" and "Blue Background" found in

NBDT. ITCNN has the ability to identify factors that lead to inference errors. For example, on the Cifar-100 dataset, ITCNN incorrectly detected a test image as a beetle in Fig. 3. The red node responsible for this error is only corrected by training the nodes within the red dashed line. Grey nodes that are not activated during inference do not require any modification due to the parameter isolation of the tree structure. Therefore, ITCNN can operate at high speed, as explained in Table 1. These findings suggest that ITCNN is an effective tool for improving models and can mitigate the need for extensive parameter modifications.

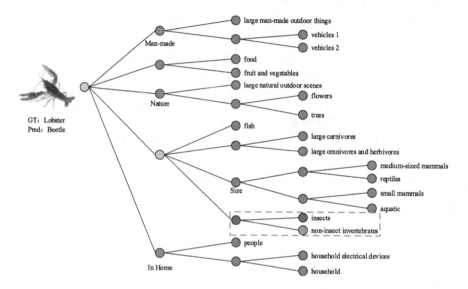

Fig. 3. During the inference process, ITCNN identifies nodes that yield inaccurate results. (Color figure online)

3.2   Experiment on CIL

We evaluate the performance trends of processing algorithms in Cifar-100, as depicted in Fig. 4, compared to LwF [11], iCaRL [15], EEIL [1], UCIR [5], PASS [25], and Tree-CNN [16]. Our analysis shows that ITCNN stands out among all stages by striking a balance between stability and plasticity. As incremental stages become more challenging, ITCNN proves its resilience. Despite their complex structure, ITCNN-Cell and ITCNN-NAS exhibit superior capabilities, indicating that the models extract generalized features that aid classification.

We compare ITCNN with other methods by Training Effort [16] that aims to count the weight updates throughout each training phase defined as

$$TrainingEffort = \sum_{network} N_{weight} \times N_{sample} \tag{12}$$

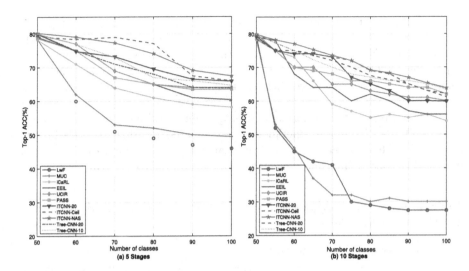

Fig. 4. Classification accuracy on CIFAR-100, which contains the complete curves.

where N_{weight} is total number of weights total number and N_{sample} is total number of training samples. To accurately measure computational cost in a constant batch size and number of training periods, we compute the product of the number of weights and training samples. This approach involves summing the training efforts of each node for ITCNN, while for other methods that use only one neural network, the number of weights in retrained layers is multiplied by the total training samples for the learning phase. This calculation allows a more

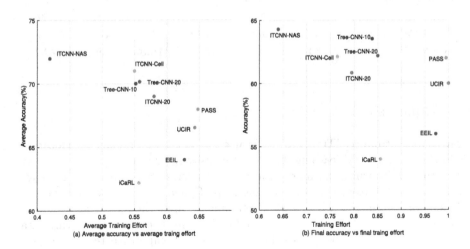

Fig. 5. The performance of ITCNN compared with other incremental learning methods. (a) Average accuracy vs average training effort over learning stages of ITCNN and other methods. (b) Final accuracy vs final training effort over all 100 classes of CIFAR-100.

precise estimate of the overall training effort for different approaches. Figure 5 shows the variation in accuracy observed during incremental learning (10 stages), comparing ITCNN performance against other methods on Cifar-100. Among all approaches, ITCNN-NAS shows the best performance in terms of accuracy and computational efficiency. Nevertheless, ITCNN-20 and ITCNN-Cell also obtain competitive results.

Table 2 presents the performance of algorithms on ImageNet100 and ImageNet1000. ITCNN shows superior results compared to other methods in this assessment. These observations confirm that our proposed approach has great potential to solve significant challenges in large-scale image recognition tasks.

Table 2. Performance comparison between ITCNN and other algorithms on ImageNet100 and ImageNet1000.

New classes per stage	ImageNet100				ImageNet1000	
	50 stages	25 stages	10 stages	5 stages	10 stages	5 stages
	1	2	5	10	50	100
iCaRL [15]	54.97	54.56	60.90	65.56	46.72	51.36
BiC [22]	46.49	59.65	65.14	68.97	44.31	45.72
UCIR (NME) [5]	55.44	60.81	65.83	69.07	59.92	61.56
UCIR(CNN) [5]	57.25	62.94	70.71	71.04	61.28	64.34
UCIR(CNN)+DDE [5]	–	–	70.20	72.34	**65.77**	67.51
Mnemonics [13]	–	69.74	71.37	72.58	63.01	64.54
PODNet (CNN) [3]	62.48	68.31	74.33	75.54	64.13	66.95
PODNet (CNN)+DDE [3]	–	–	**75.41**	76.71	64.71	66.42
GeoDL [19]	–	71.72	73.55	73.87	64.46	65.23
ITCNN-20	64.61	66.87	70.18	73.17	63.81	65.60
ITCNN-Cell	64.19	67.64	71.12	76.71	64.81	67.45
ITCNN-NAS	**64.72**	**70.78**	73.35	**76.85**	64.91	**67.57**

4   Conclusion

This paper introduces ITCNN, which is an effective model for image classification and class incremental learning. ITCNN can easily incorporate new data into its structure and learning parameters, using mechanisms such as residual modules, ESA, and ITDA to achieve superior model performance with fewer parameters. To support continuous learning while retaining the ability to recognize previously learned classes, ITCNN evolves in a tree-like fashion. By organizing existing data into feature-driven superclasses, ITCNN achieves competitive results in CIL. However, due to the specificity of the tree structure, some parameters are not used in the inference process, resulting in massive model parameters. In future developments, ITCNN can be applied to other areas of incremental learning.

References

1. Castro, F.M., Marín-Jiménez, M.J., Guil, N., Schmid, C., Alahari, K.: End-to-end incremental learning. In: Ferrari, V., Hebert, M., Sminchisescu, C., Weiss, Y. (eds.) ECCV 2018. LNCS, vol. 11216, pp. 241–257. Springer, Cham (2018). https://doi.org/10.1007/978-3-030-01258-8_15
2. Criminisi, A., Shotton, J.: Decision Forests for Computer Vision and Medical Image Analysis, p. 1. Springer, London (2013). https://doi.org/10.1007/978-1-4471-4929-3
3. Douillard, A., Cord, M., Ollion, C., Robert, T., Valle, E.: PODNet: pooled outputs distillation for small-tasks incremental learning. In: Vedaldi, A., Bischof, H., Brox, T., Frahm, J.-M. (eds.) ECCV 2020. LNCS, vol. 12365, pp. 86–102. Springer, Cham (2020). https://doi.org/10.1007/978-3-030-58565-5_6
4. Frosst, N., Hinton, G.: Distilling a neural network into a soft decision tree. Comput. Res. Repository, 1–8 (2017)
5. Hou, S., Pan, X., Loy, C.C., Wang, Z., Lin, D.: Learning a unified classifier incrementally via rebalancing. In: Proceedings of the IEEE/CVF Conference on Computer Vision and Pattern Recognition, pp. 831–839 (2019)
6. Ji, R., Wen, L., Zhang, L., Du, D., Huang, F.: Attention convolutional binary neural tree for fine-grained visual categorization. In: 2020 IEEE/CVF Conference on Computer Vision and Pattern Recognition, pp. 10465–10474 (2020)
7. Kirkpatrick, J., et al.: Overcoming catastrophic forgetting in neural networks. Proc. Natl. Acad. Sci. **114**(13), 3521–3526 (2017)
8. Kontschieder, P., Kohli, P., Shotton, J., Criminisi, A.: GeoF: geodesic forests for learning coupled predictors. In: 2013 IEEE Conference on Computer Vision and Pattern Recognition, pp. 65–72 (2013)
9. Lai, Z., Xu, Y., Yang, J., Tang, J., Zhang, D.: Sparse tensor discriminant analysis. Trans. Imgage Process. **22**(10), 3904–3915 (2013)
10. Laptev, D., Buhmann, J.M.: Convolutional decision trees for feature learning and segmentation. Pattern Recogn. 95–106 (2014)
11. Li, Z., Hoiem, D.: Learning without forgetting. IEEE Trans. Pattern Anal. Mach. Intell. 1 (2017)
12. Lin, H., Hu, Y., Chen, S., Yao, J., Zhang, L.: Fine-grained classification of cervical cells using morphological and appearance based convolutional neural networks. IEEE Access **7**, 71541–71549 (2019)
13. Liu, Y., Su, Y., Liu, A.A., Schiele, B., Sun, Q.: Mnemonics training: Multi-class incremental learning without forgetting. In: Proceedings of the IEEE/CVF Conference on Computer Vision and Pattern Recognition, pp. 12245–12254 (2020)
14. Prabhu, A., Torr, P.H.S., Dokania, P.K.: GDumb: a simple approach that questions our progress in continual learning. In: Vedaldi, A., Bischof, H., Brox, T., Frahm, J.-M. (eds.) ECCV 2020. LNCS, vol. 12347, pp. 524–540. Springer, Cham (2020). https://doi.org/10.1007/978-3-030-58536-5_31
15. Rebuffi, S.A., Kolesnikov, A., Sperl, G., Lampert, C.H.: ICARL: incremental classifier and representation learning. In: 2017 IEEE Conference on Computer Vision and Pattern Recognition, pp. 5533–5542 (2017)
16. Roy, D., Panda, P., Roy, K.: Tree-CNN: a hierarchical deep convolutional neural network for incremental learning. Neural Netw. 148–160 (2018)
17. Sarwar, S.S., Panda, P., Roy, K.: Gabor filter assisted energy efficient fast learning convolutional neural networks. In: 22nd IEEE/ACM International Symposium on Low Power Electronics and Design, pp. 1–6 (2017)

18. Sarwar, S.S., Ankit, A., Roy, K.: Incremental learning in deep convolutional neural networks using partial network sharing. IEEE Access **8**, 4615–4628 (2017)
19. Simon, C., Koniusz, P., Harandi, M.: On learning the geodesic path for incremental learning. In: Proceedings of the IEEE/CVF Conference on Computer Vision and Pattern Recognition, pp. 1591–1600 (2021)
20. Tanno, R., Arulkumaran, K., Alexander, D.C., Criminisi, A., Nori, A.: Adaptive neural trees. 2018 International Conference on Machine Learning, pp. 6166–6175 (2018)
21. Wan, A., et al.: NBDT: neural-backed decision trees. International Conference on Learning Representations (2020)
22. Wu, Y., et al.: Large scale incremental learning. In: Proceedings of the IEEE/CVF Conference on Computer Vision and Pattern Recognition, pp. 374–382 (2019)
23. Yan, T., Liu, C., Zhao, W.D.: Incremental tensor discriminant analysis for image detection. In: Advanced Materials Research, vol. 710, pp. 579–583. Trans Tech Publ (2013)
24. Yu, L., et al.: Semantic drift compensation for class-incremental learning. In: Proceedings of the IEEE/CVF Conference on Computer Vision and Pattern Recognition, pp. 6982–6991 (2020)
25. Zhu, F., Zhang, X.Y., Wang, C., Yin, F., Liu, C.L.: Prototype augmentation and self-supervision for incremental learning. In: Proceedings of the IEEE/CVF Conference on Computer Vision and Pattern Recognition, pp. 5871–5880 (2021)

Periodic-Aware Network for Fine-Grained Action Recognition

Senzi Luo[1], Jiayin Xiao[1], Dong Li[1(✉)], and Muwei Jian[2]

[1] School of Automation, Guangdong University of Technology,
510006 Guangzhou, Guangdong, China
`dong.li@gdut.edu.cn`
[2] School of Computer Science and Technology, Shandong University of Finance
and Economics, 250014 Jinan, Shandong, China

Abstract. Recently, skeleton-based action recognition has gained increasing attention and achieved remarkable results in coarse-grained action recognition. Despite the positive results shown in these attempts, they are less effective in scenarios that require a detailed comparison between fine-grained classes, e.g. different moves during a vault. In such scenarios, existing methods make it hard to distinguish subtle differences between actions with different numbers of repetitions. In this article, to solve the above problem, we introduce periodicity into fine-grained action classification and propose a novel network architecture named periodic-aware network (PAN) to distinguish fine-grained actions with different numbers of repetitions. Firstly, a periodicity feature extraction module (PFEM) is proposed to capture periodicity information and extract periodicity features of different levels. Then, a periodicity fusion module (PFM) is proposed to fuse periodicity features and spatiotemporal features. We apply multiple periodicity fusion modules to fuse different levels of features. Finally, the results are obtained by classifying the fusion features. Extensive experiments on two fine-grained skeleton-based action recognition datasets, namely FineGym and Diving48, show that our proposed method outperforms previous skeleton-based action recognition methods.

Keywords: Fine-grained action recognition · Periodicity · skeleton

S. Luo and J. Xiao—These authors contributed equally to this work and should be considered co-first authors. This work was supported by the Guangdong Basic and Applied Basic Research Foundation No. 2021A1515011867, National Natural Science Foundation of China (NSFC) (61976123); Taishan Young Scholars Program of Shandong Province; and Key Development Program for Basic Research of Shandong Province (ZR2020ZD44).

Supplementary Information The online version contains supplementary material available at https://doi.org/10.1007/978-981-99-8543-2_9.

Q. Liu et al. (Eds.): PRCV 2023, LNCS 14432, pp. 105–117, 2024.
https://doi.org/10.1007/978-981-99-8543-2_9

1   Introduction

Action recognition is an important task in video understanding that involves recognizing human actions based on visual cues. In recent years, several modalities have been explored to represent features for action recognition, including RGB frames [1], optical flows [2], audio waves [3], and human skeletons [4,5]. Recently, skeleton-based action recognition has gained increasing attention due to its high-level representation, robust adaptability to dynamic circumstances, and complicated backgrounds.

Previous skeleton-based methods [4–9] achieved remarkable results in coarse-grained action recognition by capturing discriminating spatial structures and temporal motion patterns, however, their performance remains far from being satisfactory in distinguishing between fine-grained actions. Fine-grained analysis capability is crucial to distinguish between different subcategories of actions for some scenarios. For example, in diving, judges cannot score or further analyze athletes based solely on coarse-grained categories such as "twist" or "somersault". A detailed comparison between fine-grained classes is needed.

The main reason for such performance degradation lies in the difficulty of distinguishing subtle differences between different numbers of repetitions. For example, relying on spatiotemporal information, previous skeleton-based methods are easy to distinguish events like "switch leap" and "split leap". However, they have difficulty distinguishing the subcategories of the above events, such as "switch leap with 1 turn" and "switch leap with 2 turns". Different numbers of repetitive actions have local and global patterns, and these actions only have small variations in time and space. Therefore, it is very difficult to distinguish these action categories only through temporal and spatial information. Although the periodicity information of a motion can effectively represent the number of repetitions of the motion, it is not easy to discover the periodicity information of a motion in high-dimensional data when effective representations are lacking.

Based on the above observation, we aim to distinguish the fine-grained actions by representing and incorporating periodicity information of actions with their spatiotemporal features. Specifically, we propose a novel architecture periodic-aware network (PAN) for addressing the aforementioned problem. According to Fig. 1, our architecture contains three parts: general 3D-CNN backbone, periodicity feature extraction module (PFEM), and periodicity fusion module (PFM). High-dimensional features have significant noise, deviating it far from true periodicity. Therefore, we first use temporal self-similarity matrices (TSM) to represent the periodicity information of actions, which exhibit strong robustness to large viewpoint changes and serve as an information bottleneck (IB) within the network. It also provides regularization for the network. Then, we extract periodicity features from the constructed TSM and fuse them into low-level detailed and high-level semantic features in the 3D-CNN, allowing the features of each level of the network to perceive periodicity information. Inspired by Squeeze and Excitation (SE) module [10], we propose a periodicity fusion module (PFM) to integrate the periodicity and spatiotemporal information. It combines both spatiotemporal information in 3DCNN and the periodicity information of actions to distinguish fine-grained actions with different numbers of repetitions. To summarize, the main contributions of this article are as follows.

1) We first introduce periodicity into fine-grained action classification. Specifically, we propose an architecture named periodic-aware network (PAN), which can effectively distinguish the subcategories with different numbers of repetitions. Evaluations on two fined-grained benchmarks have demonstrated the superiority of the proposed PAN compared with existing skeleton-based action recognition methods.

2) We design a periodicity feature extraction module (PFEM) to represent periodicity information by TSM and extract periodicity features from the constructed TSM. These features are used to fuse into the low-level detailed and high-level semantic features of the 3D-CNN, enabling features of each level of the network to perceive periodic information.

3) We proposed a novel periodicity fusion module (PFM) to fuse periodicity features and spatiotemporal features by a two-step squeeze and excitation process.

2   Related Work

2.1   Skeleton-Based Action Recognition

The skeleton has gained popularity in action recognition missions due to its advantages over traditional RGB-based approaches, including robustness to lighting changes, background complexity, and visual diversity [11]. In recent years, research on skeleton-based action recognition mainly focused on three main directions: RNN-based methods [12], CNN-based methods [5], and GCN-based methods [4,6,7,13,14].

In CNN-based methods, the skeleton data is still treated as image features. Duan et al. [5] proposed a new framework for skeleton-based action recognition called PoseConv3D, which effectively learns spatiotemporal features. This framework uses a 3D heatmap volume as the base representation of human skeletons and outperforms GCN-based methods in various datasets in terms of robustness and efficiency.

2.2   Periodicity Estimation of Videos

Periodicity estimation aims to analyze and detect repetitive patterns in object motions in both temporal and spatial domains [15]. This technique is widely used for periodicity detection [16] and periodicity counting [17] in videos.

Temporal Self-similarity Matrices (TSMs) are robust and useful for representing action recognition based on autocorrelation analysis. TSMs have been widely used in periodicity detection [16,18] and periodicity counting [17]. Debidatta et al. [16] showed that TSMs can effectively capture the temporal dynamics of repetitive actions.

2.3   Squeeze and Excitation Module

Hu et al. [10] introduced a squeeze-and-excitation (SE) module to enhance the learning of channel-wise features by explicitly modeling channel-wise dependencies. The SE module utilized two fully connected (FC) layers in a squeeze-and-unsqueeze manner then applied a Sigmoid activation for exciting essential channelwise features. To combine both periodicity and spatiotemporal information, we propose a periodicity fusion module in a squeeze-and-unsqueeze manner without FC layers and Sigmoid activation.

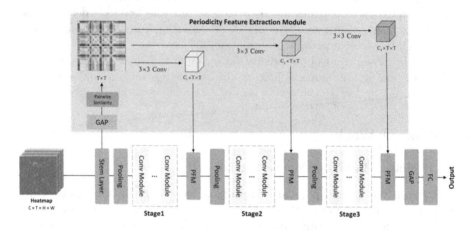

Fig. 1. Periodic-aware network Architecture instantiates with a general 3D-CNN architecture. The stem layer quickly downsamples the input feature maps with convolutions of a quite large kernel size. The module PFM is proposed to combine both periodicity and spatiotemporal information.

3   Method

In this section, we introduce our PAN architecture. According to Fig. 1, the proposed PAN contains three main steps:

1) **Capture spatiotemporal information.** Our method adapts 3D-CNN as the backbone and uses 3D heatmap volumes as the base representation of human skeletons.

2) **Extract periodicity information.** We propose PFEM to represent and extract the periodicity information from videos. The module can capture periodicity information of actions, which is neglected in the 3D-CNN backbone. These features are used to fuse into the low-level detailed and high-level semantic features of the 3D-CNN, enabling features of each level of the network to perceive periodic information.

3) **Combine both information.** After obtaining different levels of periodic features, we need to integrate them into 3D-CNN architecture. We propose

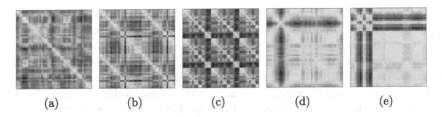

(a)   (b)   (c)   (d)   (e)

Fig. 2. Diversity of temporal self-similarity matrices in gymnastic videos (Yellow means high similarity, and blue means low similarity). (a) reflects one turn on one leg, free leg optional below horizontal, (b) reflects two turns on one leg, free leg optional below horizontal, (c) reflects three turns on one leg, free leg optional below horizontal, (d) reflects salto forward stretched with 1 twist, (e) reflects salto forward stretched with 2 twist (Color figure online)

PFM for such integration by a two-step squeeze and excitation process. Lastly, the architecture outputs the probability of each action by classifying the fusion features.

The following symbols are used: B (batch size), C (channels), T (temporal length), H (height of the spatial domain), and W (width of the spatial domain). For convenience, we will omit B.

3.1 3D-CNN Backbone

Following PoseConv3D [5], we adopt the 3D heatmap volume as input by representing the 2D skeleton data as a joint heatmap. First, we receive 2D skeleton coordinate triplets (x_k, y_k, c_k) extracted by 2D Top-Down pose estimators, then joint heatmaps are composed by using K Gaussian maps centered at every joint in each frame:

$$J_{kij} = e^{-\frac{(i-x_k)^2+(j-y_k)^2}{2*\sigma^2}} * c_k \tag{1}$$

where δ controls the variance of Gaussian maps, K is the number of joints, and (x_k, y_k) and c_k are respectively the location and confidence score of the k-th joint. Finally, The 3D heatmap volume is obtained by stacking all heatmaps along the temporal dimension, resulting in the size of $K \times T \times H \times W$. After receiving the 3D heatmap volume, two techniques, named Subject Centered Cropping [5] and Uniform Sampling strategy [19] are applied to reduce the redundancy of 3D heatmap volumes.

Based on 3D heatmap volume input, we adapt two 3D-CNNs backbones to extract spatiotemporal information: Slowonly [20], X3D [21]. In experiments, we use SlowOnly as the default backbone due to its simplicity and good recognition performance.

3.2 Periodicity Feature Extraction Module

PFEM can be divided into two successive steps: constructing TSM and extracting periodicity features.

1) **Constructing Temporal Self-similarity Matrix.** The 3D-based CNNs implement spatiotemporal convolution operation, fusing temporal information from each frame. Hence we use shallow features of the network after dimensionality reduction to construct TSM. According to Fig. 1. Given intermediate feature maps $F \in \mathbb{R}^{C \times T \times H \times W}$ from 3D-CNN architecture, we first transform F to F_{tr} by stem layers, which work as a compression mechanism over the initial video to reduce computational complexity. Then we reduce the dimensional of F_{tr} using Global Average Pooling (GAP) over the spatial dimensions and obtain embedding $x_i \in \mathbb{R}^{C \times 1 \times 1}$ corresponding to T frames. By dimensionality reduction, we don't need to track the region of interest as done explicitly in prior methods [22] and focus on temporal information of given spatiotemporal features.

To obtain periodic information, here we treat sequences of x_i for T frames as multivariate series. Given the embedding of frame i and frame j, here we use $f(x_i, x_j)$ to compute the similarity of two multivariate series, where $f(.)$ is the similarity function. Then, we construct the TSM by computing all pairwise similarities of sequences of x_i, where $S_{ij} = f(x_i, x_j)$. Referring to RepNet [16], we use the negative of the squared euclidean distance as the similarity function:

$$f(a, b) = -\|a - b\|^2. \tag{2}$$

After constructing a self-similarity matrix, the row-wise softmax operation is applied. TSM represents all the periodic information of T frames embedding. Through such representation, model interpretability is baked into the network architecture. We provide visualization for constructed TSM. Notably, we don't apply uniform sampling for input data. According to Fig. 2, our constructed TSM can visualize the periodicity information of the input video. TSMs make the model temporally interpretable which brings further insights into the predictions made by the model.

2) **Extracting Periodicity Features.** To extract the periodicity features in TSM, we first implement 2D convolution on constructed TSM, followed by the ReLU activation function, the operation can be described as:

$$F_p = \text{ReLU}(\mathbf{K}_1 * S), \tag{3}$$

where $\mathbf{K}_1$ is a 3×3 2D convolution layer, $F_p \in \mathbb{R}^{C \times T \times T}$ is the extracted feature of TSM. To enable low-level detailed and high-level semantic features in the network to perceive periodic features, we extract periodicity features with different channel numbers. Each periodicity feature corresponds to different levels of semantic information in the network. Specifically, for SlowOnly, we use the feature activations output by each stage's last residual block. We denote the output of the last residual blocks as $\{P_2, P_3, P_4\}$ for conv2, conv3, and conv4 outputs, and note that they have strides of $\{2, 4, 8\}$ pixels with respect to the input video. The channels of periodicity features is $\{C_2, C_3, C_4\}$, corresponding to channels of $\{P_2, P_3, P_4\}$ respectively. These features are used to fuse into the low-level detailed and high-level semantic features of the 3D-CNN, enabling features of each level of the network to perceive periodic information.

3.3   Periodicity Fusion Module

In this section, we propose PFM to combine both periodicity and spatiotemporal information. Figure 3 describes the architecture of PFM. PFM receives one level of periodicity features as input, which we denote as $\boldsymbol{F}_p$. PFM gets a global embedding from it and uses this embedding to recalibrate the spatiotemporal features. Inspired by the SE module [10], this is done by a two-step squeeze and excitation process described below.

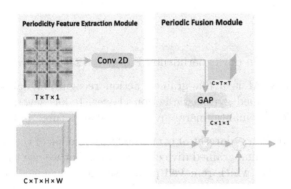

Fig. 3. Architecture of periodicity fusion module (PFM)

1) **Squeeze.** Periodicity features contain temporal and spatial information. Here, we use squeeze operations to generate a global feature descriptor from periodicity features. Different periodic movements often have different speeds and involve different joints. In order to alleviate the influence of temporal variation and joint variation, we reduce the dimensional of $\boldsymbol{F}_p$ using GAP over the different dimensions. Specifically, we investigate two different schemes. First, we apply GAP on two temporal dimensions:

$$\boldsymbol{F}^* = \frac{1}{T \times T} \sum_{i=1}^{T} \sum_{j=1}^{T} \boldsymbol{F}_p[:, i, j], \tag{4}$$

Second, apply GAP on both temporal and channel dimensions:

$$\boldsymbol{F}^* = \frac{1}{C \times T} \sum_{i=1}^{C} \sum_{j=1}^{T} \boldsymbol{F}_p[i, j, :]. \tag{5}$$

We compare the performance of the two proposed squeeze schemes in the experiment.

2) **Excitation.** The function of this unit is to generate excitation signals, which can be used to recalibrate the input features. To preserve the information on periodicity features, we don't implement any convolution and activation

functions on squeezed features. Here, F^* acts as a map for element-wise multiplication to the original input F. To maintain basic performance, we add a residual connection at the end, following the approach used in [23]. The final output can be expressed as:

$$\hat{F} = F \otimes F^* + F, \tag{6}$$

where $\otimes$ denotes element-wise multiplication.

4   Experiment

4.1   Datasets

We use two datasets in our experiments: FineGym [24] and Diving48 [19].

FineGym. FineGYM is a fine-grained action recognition dataset with 29K videos of 99 fine-grained gymnastic action classes. It has several distinguished features: multi-level semantic hierarchy, temporal structure, and high quality.

Diving48. Diving48 is a dataset of over 18,000 video clips of competitive diving actions, spanning 48 fine-grained dive classes. Note that we use the Diving48-V2 released in Oct 2020, which removed poorly segmented videos and cleaned up annotations.

4.2   Implementation Details

Following [5], we use the Top-Down approach for pose extraction: the detector is Faster-RCNN [25] with the ResNet50 backbone, and the pose estimator is HRNet [26] pre-trained. Here we use Mean-Top-1 accuracy in PoseConv3D [5] as a metric for all datasets. We implement our model in the widely used MMAction2 framework by OpenMMLab [27]. We use 2 Nvidia GTX 3060 GPUs for the model training and apply the stochastic gradient descent (SGD) algorithm with Nesterov momentum (0.9) as the optimizer. The input size for all 3D-CNN experiments is $T \times H \times W = 48 \times 56 \times 56$. We use automatic mixed precision (amp) to train all models in this article. In our experiment, we set different hyperparameters for the GCN-based method and the CNN-based method. All models are trained with the CosineAnnealing LR scheduler and cross-entropy losses. The parameter is described below:

For GCN-based methods, we follow the setting in [28]. Specifically, we set the initial learning rate to 0.25, and batch size to 32. We train DGST-GCN [7] for 240 epochs and other GCN-based models for 160 epochs. For the optimizer, we set the momentum to 0.9, and weight decay to 5×10^{-4}. The learning rate decay is set to 0.005.

For CNN-based methods, we set the initial learning rate to 0.05, and batch size to 32. For the optimizer, we set the momentum to 0.9, and weight decay to 5×10^{-4}. The learning rate decay is set to 0.003. All models are trained for 240 epochs.

4.3   Ablation Study

Different Architectures. To validate the effectiveness of our architecture, we adapted two backbones from PoseConv3D [5] as well as their variant:

Table 1. Performance of different architectures

Architecture	FineGym		Diving48	
	Top1 (%)	Top5 (%)	Top1 (%)	Top5 (%)
X3D-s [5]	92.84	99.69	49.85	85.58
PAN-X3D-s	**93.22**	**99.74**	**50.91**	**86.29**
SlowOnly [5]	95.36	**99.79**	55.94	87.01
PAN-SlowOnly	**95.65**	**99.79**	**57.66**	**88.12**

X3D-s [21]. X3D is an advanced 3D-CNN for action recognition. Compared to traditional convolutional models, X3D replaces vanilla convolutions with depth-wise convolutions, enabling the model to achieve high recognition accuracy with fewer parameters and FLOPs. Following X3D-s in [5], we remove the original first stage and convolution layers from each stage uniformly by changing the hyper-parameter γ_d from 2.2 to 1.

SlowOnly [20]. SlowOnly is a widely used 3D-CNN for RGB-based action recognition. This is achieved by converting the ResNet layers in the last two stages of the model from 2D to 3D. Referring to [5], we apply SlowOnly to skeleton-based action recognition, we modify the network by reducing its channel width to half (from 64 to 32) and removing the original first stage from the network architecture.

   We evaluate different backbone architectures on the Diving48 and Finegym benchmarks. Table 1 shows that PAN can improve the performance of the original 3D-CNN backbone architecture.

Different Stages Fusion of PFEM. By extracting periodicity features and integrating periodicity features into different stages of SlowOnly, We explore the importance of enabling each level of a network to perceive periodic information. Specifically, we extract periodicity features and add PFM to the intermediate stages: stage1, stage2, stage3, and report the results in Table 2. We observed that the network performance gradually improves as it perceives periodic information from low to high levels. These experimental results indicate that integrating periodicity features at various levels of the network is effective.

Squeeze Methods. We investigated the impact of different squeeze operations on the performance of a model. Specifically, we employed two different squeeze operations: 1) retaining the channel dimension while compressing the other dimensions, and 2) retaining one of the time dimensions while compressing the other dimensions. The results are shown in Table 3. Regardless of which squeeze operation was used, the performance of the network improved significantly. These results suggest that incorporating periodic features can significantly enhance model performance. The influence of periodic information on the channel dimension was more pronounced, suggesting that retaining the channel dimension was more effective.

Table 2. Different stages fusion of periodicity feature extraction module

	Stage1	Stage2	Stage3	Mean-Top1(%)
SlowOnly				46.72
	✓			48.40
	✓	✓		49.44
PAN	✓	✓	✓	**50.39**

Table 3. Different squeeze methods for periodicity fusion module

Method	Mean-Top1 (%)	Top1 (%)
Base	46.72	55.94
Channel	**50.39**	**57.66**
Temporal	49.54	56.95

Table 4. Performance Comparison of Different Methods in FineGym & Diving48 benchmarks

Model	FineGym			Diving48-V2		
	Mean-Top1 (%)	Top1 (%)	Top5 (%)	Mean-Top1 (%)	Top1 (%)	Top5 (%)
ST-GCN [4]	86.40	90.62	99.34	34.84	39.49	76.65
ST-GCN++ [28]	90.03	93.12	99.57	40.89	46.65	81.62
CTR-GCN [14]	90.46	93.15	99.59	38.34	45.63	78.73
MS-AAGCN [29]	87.58	91.35	99.33	35.25	43.10	75.43
DG-STGCN [7]	90.60	92.90	99.59	35.26	41.32	75.38
MS-G3D [9]	90.45	93.56	99.62	37.63	45.69	79.59
Pose-C3D-s [5]	92.07	94.85	**99.79**	41.66	52.03	86.19
Pose-X3D-s [5]	88.75	92.84	99.69	41.26	49.85	85.58
Pose-SlowOnly [5]	93.22	95.36	**99.79**	46.72	55.94	87.01
PAN-SlowOnly (ours)	**93.91**	**95.65**	**99.79**	**50.39**	**57.66**	**88.12**

4.4   Comparison with State-of-the-Art Methods

As shown in Table 4, we compare our results with the prior skeleton-based models over FineGym and Diving48 datasets. The methods used for comparison include GCN-based methods, including ST-GCN [4], ST-GCN++ [28], CTR-GCN [14], MS-AAGCN [29], DG-STGCN [7], and MS-G3D [9]. For 3D-CNN-based method, our model architecture is compared with PoseConv3D that instantiated with three backbones: SlowOnly [20], X3D [21] and C3D [22]. For a fair comparison, we apply the extracted 2D skeleton for FineGym and Divin48. All GCN-based methods directly take the extracted coordinate triplets (x, y, c) as inputs. All 3D-CNN-based methods (including PoseConv3D with different backbones and PAN) take pseudo heatmap volumes of shape $48 \times 56 \times 56$ generated from the coordinate triplets as inputs and report the accuracy obtained by 10-clip testing. We choose SlowOnly [5] as the backbone of PAN. Compared with previous skeleton-based methods, our method achieves state-of-the-art performance on FineGym and Diving48. The superiority of our method can be attributed to its ability to represent and integrate the periodicity information of actions. In contrast, other methods solely rely on utilizing the spatiotemporal information from the input feature maps.

5   Conclusion

In this paper, we propose periodic-aware network (PAN), a novel architecture to distinguish fine-grained actions with different numbers of repetitions. To the best of our knowledge, our work is the first to introduce periodicity into fine-grained action classification and outperforms previous skeleton-based action recognition methods on two fine-grained skeleton-based action recognition benchmarks.

References

1. Herzig, R., et al.: Object-region video transformers. In: Proceedings of the IEEE/CVF Conference on Computer Vision and Pattern Recognition, pp. 3148–3159 (2022)
2. Simonyan, K., Zisserman, A.: Two-stream convolutional networks for action recognition in videos. In: Advances in Neural Information Processing Systems, vol. 27 (2014)
3. Xiao, F., Lee, Y.J., Grauman, K., Malik, J., Feichtenhofer, C.: Audiovisual slowfast networks for video recognition. arXiv preprint arXiv:2001.08740 (2020)
4. Yan, S., Xiong, Y., Lin, D.: Spatial temporal graph convolutional networks for skeleton-based action recognition. In: Proceedings of the AAAI Conference on Artificial Intelligence, vol. 32 (2018)
5. Duan, H., Zhao, Y., Chen, K., Lin, D., Dai, B.: Revisiting skeleton-based action recognition. In: Proceedings of the IEEE/CVF Conference on Computer Vision and Pattern Recognition, pp. 2969–2978 (2022)
6. Kipf, T.N., Welling, M.: Semi-supervised classification with graph convolutional networks. arXiv preprint arXiv:1609.02907 (2016)
7. Ye, F., Pu, S., Zhong, Q., Li, C., Xie, D., Tang, H.: Dynamic GCN: context-enriched topology learning for skeleton-based action recognition. In: Proceedings of the 28th ACM International Conference on Multimedia, pp. 55–63 (2020)
8. Pan, H., Bai, Y., He, Z., Zhang, C.: AAGCN: adjacency-aware graph convolutional network for person re-identification. Knowl.-Based Syst. **236**, 107300 (2022)
9. Liu, Z., Zhang, H., Chen, Z., Wang, Z., Ouyang, W.: Disentangling and unifying graph convolutions for skeleton-based action recognition. In: Proceedings of the IEEE/CVF Conference on Computer Vision and Pattern Recognition, pp. 143–152 (2020)

10. Hu, J., Shen, L., Sun, G.: Squeeze-and-excitation networks. In: Proceedings of the IEEE Conference on Computer Vision and Pattern Recognition, pp. 7132–7141 (2018)

11. Yue, R., Tian, Z., Du, S.: Action recognition based on RGB and skeleton data sets: a survey. Neurocomputing (2022)

12. Si, C., Chen, W., Wang, W., Wang, L., Tan, T.: An attention enhanced graph convolutional LSTM network for skeleton-based action recognition. In: Proceedings of the IEEE/CVF Conference on Computer Vision and Pattern Recognition, pp. 1227–1236 (2019)

13. Zhang, P., Lan, C., Zeng, W., Xing, J., Xue, J., Zheng, N.: Semantics-guided neural networks for efficient skeleton-based human action recognition. In: Proceedings of the IEEE/CVF Conference on Computer Vision and Pattern Recognition, pp. 1112–1121 (2020)

14. Chen, Y., Zhang, Z., Yuan, C., Li, B., Deng, Y., Hu, W.: Channel-wise topology refinement graph convolution for skeleton-based action recognition. In: Proceedings of the IEEE/CVF International Conference on Computer Vision. pp. 13359–13368 (2021)

15. Cutle, R., Davis, L.: Robust real-time periodic motion detection. Anal. Appl. IEEE Comput. Soc. **22**(8), 781–796 (2000)

16. Dwibedi, D., Aytar, Y., Tompson, J., Sermanet, P., Zisserman, A.: Counting out time: class agnostic video repetition counting in the wild. In: Proceedings of the IEEE/CVF Conference on Computer Vision and Pattern Recognition. pp. 10387–10396 (2020)

17. Jacquelin, N., Vuillemot, R., Duffner, S.: Periodicity counting in videos with unsupervised learning of cyclic embeddings. Pattern Recogn. Lett. **161**, 59–66 (2022)

18. Karvounas, G., Oikonomidis, I., Argyros, A.: Reactnet: temporal localization of repetitive activities in real-world videos. arXiv preprint arXiv:1910.06096 (2019)

19. Li, Y., Li, Y., Vasconcelos, N.: RESOUND: towards action recognition without representation bias. In: Ferrari, V., Hebert, M., Sminchisescu, C., Weiss, Y. (eds.) ECCV 2018. LNCS, vol. 11210, pp. 520–535. Springer, Cham (2018). https://doi.org/10.1007/978-3-030-01231-1_32

20. Feichtenhofer, C., Fan, H., Malik, J., He, K.: Slowfast networks for video recognition. In: Proceedings of the IEEE/CVF International Conference on Computer Vision, pp. 6202–6211 (2019)

21. Feichtenhofer, C.: X3d: Expanding architectures for efficient video recognition. In: Proceedings of the IEEE/CVF Conference on Computer Vision and Pattern Recognition, pp. 203–213 (2020)

22. Dwibedi, D., Tompson, J., Lynch, C., Sermanet, P.: Learning actionable representations from visual observations. In: 2018 IEEE/RSJ International Conference on Intelligent Robots and Systems (IROS), pp. 1577–1584. IEEE (2018)

23. He, K., Zhang, X., Ren, S., Sun, J.: Deep residual learning for image recognition. In: Proceedings of the IEEE Conference on Computer Vision and Pattern Recognition, pp. 770–778 (2016)

24. Shao, D., Zhao, Y., Dai, B., Lin, D.: FineGYM: a hierarchical video dataset for fine-grained action understanding. In: Proceedings of the IEEE/CVF Conference on Computer Vision and Pattern Recognition, pp. 2616–2625 (2020)

25. Ren, S., He, K., Girshick, R., Sun, J.: Faster R-CNN: towards real-time object detection with region proposal networks. In: Advances in Neural Information Processing Systems, vol. 28 (2015)

26. Sun, K., Xiao, B., Liu, D., Wang, J.: Deep high-resolution representation learning for human pose estimation. In: Proceedings of the IEEE/CVF Conference on Computer Vision and Pattern Recognition, pp. 5693–5703 (2019)
27. Contributors, M.: Openmmlab's next generation video understanding toolbox and benchmark (2020). https://github.com/open-mmlab/mmaction2
28. Duan, H., Wang, J., Chen, K., Lin, D.: Pyskl: towards good practices for skeleton action recognition. In: Proceedings of the 30th ACM International Conference on Multimedia, pp. 7351–7354 (2022)
29. Shi, L., Zhang, Y., Cheng, J., Lu, H.: Skeleton-based action recognition with multi-stream adaptive graph convolutional networks. IEEE Trans. Image Process. **29**, 9532–9545 (2020)

Learning Domain-Invariant Representations from Text for Domain Generalization

Huihuang Zhang[1], Haigen Hu[1(✉)], Qi Chen[1], Qianwei Zhou[1], and Mingfeng Jiang[2(✉)]

[1] Zhejiang University of Technology, Hangzhou, China
hghu@zjut.edu.cn
[2] Zhejiang Sci-Tech University, Hangzhou, China
m.jiang@zstu.edu.cn

Abstract. Domain generalization (DG) aims to transfer the knowledge learned in the source domain to the unseen target domain. Most DG methods focus on studying how to learn domain-invariant representations that remain invariant across different domains. For humans, we tend to use the same word or text to describe images from different domains but of the same category. Therefore, text can be considered a natural domain-invariant representation. Inspired by this, this paper studies how to introduce text representations into domain generalization tasks. Specifically, the text representations generated by CLIP text encoder are used to guide the image representation learning of the visual model. To alleviate domain bias and weak discriminability caused by CLIP representations, a joint loss is proposed by combining the text representation regularization loss with standard image-level supervised loss. The proposed method is simple yet efficient, and can achieve competitive performance compared with the existing state-of-the-art methods on five standard DG datasets.

Keywords: Domain Generalization · Transfer learning · CLIP

1 Introduction

Deep learning models are designed based on the assumption that data is independently and identically distributed (IID) [5]. However, when the data distribution during training differs from that during testing (i.e. Out-Of-Distribution), these models will suffer significant performance degradation. In many real-world problems, training and testing datasets are collected under different scenarios, which will lead to the performance degradation of the model on OOD. Domain Generalization (DG) aims to solve this problem by training a model with strong generalization on multi-source domains, and test in target domain.

Existing DG methods mostly focus on learning domain-invariant representations by reducing differences between multi-source domains in the feature space

© The Author(s), under exclusive license to Springer Nature Singapore Pte Ltd. 2024
Q. Liu et al. (Eds.): PRCV 2023, LNCS 14432, pp. 118–129, 2024.
https://doi.org/10.1007/978-981-99-8543-2_10

[24]. Humans' use of consistent language across any domains (Fig. 1) suggests that text/language possesses naturally domain-invariant representation. Based on the above motive, this paper introduces the CLIP [17] text representations, which connects the visual and natural language domains, to guide model training. By pre-training with 400M images, CLIP's text representation will be excellent domain-invariant representations. CLIP maximizes the similarity between image and text representations during pre-training (Fig. 1). It has exhibited strong generalization in zero-shot experiments across different domains. This paper aims to explore the source of CLIP's generalization ability and maximize its potential in downstream domain generalization tasks. Our experiments demonstrate that CLIP text representation is the key to its strong generalizability and its effectiveness in guiding DG tasks. In our experiments (Sect. 4), we observed that replacing the image encoder of CLIP with the **ImageNet** [18] pre-trained model still resulted in significant improvement on DG tasks when using the text representations generated by the CLIP text encoder.

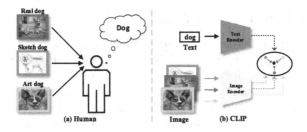

Fig. 1. An illustration of language generality. (a) The human use the same word or text to describe the same object in different domains. (b) The CLIP text encoder converts words into text repesentations that can be matched with objects of the same category in different domains.

Overall, We propose a simple yet effective domain generalization method called text regularization. We discovered that CLIP representations have limited discriminability and tend to bias certain domains. Therefore, we recommend using CLIP text representations in conjunction with standard supervised methods rather than relying solely on text representation. It's the optimal way to use CLIP text representation. Section 3.3 provides an illustration of the limitations of using text representations alone. And our experimental results demonstrate that the optimal training approach involves combining CLIP text representations with standard image-level supervision.

Our contribution is as follows: (1) A simple yet effective DG method is proposed by using CLIP text regularization. This method utilizes pre-generated text representations, requiring no additional resources during the training phase. (2) Our analysis reveals the problems of domain bias and weak discriminability in CLIP text representations. We show that the optimal training approach involves combining text regularization loss with standard supervised loss. (3) The proposed method can achieve competitive performance compared with the existing SOTA method on 5 standard DG datasets.

2   Related Work

2.1   Domain Generalization

Most Domain Generalization (DG) methods can be grouped into three broad categories. (1) Data preprocessing: data augmentation and data generation. Data augmentation is used to increase the amount of data for better generalization by copying, transforming, and adding minor alterations from existing data [26]. Compared with data augmentation, data generation is to generate samples of different styles to help Deep Neural Networks (DNN) more robust [12]. (2) Learning strategies: ensemble learning [3], meta-learning [25], robust optimization [9], et al. (3) Representation learning: it is widely in DG method, including feature disentanglement and domain-invariant representation learning. Feature disentanglement separates features into domain-shared and domain-specific components, enhancing generalization [2]. Domain-invariant representation learning includes adversarial learning [11], et al.

2.2   CLIP in Domain Generalization

CLIP is a multi-modal vision and language model by learning visual concepts from natural language supervision. GVRT [14] is the first work to use CLIP's text encoder for DG, but it is limited to datasets with detailed text descriptions. This restricts its applicability to traditional DG datasets without such information. MIRO [4] addresses this by using CLIP image encoder as the original pretrained model to fine-tune the training model in learning domain-invariant representations. However, MIRO requires additional computational resources since the oracle model performs forward propagation during training to obtain image representations.

The proposed method avoids the need for detailed text descriptions by directly using single category words, making it applicable to traditional universal DG datasets. Additionally, the generation of category text representations using the CLIP text encoder can be performed prior to training, minimizing the computational resources required during the training phase.

3   Method

3.1   Problem Formulation

In DG task, K labeled source domains $D_s = \{D_1, D_2, \ldots D_K\}$ ($D_k = \{(x_i^k, y_i^k)\}_{i=1}^{N_k}$) are given for training, where N_k is the number of labeled samples in the domain D_k. An unseen target domain $D_t = \{(x_i^t, y_i^t)\}_{i=1}^{N_t}$ is given. The goal of DG is to train a model using the labeled data from the source domains D_s in such a way that it performs well on the target domain D_t. The domains D_s and D_t share C categories, and their corresponding labels are denoted as $Y_s = Y_t = \{1, 2, \ldots, C\}$. Moreover, each category is associated with a text representation, defined as $\{T_1, T_2, \ldots, T_C\}$. The image x and text t are input to the image encoder and text encoder to generate the image representation z^{image} and the text representation z^{text}, respectively.

$$z^{image}, z^{text} = f_{image}(x), f_{text}(t) \qquad (1)$$

In this work, we select the pre-trained model of CLIP/ImageNet and the CLIP text encoder as the image encoder $f_{image}(\cdot)$ and $f_{text}(\cdot)$, respectively. The goal is to obtain the text representation z^{text} without fine-tuning the text encoder $f_{text}(\cdot)$. To achieve this, we can pre-generate category text representations by inputting all category words from the dataset beforehand. Hence, the pre-generated z^{text} can be saved in advance and reused during the training phase.

3.2   Text Regularization

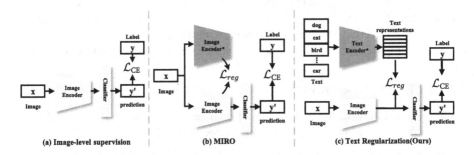

(a) Image-level supervision   (b) MIRO   (c) Text Regularization(Ours)

Fig. 2. An illustration of the architectures. $*$ indicates that the model'parameters is frozen. (a)Standard image-level supervision. (b) MIRO [4] uses oracle representations of pre-trained image encoder to guide model learning. (c)Text Regularization(Ours), using CLIP text representations to guide model learning.

In the training phase, we followed the contrast learning method used in the CLIP pre-training phase, i.e., maximizing the cosine similarity between the image representation z_i^{image} and its corresponding text representation z_i^{text}, while minimizing the similarity between z_i^{image} and other text representations z_j^{text} in a batch N_{batch}, given as follows:

$$s_{ij} = \frac{z_i^{image} \cdot z_j^{text}}{|z_i^{image}||z_j^{text}|} \qquad (2)$$

$$\mathcal{L}_{reg} = - \sum_{i=1}^{N_{batch}} \log \frac{exp(s_{ii})}{\sum_{j=1}^{N_{batch}} exp(s_{ij})} \qquad (3)$$

According to the regularization term $\mathcal{L}_{reg}$, we constrain the image encoder representation learning by maximizing the similarity of image text pairs. The total joint loss by integrating the standard Image-level(i.e. image label y_i) supervised loss $\mathcal{L}_{CE}$ is defined as follows.

$$\mathcal{L}_{CE} = -\sum_{i=1}^{N_{batch}} \log \frac{exp(z_i^{image} \cdot w_+)}{\sum_{j=1}^{C} exp(z_i^{image} \cdot w_j)} \tag{4}$$

$$\mathcal{L}_{total} = \mathcal{L}_{CE} + \lambda \mathcal{L}_{reg} \tag{5}$$

where w_+ denotes the target class of z_i^{image} and w is a learnable parameter of the classifier, λ is a hyperparameter. In the testing phase, the predicted probability of the classifier is integrated with the cosine similarity of the image-text representations, and the final result is regarded as the image classification results. Compared to MIRO, our method uses pre-generated text representations instead of regenerating them during training. This results in a faster forward propagation, taking only half the time of MIRO (Fig. 2b).

In theory, it is possible to train the image encoder only using $\mathcal{L}_{reg}$ and remove the classifier ($\mathcal{L}_{CE}$), which we called text supervision. In Sect. 3.3, we experimentally demonstrate that relying solely on text representations (z^{text}) for training results in bias to certain domains and weak discriminability of representations.

3.3   CLIP Representations

In this subsection, we address the issue of domain bias and weak discriminability in representations when using CLIP text supervision alone.

Domain Bias. CLIP is pre-trained on a dataset collected from the Internet, which suggests that its representations are likely to align well with the Internet data distribution. However, relying solely on CLIP's text representations for supervised training may introduce bias towards the Internet distribution. We analyze potential biases in CLIP text representations using the diverse Domain-Net dataset, consisting of six different domains.

Table 1 illustrates the image classification results on 6 domains using text supervised training (Text) and image-level supervised training (Baseline). It shows the models trained with text supervision perform better than image-level supervision on (*infograph, painting, readworld, sketch*), which confirms our previous hypothesis that CLIP text representations more biased or familiar with certain domains and poor performance in certain domains(like *clipart*). Since

Table 1. Accuracy comparisons between two supervised fine-tune on DomainNet(6 domains), where *Baseline* indicates image-level supervision ($\mathcal{L}_{CE}$) and *Text* indicates text representation supervision ($\mathcal{L}_{reg}$).

Method	clip	info	paint	quick	real	sketch	Avg.
Baseline	**61.8**	20.1	47.8	12.6	60.9	51.1	42.4
Text	57.7	**22.5**	**51.8**	12.8	**64.2**	**51.2**	43.4

CLIP does not open source its pre-training dataset, we cannot validate whether its domain distribution is consistent with our analysis on DomainNet. But at least it can be demonstrated that CLIP text representations bias some domains.

Table 2. Accuracy comparisons between two pre-trained image encoder under different supervised training methods on OfficeHome which contains 4 domains (**Art**, **Clipart**, **Product**, **RealWorld**). DG tasks are evaluated by leave-one-out cross-validation, where averaging all cases that use a single domain as the target (test) domain and the others as the source (train) domains.

Method	Image Encoder	Domain	A	C	P	R	Avg.
Baseline (Group I)	ImageNet	Target	59.4	51.8	74.2	76.2	65.4
		Source	82.9	82.6	78.3	80.1	**81.0**
	CLIP	Target	61.3	52.2	74.3	76.6	66.1
		Source	76.8	81.2	73.0	76.5	76.9
Text (Group II)	ImageNet	Target	63.7	52.7	73.5	77.9	66.9
		Source	82.9	83.0	80.0	80.4	**81.6**
	CLIP	Target	69.9	54.0	76.7	79.1	69.9
		Source	71.0	77.4	69.8	69.2	71.9

Weak Discriminability. Table 2 shows that CLIP text representation is equally effective for the **ImageNet** pre-trained model. Moreover, the two pre-trained models perform similarly on the target domain (OOD evaluation). However, with the poor results of CLIP on source domains, we speculate that the discriminability of CLIP representations maybe weak. To support the above hypothesis, experiments and visualizations are conducted. We analyze the discriminability of CLIP text representations (Fig. 3) and image representations (Fig. 4) respectively. The high similarity between cats and dogs (bus and car) indicates that CLIP text representations have complete semantic information. Since the CLIP text encoder is trained to embed closely related concepts near one another(for example, *cat* and *dog* belong to the same concept of *animal* or *pet*, so the two are close to each other and far from *bus*). Therefore, we can speculate that CLIP generates generalized representations rather than discriminative representations.

The image representations of discriminability are further compared between ImageNet pre-trained model and CLIP pre-trained image encoder. Figure 4 shows the similarity between the two pre-trained image encoders (i.e., CLIP and ImageNet) and two supervised training methods (image-level and text) on OfficeHome. The OfficeHome dataset contains 65 categories, so we can get 65 category representations z_c^{image} with a similarity matrix of size (65×65). Specifically, all images of the same category are input to f_{image} to generate image vectors z_i^{image}, and then all vectors are averaged as the image representation of that category z_c^{image}, given as Eqs. 6 and 7.

$$z_i^{image} = f_{image}(x_i) \tag{6}$$

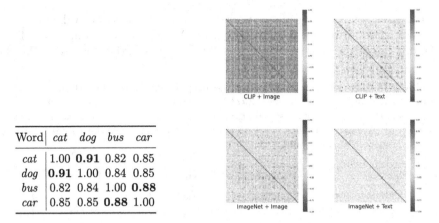

Fig. 3. The cosine similarity of CLIP text representations.

Word	cat	dog	bus	car
cat	1.00	**0.91**	0.82	0.85
dog	**0.91**	1.00	0.84	0.85
bus	0.82	0.84	1.00	**0.88**
car	0.85	0.85	**0.88**	1.00

Fig. 4. The similarity matrix of image representations on OfficeHome.

$$z_c^{image} = Average(\sum_{i=1}^{N_c} z_i^{image}) \tag{7}$$

where N_c denotes the total number of samples in category c. From Fig. 4, it can be observed that the image representations of CLIP have higher similarities between different categories under the same supervised method, i.e., weaker discriminability than ImageNet pre-trained. Surprisingly, the text supervision method enables the image encoder (either CLIP-pre or ImageNet-pre) to generate representations with stronger discriminability than the image-level supervision. To further enhance the discriminability of the representations, we consider the incorporation of image-level supervised loss ($\mathcal{L}_{CE}$). Interestingly, the representations of the ImageNet pre-trained model are consistently more discriminative than CLIP pre-trained, and it is related to the ImageNet pre-training approach, i.e., image-level supervised training.

Based on the analysis of these two CLIP representation problems, we conclude that the combination of text supervised loss ($\mathcal{L}_{reg}$) and image-level supervised loss ($\mathcal{L}_{CE}$) leads to better performance. In our experiments, the results are consistent with the above conclusion.

4   Experiments and Results

4.1   Datasets and Experimental Settings

Datasets. A series of comparison experiments are conducted on 5 datasets, such as PACS [10], Office-Home [22], DomainNet [16], VLCS [6] and TerraIncognita [1].

Evaluation Protocols. The DomainBed framework [7] is used for testing on the DG task. All results are evaluated by leave-one-out cross-validation, where averaging all cases that use a single domain as the target (test) domain and the others as the source (train) domains.

Implementation Details. If not specified, the default ResNet50 (**ImageNet-1K** pre-trained) is used. For the ImageNet pre-trained and the CLIP pre-trained model, the learning rates are 5e−5 and 1e−6, respectively. The batch size per domain is 32. λ is set to 1 by default. In our study, we trained the PACS, OH, Tec, and VLCS for a total of 5,000 iterations, and the DN for 15,000 iterations. The proposed method can utilize the SWAD [3] ensemble algorithm. If not mentioned, the following experimental results are obtained by the SWAD ensemble. All experiments are conducted with NVIDIA Quadro RTX A6000 graphics cards.

4.2   Comparison with Existing DG Methods

As shown in Tables 3, the proposed method outperform other DG methods (The results of the comparison are all from [4]) on five DG datasets. For further comparison, the proposed method is compared with other SOTA DG methods with SWAD, and outperforms the SOTA DG methods (MIRO [4] and PCL [24]). We also conducted a zero-shot test using the original CLIP model.

Table 3. Average accuracy comparison between our method and other DG methods on 5 datasets. All results are from [3]. ‡ denotes the reproduced result.

Method	PACS	VLCS	OH	TI	DN	Avg.
CLIP zero-shot	91.1	77.6	67.1	21.2	44.7	60.3
MixStyle [26]	85.2	77.9	60.4	44.0	34.0	60.3
ARM [25]	85.1	77.6	64.8	45.5	35.5	61.7
VREx [9]	84.9	78.3	66.4	46.4	33.6	61.9
CDANN [11]	82.6	77.5	65.8	45.8	38.3	62.0
Fish [19]	85.5	77.8	68.6	45.1	42.7	63.9
ERM [21]	84.2	77.3	67.6	47.8	44.0	64.2
SagNet [15]	86.3	77.8	68.1	48.6	40.3	64.2
SelfReg [8]	85.6	77.8	67.9	47.0	42.8	64.2
CORAL [20]	86.2	78.8	68.7	47.6	41.5	64.6
mDSDI [2]	86.2	79.0	69.2	48.1	42.8	65.1
MIRO [4]	85.4	79.0	**70.5**	50.4	**44.3**	65.9
Ours	**87.2**	**80.3**	70.4	**50.9**	44.0	**66.6**
Combined with SWAD [3]						
ERM+SWAD	88.1	79.1	70.6	50.0	46.5	66.9
CORAL+SWAD	88.3	78.9	71.3	51.0	46.8	67.3
PCL [24]+SWAD	**88.7**	77.7‡	71.6	52.1	**47.7**	67.6
MIRO+SWAD	88.4	79.6	72.4	52.9	47.0	68.1
Ours+SWAD	88.5	**80.4**	**72.7**	**53.6**	47.6	**68.6**

4.3   Ablation Study

A series of ablation studies are conducted to validate the effectiveness of the proposed method and improvements for the mentioned problems (Sect. 3.3).

Table 4. Loss ablation study on **ImageNet** pre-trained on OfficeHome.

$\mathcal{L}_{CE}$	$\mathcal{L}_{reg}$	A	C	P	R	Avg.
✓		59.4	51.8	74.2	76.2	65.4 (+0.0)
	✓	63.7	52.7	73.5	77.9	66.9 (+1.5)
✓	✓	**67.1**	**56.5**	**77.8**	**80.4**	**70.4** (+5.0)

Table 5. Loss ablation study on **CLIP** pre-trained on OfficeHome.

$\mathcal{L}_{CE}$	$\mathcal{L}_{reg}$	A	C	P	R	Avg.
✓		61.3	52.2	74.3	76.6	66.1 (+0.0)
	✓	69.9	54.0	76.7	79.1	69.9 (+3.8)
✓	✓	**69.3**	**56.5**	**80.0**	**80.7**	**71.6** (+5.5)

Effectiveness of Text Representation. In this work, CLIP text representation is introduced to guide image encoder representation learning. The comparison results by adopting image-level supervision($\mathcal{L}_{CE}$), text supervision ($\mathcal{L}_{reg}$), and text regularization ($\mathcal{L}_{reg} + \mathcal{L}_{CE}$) are illustrated in Tables 4 and 5. Text supervision performs significantly better than image-level supervision, and text regularization achieves the best performance, verifying the effectiveness of the text representations, which can provide a significant improvement for both ImageNet pre-trained image encoder and CLIP pre-trained image encoder.

Table 6. Ablation study about Domain bias on DomainNet.

Method	clip	info	paint	quick	real	sketch	Avg.
Baseline	61.8	20.1	47.8	12.6	60.9	51.1	42.4 (+0.0)
Text	57.7	**22.5**	**51.8**	12.8	**64.2**	51.2	43.4 (+1.0)
Ours	**62.6**	21.8	50.8	**14.9**	61.2	**52.7**	**44.0**(+1.6)

Table 7. Ablation study about the discriminability of representation on OfficeHome.

Method	Domain	A	C	P	R	Avg.
Baseline	Target	61.6	51.9	74.2	76.4	66.0
	Source	61.3	52.2	74.3	76.6	66.1
Text	Target	67.8	48.8	77.5	78.7	68.2
	Source	67.8	49.2	77.6	78.7	68.3
Ours	Target	69.3	56.9	80.3	80.6	**71.8**
	Source	73.7	79.2	72.1	72.0	**74.3**

Impact of Domain Bias and Weak Discriminability. For Domain bias and weak discriminability of CLIP text representations, joint text supervision and image-level supervision are the optimal way to utilize CLIP text representations. Table 6 shows the joint loss can alleviate the domain bias of CLIP text representations and obtain better average accuracy. Table 7 and Fig. 5 respectively show that the proposed method obtains better accuracy and larger inter-class distances on the source domains, which means that the representations are more discriminable. As the discriminability of the representation rises, the performance on the target domain also is improved.

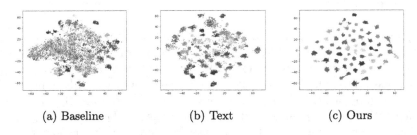

| (a) Baseline | (b) Text | (c) Ours |

Fig. 5. t-SNE [13] visualization on representations of source domains.

Effectiveness Under Various Pre-trained Models. To verify the generalizability of the proposed method, the image encoder is replaced with a convolutional neural network (ResNet50) and Transformer network (ViT-B/16). From Table 8, the proposed method outperforms MIRO (SOTA) and has a significant improvement on both ImageNet pre-trained model and CLIP pre-trained model.

Table 8. Comparison with different pre-training model, methods, and backbones.

Pre-trained	Backbone	Method	PACS	VLCS	OH	TI	DN	Avg.
ImageNet	ResNet	ERM	84.2	76.9	65.4	47.9	42.5	63.4 (+0.0)
		MIRO [4]	85.4	79.0	**70.5**	50.4	**44.3**	65.9 (+2.5)
		Ours	**87.2**	**80.3**	70.4	**50.9**	44.0	**66.6** (+3.2)
	ViT	ERM	83.4	79.3	71.5	42.2	47.5	64.8 (+0.0)
		Ours	**86.2**	**79.6**	**72.8**	**45.8**	**48.2**	**66.5** (+1.7)
CLIP	ResNet	ERM	86.4	80.4	66.1	23.4	16.9	54.6 (+0.0)
		MIRO [4]	76.6	78.9	59.5	49.0	42.0	61.2 (+6.6)
		Ours	**91.3**	**82.8**	**71.6**	49.2	**44.6**	67.1 (+13.5)
	ViT	ERM	93.8	82.2	72.9	48.3	30.3	65.5 (+0.0)
		MIRO [4]	95.6	82.2	**82.5**	54.3	54.0	73.7 (+8.2)
		Ours	**95.9**	**83.0**	82.3	**56.5**	**57.9**	**75.1** (+10.6)
Combined with SWAD								
ImageNet	ResNet	SWAD	88.1	79.1	70.6	50.0	46.5	66.9 (+0.0)
		Ours	**88.5**	**80.4**	**72.7**	**53.6**	**47.6**	**68.6** (+1.7)
	ViT	SWAD	89.5	**80.1**	75.8	45.9	50.5	67.4 (+0.0)
		Ours	**89.9**	**80.1**	**75.9**	**46.6**	**51.3**	**67.8** (+0.4)
CLIP	ResNet	SWAD	85.8	80.4	45.4	23.7	16.5	50.4 (+0.0)
		Ours	**91.2**	**82.1**	**71.6**	**40.4**	**45.1**	**66.1** (+15.7)
	ViT	SWAD	93.8	82.2	72.9	48.7	29.9	65.5 (+0.0)
		Ours	**95.6**	**82.3**	**83.4**	**56.8**	**58.3**	**75.3** (+9.8)

5   Conclusions

This paper proposes a method for domain generalization using CLIP text representations. By combining text regularization loss with standard image-level supervised loss during **ImageNet/CLIP** image encoder training, it addresses domain bias and weak discriminability issues associated with CLIP representations. This approach enhances domain-invariant representations while requiring minimal training resources, as text representations can be generated prior to training. Experiments on five domain generalization datasets demonstrate its competitive performance compared to state-of-the-art methods

Limitation. When the downstream tasks contain objects rarely seen in CLIP pre-trained datasets (like medical images), the proposed method will be limited. But there are more and more CLIP-like works such as MedCLIP [23] (a medical image-text model). And as these works increasing, the limitation will be solved.

Acknowledgments. This work was supported in part by National Natural Science Foundation of China (Grant Nos. 62373324, 62271448 and U20A20171), in part by Zhejiang Provincial Natural Science Foundation of China (Grant Nos. LGF22F030016 and LY21F020027), and in part Key Programs for Science and Technology Development of Zhejiang Province (2022C03113).

References

1. Beery, S., Van Horn, G., Perona, P.: Recognition in terra incognita. In: Ferrari, V., Hebert, M., Sminchisescu, C., Weiss, Y. (eds.) ECCV 2018. LNCS, vol. 11220, pp. 472–489. Springer, Cham (2018). https://doi.org/10.1007/978-3-030-01270-0_28
2. Bui, M.H., Tran, T., Tran, A., Phung, D.: Exploiting domain-specific features to enhance domain generalization. In: Advances in Neural Information Processing Systems, vol. 34, pp. 21189–21201 (2021)
3. Cha, J., et al.: Swad: domain generalization by seeking flat minima. In: Advances in Neural Information Processing Systems, vol. 34, pp. 22405–22418 (2021)
4. Cha, J., Lee, K., Park, S., Chun, S.: Domain generalization by mutual-information regularization with pre-trained models. In: Avidan, S., Brostow, G., Cissé, M., Farinella, G.M., Hassner, T. (eds.) ECCV 2022. LNCS, vol. 13683, pp. 440–457. Springer, Cham (2022). https://doi.org/10.1007/978-3-031-20050-2_26
5. Dosovitskiy, A., et al.: An image is worth 16x16 words: transformers for image recognition at scale. In: ICLR (2021)
6. Fang, C., Xu, Y., Rockmore, D.N.: Unbiased metric learning: on the utilization of multiple datasets and web images for softening bias. In: Proceedings of the IEEE International Conference on Computer Vision, pp. 1657–1664 (2013)
7. Gulrajani, I., Lopez-Paz, D.: In search of lost domain generalization. In: International Conference on Learning Representations (2020)
8. Kim, D., Yoo, Y., Park, S., Kim, J., Lee, J.: Selfreg: self-supervised contrastive regularization for domain generalization. In: Proceedings of the IEEE/CVF International Conference on Computer Vision, pp. 9619–9628 (2021)
9. Krueger, D., et al.: Out-of-distribution generalization via risk extrapolation (rex). In: International Conference on Machine Learning, pp. 5815–5826. PMLR (2021)

10. Li, D., Yang, Y., Song, Y.Z., Hospedales, T.M.: Deeper, broader and artier domain generalization. In: Proceedings of the IEEE International Conference on Computer Vision, pp. 5542–5550 (2017)
11. Li, H., Pan, S.J., Wang, S., Kot, A.C.: Domain generalization with adversarial feature learning. In: Proceedings of the IEEE Conference on Computer Vision and Pattern Recognition, pp. 5400–5409 (2018)
12. Li, L., et al.: Progressive domain expansion network for single domain generalization. In: Proceedings of the IEEE/CVF Conference on Computer Vision and Pattern Recognition, pp. 224–233 (2021)
13. Van der Maaten, L., Hinton, G.: Visualizing data using t-SNE. J. Mach. Learn. Res. **9**(11) (2008)
14. Min, S., Park, N., Kim, S., Park, S., Kim, J.: Grounding visual representations with texts for domain generalization. In: Avidan, S., Brostow, G., Cissé, M., Farinella, G.M., Hassner, T. (eds.) ECCV 2022. LNCS, vol. 13697, pp. 37–53. Springer, Cham (2022). https://doi.org/10.1007/978-3-031-19836-6_3
15. Nam, H., Lee, H., Park, J., Yoon, W., Yoo, D.: Reducing domain gap by reducing style bias. In: Proceedings of the IEEE/CVF Conference on Computer Vision and Pattern Recognition, pp. 8690–8699 (2021)
16. Peng, X., Bai, Q., Xia, X., Huang, Z., Saenko, K., Wang, B.: Moment matching for multi-source domain adaptation. In: Proceedings of the IEEE/CVF International Conference on Computer Vision, pp. 1406–1415 (2019)
17. Radford, A., et al.: Learning transferable visual models from natural language supervision. In: International Conference on Machine Learning, pp. 8748–8763. PMLR (2021)
18. Russakovsky, O., et al.: Imagenet large scale visual recognition challenge. Int. J. Comput. Vision **115**, 211–252 (2015)
19. Shi, Y., et al.: Gradient matching for domain generalization. arXiv preprint arXiv:2104.09937 (2021)
20. Sun, B., Saenko, K.: Deep CORAL: correlation alignment for deep domain adaptation. In: Hua, G., Jégou, H. (eds.) ECCV 2016. LNCS, vol. 9915, pp. 443–450. Springer, Cham (2016). https://doi.org/10.1007/978-3-319-49409-8_35
21. Vapnik, V.: The Nature of Statistical Learning Theory. Springer, New York (1999). https://doi.org/10.1007/978-1-4757-3264-1
22. Venkateswara, H., Eusebio, J., Chakraborty, S., Panchanathan, S.: Deep hashing network for unsupervised domain adaptation. In: Proceedings of the IEEE Conference on Computer Vision and Pattern Recognition, pp. 5018–5027 (2017)
23. Wang, Z., Wu, Z., Agarwal, D., Sun, J.: Medclip: contrastive learning from unpaired medical images and text. arXiv preprint arXiv:2210.10163 (2022)
24. Yao, X., et al.: PCL: proxy-based contrastive learning for domain generalization. In: Proceedings of the IEEE/CVF Conference on Computer Vision and Pattern Recognition, pp. 7097–7107 (2022)
25. Zhang, M., Marklund, H., Dhawan, N., Gupta, A., Levine, S., Finn, C.: Adaptive risk minimization: a meta-learning approach for tackling group distribution shift (2020)
26. Zhou, K., Yang, Y., Qiao, Y., Xiang, T.: Domain generalization with mixstyle. In: International Conference on Learning Representations (2020)

TSTD:A Cross-modal Two Stages Network with New Trans-decoder for Point Cloud Semantic Segmentation

Zhao Gao[1], Li Yan[1,2], Hong Xie[2,3(✉)], Pengcheng Wei[2], Hao Wu[2], and Jian Wang[2]

[1] School of Computer Science, Wuhan University, Wuhan 430072, China
gaozzz@whu.edu.cn, lyan@sgg.whu.edu.cn
[2] School of Geodesy and Geomatics, Wuhan University, Wuhan 430079, China
{wei.pc,2021202140055,wj_sgg}@whu.edu.cn
[3] Hubei Luojia Laboratory, Wuhan University, Wuhan 430079, China
hxie@sgg.whu.edu.cn

Abstract. In recent years, exploring integrated heterogeneous features architecture has become one of the hot spots in 3D point cloud understanding. However, the efficacy of end-to-end training in enhancing the precision of multi-view fusion for point cloud segmentation and its flexibility remain limited. Furthermore, it is worth highlighting that prior studies have consistently employed encoder-decoder architectures, predominantly emphasizing the refinement of encoder designs, while relatively neglecting the significance of decoders which ultimately leads to increased computing costs. In this study, we present our novel TSTD model, which exhibits remarkable efficacy and efficiency, addressing the constraints encountered in prior research. Diverging from existing approaches that exclusively employ either geometry or RGB data for semantic segmentation, our proposed methodology incorporates both modalities within a unified, two-stage network architecture. This integrative approach enables the effective fusion of heterogeneous data features, leading to notable enhancements in semantic segmentation outcomes. Moreover, we have devised an innovative and efficient decoder utilizing a lightweight transformer module. This novel design further enhances the decoding process, resulting in improved performance and effectiveness. The performance of our model, TSTD, demonstrates strong results with an mIoU of 72.5% on the ScanNet v2 validation set and 67.6% on the test set. Notably, TSTD outperforms the current leading cross-modal point cloud semantic segmentation method CMX by a significant margin of 6.3% mIoU. It also reduces Flops by 31.4% compared to Point Transformer. Extensive experiments confirm that TSTD achieves state-of-the-art performance in cross-model point cloud semantic segmentation.

This work was supported in part by the Open Fund of Hubei Luojia Laboratory under Grant 220100053 and in part by the Science and Technology Major Project of Hubei Province under Grant 2021AAA010.

Supplementary Information. The supplementary materials are available at https://github.com/Whu-gaozhao/TSTD-Supp.

Keywords: cross-modal · semantic segmentation · two-stage · decoder

1 Introduction

Semantic scene segmentation plays a crucial role in numerous applications as it facilitates the understanding of visual data. As a convention, contemporary techniques for point cloud semantic segmentation frequently employ a solitary point cloud as their data source. The Point Transformer [30] utilizes point coordinates as inputs, with self-attention and group vector attention mechanisms being incorporated individually to enhance the understanding of 3D point clouds. In contrast, 3DMV [4] presents a joint 2D-3D convolutional framework that learns from both 2D RGB input and 3D geometry. Their findings suggest that multi-view RGB images could potentially improve the accuracy of point cloud semantic segmentation. In this work, our research aims to introduce a novel two-stage feature fusion framework that can effectively integrate heterogeneous data across multiple modes, with the goal of leveraging multi-view images to improve point cloud understanding. In addition, we also propose a novel single-head transformer decoder that incorporates an additional loss function as a constraint to improve context information utilization.

The PointNet [15] model utilizes a shared function, comprised of fully connected layers and max-pooling operations, to extract features from point clouds. However, such an approach can only represent individual point features, which may not be sufficient to effectively describe an object's features. Given the irregularity and sparsity of point clouds, we propose to address these challenges by incorporating transformer architecture and multi-view image data sources.

Furthermore, the two-stage network architecture offers greater flexibility and facilitates better merging of 2D image features with 3D point cloud features. However, previous end-to-end framework such as 3DMV [4] and MVPNet [23] fail to significantly improve the accuracy of semantic segmentation in multi-view fusion. To this end, we introduce a novel two-stage network architecture that can effectively integrate heterogeneous data features. Moreover, it is worth noting that both studies continue to adopt a U-Net style encoder-decoder architecture, with more emphasis placed on the design of the encoder while the importance of the decoder is relatively overlooked. Typically, these frameworks incorporate a complex transformer encoder to enhance performance, but this choice comes at the expense of incurring significant computational costs. To solve this problem, we design a simple but effective decoder based on a light weighted transformer module.

Our main contributions can be summarized as follows:

- A novel two-stage network architecture is designed to effectively integrate heterogeneous data features for enhancing the feature representation ability of point cloud.
- We propose a simple yet effective decoder based on a lightweight transformer module, which incorporates an additional loss function as a constraint to improve context information utilization.

- In the end, our approach outperforms CMX by 6.3% mIoU on the ScanNet v2 test set and achieves state-of-the-art performance in cross-modal point cloud semantic segmentation.

2   Related Works

2.1   Image Transformers

Due to Vis Transformer(ViT) [12] is scalable and simple and effective in vision tasks, it has emerged as a prominent trend in 2D image understanding [18]. Building upon this momentum, EDFT [26] and SegFormer [25] have presented innovative and influential semantic segmentation frameworks that integrate Transformers with MLP decoders. EDFT [26] introduces a novel and efficient depth fusion transformer network to address the challenges associated with memory consumption. Additionally, Segdecoder [18] proposes an efficient self-attention mechanism that enables the operation of the transformer block in a sequence of shifted windows, further enhancing memory efficiency.

2.2   Point Cloud Transformer

Transformer-based networks are categorized as point-based networks that are used for understanding point clouds. Notably, during the rise of research on vision transformers, Zhao et al. [30] and Guo et al. [7] independently conducted explorations on applying attention mechanisms to point cloud understanding. Guo et al. proposed PCT [7], which directly applies global attention to the point cloud. However, similar to ViT [5], their approach faces limitations in terms of memory consumption and computational complexity. On the other hand, Point Transformer has achieved significant advancements in various point cloud understanding tasks and has attained state-of-the-art performance in several competitive challenges. In this work, we introduce a novel two-stage network architecture and a novel method that can effectively integrate heterogeneous data features. Meanwhile, we propose a simple yet effective decoder based on a lightweight transformer module, which incorporates an additional loss function as a constraint to improve context information utilization.

2.3   Joint 2D-3D Network

Semantic segmentation on 3D point clouds has gained significant attention and has been extensively studied by the use of end-to-end neural network [4,11] approaches. Although it is a relatively new task, its relevance extends to various applications, particularly in the field of autonomous driving, where a comprehensive understanding of spatial semantics is crucial. To facilitate the development of joint 2D-3D networks, several datasets and benchmarks have been introduced. One notable dataset is ScanNet [3], which includes a task for 3D semantic segmentation. The dataset provides ground truth annotations for training, validation, and testing directly on the 3D reconstructions. It consists of 2.5k RGB-D frames, where the 2D annotations are obtained by rendering 3D-to-2D projections. CMX [28] explores segmentation of cross-model sensing data combinations and achieves remarkable performance on ScanNet v2 dataset.

3   Method

The overall architecture of the joint 2D-3D framework, along with the networks employed in the two-stage architecture, is presented in Sect. 3.1. Subsequently, in Sect. 3.2, the projection method is introduced, demonstrating the effective integration of heterogeneous data features by incorporating 2D semantic information into the preparation of point clouds. Additionally, the novel trans-decoder is proposed in Sect. 3.3.

3.1   Overall Architecture

An overview of our architecture is shown in Fig. 1. Our network is composed of a 3D stream and A 2D stream that are combined in a joint 2D-3D network architecture. The phased joint segmentation network comprises three core components: the training of a 2D image segmentation network, the preparation of a point cloud with 2D semantic information, and the training of a 3D point cloud semantic segmentation network.

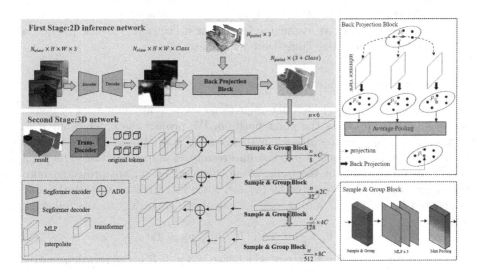

Fig. 1. Network overview: our architecture consists of a 2D and a 3D components. The 2D component takes several RGB multi-view images as input, from which it learns features. The output of the 2D network is the per-pixel uncertainty for each category. After the 3D encoder processes the input point cloud with 2D semantic information, the extracted features are utilized in the 3D decoder with attention loss and segmentation loss.

Training and Inference of the 2D Image Segmentation Network. We have utilized a 2D network called SegFormer [25] to learn the features, as it is known for its simplicity, fast run-time and memory-efficient operations. 2D stream is asked to predict meaningful semantic features for an RGB image segmentation task. The ultimate objective of the 2D component is to achieve per-pixel uncertainty for each category, which is utilized to prepare point cloud with 2D semantic information.

Preparation of a Point Cloud with 2D Semantic Information. In order to obtain semantically enhanced point clouds, both training and reasoning of the 2D network are necessary to process multi-view images. The process is shown at the top left of Fig. 1. A detailed exposition on the projection method will be presented in Sect. 3.2. The process of preparing point cloud with 2D semantic information is showed at the top right of Fig. 1.

Training of a 3D Point Cloud Semantic Segmentation Network. We have utilized simplified Point Transformer due to hardware limitations. Compared with the original Point Transformer, the downsampling ratio of the first stage is changed from 4 times to 8 times, and only the Point Transformer network layer is added to the Pointnet++ decoder. The simplified Point Transformer network structure is shown in Fig. 1. We introduce a novel decoder based on a lightweight transformer module which incorporates an additional loss function as a constraint to improve context information utilization. We will provide a detailed explanation of the Trans-Decoder in Sect. 3.3.

3.2   2D-3D Backprojection

Let $P = p_i|p_i = (x_i, y_i, z_i)$ be a 3D point cloud scene containing a set of points where $p_i \in \mathbb{R}^3$ represents the point position. We project the point cloud $P = p_i|p_i = (x_i, y_i, z_i)$ onto the image I with size of $H \times W$. Image I has a total of C channels, each of which stores the uncertainty of a specific class. The pixel coordinates of a point in the point cloud, denoted by p_i, are projected onto the image plane and represented by (u_i, v_i), while its corresponding depth value is denoted by d_i. (u_i, v_i) and d_i can be calculated according to the following formula:

$$d_i(u_i, v_i, 1)^T = K(R|T)(x_i, y_i, z_i, 1)^T \tag{1}$$

where K is camera intrinsic matrix which size is 3×3. R is the rotation matrix in camera extrinsic. T is the translation matrix in camera intrinsic matrix.

By sequentially projecting all the selected 2D images onto the corresponding point cloud and then aggregating the resulting features, we obtain the multi-view point cloud features F with a size of $n_{view} \times n_{point} \times (3 + C)$. n_{view} represents the number of multi-view. n_{point} is the number of point cloud. C is the number of class. The multi-view point cloud feature F is formulated as follows,

$$F(i, j, k < 3) = (x_i, y_i, z_i) \tag{2}$$

$$F(i, j, k \geq 3) = \begin{cases} I_i(u_j, v_j), & if(u_j, v_j, d_j) \in S \\ 0, & else \end{cases} \tag{3}$$

$$S = \begin{cases} 0 \leq u_j \leq H \\ 0 \leq v_j \leq W \\ d_j \geq 0 \\ |d_j - D_j| < \delta \end{cases} \tag{4}$$

where $F(i, j, k)$ represents the value of feature F in the k-th channel of j-th point of the i-th image. $I_i(u_j, v_j)$ is the value of the position (u_j, v_j) in the i-th image, and (u_j, v_j) represents the projection coordinate of the point p_i onto the i-th image. S represents the constraint on pixel coordinates and pixel depth. D_j represents the value at pixel coordinates (u_j, v_j) in the original depth map. δ is the size of the depth constraint in millimeters.

To aggregate the multiple-view features, an average pooling method is adopted in this study.

$$F(j, k) = \frac{\sum_i F_i(j, k)}{n_{valid}} \tag{5}$$

where n_{valid} represents the number of views from which the value 0 has been removed.

3.3 Trans-Decoder

Subsequent to acquiring these features from the encoder, we feed them into the proposed Trans-Decoder model to enhance the representations and predict the final segmentation mask M, which has a dimensionality of $B \times C \times N_{cls}$, where N_{cls} corresponds to the number of categories. The Trans-Decoder model proposed in this study encompasses the following principal steps. Initially, in line with the Point Transformer approach, MLP layers and up-sampling techniques are applied to standardize the

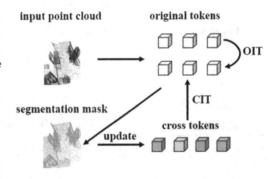

Fig. 2. OIT means original information tokens. CIT means cross information tokens.

channel dimension and feature scale of the multi-stage features. Subsequently, the fused feature is flattened back to the original token sequence $X_{original}$, which has a size of $N \times C$, where N denotes the number of points and represents the length of $X_{original}$. A visual depiction of the original tokens and cross tokens is presented in Fig. 2 for illustrative purposes. The original tokens from the current point cloud and the introduced learnable external tokens can easily interact with each other for multi-level context feature extraction in a transformer-based framework. Next, we perform feature extraction on the original and cross contexts as follows,

$$Y_{original} = \Upsilon_{original}(X_{original}) \tag{6}$$

$$Y_{cross} = \Upsilon_{cross}(X_{original}, X_{cross}) \tag{7}$$

where $\Upsilon_{original}$ and Υ_{cross} denote the original context extraction module and cross context extraction module, respectively. $Y_{original}$ and Y_{cross} are their outputs. The original context extraction module and cross context extraction module are shown in the Supplementary material. $X_{cross} \in \mathbb{R}^{N_{cls} \times C}$ is a learnable cross token sequence. Subsequently, the augmented feature $F_{aug} \in \mathbb{R}^{N \times C}$ is obtained through a two-step process involving a feature fusion operation P and a reshape operation,

$$F_{aug} = RESHAPE(P(X_{original}, Y_{original}, Y_{cross})) \tag{8}$$

In our proposed methodology, denoted in the present study, the element-wise addition operation is employed to establish the value of P. Consequently, the resulting segmentation mask, denoted as M, is determined in the following manner.

$$M = Upsample(N)H(F_{aug}) \tag{9}$$

where H denotes the classification head. The overall architecture of Trans-Decoder is showed in Fig. 3.

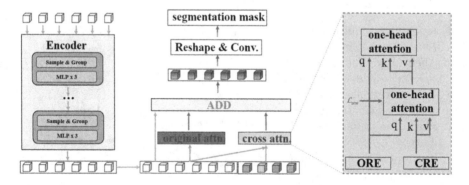

Fig. 3. The overall architecture of Trans-Decoder. The Trans-Decoder model comprises two pivotal components, namely original attention and cross attention. The attention loss $\mathcal{L}$ is designed to constrain on X_{cross}. ORE refers to the original embedding operation, while CRE denotes the cross embedding operation.

Loss Function. X_{cross} can be dynamically updated during the learning process, as it forms an integral part of the network parameter. To accomplish category specialization, an additional cross-entropy loss is applied to the mid output M_{mid} arising from $Attn_{mid} \in \mathbb{R}^{N \times N_{cls}}$,

$$M_{mid} = Upsample(N)(RESHAPE(Attn_{mid})) \tag{10}$$

$$\mathcal{L}_{attn} = \frac{1}{N} \sum_n \mathcal{L}_{ce}(M_{mid}^{[n,*]}, \Diamond \mathcal{GT}^{[n]}) \tag{11}$$

where $\Diamond$ represents the conversion of the ground truth class label stored in $\mathcal{GT}$ into a one-hot format, The summation symbol $\sum_n$ indicates that the summation is performed over all the points of $\mathcal{GT}$. The cross-entropy loss, denoted as $\mathcal{L}_{ce}$, is employed for this purpose. Analogously, the final segmentation loss, denoted as $\mathcal{L}seg$, is obtained by utilizing the mask prediction M.

$$\mathcal{L}_{seg} = \frac{1}{N} \sum_n \mathcal{L}_{ce}(M^{[n,*]}, \Diamond \mathcal{GT}^{[n]}) \tag{12}$$

$$\mathcal{L}_{total} = \mathcal{L}_{seg} + \alpha \mathcal{L}_{attn} \tag{13}$$

where α is the hyperparameters to balance the losses. By default, we empirically establish the value of α as 0.4 based on empirical observations and experimentation.

Efficient Self-attention. To address the computational challenges arising from transformer, we introduce a reduction ratio denoted as R. By utilizing a convolutional operation with an $R \times R$ kernel and stride R, we can effectively decrease the scale of the matrices K and V in the equations. This reduction in scale helps alleviate the computational load associated with attention operations, enabling more efficient processing while maintaining the integrity of the model's performance.

4   Experiments

In order to assess the efficacy of the proposed methodology, we perform experimental evaluations on ScanNet v2 datasets for the task of semantic segmentation. The training strategy and ablation experiments are shown in the Supplementary material.

4.1   Dataset and Metric

In our study on point cloud semantic segmentation, we performed experiments on the ScanNet v2 dataset. For evaluating the performance of our approach, we utilized the mean Intersection over Union (mIoU) as the primary evaluation metric.

Table 1. Compared with the single-modal method on ScanNet v2.

Method	Input	Val	Test
PointNet++ [16]	point	53.5	55.7
PointCNN [11]	point	-	45.8
PanopticFusion [14]	point	-	52.9
SegGCN [10]	point	-	58.9
UPB [1]	point	-	63.4
MVPNet [23]	point	-	64.1
PointASNL [27]	point	63.5	66.6
PCT [7]	point	58.9	-
RandLA-Net [9]	point	-	64.5
KPConv [21]	point	69.2	68.6
PointConv [24]	point	61.0	66.6
SparseConvNet [6]	voxel	69.3	72.5
Point Transformer	point	70.6	-
MinkowskiNet [2]	voxel	72.1	73.6
TSTD(Ours)	point	**72.5**	67.6

Table 2. Compared with the cross-modal method on ScanNet v2.

Method	Model	Test
PSPNet [29]	RGB	47.5
AdapNet++ [22]	RGB	50.3
3DMV(2d-porj) [26]	RGB-D	49.8
SegFusion [13]	RGB-D	51.5
FuseNet [8]	RGB-D	53.5
SSMA [22]	RGB-D	57.7
GRBNet [17]	RGB-D	59.2
MCA-Net [20]	RGB-D	59.5
DMMF [19]	RGB-D	59.7
CMX [28]	RGB-D	61.3
TSTD(Ours)	RGB-D	**67.6**

4.2 Performance Comparison

The results of our TSTD model, in comparison to previous methods, are presented in Table 1, showcasing the performance on the ScanNet v2 dataset. Notably, TSTD significantly outperforms Point Transformer by 1.94% mIoU on the ScanNet v2 validation set. As shown in Table 2, RGB-D methods typically achieve superior performance compared to RGB-only methods. The superiority of TSTD over RGB methods is evident, as it achieves a remarkable top mean Intersection over Union (mIoU) of 67.6% among RGB-D methods. While some methods, like CMX, attain higher scores on the ScanNetV2 leaderboard by solely utilizing 2D data and exploiting the complementary information within RGB-D modalities, our approach surpasses these scores. Our method leverages the power of joint 2D and 3D reasoning by incorporating 2D semantic information from multiple viewpoints, which contributes to the improved performance observed in our results. The qualitative results of point cloud segmentation are depicted in Fig. 4.

4.3 Ablation Experiment

A series of ablation studies are conducted to assess the effectiveness of each module within our design. The results of these ablation studies are reported on the ScanNet v2 validation set. The main ablation result is shown in Table 3, and other ablation study results are shown in Supplementary material.

Two Stages Strategy and Trans-decoder. In this section, we use Mit-B4 as 2D network backbone in order to effectively obtain best experiment results. We ablate different components in our TSTD: two stages strategy and Trans-Decoder and the results are illustrated in Table 3. We introduce a multi-view PCT (MVE_PCT) which based on PCT and recurrent simplified version Point Transformer. Benefiting from two stages, MVE_PCT result increased from 58.95% to 69.09%, and Point Transformer result increased from 56.86% to 69.77%. Furthermore, benefiting from two stages strategy and Trans-Decoder, MVE_PCT result increased from 58.95% to 70.20%, and Point Transformer result increased from 56.86% to 72.54%. It should be noted that we did not reproduce Point Transformer perfectly, only with network reproduction based on the adaptation of hardware devices.

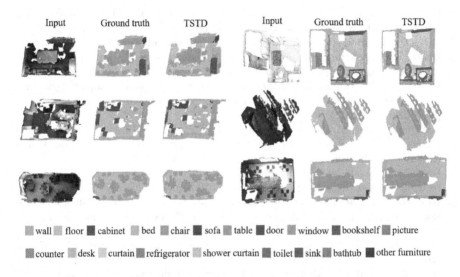

Input   Ground truth   TSTD   Input   Ground truth   TSTD

wall floor cabinet bed chair sofa table door window bookshelf picture
counter desk curtain refrigerator shower curtain toilet sink bathtub other furniture

Fig. 4. Visualization results on ScanNet v2.

Table 3. Different components ablation

ID	MVE_PCT	Point Transformer	Two Stages	Trans-Decoder	mIoU(%)
(1)	✓				58.95
(2)	✓		✓		69.09
(3)	✓		✓	✓	70.20
(4)		✓			56.86
(5)		✓	✓		69.77
(6)		✓	✓	✓	**72.54**

5 Conclusion

We propose Two Stages Network Design and Trans-Decoder (TSTD), a novel two-stage network architecture and a simple yet effective decoder based on a lightweight transformer module. Our work can effectively integrate heterogeneous data features and make several non-trivial improvements upon Point Transformer, including the two-stage strategy, the novel Trans-Decoder, and an additional loss function. Our TSTD outperforms CMX by 6.3% mIoU on the ScanNet v2 test set and achieves state-of-the-art performance in cross-modal point cloud semantic segmentation.

References

1. Chiang, H.Y., et al.: A unified point-based framework for 3d segmentation. In: 2019 International Conference on 3D Vision (3DV), pp. 155–163. IEEE (2019)
2. Choy, C., et al.: 4d spatio-temporal convnets: minkowski convolutional neural networks. In: Proceedings of the IEEE/CVF Conference on Computer Vision and Pattern Recognition, pp. 3075–3084 (2019)
3. Dai, A., et al.: Scannet: richly-annotated 3d reconstructions of indoor scenes. In: Proceedings of the IEEE Conference on Computer Vision and Pattern Recognition, pp. 5828–5839 (2017)
4. Dai, A., Nießner, M.: 3dmv: joint 3d-multi-view prediction for 3d semantic scene segmentation. In: Proceedings of the European Conference on Computer Vision (ECCV), pp. 452–468 (2018)
5. Dosovitskiy, A., et al.: An image is worth 16x16 words: transformers for image recognition at scale. arXiv preprint arXiv:2010.11929 (2020)
6. Graham, B., et al.: 3d semantic segmentation with submanifold sparse convolutional networks. In: Proceedings of the IEEE Conference on Computer Vision and Pattern Recognition, pp. 9224–9232 (2018)
7. Guo, M.H., et al.: Pct: point cloud transformer. Comput. Visual Media **7**, 187–199 (2021)
8. Hazirbas, C., Ma, L., Domokos, C., Cremers, D.: FuseNet: incorporating depth into semantic segmentation via fusion-based CNN architecture. In: Lai, S.-H., Lepetit, V., Nishino, K., Sato, Y. (eds.) ACCV 2016. LNCS, vol. 10111, pp. 213–228. Springer, Cham (2017). https://doi.org/10.1007/978-3-319-54181-5_14
9. Hu, Q., et al.: Randla-net: efficient semantic segmentation of large-scale point clouds. In: Proceedings of the IEEE/CVF Conference on Computer Vision and Pattern Recognition, pp. 11108–11117 (2020)
10. Lei, H., et al.: Seggcn: efficient 3d point cloud segmentation with fuzzy spherical kernel. In: Proceedings of the IEEE/CVF Conference on Computer Vision and Pattern Recognition, pp. 11611–11620 (2020)
11. Li, Y., et al.: Pointcnn: convolution on x-transformed points. In: Advances in neural information processing systems 31 (2018)
12. Liu, Z., et al.: Swin transformer: hierarchical vision transformer using shifted windows. In: Proceedings of the IEEE/CVF International Conference on Computer Vision, pp. 10012–10022 (2021)
13. Menini, D., Kumar, S., et al.: A real-time online learning framework for joint 3d reconstruction and semantic segmentation of indoor scenes. IEEE Robot. Autom. Lett. **7**(2), 1332–1339 (2021)

14. Narita, G., et al.: Panopticfusion: online volumetric semantic mapping at the level of stuff and things. In: 2019 IEEE/RSJ International Conference on Intelligent Robots and Systems (IROS), pp. 4205–4212. IEEE (2019)

15. Qi, C.R., et al.: Pointnet: deep learning on point sets for 3d classification and segmentation. In: Proceedings of the IEEE Conference on Computer Vision and Pattern Recognition, pp. 652–660 (2017)

16. Qi, C.R., et al.: Pointnet++: deep hierarchical feature learning on point sets in a metric space. In: Advances in Neural Information Processing Systems 30 (2017)

17. Qian, Y., Deng, L., et al.: Gated-residual block for semantic segmentation using rgb-d data. IEEE Trans. Intell. Transp. Syst. **23**(8), 11836–11844 (2021)

18. Shi, B., et al.: A transformer-based decoder for semantic segmentation with multi-level context mining. In: ECCV 2022, Part XXVIII, pp. 624–639. Springer, Cham (2022). https://doi.org/10.1007/978-3-031-19815-1_36

19. Shi, W., Xu, J., et al.: Rgb-d semantic segmentation and label-oriented voxelgrid fusion for accurate 3d semantic mapping. IEEE Trans. Circuits Syst. Video Technol. **32**(1), 183–197 (2021)

20. Shi, W., Zhu, D., et al.: Multilevel cross-aware rgbd indoor semantic segmentation for bionic binocular robot. IEEE Trans. Med. Robot. Bionics **2**(3), 382–390 (2020)

21. Thomas, H., et al.: Kpconv: flexible and deformable convolution for point clouds. In: Proceedings of the IEEE/CVF International Conference on Computer Vision, pp. 6411–6420 (2019)

22. Valada, A., et al.: Self-supervised model adaptation for multimodal semantic segmentation. Int. J. Comput. Vision **128**(5), 1239–1285 (2020)

23. Wang, J., Sun, B., Lu, Y.: Mvpnet: multi-view point regression networks for 3d object reconstruction from a single image. In: Proceedings of the AAAI Conference on Artificial Intelligence, vol. 33, pp. 8949–8956 (2019)

24. Wu, W., et al.: Pointconv: deep convolutional networks on 3d point clouds. In: Proceedings of the IEEE/CVF Conference on Computer Vision and Pattern Recognition, pp. 9621–9630 (2019)

25. Xie, E., et al.: Segformer: simple and efficient design for semantic segmentation with transformers. Adv. Neural. Inf. Process. Syst. **34**, 12077–12090 (2021)

26. Yan, L., et al.: Efficient depth fusion transformer for aerial image semantic segmentation. Remote Sens. **14**(5), 1294 (2022)

27. Yan, X., et al.: Pointasnl: robust point clouds processing using nonlocal neural networks with adaptive sampling. In: Proceedings of the IEEE/CVF Conference on Computer Vision and Pattern Recognition, pp. 5589–5598 (2020)

28. Zhang, J., et al.: Cmx: cross-modal fusion for rgb-x semantic segmentation with transformers. arXiv preprint arXiv:2203.04838 (2022)

29. Zhao, H., et al.: Pyramid scene parsing network. In: Proceedings of the IEEE Conference on Computer Vision and Pattern Recognition, pp. 2881–2890 (2017)

30. Zhao, H., et al.: Point transformer. In: Proceedings of the IEEE/CVF International Conference on Computer Vision, pp. 16259–16268 (2021)

NeuralMAE: Data-Efficient Neural Architecture Predictor with Masked Autoencoder

Qiaochu Liang, Lei Gong$^{(\boxtimes)}$, Chao Wang, Xuehai Zhou, and Xi Li

School of Computer Science and Technology, University of Science and Technology of China, Hefei, China
liangqc@mail.ustc.edu.cn, {leigong0203,cswang,xhzhou,llxx}@ustc.edu.cn

Abstract. Predictor-based Neural Architecture Search (NAS) offers a promising solution for enhancing the efficiency of traditional NAS methods. However, it is non-trivial to train the predictor with limited architecture evaluations for efficient NAS. While current approaches typically focus on better utilizing the labeled architectures, the valuable knowledge contained in unlabeled data remains unexplored. In this paper, we propose a self-supervised transformer-based model that effectively leverages unlabeled data to learn meaningful representations of neural architectures, reducing the reliance on labeled data to train a high-performance predictor. Specifically, the predictor is pre-trained with a masking strategy to reconstruct input features in both latent and raw data spaces. To further enhance its representative capability, we introduce a multi-head attention-masking mechanism that guides the model to attend to different representation subspaces from both explicit and implicit perspectives. Extensive experimental results on NAS-Bench-101, NAS-Bench-201 and NAS-Bench-301 demonstrate that our predictor requires less labeled data and achieves superior performance compared to existing predictors. Furthermore, when combined with search strategies, our predictor exhibits promising capability in discovering high-quality architectures.

Keywords: Neural architecture search · Masked autoencoder · Transformer

1 Introduction

Neural Architecture Search (NAS) has emerged as a prominent technique for automating the design of deep learning models, demonstrating superior performance over hand-crafted methods in several domains [6,15]. However, some conventional NAS methods [25] require enormous computation resources to train each candidate architecture from scratch for evaluation, making them impractical for real-world applications [18]. To address these issues, predictor-based NAS approaches [9] have emerged, which aim to learn predictors that can estimate the performance of unseen architectures directly to reduce the estimation time.

© The Author(s), under exclusive license to Springer Nature Singapore Pte Ltd. 2024
Q. Liu et al. (Eds.): PRCV 2023, LNCS 14432, pp. 142–154, 2024.
https://doi.org/10.1007/978-981-99-8543-2_12

However, one of the key challenges with such predictor-based NAS is that building a reliable predictor requires plenty of training samples (e.g., architecture-accuracy/latency pairs), which are very expensive to obtain in practice. For instance, BRP-NAS [3] uses 900 samples to train latency predictors for each device, which is prohibitively costly for the real-world deployment of a model to different hardware devices. Thus, how to efficiently evaluate architectures with limited labeled training data has become an essential problem. Some researchers attempt to improve predictor structures to extract more information from the available labeled data, such as using graph-based models [19] and transformer-based methods [11]. Others focus on learning the relative ranking of different architectures rather than estimating absolute performance values, as seen in ReNAS [21]. While most previous works have aimed to make better use of labeled data, they have not fully exploited the useful information contained in easily accessible unlabeled data (e.g., untrained architectures).

Therefore, inspired by the success of generative self-supervised learning [1, 4,5] in diverse fields, we propose a data-efficient masked self-supervised neural predictor (NeuralMAE) that can learn transferable features of untrained architectures during pre-training, reducing the need for labeled data in training a high-performance predictor. Specifically, the main goal of the pretext task is to reconstruct the randomly masked input features, allowing the model to learn more informative representations of architectures with signals acquired from the data itself. However, unlike masked images or masked sentences, the masked structure of an architecture is less predictable, making it challenging to directly reconstruct the masked architecture based on the input data of nearby architecture. To enhance the self-supervision mechanism, we combine the prediction of masked features in the latent space, which provides more informative targets, with reconstruction in the raw data space.

To further improve our predictor, we design a multi-head attention-masking mechanism based on different graph priors to incorporate the structural information of architectures into the vanilla Transformer [17]. After pre-training, the model can be fine-tuned with minimal labeled data to predict different architecture attributes (e.g., accuracy or latency), enabling the predictor to be coupled with any search strategies for efficient NAS.

The main contributions of this paper can be summarized as follows:

- We introduce generative self-supervised learning to neural architecture performance predictor and further improve the reconstruction mechanism, enhancing the robustness of the predictor with limited labeled data.
- We propose a novel multi-head attention-masking mechanism for transformers, encoding topological priors of architectures from different representation subspaces into the predictor.
- We conduct extensive experiments to demonstrate the superiority of our method on various NAS benchmarks.

2 Related Work

2.1 Neural Architecture Performance Predictors

Neural architecture performance predictors play an important role in reducing the evaluation cost for efficient NAS. Given the inherent nature of neural architectures as directed acyclic graphs (DAGs), some works have explored the encoding of architectures using diverse graph neural networks (GNNs) [3,19] to predict their performance. Building upon the success of transformer models in various domains, TNASP [11] introduced transformers into architecture encoding for the first time and outperformed previous methods. However, the above works focus on the fully supervised way to train performance predictors, neglecting the potential value of the vast amount of unlabeled architectures. To reduce the dependence on labeled data, GMAE [7] developed a masked autoencoder enhanced predictor using a GNN model, which is closer to our work. Nevertheless, the key difference is that we employ a more expressive transformer-based model and further improve the feature reconstruction mechanism, leading to improved performance in our method.

2.2 Generative Self-supervised Learning

Generative self-supervised learning, which aims to reconstruct the missing parts of the masked input data, has shown effectiveness in pre-training models that can be generalized well to various downstream tasks. Inspired by the achievements of masked autoencoders (MAEs) in natural language processing (NLP) [1] and computer vision (CV) [4], some researchers have extended their exploration of these techniques to handle plenty of unlabeled graph data, focusing on objectives related to structural [16] or feature reconstruction [5].

3 Method

3.1 Overall Framework

Following the previous NAS works [24], our architecture space is also based on cells. Each cell can be considered as a DAG $\mathcal{G} = \{\mathcal{V}, \mathcal{E}\}$, where $\mathcal{V}$ denotes a set of N nodes that represent the operations of an architecture, and $\mathcal{E}$ is a set of edges that connect the nodes. Figure 1 shows the overview of our proposed NeuralMAE, which consists of two phases: pre-training and fine-tuning. Firstly, the model is pre-trained using the improved feature reconstruction mechanism to extract meaningful representations from unlabeled architectures (Sect. 3.2). Then the pre-trained encoder is fine-tuned with limited labeled data for transferring to various performance prediction tasks (Sect. 3.3). The backbones of the encoder and decoder are transformers with a specifically designed multi-head attention-masking mechanism (Sect. 3.4).

3.2   Pre-training

The pre-training phase can be mainly divided into three parts: data masking, latent feature prediction and feature reconstruction. The goal of the pretext task is to reconstruct the features of masked nodes in both latent feature and raw data spaces, based on the visible ones.

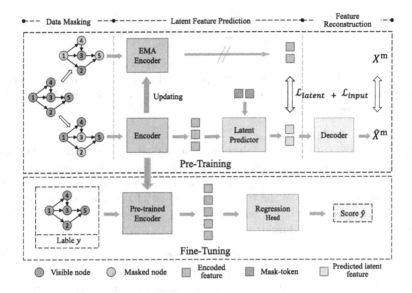

Fig. 1. The overview of NeuralMAE.

Data Masking. We randomly mask a subset of nodes in $\mathcal{G}$ with mask tokens under a specified masking ratio r, leading to the division of nodes into visible nodes $\mathcal{V}^v$ and masked nodes $\mathcal{V}^m$.

Latent Feature Prediction. After data masking, the masked input data is mapped to the latent feature space by the encoder. Notably, the encoder only takes the visible nodes as input to ensure consistency between the pretext task and the downstream task. It first embeds the operations of nodes via an embedding layer, followed by the addition of positional embeddings. The combined embeddings are then passed through a series of self-attention transformer blocks to generate the representations of visible nodes.

Following the encoder, a latent feature predictor is designed to predict the latent features of masked nodes based on the representations of the visible ones output from the encoder, leveraging high-level node correlations that are difficult to observe directly in raw data space. The predictor consists of a sequence of cross-attention transformer blocks, where the mask tokens serve as queries and the concatenation of visible node features and the previous transformer block's output serve as both keys and values. To guide the prediction of masked nodes effectively, we employ a momentum encoder, with the same architecture as the encoder but distinct weights, to produce latent features as prediction targets. The parameters of the momentum encoder are updated using an exponential moving average of the encoder's parameters and are detached from the gradient back-propagation. The encoder is trained to match the output of the momentum encoder on masked nodes by minimizing the following scaled cosine error:

$$\mathcal{L}_{\text{latent}} = \frac{1}{|\mathcal{V}^m|} \sum_{i=1}^{|\mathcal{V}^m|} \left(1 - \frac{\langle \widehat{z}_i^m, z_i^m \rangle}{\|\widehat{z}_i^m\| \cdot \|z_i^m\|}\right)^{\gamma} \tag{1}$$

where $|\mathcal{V}^m|$ is the number of masked nodes, and $\langle \cdot, \cdot \rangle$ is an inner product operation. $\widehat{z}_i^m \in \mathbb{R}^d$ and $z_i^m \in \mathbb{R}^d$ refer to the predicted and target feature vectors of i-th masked node, where d is the dimension of latent features. γ is a scale factor and is set to 2 empirically.

Feature Reconstruction. The decoder is a shallow transformer network followed by a linear projection, which aims to reconstruct the input operation types of masked nodes given the predicted latent representations. We apply the typical cross-entropy loss as the reconstruction loss of raw data space. By combining the losses from the latent feature and raw data spaces, we create a unified pre-training optimization target:

$$\mathcal{L} = \mathcal{L}_{\text{input}} + \lambda \mathcal{L}_{\text{latent}} \tag{2}$$

where λ is 1 in our experiments. The latent loss is designed to learn the latent representations, while the input loss aims to reconstruct the original input data.

3.3 Fine-Tuning

After pre-training, the encoder is able to generate informative representations of neural architectures, which can be leveraged to improve downstream prediction tasks. In the fine-tuning stage, only the encoder with pre-trained parameters is preserved, followed by an average pooling readout layer and a single linear projection layer as a regression head to construct the final predictor. The predictor is then fine-tuned using limited labeled data through end-to-end training.

Given that the accurate relative ranking of architectures holds more significance than the absolute performance values for NAS tasks, we incorporate predicted ranking into the optimization process instead of relying solely on mean squared error (MSE) between predicted and actual performance values. The whole loss function for fine-tuning is a weighted sum defined as follows:

$$\mathcal{L} = \mathcal{L}_{\text{mse}} + \lambda \mathcal{L}_{\text{ranking}} \tag{3}$$

where λ is a balancing coefficient and is set to 2. $\mathcal{L}_{\text{ranking}}$ is a pairwise ranking-based loss function, leveraging the inherent ordering of labels to optimize the consistency of rankings, formulated as:

$$\mathcal{L}_{\text{ranking}} = \sum_{\substack{i,j \in S \\ y_i > y_j}} max(0, m - (\widehat{y}_i - \widehat{y}_j)) \tag{4}$$

where S is the pairwise architecture set, and the margin m is set to 0.1 empirically. $\hat{y}_i$ and y_i correspond to the predicted performance and ground-truth performance of i-th architecture, respectively.

3.4   Multi-head Attention-Masked Transformer

To further enhance the representative capability of our predictor, we employ transformer models, which are known for their expressive power, as backbones of the encoder and decoder. However, the original Transformer lacks consideration for the structural information of a graph, as it allows nodes to attend to all other nodes in a graph. Inspired by Graphormer [23], we incorporate the centrality encoding where the node degrees are encoded as the positional embeddings and added to the input features to capture the node importance in the graph.

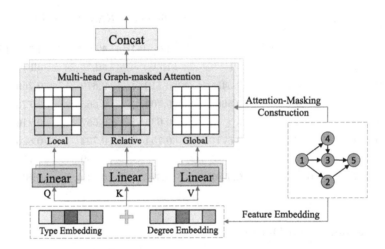

Fig. 2. The proposed multi-head attention-masking mechanism for encoder backbone.

To make the representation more distinctive and informative, we propose a novel multi-head attention-masking mechanism, as shown in Fig. 2. Specifically, the mechanism incorporates the structural graph information into the model through the utilization of three types of mask matrices from different perspectives: local, relative and global. By grouping the heads into three sets and introducing the corresponding matrix as a bias term to assist in the calculation of attention scores, our mechanism ensures that the heads explicitly attend to different subspaces with graph priors.

These three different mask matrices essentially enable the gradual expansion of the receptive field of the transformer, guiding the extraction of valuable information based on the graph structure. The local mask matrix, derived from the adjacency matrix, limits the receptive field of each node to its neighbors, capturing crucial natural relation information within the graph. The relative mask matrix is computed based on the shortest path distance of each node pair, enabling adaptive attention between nodes according to their spatial relations. Conversely, the global mask matrix, being a zero matrix, imposes no structural constraints and allows the model to learn implicit correlations between nodes without any prior assumptions.

4   Experiments

We first introduce the implementation details in Sect. 4.1. Then, in Sect. 4.2 - 4.4, we validate the superiority of our NeuralMAE in prediction tasks and combine the predictor with a search strategy to verify our approach in actual NAS tasks on different search spaces. Additionally, extensive ablation studies are conducted in Sect. 4.5 to evaluate the effect of each component of our model.

Table 1. Kendall's Tau of different predictors on NAS-Bench-101.

Training Samples	100 (0.02%)	172 (0.04%)	424 (0.1%)	424 (0.1%)	4236 (1%)
Test Samples	all	all	100	all	all
NP [19]	0.391	0.545	0.710	0.679	0.769
NAO [12]	0.501	0.566	0.704	0.666	0.775
ReNAS [21]	-	-	0.634	0.657	0.816
TNASP [11]	0.613	0.671	0.754	0.722	0.820
GMAE [7]	-	-	-	0.739	0.778
NeuralMAE	**0.651**	**0.716**	**0.787**	**0.756**	**0.852**

4.1   Implementation Details

In our model, the encoder consists of 4 transformer layers, while both the latent predictor and decoder have 2 layers each. The default masking ratio is set to 0.4, and a sensitivity analysis on different masking ratios is conducted in our ablation study. The hidden dimension size is 128 and the number of attention heads in the transformer layers is 8. During pre-training, all models are trained for 200 epochs with a batch size of 2048. For fine-tuning, the maximum epoch is set to 300, and the batch size is dynamically adjusted based on the number of labeled data. We employ the AdamW optimizer with a learning rate of $1e-3$ and apply a cosine decay strategy with warmup to schedule the learning rate.

4.2   Experiments on NAS-Bench-101

NAS-Bench-101. NAS-Bench-101 [24] is a typical NAS benchmark that provides the accuracy data for 423,624 unique cell-based architectures on CIFAR-10. Each cell has a maximum of 7 nodes and 9 edges, where nodes represent different candidate operations and edges indicate the connections between the nodes.

Results Comparison. Following the settings in [11], we select 0.02%, 0.04%, 0.1% and 1% of the whole data as the training set for fine-tuning. To assess the performance of different predictors, we use the entire dataset as the test set to calculate Kendall's Tau (KTau) as the evaluation metric. KTau is a suitable metric for evaluating the accuracy of predictive rankings, with a value closer to 1 indicating a stronger correspondence between the predicted rankings and the ground-truth rankings.

Table 1 presents KTau of different predictors on various training set sizes. We can see that our NeuralMAE outperforms other predictors in all settings, especially when the training data is limited, which is precisely the scenario we are most concerned about. The experimental results indicate that the prior representations obtained from pre-training facilitate improved performance of the predictor even with an extremely limited number of samples.

Table 2. Kendall's Tau of different predictors on NAS-Bench-201.

Method	Dataset	Training Samples				
		78 (0.5%)	156 (1%)	469 (3%)	781 (5%)	1563 (10%)
NP [19]	CIFAR-10	0.343	0.413	0.584	0.634	0.646
NAO [12]	CIFAR-10	0.467	0.493	0.470	0.522	0.526
TNASP [11]	CIFAR-10	0.565	0.594	0.642	0.690	0.726
NeuralMAE	CIFAR-10	**0.629**	**0.717**	**0.806**	**0.837**	**0.879**
NeuralMAE	CIFAR-100	0.656	0.715	0.806	0.831	0.873
NeuralMAE	ImageNet-16-120	0.604	0.687	0.779	0.814	0.866

Table 3. Spearman's ranking correlation of different latency predictors for various devices on NAS-Bench-201.

Method	Training Samples	Device		
		FPGA	Eyeriss	Raspi4
FLOPs	-	0.937	0.542	0.872
BRP-NAS [3]	900	0.801	0.811	0.847
NeuralMAE	50	**0.998**	**0.997**	**0.931**

4.3 Experiments on NAS-Bench-201

NAS-Bench-201. NAS-Bench-201 [2] is also a cell-based search space including 15,625 different architectures and provides the accuracy data for all architectures on CIFAR-10, CIFAR-100, and ImageNet-16-120. All cells in NAS-Bench-201 consist of 4 nodes and 6 edges, where each edge represents an operation. We convert each cell into an equivalent "operation on nodes" (OON) form to align with the representation format used in NAS-Bench-101.

Results Comparison. Similar to the settings in [11], we employ 0.5%, 1%, 3%, 5% and 10% of the whole dataset for fine-tuning, and the performance of different predictors is evaluated using KTau on the whole dataset. As shown in Table 2, NeuralMAE is still superior to other models regardless of the amount of training data available, indicating its effectiveness in diverse search spaces.

Additionally, HW-NAS-Bench [8] provides the measured hardware performance for models in NAS-Bench-201, allowing us to extend our experiments beyond accuracy prediction and verify the capability of NeuralMAE in latency prediction tasks. For more details regarding the hardware devices, please refer to HW-NAS-Bench [8]. We use Spearman's ranking correlation to evaluate the ranking accuracy and consider two baselines: a predictor based on the number of FLOPs and the latency predictor from BRP-NAS [3]. As shown in Table 3, our method achieves a Spearman correlation close to 1 even with a limited training set of 50 samples per device, demonstrating that the representations acquired from the pre-trained encoder are generalized for architecture evaluation, enabling the estimation of various hardware metrics for efficient hardware-aware NAS.

Table 4. Test accuracy (%) of searched architectures on NAS-Bench-201. "#Queries" denotes the number of queries.

Method	#Queries	Dataset		
		CIFAR-10	CIFAR-100	ImageNet-16-120
Random Search	50	94.03	72.09	45.75
BANANAS [20]	150	94.31	73.19	46.31
ReNAS [21]	90	94.22	72.88	46.42
NeuralMAE	50	**94.34**	**73.21**	**46.48**
optimal	-	94.37	73.51	47.31

Table 5. Comparison with other methods on NAS-Bench-301. The maximum number of queries is set to 100 in search.

Method	Test Acc. (%)	KTau	
		100	200
Random Search	94.06	-	-
BANANAS [20]	94.65	-	-
CATE-DNGO-LS [22]	94.72	-	-
GMAE [7]	94.85	0.599	0.630
NeuralMAE	**94.87**	**0.787**	**0.823**

To explore the impact of incorporating NeuralMAE into the actual NAS task, we combine our predictor with an evolutionary algorithm [13] as the search strategy. The test accuracy of the best architecture discovered using different methods on NAS-Bench-201 is reported in Table 4, with the last row indicating the oracle where the performance of all architecture is known. Experimental results show that our predictor can effectively collaborate with the search strategy, enabling the discovery of superior architectures with a limited number of queries. This highlights our significant advantage in terms of query utilization.

4.4 Experiments on NAS-Bench-301

NAS-Bench-301. NAS-Bench-301 [14] is a surrogate NAS benchmark on the DARTS [10] search space that contains more than 10^{18} architectures and provides estimated accuracy values for each architecture on CIFAR-10. Architectures in NAS-Bench-301 are constructed using normal cells and reduced cells, with each containing 7 nodes and 14 edges. Following the settings in CATE [22], we employ the same cell for both the normal and reduction cell.

Results Comparison. To verify the generalization ability of NeuralMAE in larger search spaces, we randomly sample 1,000,000 architectures from NAS-Bench-301 for pre-training and use 100 and 200 training samples for fine-tuning. Table 5 reports KTau for all samples in different training set sizes, indicating that our method can achieve better performance even with limited training data in larger search spaces. We also compare the best accuracy of architectures selected by different methods after 100 queries, as shown in Table 5, which further confirms the advantages of our method in larger search spaces.

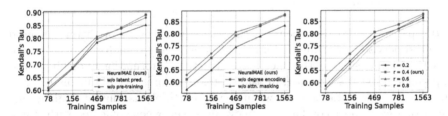

Fig. 3. The ablation studies of pre-training (left), structural encoding (middle) and masking ratio (right) on NAS-Bench-201.

4.5 Ablation Study

All ablation studies are conducted on NAS-Bench-201 and use KTau as the evaluation metric. The results are shown in Fig. 3.

Impacts of Pre-training. As shown in Fig. 3, we analyze the impacts of the enhanced self-supervision mechanism, where "w/o latent pred." means that the target of latent feature prediction is removed from pre-training, and "w/o pre-training" denotes that the predictor is trained from scratch without any pre-training. Our method achieves a higher KTau with extremely few training samples, proving that our predictor can learn better representations from unlabeled data and effectively transfer them to prediction tasks.

Impacts of Structural Encoding. We compare the results obtained without degree embedding or attention masking to verify the effectiveness of the proposed multi-head attention-masking mechanism. It can be seen that both strategies contribute to performance improvement. However, the attention-masking mechanism provides greater benefits, as evidenced by the larger drop in KTau whenss the component is removed. This is because compressing the graph structure into fixed-size vectors may result in information loss as the number of layers increases. Therefore, incorporating different structure priors to assist in attention matrix calculations during the computation process proves to be highly effective in mitigating this issue.

Impacts of Masking Ratio. Figure 3 also shows the influence of different masking ratios r. Notably, a masking ratio of 0.4 demonstrates the optimal performance, while using masking ratios that are excessively high or low introduces challenges or oversimplification in the pretext task, leading to suboptimal feature representations and a subsequent decline in performance.

5   Conclusion

In this paper, we propose a neural architecture performance predictor enhanced by self-supervised pre-training and design a specialized multi-head attention-masking mechanism to better encode the structural information of architectures. Experimental results demonstrate the superior performance of our predictor in different prediction tasks, even when the labeled data is limited, highlighting our advantage in terms of data-efficiency. Moreover, when combined with search strategies, our predictor can further improve the efficiency of predictor-based NAS. In the future, we will further explore how to bridge the gap between the pretext and downstream tasks to mitigate performance degradation during task transfer.

Acknowledgments. This work was supported in part by the National Key R&D Program of China under Grants 2022YFB4501603, in part by the National Natural Science Foundation of China under Grants 62102383, 61976200, and 62172380, in part by Jiangsu Provincial Natural Science Foundation under Grant BK20210123, in part by Youth Innovation Promotion Association CAS under Grant Y2021121, and in part by the USTC Research Funds of the Double First-Class Initiative under Grant YD2150002005.

References

1. Devlin, J., Chang, M.W., Lee, K., Toutanova, K.: Bert: pre-training of deep bidirectional transformers for language understanding. arXiv preprint arXiv:1810.04805 (2018)
2. Dong, X., Yang, Y.: Nas-bench-201: extending the scope of reproducible neural architecture search. arXiv preprint arXiv:2001.00326 (2020)
3. Dudziak, L., Chau, T., Abdelfattah, M., Lee, R., Kim, H., Lane, N.: Brp-nas: prediction-based nas using gcns. Adv. Neural. Inf. Process. Syst. **33**, 10480–10490 (2020)
4. He, K., Chen, X., Xie, S., Li, Y., Dollár, P., Girshick, R.: Masked autoencoders are scalable vision learners. In: Proceedings of the IEEE/CVF Conference on Computer Vision and Pattern Recognition. pp. 16000–16009 (2022)
5. Hou, Z., et al.: Graphmae2: a decoding-enhanced masked self-supervised graph learner. In: Proceedings of the ACM Web Conference 2023, pp. 737–746 (2023)
6. Howard, A., et al.: Searching for mobilenetv3. In: Proceedings of the IEEE/CVF International Conference on Computer Vision, pp. 1314–1324 (2019)
7. Jing, K., Xu, J., Li, P.: Graph masked autoencoder enhanced predictor for neural architecture search. In: Thirty-First International Joint Conference on Artificial Intelligence, vol. 4, pp. 3114–3120 (2022)
8. Li, C., et al.: Hw-nas-bench: Hardware-aware neural architecture search benchmark. arXiv preprint arXiv:2103.10584 (2021)
9. Liu, C., et al.: Progressive neural architecture search. In: Proceedings of the European conference on computer vision (ECCV), pp. 19–34 (2018)
10. Liu, H., Simonyan, K., Yang, Y.: Darts: Differentiable architecture search. arXiv preprint arXiv:1806.09055 (2018)
11. Lu, S., Li, J., Tan, J., Yang, S., Liu, J.: Tnasp: a transformer-based nas predictor with a self-evolution framework. Adv. Neural. Inf. Process. Syst. **34**, 15125–15137 (2021)
12. Luo, R., Tian, F., Qin, T., Chen, E., Liu, T.Y.: Neural architecture optimization. In: Advances in Neural Information Processing Systems 31 (2018)
13. Real, E., Aggarwal, A., Huang, Y., Le, Q.V.: Regularized evolution for image classifier architecture search. In: Proceedings of the AAAI Conference on Artificial Intelligence, vol. 33, pp. 4780–4789 (2019)
14. Siems, J., Zimmer, L., Zela, A., Lukasik, J., Keuper, M., Hutter, F.: Nas-bench-301 and the case for surrogate benchmarks for neural architecture search. arXiv preprint arXiv:2008.09777 (2020)
15. Tan, M., Le, Q.: Efficientnet: Rethinking model scaling for convolutional neural networks. In: International Conference on Machine Learning, pp. 6105–6114. PMLR (2019)
16. Tan, Q., et al.: S2gae: self-supervised graph autoencoders are generalizable learners with graph masking. In: Proceedings of the Sixteenth ACM International Conference on Web Search and Data Mining, pp. 787–795 (2023)
17. Vaswani, A., et al.: Attention is all you need. In: Advances in Neural Information Processing Systems 30 (2017)
18. Wang, C., Gong, L., Li, X., Zhou, X.: A ubiquitous machine learning accelerator with automatic parallelization on fpga. IEEE Trans. Parallel Distrib. Syst. **31**(10), 2346–2359 (2020)

19. Wen, W., Liu, H., Chen, Y., Li, H., Bender, G., Kindermans, P.J.: Neural predictor for neural architecture search. In: Computer Vision–ECCV 2020: 16th European Conference, Glasgow, UK, August 23–28, 2020, Proceedings, Part XXIX, pp. 660–676. Springer (2020)
20. White, C., Neiswanger, W., Savani, Y.: Bananas: Bayesian optimization with neural architectures for neural architecture search. In: Proceedings of the AAAI Conference on Artificial Intelligence, vol. 35, pp. 10293–10301 (2021)
21. Xu, Y., et al.: Renas: relativistic evaluation of neural architecture search. In: Proceedings of the IEEE/CVF Conference on Computer Vision and Pattern Recognition, pp. 4411–4420 (2021)
22. Yan, S., Song, K., Liu, F., Zhang, M.: Cate: computation-aware neural architecture encoding with transformers. In: International Conference on Machine Learning, pp. 11670–11681. PMLR (2021)
23. Ying, C., et al.: Do transformers really perform badly for graph representation? Adv. Neural. Inf. Process. Syst. 34, 28877–28888 (2021)
24. Ying, C., Klein, A., Christiansen, E., Real, E., Murphy, K., Hutter, F.: Nas-bench-101: towards reproducible neural architecture search. In: International Conference on Machine Learning, pp. 7105–7114. PMLR (2019)
25. Zoph, B., Le, Q.V.: Neural architecture search with reinforcement learning. arXiv preprint arXiv:1611.01578 (2016)

Co-regularized Facial Age Estimation with Graph-Causal Learning

Tao Wang[1], Xin Dong[1], Zhendong Li[1,2,3(✉)], and Hao Liu[1,2,3]

[1] School of Information Engineering, Ningxia University, Yinchuan 750021, China
wangtao@stu.nxu.edu.cn
[2] Collaborative Innovation Center for Ningxia Big Data and Artificial Intelligence
Co-founded by Ningxia Municipality and Ministry of Education,
Yinchuan 750021, China
[3] Key Laboratory of the Internet of Water and Digital Water Governance of the
Yellow River in Ningxia, Yinchuan 750021, China
{lizhendong,liuhao}@nxu.edu.cn

Abstract. In this paper, we present a graph-causal regularization (GCR) for robust facial age estimation. Existing label facial age estimation methods often suffer from overfitting and overconfidence issues due to limited data and domain bias. To address these challenges and leveraging the chronological correlation of age labels, we propose a dynamic graph learning method that enforces causal regularization to discover an attentive feature space while preserving age label dependencies. To mitigate domain bias and enhance aging details, our approach incorporates counterfactual attention and bilateral pooling fusion techniques. Consequently, the proposed GCR achieves reliable feature learning and accurate ordinal decision-making within a globally-tuned framework. Extensive experiments under widely-used protocols demonstrate the superior performance of GCR compared to state-of-the-art approaches.

Keywords: Facial age estimation · Causal learning · Graph convolutional networks

1 Introduction

Facial age estimation is a critical task in various domains, including minor protection [16], personalized recommendation [7] and social media, as it involves predicting the age value of a given facial image. However, the availability of labeled data for training age estimation models is restricted due to privacy policies. Consequently, the aging dataset exhibits significant challenges, including extremely sparse sample sizes at certain ages and notable domain (*i.e.* dataset) biases, such as variations in ethnicity, facial expressions, and background session.

Supplementary Information The online version contains supplementary material available at https://doi.org/10.1007/978-981-99-8543-2_13.

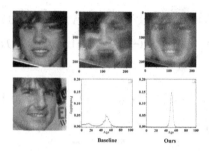

Fig. 1. The insight of GCR. It can locate precise aging related regions and capture texture details with the BCP module. The ODM further guarantees the ordinal prediction. (Better viewed in color file)

These factors contribute to issues like overfitting and overconfidence, leading to performance bottlenecks and limited generalization capabilities of the age estimation models.

In the context of numerous facial age estimation approaches, early literature [1,10,11] mainly focuses on aging feature descriptors, which struggle to collect the discriminative feature for robust age classification or regression. However, they overlook the age label ambiguity derived from progressive aging process, *i.e.* ordinal relation, and the severe dataset bias such as ethnicity and background variance badly cause the biased learning. Recently, a series of label distribution learning (LDL) based approaches [5,8,13] are developed for cross-age ambiguity exploitation, which converts the label of each image into a discrete label distribution. Although several variants have been proposed, such as [13] for data-driven label distribution learning and [5] for predicted label distribution adjustment, it is still challenging to model the non-stationary aging process and provide enough learning supervision with limited labeled data. Thus, the above approaches lead to the apparent overfitting and overconfidence problems, *i.e.* performance bottleneck and poor generalization. To address these problems, a generic solution is to devise effective co-regularizations both for reliable feature learning and cross-age ambiguity modeling.

Recent age estimation methods [5,8,13] have demonstrated remarkable performance on independent and identically distribution (IID) data collected from visible environments. However, they often struggle to generalize effectively to out-of-distribution (OOD) data obtained from unseen environments. One possible explanation for this limitation is that existing learning models tend to capture spurious correlations between environmental changes and age [31]. In other words, these models may rely on internal (*e.g.* ethnicity) or external factors (*e.g.* illumination) rather than age-related information itself. This susceptibility to environmental variations can pose significant challenges for age estimation systems deployed in real-world scenarios, raising concerns about their long-term reliability. To address this issue, it is crucial to leverage causal learning methods to uncover the underlying causal relationships instead of relying on superficial associations.

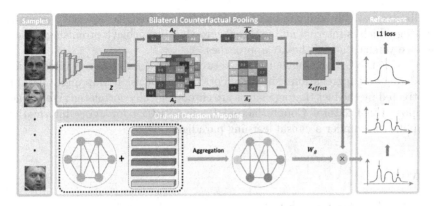

Fig. 2. The framework of our GCR. In ODM (green box), we initialize the ordinal graph edges with age label differences and node embeddings with normal distributions based on corresponding ages. GCN is then used to aggregate cross-age information and transform the ordinal graph into graph embeddings. In BCP (blue box), we employ ConvNet for feature extraction and perform counterfactual attention maximization in channel and spatial dimensions using a two-branch approach. The yellow box represents the refinement process of predicted scores. (Color figure online)

To this end, we propose a Graph-Causal Regularization (GCR) approach for stable facial age estimation especially in overfitting and overconfidence problems. The basic idea of our GCR is to implicitly reduce the solution space of model parameters both from the age prediction and feature learning process through effective co-regularization. This approach is composed of Ordinal Decision Mapping (ODM) and Bilateral Counterfactual Pooling (BCP), which is illustrated in Fig. 1. Specifically, motivated by the fact that age labels are chronologically correlated, it is desired to utilize this ordinal relation as label-level regularization to relieve the data limitation problem. We first develop a dynamic ordinal graph to model the underlying structure of age labels. Then we utilize it as the feature-to-label decision mapping regularization. In the BCP module, to remove the confounding effects on fine-grained aging feature from domain bias, we use the average treatment effect (ATE) estimation to quantify the attention and maximize the difference between attention and its counterpart to learn effective attention regularization. This can establish the causal correlation between feature and age label. We then integrate bilateral spatial and channel pooling into this counterfactual attention. The bilateral branches are respectively for aging related regions location and discriminative aging texture recognition. The pooling fusion can further achieve aging feature amplification. Note that these two regularizers are formulated in objective function as two panel terms, which means that they are the indirect co-regularization of model parameters.

The main contributions are summarized as follows:

- We propose a unified co-regularization approach to mitigate the issues of overfitting and overconfidence in facial age estimation, which are commonly caused

by limited data and domain bias. To highlight our strength, especially on OOD setting which is internal and external zero-shot, we achieve promising performance versus state-of-the-art approaches.
- We utilize the Graph Convolutional Network (GCN) for cross-age information modeling and aggregation. The design of Ordinal Decision Mapping (ODM) is motivated by the fact that age labels are chronologically correlated. We further propose the Bilateral Counterfactual Pooling (BCP) for aging information amplification under a causal learning paradigm.

2   Method

2.1   Problem Formulation

Let $\{(x_i, y_i)\}_{i=1}^N, y_i \in \{0, 1, ..., K-1\}$ denote N images and their labels. Existing methods extract feature representation with deep convolutional networks: $z_i = f_d(x_i; W_d) = W_d^T x_i$. Then these representations are mapped into age labels by decision mapping networks (e.g. fully connected layers) and softmax layer. As for objective function, the expectation regression loss [24] is widely adopted. Besides, recent LDL methods further utilize Kullback-Leibler Divergence [9] to embed the label correlation into age prediction.

Nevertheless, these methods easily cause the overfitting and overconfidence problems. To better understand these problems, we review the empirical risk minimization (ERM) paradigm in facial age estimation, i.e. minimization of $\sum_{i=1}^N (L(y_i, h(x_i)))$. Here $h(\cdot)$ means the feature extraction and decision mapping process. This ERM approximates to expectation risk $E(L(Y, h(X)))$. But the data limitation hinders this approximation to cause overfitting. The domain bias further introduce domain shift: $P_{tr}(X, Y) \neq P_{te}(X, Y)$ to cause poor generalization.

2.2   Ordinal Decision Mapping

Decision mapping means the feature-to-label mapping in age prediction. Motivated by the chronologically correlated age labels, label correlation can help address the data limitation. But the popular Kullback-Leibler divergence [9] is a implicit, infeasible and weak supervision for ordinal relation preserving. Thus we explicitly explore the non-linear label dependencies with a dynamic graph. Then we embed the graph to mapping networks for ordinal prediction. We will demonstrate this process actually provides effective regularization on the mapping network parameters. The architecture of our proposed method is shown in Fig. 2.

Ordinal Graph Learning. We first define the structure of ordinal graph [3]. For a graph $G(V, O)$, the node set V represents age labels, and the edge set O represents the weighted difference between two nodes. For each node v_i, its stack node representations to form the graph representation is expressed as $M \in \mathcal{R}^{K \times R}$, and its adjacent matrix $B \in \mathcal{R}^{K \times K}$, where K represents the number of nodes, and R is the representation dimension. Inspired by LDL [8], we initialize the node representations with their converted normal distributions.

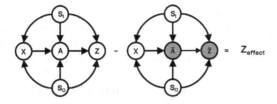

Fig. 3. Modelling the facial age estimation from causal view. The bilateral counterfactual attention utilizes average treatment estimation to ensure the reliable feature learning

Then we simply perform message aggregation with graph convolutional networks (GCN) [15]. We first define the identical matrix E and the degree matrix D conditioned on $\tilde{D}_{ii} = \sum_{j=1}^{K} B_{ij}$. Then graph convolution can be written as:

$$
\begin{aligned}
H^{(l+1)} &= f(H^{(l)}, B) \\
&= \sigma(\tilde{D}^{-\frac{1}{2}}(B + E)\tilde{D}^{-\frac{1}{2}}H^{(l)}W^{(l)}).
\end{aligned}
\tag{1}
$$

Here $H^{(l)}$ and $W^{(l)}$ represent feature and learnable weights of lth layer and we construct two-layer GCN, $\sigma(\cdot)$ is activation function. During the data-driven GCN process, the model captures age label dependencies and map the correlated age labels to the shared space. This is because GCN can force the model to aggregate aging information from neighbourhood age labels and achieve the global correlation from structural topology. Note that our ordinal graph can actually degenerate to LDL supervision if we do not perform graph convolution.

Mapping Network Regularization. The learned ordinal graph is embedded into a regularization matrix, which weights the feature to ordinal predicted scores. We rewrite the age prediction with the ordinal graph as:

$$
P_{r\_c} = softmax(ZW_gW_c),
\tag{2}
$$

where W_g represents the weight matrix learned in Eq. (1), and ZW_g means the feature regularized by ordinal relation. Z denotes aging feature whose calculation details will be further elaborated on in Sect. 2.3. Then they are mapped to high-level age labels with mapping matrix W_c. With the associative law of matrix multiplication, we combine the term W_gW_c in Eq. 2. This implies that the label-level regularization from the ordinal graph is actually an indirect but effective regularization on the decision mapping matrix. The damping effect for predicted scores can reduce the solution space from the label level and further relieve the overfitting problem due to the data limitation.

2.3 Bilateral Counterfactual Pooling

Due to the fine-grained characteristic of aging data, domain bias has confounding effects to cause domain shift. We aim to utilize the causal intervention and bilateral pooling to remove the domain bias. This can reduce the solution space in feature learning and help the stable learning.

Interventional Attention. We first utilize the structural causal model (SCM) to analyse the confounding effects from domain bias. As illustrated in Fig. 3, the aging images, aging features, attention maps and domain bias are denoted by X, Z, A and S respectively, where S_I represents internal factors (*e.g.* ethnicity) and S_O represents external factors (*e.g.* illumination). $S \rightarrow X$ means domain bias can affect (or cause) input images. $X \rightarrow A \rightarrow Z$ means attention learning for better recognition. However, they apply maximum likelihood statistic that establishes statistical rather than causal correlation. We particularly show S_I and S_O both have backdoor paths from X to Z for a spurious correlation. Thus we devise more effective attention with causal intervention, which performs backdoor adjustment in causal learning for domain bias removal.

Specifically, we learn from [23,29] to perform counterfactual intervention in attention, which quantifies the effectiveness of attention by the difference between attention and counterfacts. Unlike [29] to generate image counterfacts that lack fidelity, we only generate counterfactual attention, which samples from uniform distribution as is illustrated in Fig. 2 and is easier to optimize. We write this counterfactual attention as:

$$
\begin{aligned}
Z_{\text{effect}} &= E[Z|A = A, X = x] - E_{\tilde{A} \sim U}[Z|A = \overline{A}, X = x] \\
&= E_{\overline{A} \sim U}[Z(A = A, X = x) - Z(do(A = \overline{A}), X = x)],
\end{aligned}
\tag{3}
$$

where we denote the effect on the aging feature as Z_{effect} and U is uniform distribution The philosophy behind this is that maximization of attention difference equals to the maximization of average treatment effect estimation (ATE) in causal learning. Here $do(A = \overline{A})$ operation is backdoor adjustment, which generates attention values from uniform distribution.

Bilateral Pooling Fusion. We integrate bilateral pooling architecture into counterfactual attention for better internal and external factors removal and aging information amplification. Specifically, we split A into two branches, spatial attention A_S and channel attention A_C, and perform spatial counterfactual attention $\overline{A_S}$ and channel counterfactual attention $\overline{A_C}$ on the bilateral branches for aging related regions location and discriminative aging texture capture respectively. Their refined feature X_S and X_C with counterfactual attention as:

$$
X_S = X \odot (A_S - \overline{A_S}),
\tag{4}
$$

$$
X_C = X * (A_C - \overline{A_C}).
\tag{5}
$$

Here $\odot$ and * represent Hadamard product and channel-wise multiplication respectively. A_S and A_C are implemented with spatial and channel attention from CBAM [32]. We then propose the attentive feature pooling as

$$
\begin{aligned}
Z &= P(X_S, X_C) \\
&= concat([g(X_S \odot X_C^1), g(X_S \odot X_C^2), ..., g(X_S \odot X_C^M)]).
\end{aligned}
\tag{6}
$$

Here $g(\cdot)$ means the average pooling operation and M is the channel number of X_C. This pooling fusion can harmony with the spatial and channel dimension

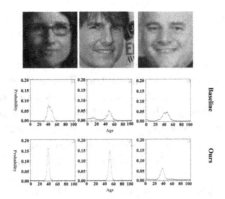

Fig. 4. The predicted scores generated by baseline and GCR in the UTKFace dataset. (Better viewed in color file.)

variations, and transform fine-grained aging information into high-order amplified feature representation. Note that the proposed bilateral attentive pooling can be viewed as an effective self attention.

Overall Objective Function. The overall loss function for the proposed method are written as:

$$L = L_r(\hat{p}, y) + \beta L_r(\hat{p}_{\text{effect}}, y) + \sum_{E_{i,j} \in W_C} ||W_i, W_j||. \tag{7}$$

Here L_r is expectation regression loss [24], $\hat{p}$ and $\hat{p}_{effect}$ are original and counterfactual interventional predicted scores respectively. W_i and W_j are weights for ith and jth age nodes. The second term is corresponding to the BCP module. For ODM optimization, we utilize GCN to replace the non-convex optimization in loss function [3]. Thus the above modules can be optimized in a globally-tuned paradigm with the standard back propagation algorithm. The joint optimization of ODM and BCP can establish the causality between feature representation and age labels, achieving performance improvement and stable generalization.

3   Experiments

3.1   Datasets and Evaluation Settings

Datasets: *MORPH II* [14] has more than 55,000 images from more than 13,000 persons. The age covers from 16 to 77 years old. Like [5], we use the Setting I and Setting II for evaluation. *Chalearn LAP 2015* [6] has 4699 images from Internet. Each image is annotated by 10 volunteers based on appearance and their results are shown in normal distribution form. As for evaluation metrics, we followed previous works [5,8,18].

Evaluation Metrics: In this paper, we utilized Mean Absolute Error (MAE) as the metric for all datasets and ϵ-error as an additional metric for Chalearn

LAP dataset. These two metrics were widely adopted in most recent approaches [5,17]. We followed this metric setting for better comparisons with existing approaches.

OOD Settings: We further devised the OOD settings to validate the robustness under internal factors and external factors. The settings are based on *UTK-FACE* [36] and *Megaage-Asian* [35]. The former has more than 20,000 images with age, gender and other attributes annotations. The latter has over 40,000 images of Asian persons, containing variations of expression and illumination.

Specifically, in line with the zero-shot OOD settings in [30], we used the images of White persons in UTKFACE to train the model while testing it on images from Asian persons. We named it an internal zero-shot setting. We then collected training images from Asian persons in UTKFACE and tested the model with images from Asian persons in Megaage-Asian. The setting validates the stability under external factors and we named it an external zero-shot setting.

Table 1. Comparisons with state-of-the-art methods on the Morph II Setting I and Setting II. (* and † respectively mean pretrained weights on IMDB-WIKI and private dataset)

Setting I	MAE	Year	Setting II	MAE	Year
DRFs [26]	2.17	2018	DEX [24]	2.68*	2018
DLDL v2 [8]	1.97†	2018	AgeED [27]	2.52*	2018
BridgeNet [18]	2.38*	2019	DRFs [26]	2.91	2018
MA-SFV2 [22]	2.68	2020	DHAA [28]	2.49*	2019
AVDL [33]	**1.94***	2020	AVDL [33]	2.37*	2020
Dapher et al. [4]	2.94	2021	PML [5]	2.31	2021
PML [5]	2.15	2021	MetaAge [19]	2.23	2022
OrdCLIP [17]	2.32	2022	DCT [2]	2.28	2022
Ours	2.136	-	**Ours**	**1.985**	-

Table 2. Comparisons with existing methods on the Chalearn. (* means pretrained weights on IMDB-WIKI)

Method	$\epsilon-$error	MAE	Year
CVL_ETHZ [24]	0.282*	3.252*	2018
AgeED [27]	0.280*	3.210*	2018
ODL [21]	0.312	3.950	2019
DHAA [28]	0.265*	3.052*	2019
PML* [5]	0.243*	2.915*	2021
PML [5]	0.293	3.455	2021
Ours	**0.280**	**3.273**	-
Ours*	**0.216***	**2.977***	-

Implementation Details: We utilized MTCNN [34] to detect and align faces. Then the images were cropped into normalized size (224 × 224). For robust training and fast convergence, augmentations with horizontal flip, random rotation and color jitters were used. During the training, we adopted the pretrained Resnet-34 [12] as the backbone. We set the learning rate to 1e-3 and training epoches to 1000. To optimize the network, we employed SGD with a momentum of 0.9. The weight decay was set to 0.1. Cosine annealing algorithm with warm-up strategy was adopted for learning rate adjustment. Our models were both trained and tested on one Tesla V100 GPU in a batch size of 600.

Fig. 5. The heatmaps generated by GCR in the UTKFace.

3.2   Comparison with State-of-the-Art Methods

We compared results of GCR in MORPH II and Chalean datasets with state-of-the-art methods. As shown in Table 1, GCR achieves the best results with other comparison methods under w/o pretrained weights setting.[1] Furthermore, the GCR suppressed all methods in Morph II setting II. As shown in Table 2, in Chalearn the GCR outperforms state-of-the-art methods under all settings for ϵ-error metric. The ϵ-error metric is designed specially for appearance based dataset, which is more important than MAE in these datasets. These results demonstrate the effectiveness of GCR with ODM and BCP modules.

Table 3. Ablation study on the Morph II Setting II and Chalearn LAP 2015.

	MORPH	Chalearn	
Method	MAE	ϵ-error	MAE
Baseline	2.812	0.340	3.687
w/o ODM	2.607	0.319	3.500
w/o BCP	2.533	0.323	3.550
Ours	**1.985**	**0.280**	**3.273**

---

[1] GCR outperformed other methods except for DLDL v2 and AVDL in Morph II setting I. However, they were pretrained on large dataset (*i.e.* IMDB-WIKI [24]) or private dataset.

3.3 Ablation Study

We conducted ablation experiments to validate the performance improvement from each module. As shown in Table 3, when we removed any module the performance would degrade obviously. The joint optimization of ODM and BCP achieve the best result, which demonstrate that each module of the GCR had an indispensable effect on solving the overfitting and overconfidence problems.

Table 4. Performance in internal and external OOD settings.

Method	Internal zero shot	External zero shot
DLDL v2 [8]	5.650	12.270
PML [5]	4.610	10.035
Ours	**3.915**	**9.385**

3.4 Performance Under Out-of-Distribution Settings

To demonstrate the generalization ability of the GCR, we conducted experiments under the OOD settings, which has defined in Sect. 3.1. We compared the results of GCR with DLDL v2 [8] and recent PML [5] methods with MAE. For a fair comparison, we utilized their source codes and did not adjust many hyper parameters to respect the original performance. As shown in Table 4, our GCR improves the accuracy by over 0.6 in the internal zero-shot setting compared with PML, which means GCR can better cope with internal semantic disturbances such as ethnicity variance. Although the difficulty of the external zero-shot setting, our GCR suppresses the DLDL v2 and PML. These results are from the spatial and channel bilateral counterfactual pooling to precisely locate the age related regions while capturing the fine-grained aging texture. The label-level regularization from ODM further helps to address the overconfidence.

3.5 Qualitative Results

To show the qualitative results the GCR, we conducted two kinds of experiments: 1) Grad-CAM based heatmaps, 2) predicted scores. Specifically, heatmap visualizations can vividly show the regions of interests learned by models. Predicted scores visualization can judge whether the model preserves the ordinal relation. As illustrated in Fig. 5, GCR can better locate the face regions and pay more attention to eyes and nose under variable poses and expressions. In Fig. 4, GCR can predict a more precise score: 1) it has a single peak, 2) it obeys the ordinal consistency. These visual results validate the effectiveness of GCR for ordinal relation preserving and reliable feature learning, thus achieving performance improvement and stable generalization. More qualitative experiments and results can be seen in the supplementary file.

4   Conclusion

In this paper, we have proposed the graph-causal regularization (GCR) approach to address the overfitting and overconfidence problems in facial age estimation. Specifically, we have developed the dynamic graph to explore the underlying ordinal relation, which can achieve the ordinal relation preserving in predicted scores. Furthermore, we have incorporated with counterfactual attention and bilinear pooling for domain bias removal and aging details amplification, which is an effective regularization in feature space. Comprehensive experiments on three public protocols and the out-of-distributed settings show the effectiveness of our GCR. In the future, we will explore the importance of causal interventions in multi-modality tasks [20].

Acknowledgment. This work was supported in part by the National Science Foundation of China under Grants 62076142 and 62241603, in part by the National Key Research and Development Program of Ningxia under Grant 2023AAC05009, 2022BEG03158 and 2021BEB0406.

References

1. Ahonen, T., Hadid, A., Pietikäinen, M.: Face description with local binary patterns: application to face recognition. TPAMI **28**(12), 2037–2041 (2006)
2. Bao, Z., Tan, Z., Wan, J., Ma, X., Guo, G., Lei, Z.: Divergence-driven consistency training for semi-supervised facial age estimation. TIFS **18**, 221–232 (2022)
3. Cai, X., Nie, F., et al.: New graph structured sparsity model for multi-label image annotations. In: ICCV, pp. 801–808 (2013)
4. Dagher, I., Barbara, D.: Facial age estimation using pre-trained cnn and transfer learning. Multimed. Tools. Appl. **80**, 20369–20380 (2021)
5. Deng, Z., et al: PML: progressive margin loss for long-tailed age classification. In: CVPR, pp. 10503–10512 (2021)
6. Escalera, S., et al.: Chalearn looking at people 2015: apparent age and cultural event recognition datasets and results. In: ICCV Workshops, pp. 243–251 (2015)
7. Fu, Y., Guo, G., Huang, T.S.: Age synthesis and estimation via faces: a survey. TPAMI **32**(11), 1955–1976 (2010)
8. Gao, B., et al.: Age estimation using expectation of label distribution learning. In: Lang, J. (ed.) IJCAI, pp. 712–718 (2018)
9. Geng, X., Ji, R.: Label distribution learning. In: ICDM Workshops, pp. 377–383 (2013)
10. Geng, X., Yin, C., Zhou, Z.: Facial age estimation by learning from label distributions. TPAMI **35**(10), 2401–2412 (2013). https://doi.org/10.1109/TPAMI.2013.51
11. Guo, G., Mu, G., Fu, Y., Huang, T.S.: Human age estimation using bio-inspired features. In: CVPR, pp. 112–119 (2009)
12. He, K., Zhang, X., Ren, S., Sun, J.: Deep residual learning for image recognition. In: CVPR, pp. 770–778 (2016)
13. He, Z., et al.: Data-dependent label distribution learning for age estimation. TIP, pp. 3846–3858 (2017)

14. Jr., K.R., Tesafaye, T.: MORPH: a longitudinal image database of normal adult age-progression. In: FG, pp. 341–345 (2006)
15. Kipf, T.N., Welling, M.: Semi-supervised classification with graph convolutional networks. In: ICLR. OpenReview.net (2017)
16. Lanitis, A., Draganova, C., Christodoulou, C.: Comparing different classifiers for automatic age estimation. SMC **34**(1), 621–628 (2004)
17. Li, W., Huang, X., Zhu, Z., Tang, Y., Li, X., Zhou, J., Lu, J.: Ordinalclip: learning rank prompts for language-guided ordinal regression. arXiv preprint arXiv:2206.02338 (2022)
18. Li, W., Lu, J., Feng, J., Xu, C., Zhou, J., Tian, Q.: Bridgenet: a continuity-aware probabilistic network for age estimation. In: CVPR, pp. 1145–1154 (2019)
19. Li, W., Lu, J., Wuerkaixi, A., Feng, J., Zhou, J.: Metaage: meta-learning personalized age estimators. TIP **31**, 4761–4775 (2022)
20. Liu, C., Ding, H., Jiang, X.: Gres: generalized referring expression segmentation. In: CVPR, pp. 23592–23601 (2023)
21. Liu, H., Lu, J., Feng, J., Zhou, J.: Ordinal deep feature learning for facial age estimation. In: FG, pp. 157–164 (2017)
22. Liu, X., Zou, Y., Kuang, H., Ma, X.: Face image age estimation based on data augmentation and lightweight convolutional neural network. Symmetry **12**(1), 146 (2020)
23. Rao, Y., et al.: Counterfactual attention learning for fine-grained visual categorization and re-identification. In: ICCV, pp. 1005–1014 (2021)
24. Rothe, R., Timofte, R., Gool, L.V.: DEX: deep expectation of apparent age from a single image. In: ICCV Workshop, pp. 252–257 (2015)
25. Rothe, R., Timofte, R., Gool, L.V.: Ijcv. Int. J. Comput. Vis. **126**(2–4), 144–157 (2018)
26. Shen, W., Guo, Y., Wang, Y., Zhao, K., Wang, B., Yuille, A.L.: Deep regression forests for age estimation. In: CVPR, pp. 2304–2313 (2018)
27. Tan, Z., et al.: Efficient group-n encoding and decoding for facial age estimation. TPAMI, pp. 2610–2623 (2018)
28. Tan, Z., etal.: Deeply-learned hybrid representations for facial age estimation. In: IJCAI, pp. 3548–3554 (2019)
29. Vermeire, T., Martens, D.: Explainable image classification with evidence counterfactual. CoRR **abs/2004.07511** (2020)
30. Wang, T., Zhou, C., Sun, Q., Zhang, H.: Causal attention for unbiased visual recognition. CoRR **abs/2108.08782** (2021)
31. Wang, X., Saxon, M., Li, J., Zhang, H., Zhang, K., Wang, W.Y.: Causal balancing for domain generalization. arXiv preprint arXiv:2206.05263 (2022)
32. Woo, S., Park, J., et al.: CBAM: convolutional block attention module. In: ECCV, pp. 3–19 (2018)
33. Zhang, C., Liu, S., Xu, X., Zhu, C.: C3AE: exploring the limits of compact model for age estimation. In: CVPR, pp. 12587–12596 (2019)
34. Zhang, K., Zhang, Z., Li, Z., Qiao, Y.: Joint face detection and alignment using multitask cascaded convolutional networks. IEEE Signal Process. Lett. **23**(10), 1499–1503 (2016)
35. Zhang, Y., Liu, L., et al.: Quantifying facial age by posterior of age comparisons. In: BMVC (2017)
36. Zhang, Z., Song, Y., Qi, H.: Age progression/regression by conditional adversarial autoencoder. In: CVPR, pp. 4352–4360 (2017)

Online Distillation and Preferences Fusion for Graph Convolutional Network-Based Sequential Recommendation

Youhui Cheng[1], Jianping Gou[2(✉)], and Weihua Ou[3]

[1] School of Computer Science and Communication Engineering,
Jiangsu University, Jiangsu 212013, China
[2] College of Computer and Information Science, College of Software,
Southwest University, Chongqing 400715, China
cherish.gjp@gmail.com
[3] School of Big Data and Computer Science, Guizhou Normal University,
Guiyang, Guizhou 550025, China

Abstract. Sequential recommendations make an attempt to predict the next item that a user will interact with based on their historical behavior sequence. Recently, considering the relationship learning ability of graph convolutional network (GCN), a number of GCN-based sequence recommendation models have emerged. However, in real-world applications, sparse interactions are common, with early and current short-term preferences playing diverse roles in sequential recommendation. As a result, vanilla GCNs fail to investigate the explicit relationship between these early and current short-term preferences. To address the above limitations, we propose a scheme of Online Distillation and Preferences Fusion for GCN-based sequential recommendation (ODPF). Specifically, our approach performs online distillation among multiple networks to learn item feature representations. To distinguish between early and recent short-term preferences, we divide each sequence into two subsequences and construct two graphs separately. On this basis, a fusion network is introduced to capture more accurate preferences by fusing these two types of preferences. Experimental evaluations conducted on two public datasets demonstrate that our proposed method outperforms recent state-of-the-art methods in terms of recommendation precision.

Keywords: Graph Convolutional Networks · Sequential Recommendation · Online Distillation · Feature Fusion

1 Introduction

Sequential recommendation predicts the users' next behavior based on their historical interaction sequences. Since the historical sequences are arranged in chronological order, in order to improve the effectiveness of sequential recommendation, it is essential to correctly capture user preferences [3]. Specifically,

© The Author(s), under exclusive license to Springer Nature Singapore Pte Ltd. 2024
Q. Liu et al. (Eds.): PRCV 2023, LNCS 14432, pp. 167–178, 2024.
https://doi.org/10.1007/978-981-99-8543-2_14

early works are based on matrix factorization and human-designed rules to process chronological item sequences [12], aiming to capture long-term preferences and temporal information. However, this strategy demands strong dependency assumptions about user behaviors. With the advancement of deep learning, many neural network-based methodologies have been proposed, including recurrent neural network (RNN)-based methods [10,15], and convolutional neural network (CNN)-based methods [16,20]. Specifically, GRU4Rec [15] and HRNN [10] utilize recurrent neural networks to learn long- and short-term information from the sequences. However, they cannot effectively capture users' dynamic interests. CNN-based models, for example, Caser [16], uses convolutional layers to extract different levels of patterns from the sequences. However, the above mentioned approaches focus more on recent user behaviour and fail to take full advantage of the early user behavior to capture user preferences more accurately. Recently, due to the superiority of GCN in handling non-Euclidean spatial data. GCN-based methods for recommendation tasks have been popular. GCMC [2], NGCF [18], LightGCN [7] have achieved excellent performance in traditional recommendations by exploiting the property of GCNs to iteratively aggregate hidden features from neighborhoods. Furthermore, various GCN-based models have also been devised for sequential recommendation scenarios [3,4,9,13] to improve the precision of sequential recommendations.

Despite the encouraging performance achieved by GCN-based approaches, there are still two major challenges that have not been well addressed. Firstly, GCN aggregates neighbour information via utilizing graph topology structure and message passing, but the practical data often exhibit sparsity, which increases the difficulty of obtaining accurate item feature representations. Second, current GCN-based models usually use shallow networks with two to three layers, which leads to their inability to extract higher-order neighborhood information [22]. This makes these models ignore the associations between early and recent short-term preferences in sequential recommendation, where short-term preferences refer to the user's interest in certain similar items within a short period of time [9].

To tackle these two challenges, we present a novel scheme called Online Distillation and Preferences Fusion (ODPF) to learn items feature representations and capture user preferences. We first divide the sequence into two subsequences: the source sequence and the target sequence, then we convert these two subsequences into item-item graphs, separately. We extract user's early short-term preferences and recent short-term preferences from the two graphs, respectively. A fusion network is then used to fuse the above two types of preferences. In addition, inspired by collaborative learning (CL) [1], an online knowledge distillation approach to transfer knowledge across multiple student networks without teacher network. And as a collaborative learning method, KDCL [6] integrates the logits of multiple networks into a teacher network, which serves as a guidance for individual networks. Motivated by the idea behind KDCL, our approach involves collaborative learning among the latent features obtained from several graph convolutional networks (GCNs). We acquire ensemble features through the collaboration and subsequently utilize these ensemble features to guide the latent features of the individual GCNs.

The main contributions of this paper can be summarised as follows: 1) We divide a sequence into two subsequences and construct the source and target graphs, allowing us to capture early and recent short-term preferences separately. Then we bridge early and recent short-term preferences in a sequence from the perspective of feature fusion to alleviate the issue that GCN can not effectively extract higher-order dependencies; 2) We introduce collaborative learning to transfer feature knowledge between several randomly initialized networks to acquire more accurate latent features of items; and 3) We conduct extensive experiments on two popular datasets to evaluate the proposed method, ODPF. The experimental results clearly demonstrate that our method outperforms recent state-of-the-art recommendation methods across various evaluation metrics.

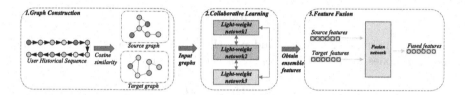

Fig. 1. The main ODPF framework for sequential recommendation.

2   Method

The main proposed ODPF framework for sequential recommendation is shown in Fig. 1. This section presents the details of our proposed ODPF.

2.1   Graph Construction

We construct two graphs to capture early short-term preferences and recent short-term preferences separately. Specifically, by dividing the sequence into two subsequences, we use the first subsequence as the source sequence to capture early short-term preferences and the second subsequence as the target sequence to capture recent short-term preferences. We then construct two graphs accordingly based on the above-mentioned two sequences, that is, source graph and target graph, separately. Moreover, in order to facilitate feature fusion of the two graphs, the two subsequences are always in the same length.

Let U and V denote the set of users and items, respectively. Let the historical interaction sequences set be $L = \{L_1, L_2, \cdots, L_n\}$, where n is the number of users in U. For example, $L_1 = \{v_1, v_2, \cdots, v_m\}$ should be divided into two subsequences: $L_1^{source} = \{v_1, v_2, \cdots, v_{m/2}\}$ and $L_1^{target} = \{v_{m/2+1}, v_{m/2+2}, \cdots, v_m\}$. Then we will use L_1^{source} and L_1^{target} to construct two graphs: source graph G_1^{source} and target graph G_1^{target}. Take the construction process of G_1^{source} as an example: when we construct a graph, following [19], we use cosine similarity to metric similarities between items, a similarity matrix M is calculated as follows:

$$M_{ij} = \begin{cases} cos(v_i, v_j), & i \neq j \\ 0, & i = j \end{cases}, \tag{1}$$

where v_i and v_j are the features of nodes i and j, respectively. Then we rank its elements M_{ij} in descending order and take the top $k\%$ of the elements as edges to construct the graph. Additionally, we add self-loop for all nodes. In this way, we add edges between similar items, reducing computational consumption and limiting the effect of noise. k controls the sparsity of the graph. Since it is extremely important to choose an appropriate k. We determine the number of edges based on the graph sparsification criteria [5]:

$$q = p \log_2 p, \tag{2}$$

where q is the number of edges and p is the number of nodes. Therefore, the number of nodes in G_1^{source} is $p = m/2$, k is calculated as follows:

$$k = \frac{p \log_2 p}{p^2} = \frac{\log_2 p}{p}, \tag{3}$$

where the value of k varies with the sequence length.

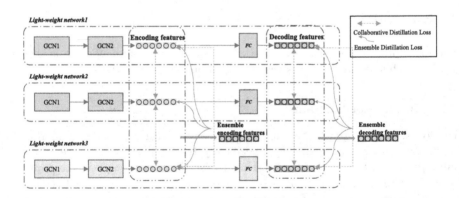

Fig. 2. The main collaborative learning pipeline across multiple networks. Specifically, knowledge distillation is performed 1) between features of individual networks; and 2) between ensemble features and network features.

2.2   Collaborative Learning

After source graph and target graph are constructed, we input the graphs into the feature learning network. This component can be summarized as collaborative learning with multiple networks. We employ multiple networks with different initializations to obtain more information, and utilize collaborative learning to improve network performance for more accurate item hidden features. The framework is shown in Fig. 2. The specific training process is as follows.

First, we obtain the features of items based on their own properties $H = \{h_1, h_2, \cdots, h_{m/2}\}$, then these item features are used as the corresponding node initial features in graph. Next, we generate the latent features of node i by applying the following aggregation function on the two graph convolution layers:

$$h_i^{(l+1)} = \sigma \left(b_n^l + \sum_{j \in N(i)} \frac{1}{c_{ij}} h_j^{(l)} W_n^l \right), \tag{4}$$

where $N(i)$ is the one-hop neighbor of node i, h_j^l is the features of node j in the former layer, and $c_{ij} = \sqrt{|N(i)|}\sqrt{|N(j)|}$ is the product of the square root of the node degrees between node i and its neighbor node j. W_n^l and b_n^l are two trained parameters. σ is rectified linear unit (ReLU), a nonlinear activation function. For network n, $E_n \in R^{\frac{m}{2} \times d}$ is the encoded features of the two-layer GCN, and the decoder utilizes the encoded features E_n as input to generate decoder features D_n through a fully connected layer.

$$D_n = \sigma \left(W_n E_n + b_n \right), \tag{5}$$

where W_n and b_n are the trained parameters, activation function is ReLU. After that, we use the averaging method to obtain the ensemble encoded features E of several networks and E is the output of this part:

$$E = \frac{1}{n} \sum_{i=1}^{n} E_i, \tag{6}$$

To further boost the feature extraction capability of the light-weight networks and reduce the noise impact of a single network during feature learning, we perform collaborative learning between encoded features. Besides, we view E as the teacher then transferring this integrated knowledge to the light-weight networks by means of teacher-student learning:

$$L_{\text{coe}} = \frac{1}{n(n-1)} \sum_{i=1, i \neq j}^{n} \text{MSE}\left(E_i, E_j\right) + \frac{1}{n} \sum_{i=1}^{n} \text{MSE}\left(E, E_i\right), \tag{7}$$

Similarly, we get the loss for ensemble decoded features L_{cod}. The overall loss function is then defined as follows:

$$L_1 = \alpha L_{\text{coe}} + (1 - \alpha)L_{\text{cod}} + L_{\text{recon}}, \tag{8}$$

where α is the hyper-parameter employed to control the importance of encoded features loss and decoded features loss. L_{recon} is the reconstruction loss between decoded features and initial features:

$$L_{\text{recon}} = \text{MSE}(D, H), \tag{9}$$

2.3   Feature Fusion

In this subsection, we introduce the feature fusion to establish a connection between the source graph and target graph. The main feature fusion pipeline is presented in Fig. 3, and the detailed fusion process is described below.

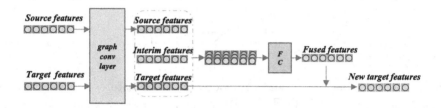

Fig. 3. The main feature fusion pipeline.

First, encoded features E_s and E_t are input to one graph convolutional layer to get features F_s and F_t, as in Eq. (4), respectively. Next, we perform feature fusion on F_s and F_t. The most straightforward is to combine them directly. However, this approach ignores the correlation knowledge between different short-term preferences. Since we have obtained the accurate item features in the Collaborative Learning, the simplified fusion strategy [17] is adapted into Feature Fusion. We derive an intermediate feature before performing the fusion operation, which is a transition from F_s to F_t, i.e., we have

$$F_i = A_{\text{att}} F_s W, \qquad (10)$$

where $W \in R^{d \times d}$ is a trainable weight matrix. $A_{att} \in R^{\frac{m}{2} \times \frac{m}{2}}$ is a transfer matrix.

We employ the attention mechanism to adaptively integrate knowledge from various early short-term preferences because for a short-term preference on the target graph, the importance of distinct early short-term preferences on the source graph is different:

$$a_{ij} = \frac{\exp\left(\cos\left(v_i, v_j\right)\right)}{\sum_j \exp\left(\cos\left(v_i, v_j\right)\right)}, \qquad (11)$$

where v_i is the feature of node i in target graph, v_j is the feature of node j in source graph, and $cos(v_i, v_j)$ is the cosine similarity between v_i and v_j. Then, we have the feature fusion process as follows:

$$F_t = F_t + \sigma\left(\text{FC}\left(F_t, F_i, F_s\right)\right), \qquad (12)$$

where FC represents the fully connected layer in fusion network, (F_t, F_i, F_s) means concatenation of three node features. σ is the ReLU function. After obtaining the final node features according to Eq. (12), we use Bayesian personalized ranking loss L_{BPR} to optimize the node features and train a user features

$U = \{u_1, u_2, \cdots, u_n\}$. L_{BPR} aims to make users score positive items higher than the scores of negative items:

$$L_{\text{BPR}} = -\frac{1}{n} \sum_{i=1}^{n} \sum_{j=1}^{N} \sum_{k=1}^{M} \log (r_j - r_k), \tag{13}$$

where n is the number of users, N indicates the number of positive items, M indicates the number of negative items, r_j and r_k represent the scores of user i for positive item j and negative item k, respectively:

$$\begin{cases} r_j = u_i \odot f_j^+ \\ r_k = u_i \odot f_k^- \end{cases}, \tag{14}$$

where $\odot$ indicates dot product. u_i, f_j^+, f_k^- are three elements of a sample triplet (u_i, f_j^+, f_k^-), u_i is a trained user preferences of user i, f_j^+ is the features of positive item j and f_k^- is the features of negative item k.

After training, the user features U are used to calculate the user's ratings of the items. Then we rank the ratings and selected the top few items as the recommended results.

Table 1. Statistics of the Datasets

Dataset	Users	Items	Ratings	Average Sequence Length	Sparsity
MLs-100K	942	1350	99249	105.5	92.20%
MLs-1M	6036	3416	999513	165.6	95.15%

3 Experiment

3.1 Experimental Setup

To validate the effectiveness of our proposed ODPF, we use two real-world datasets Movielens-1M (MLs-1M) and Movielens-100K (MLs-100K) for experiments. The summary of processed datasets are shown in Table 1. We compare ODPF with recent state-of-the-art methods: BPR [11], GCMC [2], NGCF [18], LightGCN [7], NextItNet [20], SINE [14], and CORE [8]. Among them, BPR, GCMC, NGCF, and LightGCN are non-sequential models; NextItNet, SINE, and CORE are sequential models. These comparative experiments are implemented via RecBole 2.0 [21]. We split each user sequence into training sequence (80%), validation sequence (10%), test sequence (10%). All competitive models are trained on GeForce RTX 3090 GPU. To evaluate the performances of our proposed model and competitive models, we adopt recall ratio(RECALL@N) and normalized discounted cumulative gain (NDCG@N). In our experiments, we set two values for N, that is, 5 and 10. For both metrics, a larger value indicates better performances.

In the two Movielens datasets we used, the attributes include movie title, release date, and genres. The feature dimension of items is 320 in our paper. In collaborative learning, the two graph convolutional layers shaps are 320×128 and 128×64, respectively. Correspondingly, the dimension of the fully connected layer that acts as a decoder is 64×320. We set the initial learning rate as 0.01 and $\alpha = 0.5$. In feature fusion, we use the graph convolutional layer with the hidden size of 64×320, and the fully connected layer with the dimension 320×320. The initial learning rate is 0.1.

Table 2. Recommendation performance (%) on MLs-100K

	N = 5		N = 10	
	RECALL	NDCG	RECALL	NDCG
BPR-[2012]	7.59	12.23	12.17	12.96
GCMC-[2017]	9.57	15.35	15.36	16.06
NGCF-[2019]	11.66	17.23	18.16	18.81
LightGCN-[2020]	11.89	18.43	18.96	19.07
NextItNet-[2019]	10.52	12.47	17.79	14.42
SINE-[2021]	11.14	13.04	17.77	14.74
CORE-[2022]	10.84	17.37	17.51	17.96
ODPF (Ours)	**12.42**	**18.84**	**19.49**	**19.54**

3.2   Experimental Results

Tables 2–3 illustrates the comparisons with six comparative models on three datasets. During training, we use the number of light-weight networks is $n = 3$. From Tables 2–3, we draw the following conclusions:

First, the proposed ODPF obtains a better recall rate than all other baselines and a better NDCG than most of the baselines on three datasets with different

Table 3. Recommendation performance(%) on MLs-1M

	N = 5		N = 10	
	RECALL	NDCG	RECALL	NDCG
BPR-[2012]	8.64	15.49	14.56	16.32
GCMC-[2017]	9.67	18.96	16.56	19.47
NGCF-[2019]	10.3	19.26	17.44	19.87
LightGCN-[2020]	11.51	20.41	18.49	20.75
NextItNet-[2019]	11.28	19.86	19.73	21.41
SINE-[2021]	10.45	17.85	16.93	18.52
CORE-[2022]	11.4	19.11	20.19	21.46
ODPF (Ours)	**12.24**	**20.75**	**20.92**	**22.19**

sparsity. For RECALL@10 and NDCG@10 on MLs-1M dataset, ODPF surpasses the best in the comparison methods by 3.62% and 3.4%, respectively. On MLs-100K, these two values are 2.79% and 2.46%, respectively. This is possibly due to the following two reasons: 1) we perform collaborative learning among light-weight networks such that they can learn additional information each other; and 2) through a feature fusion network, we fuse the features of the source graph into the target graph. This helps to establish a correlation between early short-term preferences and recent short-term preferences.

Second, among the baselines, traditional models such as PMF and BPR do not perform well on the three datasets. The influence of the length of user interaction sequences is much greater for other sequential models than for non-sequential models. Also its effect on NGCG is larger than that on RECALL. According to Table 1, we selected the experimental results of the two datasets with the largest gap in the average user interaction sequence length for observation, that is, MLs-100K and MLs-1M. For three sequential models, the average NDCG@5 and NDCG@10 on MLs-1M and MLs-100K are 18.94%, 20.46% and 14.29%, 15.71%, respectively. For four non-sequential models, the average NDCG@5 and NDCG@10 on MLs-1M and MLs-100K are 18.52%, 19.1% and 15.81%, 16.73%, respectively. From MLs-1M to MLs-100K, sequential and non-sequential models varied by -24.55% and -23.22% on NDCG@5 and NDCG@10, respectively. For non-sequential models, that is, -14.63% and -12.41%. Since our method does not directly take the whole sequence as input, and divides the sequence into two separate parts for training, our method is more robust to sequence length.

3.3   Ablation Studies

Feature Fusion: To verify the effectiveness of feature fusion, we set up two variants: **1) ODPF-NODI**: This variant does not divide the sequence and constructs the graph using the entire sequence, without using feature fusion; **2) ODPF-NOFF**: The other steps of this variant are the same as ODPF, but feature fusion is not performed. From Fig. 4 we find that ODPF outperforms ODPF-NODI and ODPF-NOFF, this demonstrates the effectiveness of feature fusion. Specifically, ODPF-NODI surpasses ODPF-NOFF on all three datasets because ODPF-NODI constructs the graph with the entire sequence,

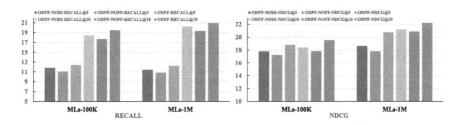

Fig. 4. Results of ablation studies on two datasets

which contains more information about the user's early short-term preferences than ODPF-NOFF, which only has the half of the sequence. However, due to the deficiency of GCN in aggregating higher-order neighborhood information, ODPF-NODI cannot fully utilize users' early short-term preference information. This is the reason why the performances of ODPF is better than ODPF-NODI. ODPF exploits feature fusion to fuse the user's early short-term preferences on the source sequence with the recent short-term preferences on the target sequence, making full use of the early short-term preference information to facilitate recommendations.

Number of Light-Weight Networks: Let n denote the hyper-parameter that controls the number of light-weight networks in collaborative learning. The values of n also affects the complexity of the model, since more light-weight networks mean more training parameters. Therefore, this section will discuss the influence of this hyperparameter on the recommended performance. In particular, these n trained light-weight networks are exploited as an ensemble to obtain the features of the output of Collaborative Learning. Table 4 shows the RECALL@N and NDCG@N values of COFF with different n values from 2–4. From Table 4, we observe that the recommendation performance improves when n rises from 2 to 3. This is because the extra light-weight network improves the feature extraction ability of the model and learns user preferences more efficiently. However, when n is increased from 3 to 4, the recommended performance decreases due to the overfitting problem caused by the excessive number of small networks. For example, the RECALL@10 values when n is 2, 3 and 4 is 19.33%, 20.92% and 19.86%, respectively, on MLs-1M. So it is essential to select a appropriate n value for the model. In all experiments, we choose $n = 3$.

Table 4. Recommendation performance(%) under different n value

	N = 5		N = 10	
	RECALL	NDCG	RECALL	NDCG
MLs-100K				
$n = 2$	11.79	18.43	18.92	18.83
$n = 3$	**12.42**	**18.84**	**19.49**	**19.54**
$n = 4$	12.18	18.72	19.01	19.28
MLs-1M				
$n = 2$	11.6	20.24	19.33	21.77
$n = 3$	**12.24**	**20.75**	**20.92**	**22.19**
$n = 4$	11.93	20.52	19.86	21.92

4 Conclusion

In this paper, we propose online distillation and feature fusion that incorporates GCNs for sequential recommendation. Specifically, we divide the original sequence into source and target sequence for graph construction and then employ feature fusion to fuse the features from the source graph and target graph to establish a bond between early short-term preferences and recent short-term preferences. Also, we apply online distillation to obtain more accurate item features. The proposed method alleviates the drawback that GCNs are difficult to obtain higher-order dependencies from the feature fusion perspective. Experimental results on three popular sequential recommendation datasets verify the effectiveness of our proposed model. In future, we will study the different online distillation methods to improve the recommendation performance, especially for those session-based recommendations with shorter sequence lengths.

Acknowledgement. This work was supported in part by National Natural Science Foundation of China (Grant Nos. 61976107 and 61502208).

References

1. Anil, R., Pereyra, G., Passos, A., Ormandi, R., Dahl, G.E., Hinton, G.E.: Large scale distributed neural network training through online distillation. arXiv preprint arXiv:1804.03235 (2018)
2. Berg, R.v.d., Kipf, T.N., Welling, M.: Graph convolutional matrix completion. arXiv preprint arXiv:1706.02263 (2017)
3. Chang, J., Gao, C., Zheng, Y., Hui, Y., Niu, Y., Song, Y., Jin, D., Li, Y.: Sequential recommendation with graph neural networks. In: Proceedings of the 44th International ACM SIGIR Conference on Research and Development in Information Retrieval, pp. 378–387 (2021)
4. Dong, X., Jin, B., Zhuo, W., Li, B., Xue, T.: Improving sequential recommendation with attribute-augmented graph neural networks. In: Karlapalem, K., Cheng, H., Ramakrishnan, N., Agrawal, R.K., Reddy, P.K., Srivastava, J., Chakraborty, T. (eds.) PAKDD 2021. LNCS (LNAI), vol. 12713, pp. 373–385. Springer, Cham (2021). https://doi.org/10.1007/978-3-030-75765-6_30
5. Fung, W.S., Hariharan, R., Harvey, N.J., Panigrahi, D.: A general framework for graph sparsification. In: Proceedings of the Forty-Third Annual ACM Symposium on Theory of Computing, pp. 71–80 (2011)
6. Guo, Q., et al.: Online knowledge distillation via collaborative learning. In: Proceedings of the IEEE/CVF Conference on Computer Vision and Pattern Recognition (CVPR), June 2020
7. He, X., Deng, K., Wang, X., Li, Y., Zhang, Y., Wang, M.: Lightgcn: simplifying and powering graph convolution network for recommendation. In: Proceedings of the 43rd International ACM SIGIR conference on research and development in Information Retrieval, pp. 639–648 (2020)
8. Hou, Y., Hu, B., Zhang, Z., Zhao, W.X.: Core: simple and effective session-based recommendation within consistent representation space. In: Proceedings of the 45th International ACM SIGIR Conference on Research and Development in Information Retrieval, pp. 1796–1801 (2022)

9. Ma, C., Ma, L., Zhang, Y., Sun, J., Liu, X., Coates, M.: Memory augmented graph neural networks for sequential recommendation. In: Proceedings of the AAAI Conference on Artificial Intelligence, vol. 34, pp. 5045–5052 (2020)

10. Quadrana, M., Karatzoglou, A., Hidasi, B., Cremonesi, P.: Personalizing session-based recommendations with hierarchical recurrent neural networks. In: Proceedings of the Eleventh ACM Conference on Recommender Systems, pp. 130–137 (2017)

11. Rendle, S., Freudenthaler, C., Gantner, Z., Schmidt-Thieme, L.: Bpr: Bayesian personalized ranking from implicit feedback. arXiv preprint arXiv:1205.2618 (2012)

12. Rendle, S., Freudenthaler, C., Schmidt-Thieme, L.: Factorizing personalized Markov chains for next-basket recommendation. In: The Web Conference (2010)

13. Sheng, Z., Zhang, T., Zhang, Y., Gao, S.: Enhanced graph neural network for session-based recommendation. Expert Syst. Appl. **213**, 118887 (2023)

14. Tan, Q., Zhang, J., Yao, J., Liu, N., Zhou, J., Yang, H., Hu, X.: Sparse-interest network for sequential recommendation. In: Proceedings of the 14th ACM International Conference on Web Search and Data Mining, pp. 598–606 (2021)

15. Tan, Y.K., Xu, X., Liu, Y.: Improved recurrent neural networks for session-based recommendations. In: Proceedings of the 1st Workshop on Deep Learning for Recommender Systems (2016)

16. Tang, J., Wang, K.: Personalized top-n sequential recommendation via convolutional sequence embedding. In: Proceedings of the Eleventh ACM International Conference on Web Search and Data Mining, pp. 565–573 (2018)

17. Wang, J., Gao, F., Dong, J., Zhang, S., Du, Q.: Change detection from synthetic aperture radar images via graph-based knowledge supplement network. IEEE J. Sel. Top. Appl. Earth Observ. Remote Sensing **15**, 1823–1836 (2022)

18. Wang, X., He, X., Wang, M., Feng, F., Chua, T.S.: Neural graph collaborative filtering. In: Proceedings of the 42nd International ACM SIGIR Conference on Research and Development in Information Retrieval, pp. 165–174 (2019)

19. Wojke, N., Bewley, A.: Deep cosine metric learning for person re-identification. In: 2018 IEEE Winter Conference on Applications of Computer Vision (WACV), pp. 748–756. IEEE (2018)

20. Yuan, F., Karatzoglou, A., Arapakis, I., Jose, J.M., He, X.: A simple convolutional generative network for next item recommendation. In: Proceedings of the Twelfth ACM International Conference on Web Search and Data Mining, pp. 582–590 (2019)

21. Zhao, W.X., et al.: Recbole 2.0: towards a more up-to-date recommendation library. In: Proceedings of the 31st ACM International Conference on Information & Knowledge Management, pp. 4722–4726 (2022)

22. Zhu, H., Koniusz, P.: Simple spectral graph convolution. In: International Conference on Learning Representations (2021)

Grassmann Graph Embedding
for Few-Shot Class Incremental Learning

Ziqi Gu, Chunyan Xu$^{(\boxtimes)}$, and Zhen Cui

PCA Lab, Key Lab of Intelligent Perception and Systems for High-Dimensional
Information of Ministry of Education, School of Computer Science and Engineering,
Nanjing University of Science and Technology, Nanjing, China
{ziqigu,cyx,zhen.cui}@njust.edu.cn

Abstract. Few-shot class incremental learning (FSCIL) poses a significant challenge in machine learning as it involves acquiring new knosstab-
wledge from limited samples while retaining previous knowledge. However, the scarcity of data for new classes not only leads to overfitting
but also exacerbates catastrophic forgetting during incremental learning.
To tackle these challenges, we propose the Grassmann Graph Embedding framework for Few-shot Class Incremental Learning (GGE-FSCIL),
which effectively preserves the geometric properties and structural relationships of learned knowledge. Unlike most existing approaches that
optimize network parameters in the Euclidean space, we leverage manifold space for optimizing the incremental learning network. Specifically,
we embed the approximated characteristics of each class onto a Grassmann manifold, enabling the preservation of intra-class knowledge and
enhancing adaptability to new tasks with few-shot samples. Recognizing
that knowledge exists within interconnected relationships, we construct
a neighborhood graph on the Grassmann manifold to maintain inter-
class structure information, thereby alleviating catastrophic forgetting.
We extensively evaluated our method on CIFAR100, miniImageNet, and
CUB200 datasets, and the results demonstrate that our approach surpasses the state-of-the-art methods in few-shot class incremental learning.

Keywords: Few-shot Class Incremental Learning · Grassmann
Manifold · Graph Embedding

1 Introduction

Recent deep learning [1,2] techniques have achieved significant success in the
field of computer vision, driven by the availability of large amounts of labeled
data and powerful computational capabilities. However, it is difficult to prepare

Supplementary Information The online version contains supplementary material
available at https://doi.org/10.1007/978-981-99-8543-2_15.

Q. Liu et al. (Eds.): PRCV 2023, LNCS 14432, pp. 179–191, 2024.
https://doi.org/10.1007/978-981-99-8543-2_15

a large-scale supervised dataset in many real-world scenarios, and a model that is trained by supervised learning can only predict a finite set of classes. Therefore, a neural network that has continual learning ability can learn new classes with only a few labeled samples. Compared to machine learning systems, humans are much better at learning a new concept with just a few samples and retaining old knowledge without forgetting. The work is devoted to processing the few-shot continual/incremental learning tasks [3,4]. There are two major problems in the incremental learning tasks: i) How to learn new class knowledge with few-shot labeled samples effectively, and ii) How to avoid catastrophic forgetting of old knowledge.

In recent years, various studies [5] have been conducted on few-shot class incremental learning from different perspectives. Some approaches in few-shot class incremental learning, as proposed by some studies [5,6], are to use graph models or neural gas networks to construct topological structures that connect different learned classes, thereby facilitating knowledge retention and preventing catastrophic forgetting. The forward compatible training method [7], which utilizes virtual prototypes to compress the embedding of learned classes and predicts potential new classes, has been proposed as an effective address to learn new classes while mitigating the forgetting of previously learned ones. There are meta-learning based approaches [8] that have been proposed to improve the ability of neural networks to learn classes while reducing interference with previously learned classes in incremental learning. The utilization of self-supervised approaches [4,9] has also been explored to improve the feature extraction capability of the model by incorporating a self-supervised loss function in the training process.

In this work, we introduce a novel Grassmann Graph Embedding framework for Few-shot Class Incremental Learning (GGE-FSCIL), which enables a deep neural network to incrementally learn from a sequential stream of few-shot labeled data. In general, the model is trained with a significant amount of labeled data in the base session, but few-shot samples of unknown classes may be encountered in the incremental learning process. We propose a Grassmann Manifold Embedding (GME) module that can embed the approximated characteristics of each class onto the Grassmann manifold. By leveraging the linear subspace properties of the Grassmann manifold [10], the GME module can preserve the geometric properties, reduce the risk of over-fitting, and improve adaptability to new classes in the incremental learning process. Furthermore, considering that the knowledge acquired by humans is not independent but rather related structure relationships [11], we aim to maintain the inter-class structure information of the previously learned classes in the incremental learning. Here we propose a Graph Structure Preserving (GSP) module to alleviate the catastrophic forgetting problem on the Grassmann manifold. We first construct a neighborhood graph with various graph manifold embedding methods, and then learn these new class representations with few-shot samples by preserving the inter-class structural relationship of old knowledge. We summarize the main contributions of this work as follows:

- We present a novel Grassmann Graph Embedding framework for Few-shot Class Incremental Learning (GGE-FSCIL). We can enable the network to learn new knowledge from a few labeled data incrementally while mitigating catastrophic forgetting of previously acquired knowledge.
- By leveraging the geometric and structural properties of the graph Grassmann manifold, we introduce two modules, Grassmann Manifold Embedding and Graph Structure Preserving to address the problems.
- Our experimental results have shown that our framework outperforms the current state-of-the-art methods on three benchmark datasets: CUB200 [12], CIFAR100 [13], and miniImageNet [14].

2   Related Work

Few-Shot Class Incremental Learning: Few-shot class incremental learning has gained increasing attention in recent years, with the goal of allowing a model for continual/incremental learning from a sequence of few-shot labeled data. Several approaches have been proposed to tackle this task, which involves addressing challenges such as forgetting, over-fitting, and data imbalance. For example, Tao et al. [5] proposed the use of a neural gas network (NG) to learn and maintain a topological structure of different category features for few-shot continual learning. Zhang et al. [6] introduced a graph attention network based evolving classifier, which incorporates graph models to facilitate information propagation between classifiers. The forward compatible training (FACT) [7] was proposed to address the catastrophic forgetting problem by generating virtual classes and incorporating new classes in the forward compatible setting while preserving knowledge of old classes. Self-supervised stochastic classifiers were also proposed in [9] to address few-shot class incremental learning using a self-supervision mechanism. Further, Zhu et al. [15] proposed an incremental prototype learning scheme to achieve continual learning by fine-tuning class prototypes using known prototypes and few-shot samples from new classes to train the model, and updating the prototypes based on this. Michael et al. [16] proposed the Constrained Few-shot Class-incremental Learning (CFSCIL) method, which utilizes hyper-dimensional embedding to enable the continual learning of more classes beyond the fixed number of dimensions in the feature space.

Graph Manifold Embedding: Recently, several approaches have been proposed to research graph manifold embedding. For example, Yan et al. [17] introduced a novel supervised method, Marginal Fisher Analysis (MFA), for reducing the dimensionality of data by constructing two graphs that capture the compactness within classes and separability between classes. Wang et al. [18] introduced a new algorithm called geometry-aware graph embedding projection metric learning (GEPML) which aims to learn a projection metric that is aware of the geometry of the data manifold. GEPML first constructs interclass and intraclass similarity graphs on the Grassmann manifold to capture the local structural information of the data and also extends the Euclidean collaborative representation mechanism to the Grassmann manifold to perform graph learning in an adaptive manner. To jointly learn the embedding mapping and the similarity metric, the authors formulate the Grassmann dimensionality reduction problem with a carefully designed metric learning regularization term. Mehrtash et al. [19] propose a graph embedding-based discriminant analysis approach on the Grassmann manifold. They demonstrate that incorporating within-class and between-class similarity graphs to capture intra-class compactness and inter-class separability can effectively utilize the geometric structure of the data.

3   The Proposed Method

3.1   Problem Definition

Few-shot class incremental learning (FSCIL) aims to learn a model continuously from a sequential stream of few-labeled samples of each class, where the data of the old tasks are no longer available in the learning of new tasks. Formally, the training set, label set, and test set are defined as X, Y, and Z, respectively. Specifically, we define a sequence of labeled training datasets X_1, X_2,..., X_T to train the model, where X_t represents the training set for the t-th session, Y_t represents the corresponding label set, and T represents the total number of incremental learning sessions. In the incremental session, the number of training samples is limited, and the task is described as "N-way K-shot" for the incremental session. It means that in each incremental session, the dataset is divided into N classes, with K samples in each class. In the base session, the model is initially trained on a base task with a sufficient number of samples in the training set X_1 and corresponding label set Y_1. In the incremental learning process, each training set is constructed such that it contains no repeated class labels, i.e., $\forall i, j$ and $i \neq j$, $Y_i \cap Y_j = \varnothing$.

3.2   Overview

The proposed framework for Grassmann Graph Embedding for Few-shot Class Incremental Learning (GGE-FSCIL) is presented in Fig. 1. In general, the base network Θ_1 is initially trained on a large training dataset X_1 in Euclidean space with the classic cross-entropy loss function. However, to preserve their learned intra-class knowledge while improving their adaptability to new tasks with few-shot samples, we embed the network parameters onto the Grassmann manifold. In the base session (i.e., $t = 1$), the base network Θ_1 is embedded onto the Grassmann manifold to preserve the geometric properties of the classes and enable subsequent incremental learning. During the incremental learning process (i.e., $t \geq 2$), the model is fed with a few labeled samples of new classes.

To learn knowledge of new classes and avoid forgetting previous knowledge of learned classes, we embed the few-shot class incremental learning model parameters Θ onto the Grassmann manifold. Specifically, we designed a Grassmann Manifold Embedding (GME) module to orthogonalize the network parameters and embed Θ onto the Grassmann manifold. Therefore, the properties of the

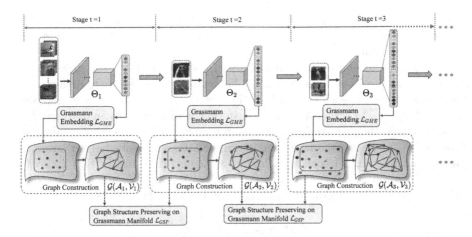

Fig. 1. The pipeline of our Grassmann Manifold Embedding for Few-shot Class Incremental Learning, where t refers to the t-th stage in the incremental learning ($t \geq 1$), and Θ_t represents the network parameter (i.e., $\Theta_1, \Theta_2, \Theta_3, \dots$). Firstly, we propose the Grassmann Manifold Embedding (GME) module to embed the network onto the Grassmann manifold. Secondly, to address the catastrophic forgetting problem, we proposed the Graph Structure Preserving (GSP) module.

parameter linear subspace can be better described, while the GME can preserve its learned intra-class knowledge and improve its adaptability to new tasks with few-shot samples. Furthermore, we not only need to focus on how to acquire new knowledge but also alleviate the problem of catastrophic forgetting. To avoid the problem, we designed the Graph Structure Preserving (GSP) module to maintain the adjacency relationships between previously learned classes, which can prevent the network from forgetting the knowledge learned. Specifically, at each learning stage t, we use the metric of the Grassmann manifold to construct a neighborhood graph $\mathcal{G}(\mathcal{A}_t, \mathcal{V}_t)$ that represents the adjacency relationship between the learned classes. Here, the $\mathcal{A}$ represents adjacency weights and the $\mathcal{V}$ represents class centers. The neighborhood graph $\mathcal{G}$ describes the correlations between the learned classes and is dynamically updated on the Grassmann manifold. Then, we use the neighborhood graph $\mathcal{G}$ to maintain the structural relationships and further preserve the adjacency relationships between already learned classes, thereby mitigating the catastrophic forgetting problem. Therefore, the approach is feasible in the incremental learning process and can preserve the geometric properties and structural relationships.

3.3 Grassmann Manifold Embedding

In incremental learning, we adopt the method of embedding the network parameters Θ onto the Grassmann manifold. The method improves the adaptability to new classes by better mitigating the model drift of the neural network. The Grassmann manifold is a special type of Riemannian manifold that embeds all

p-dimensional linear subspaces in a d-dimensional Euclidean space, and the manifold is defined as $G(d,p) = \{X \in \mathbb{R}^{d \times p} : X^T X = I\}$, where X represents any point on the manifold.

Our goal is to embed the parameters of each layer W_l of the incremental model Θ onto the Grassmann manifold. Therefore, it is necessary to orthogonalize the parameters of each layer, where $W_l \in \Theta$ and l represent the layer number, (i.e. $l \in 1, 2, 3, \dots$) Furthermore, we transform the problem into an optimization of parameter orthogonalization. Accordingly, we introduce an orthogonalization function $F(\cdot, \cdot)$ to orthogonalize the parameters W_l and embed them onto the Grassmann manifold. Recall that for an orthogonalized matrix W_l, it should satisfy the condition: $W_l^T W_l - I = 0$. Let $U = W_l^T W_l - I$, and $z \in \mathbb{R}^d$ is an arbitrary vector, the organization function $F(U, z)$ can be then defined as:

$$F(U, z) = \left| \frac{||Uz||^2}{||z||^2} - 1 \right|. \tag{1}$$

Here, $F(U, z) \le \delta_U$ and $\delta_U \in (0, 1)$ is a small value [20]. To solve for the minimum of the function $F(U, z)$, we can first define the spectral norm of matrix U as: $\alpha(U) = \sup_{z \in \mathbb{R}^d, z \ne 0} \frac{||Uz||}{||z||}$. Then, according to Eq. 1, the parameter orthogonalization problem can be transformed into minimizing the spectral norm of the matrix $\alpha(W_l^T W_l - I)$:

$$\mathcal{L}_E = \sum_{i=1}^{L} F(W_l^T W_l - I, z). \tag{2}$$

The method of embedding incremental model parameters onto the Grassmann manifold helps to preserve their learned intra-class knowledge and improve its adaptability to new tasks with few-shot samples. Therefore, we name it as Grassmann Manifold Embedding module. The learning objective consists of the classical cross-entropy loss $\mathcal{L}_{CE}$ and $\mathcal{L}_E$, and the final objective is $\mathcal{L}_{GME} = \mathcal{L}_{CE} + \beta \mathcal{L}_E$, where β is a hyper-parameter.

3.4   Graph Structure Preserving on Grassmann Manifold

Another key issue of few-shot class incremental learning is the catastrophic forgetting of learned knowledge. Recall that the knowledge in the human brain is not isolated and there exists a relational structure among different learned knowledge to maintain their connections [11]. To this end, we propose a graph structure preserving module to preserve the inter-class structure information of the previously learned classes to alleviate the catastrophic forgetting issue. Specifically, in the t-th incremental session, a neighborhood graph $\mathcal{G}(\mathcal{A}_t, \mathcal{V}_t)$ is constructed to represent the structural relationships among the learned knowledge on the Grassmann manifold, where $\mathcal{A}_t$ and $\mathcal{V}_t$ separately represent the corresponding adjacency weight matrix and node set. Here, we use the learned Grassmann embedding of the fully-connected layer parameters of the corresponding classes in the t-th session to form the graph node set C_t, i.e., $\mathcal{V}_t = \{x_1^t, x_2^t, \cdots, x_{C_t}^t\}$, where x_k^t is the learned Grassmann embedding of the k−th class and C_t is the number of all classes learned to till the t-th stage. Next, to effectively mine the inter-class structure information, based on the Grassmann embedding learned above, we can obtain the adjacency weight a_{ij} as the correlation between the two learned class embedding x_i and x_j, and $M = x_j^T x_i$:

$$a_{i,j} = ||(x_i - x_j M)^T (x_i - x_j M)||_F = ||I - M^T M||_F. \tag{3}$$

We can then use the built neighborhood graph to preserve the structural relationships of the learned classes by imposing constraints on those Grassmann embedding of the classes learned in the former sessions with the nonlinear embedding methods on the Grassmann manifold. We can then use the nonlinear embedding methods on the Grassmann manifold to avoid the knowledge forgetting the previously learned classes by using the built neighborhood graph to preserve the structural relationships of the learned classes. In each learning session t, we can conveniently save the corresponding node set $\mathcal{V}_t \in \mathbb{R}^{C_{t-1} \times C_{t-1}}$, which could then be used in the later sessions to preserve the relative positions of the Grassmann embedding of those classes learned in the t-th stage on the manifold. To be concrete, in the t-th incremental session, let Θ_t be the matrix composed of the Grassmann embedding of all the learned classes, we can also derive $\mathcal{A}_{t-1}$ according to the saved $\mathcal{V}_{t-1}$. The related positions of the first C_{t-1} classes in the $t-1$-th session could be maintained by using the nonlinear Grassmann manifold embedding methods with both Θ_t and $\mathcal{A}_{t-1}$. We next introduce three different nonlinear Grassmann manifold embedding methods that could well preserve the structural relationships of different class embedding.

1): Locally Linear Embedding (LLE): The locally linear embedding algorithm [21] is to maintain the local geometric properties on the Grassmann manifold. Let the first C_{t-1} rows which corresponds to the classes learned in the $t-1$-th session of Θ_t as Θ_t^{t-1}, the corresponding objective function is defined as:

$$\mathcal{L}_S^{(LLE)} = tr((\Theta_t^{t-1})^T (I - \mathcal{A}_{t-1})^T (I - \mathcal{A}_{t-1})(\Theta_t^{t-1})), \tag{4}$$

where $tr(\cdot)$ represents the matrix trace operator.

2): Laplacian Eigenmaps (LE): Laplacian eigenmaps algorithm [22] maintains local neighborhood information between nodes on the Grassmann manifold by constructing an undirected weighted graph. It uses the weight matrix as a penalty coefficient and then solves it using the Laplacian matrix. Similar to the LLE algorithm, in the incremental learning stage t ($t > 1$), we apply the Laplacian eigenmaps algorithm to maintain the knowledge structure, where the objective function is expressed as:

$$\mathcal{L}_S^{(LE)} = tr((\Theta_t^{t-1})^T L_{t-1}(\Theta_t^{t-1})), \tag{5}$$

where L_{t-1} is the Laplacian matrix derived from $\mathcal{A}_{t-1}$.

3): Locality Preserving Projections (LPP): Locality preserving projections algorithm [23] preserves local neighborhood information in a k-dimension manifold space ($k \ll p$), and the final objective function can be formulated as:

$$\mathcal{L}_S^{(LPP)} = tr(P^T(\Theta_t^{t-1})L_{t-1}(\Theta_t^{t-1})^T P), \qquad (6)$$

where the projection matrix P is composed of the eigen-vectors of the matrix $(\Theta_t^{t-1})L_{t-1}(\Theta_t^{t-1})^T$ w.r.t the lowest k eigen-values. In the graph structure preserving module, we optimize the Grassmann embedding of the fully connected parameters Θ_t^{t-1} using both $\mathcal{L}_S$ and $\mathcal{L}_{CE}$ and the final loss function to maintain the learned knowledge structure is defined as: $\mathcal{L}_{GSP} = \mathcal{L}_{CE} + \gamma\mathcal{L}_S$, where γ is a balancing factor.

4 Experiment

4.1 Experimental Setup

Datasets and Implementation Details: We evaluate the GGE-FSCIL approach on three public datasets. For CIFAR100 [13] and miniImageNet [14], we use the entire training data of the 60 base classes and the 40 new classes are used for eight 5-way 5-shot incremental learning tasks.

In the case of CUB200 [12], we divide the 200 classes into 100 base classes and 100 new classes, and 100 new classes are used for ten 10-way 5-shot continual learning tasks. We adopt the FSCIL approach [5] with ResNet20 as the backbone for CIFAR100 [13], and ResNet18 as the network backbone for miniImageNet [14] and CUB200 [12]. To ensure the feature extraction ability of the neural network during incremental learning, the parameters of the convolutional layers Θ_{conv} are frozen in the incremental learning sessions. We use SGD as the

Table 1. The performance of our GGE-FSCIL and state-of-the-art methods was evaluated on the CUB200 dataset [12], and the results marked with * were obtained from the authors' published code. See the supplementary material for results of other datasets.

Methods	Accuracy in each session (%) ↑											KR↑	ΔFinal	Avg↑
	1	2	3	4	5	6	7	8	9	10	11			
Ft-CNN	68.68	43.70	25.05	17.72	18.08	16.95	15.10	10.60	8.93	8.93	8.47	12.33	+52.27	22.02
NCM [24]	68.68	57.12	44.21	28.78	26.71	25.66	24.62	21.52	20.12	20.06	19.87	28.93	+40.87	32.49
iCaRL [25]	68.68	52.65	48.61	44.16	36.62	29.52	27.83	26.26	24.01	23.89	21.16	30.80	+39.58	36.67
EEIL [26]	68.68	53.63	47.91	44.20	36.30	27.46	25.93	24.70	23.95	24.13	22.11	32.19	+38.63	36.27
TOPIC [27]	68.68	62.49	54.81	49.99	45.25	41.40	38.35	35.36	32.22	28.31	26.28	38.26	+34.46	43.92
SPPR [15]	68.68	61.85	57.43	52.68	50.19	46.88	44.65	43.07	40.17	39.63	37.33	54.35	+23.41	49.32
D-DeepEMD [28]	75.35	70.69	66.68	62.34	59.76	56.54	54.61	52.52	50.73	49.20	47.60	63.17	+13.14	58.73
D-NegCosine [29]	74.96	70.57	66.62	61.32	60.09	56.06	55.03	52.78	51.50	50.08	48.47	64.66	+12.27	58.86
D-Cosine [30]	75.52	70.95	66.46	61.20	60.86	56.88	55.40	53.49	51.94	50.93	49.31	65.29	+11.43	59.36
Meta-FSCIL [8]	75.90	72.41	68.78	64.78	62.96	59.99	58.30	56.85	54.78	53.83	52.64	69.35	+8.10	61.93
CEC [6]	76.37	72.00	68.45	63.08	62.00	58.4	57.39	55.13	54.40	52.90	51.27	67.13	+10.03	61.03
CEC [6]+GGE-FSCIL	76.28	71.75	68.28	63.27	62.07	58.56	57.54	55.31	54.55	53.15	51.65	67.71	+9.45	61.13
FACT [7]	77.38	73.91	70.32	65.91	65.02	61.82	61.29	59.53	57.92	57.63	56.46	72.96	+4.20	64.29
FACT [7]+GGE-FSCIL	77.42	73.81	70.43	66.08	65.19	62.24	61.45	59.89	58.30	57.99	56.94	73.56	+4.06	64.52
GGE-FSCIL(LPP)	78.72	75.26	71.46	67.68	67.45	64.97	64.78	63.54	62.28	61.52	60.49	76.84	+0.25	67.10
GGE-FSCIL(LE)	78.72	**75.36**	**71.76**	67.87	67.60	65.10	**64.93**	63.66	62.40	**61.74**	60.71	77.12	+0.03	67.25
GGE-FSCIL(LLE)	**78.72**	75.33	71.67	**67.87**	**67.62**	**65.13**	64.91	**63.70**	**62.45**	61.72	**60.74**	77.16	–	**67.26**

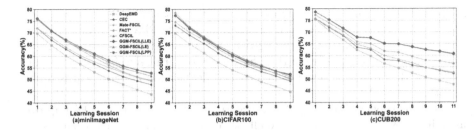

Fig. 2. The accuracy changing curves with growing incremental sessions of different methods on the three datasets.

optimizer, initialize the learning rate to 0.1, and set the batch size to 128 in the base session. At training time, we employ data augmentation techniques such as random cropping, random scaling, and random horizontal flipping. We adopt several evaluation metrics to assess the performance of our proposed method from different perspectives. "Acc_t" denotes the Top-1 accuracy in the t-th incremental session, while "Avg" represents the average score, i.e., $\sum_{t=1}^{T} Acc_t/T$. "ΔFinal" represents the difference in Acc_t between our method and the compared method in the final session, and "KR" is the knowledge retention rate, which is defined as Acc_T/Acc_1.

4.2 Comparison with State-of-the-Art Methods

We first evaluate the performance of our proposed GGE-FSCIL method on the challenging CUB200 dataset [12] and compare it with other state-of-the-art methods. Detailed results are provided in Table 1, and figures depicting the variation in accuracy with incremental learning sessions are shown in Fig. 2(c). Our proposed GGE-FSCIL method has achieved state-of-the-art performance surpassing all other compared methods by a significant margin. In particular, our method achieved the best score compared to all the other methods. Furthermore, in terms of the overall performance in the final incremental session, our GGE-FSCIL outperforms the second-best method FACT [7], by a significant margin of 4.28%. It shows that our GGE-FSCIL can maintain excellent learning capability throughout the incremental learning process. We believe that the Grassmann manifold embedding module can embed parameters onto the Grassmann manifold, preserving their learned intra-class knowledge and improving its adaptability to new classes with few-shot samples. Additionally, our method also outperforms all the compared methods by at least 2.97% on Avg, indicating that our GGE-FSCIL consistently achieves better performance across all the incremental learning sessions. Further, it demonstrates that the Grassmann manifold embedding can help mitigate model drift in incremental learning. Overall, the excellent results achieved by our GGE-FSCIL in terms of Avg and ΔFinal strongly demonstrate the effectiveness of our proposed Grassmann manifold embedding module to preserve their learned intra-class knowledge and improve its adaptability to new tasks with few-shot samples. Besides, we incorporate our proposed GGE-FSCIL mechanisms into the FACT [7] and CEC [6] frameworks. The comparative results demonstrate substantial enhancements of 0.60% and 0.58% on the KR metric. This further confirms the effectiveness of our GGE-FSCIL framework in acquiring new knowledge and mitigating catastrophic forgetting.

Our experiments show that GGE-FSCIL achieves at least a 3.89% improvement in the KR metric compared to all other methods. The KR metric measures the accuracy ratio between the final incremental session and the base session, and a higher KR value indicates less learned knowledge is forgotten during the incremental learning process. When compared with previous methods such as CEC [6] and FACT [7] which also freeze the base classifier to avoid catastrophic forgetting, our GGE-FSCIL outperforms them by a significant margin. It indicates that our proposed graph structure preserving module, which has the capability of preserving relationships, is effective in retaining the learned knowledge previously. The result confirms that the proposed graph structure preserving module is effective in mitigating the problem of catastrophic forgetting. Overall, the superior performance of our proposed GGE-FSCIL clearly confirms its superiority in learning novel class knowledge as well as mitigating the catastrophic forgetting problem. The superior performance of our approach demonstrates the effectiveness of our proposed graph structure preserving module in maintaining previously learned knowledge. Moreover, the accuracy changing curves depicted in Fig. 2(c) visually demonstrate the superiority of our GGE-FSCIL. Additionally, we have included separate accuracy changing curves for the mini-ImageNet [14] and CIFAR100 [13] datasets in Fig. 2(a) and Fig. 2(b), respectively. Observing the results, it can be seen that our proposed GGE-FSCIL also obtains the best performances on the other datasets, which further confirms the superiority of our proposed Grassmann manifold embedding module and graph structure preserving module in learning new classes knowledge while mitigating the catastrophic forgetting problem.

Effectiveness of the Proposed Modules: The detailed results of our experiments are shown in Table 2. In incremental learning sessions ($t \geq 1$), by adding the Grassmann Manifold Embedding module, the network parameters are orthogonality constrained and embedded onto the Grassmann manifold, which can preserve their learned intra-class knowledge and improve its adaptability to new classes with few-shot samples. Our proposed method achieves an improve-

Table 2. An overall ablation study of our proposed method on the CUB200 dataset [12] was conducted, which includes two modules we proposed: Grassmann Manifold Embedding (GME) and Graph Structure Preserving (GSP).

Baseline	GME	GSP	Accuracy in each session (%) ↑											KR	Avg
			1	2	3	4	5	6	7	8	9	10	11		
✓			78.49	74.94	71.05	66.21	65.93	62.97	62.78	60.78	59.34	58.46	57.15	72.82	65.28
✓	✓		78.72	75.45	71.54	67.60	67.38	64.88	64.29	63.10	62.02	61.20	60.24	76.52	66.95
✓		✓	78.49	75.19	71.37	67.06	66.93	63.90	63.29	61.77	60.72	59.99	58.97	75.13	66.15
✓	✓	✓	78.72	75.33	71.67	67.87	67.62	65.13	64.91	63.70	62.45	61.72	60.74	77.16	67.26

ment of 3.70% and 1.67% in terms of KR and Avg, respectively, compared to the baseline. It confirms the effectiveness of our Grassmann manifold embedding module. We adopt the Graph Structure Preserving module to preserve the structural relationship between the current neighborhood graph $\mathcal{G}_t$ and the previous neighborhood graph $\mathcal{G}_{t-1}$ on the Grassmann manifold as much as possible in the incremental learning process. Further, Graph Structure Persevering module leads to significant performance improvements of 0.87% and 2.31% on Avg and KR metrics, which are clearly non-trivial, respectively. The results show that our proposed Graph Structure Persevering module can effectively preserve the learned knowledge structure by constructing the neighborhood graph $\mathcal{G}$ onto the Grassmann manifold, thereby mitigating catastrophic forgetting. We believe that the structural relationships between classes can be well-preserved during the incremental learning process, which validates the effectiveness of the proposed Graph Structure Preserving module. Finally, when applying both modules, the performance of the proposed GGE-FSCIL framework could be further improved, achieving a new state-of-the-art performance.

The Shot Number in the Incremental Learning: The accuracy curves in Fig. 3(b) show that increasing the amount of data has a significant positive impact on the performance of the few-shot class incremental learning task. More data improves the class attribute distribution and enhances the accuracy of class relationships, thereby facilitating the learning of new knowledge. However, once the number of data points exceeds 15, the performance improvement becomes marginal. This indicates that obtaining an appropriate amount of newly labeled data is sufficient to achieve satisfactory performance.

Accuracy of the New Classes for Each Incremental Learning Session: We show the new class accuracy of the base model and GGE-FSCIL framework at each stage of incremental learning in Fig. 3(c). The experimental results demonstrate that compared to the base model, our proposed GGE-FSCIL framework achieves significant improvement in the new class learning accuracy

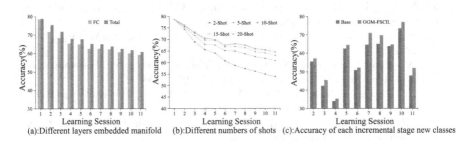

(a):Different layers embedded manifold   (b):Different numbers of shots   (c):Accuracy of each incremental stage new classes

Fig. 3. The above shows some experimental results of the GGE-FSCIL framework on the CUB200 dataset [12]. Figure 4(a) displays the experimental results of embedding all network parameters or only the fully connected layer parameters onto the Grassmann manifold. Figure 4(b) shows the impact of changes in the number of samples for each new class. Figure 4(c) presents the performance of new classes in each incremental session.

at each session. It indicates that the Grassmann Manifold Embedding module in our proposed GGE-FSCIL framework is better able to preserve the geometric properties, while the Graph Structure Preserving module can better maintain the knowledge of new classes.

Different Layers are Embedded on Grassmann Manifold: To verify the impact of embedding network parameters onto the Grassmann manifold in the incremental learning stage, we show in Fig. 3(a) the experimental results of two cases: only embedding the fully connected layer and embedding the entire model (Total) onto the Grassmann manifold. The experimental results show that embedding the entire network model onto the Grassmann manifold yields better results compared to only embedding the fully connected layer. It validates that in our Grassmann Manifold Embedding module, embedding the entire network model onto the Grassmann manifold can better learn new class knowledge.

5   Conclusion

In this paper, we present GGE-FSCIL, a novel framework for few-shot class incremental learning. GGE-FSCIL aims to preserve the geometric properties and structural relationships of classes using Grassmann Graph Embedding module. Additionally, we propose a Graph Structure Preserving module, which creates a neighborhood graph on the Grassmann manifold to maintain the evolving knowledge structure and mitigate catastrophic forgetting.

References

1. Simonyan, K., Zisserman, A.: Very deep convolutional networks for large-scale image recognition. In: CVPR (2014)
2. He, K., Zhang, X., Ren, S., Sun, J.: Deep residual learning for image recognition. In: CVPR, pp. 770–778 (2016)
3. Rebuffi, S.-A., Kolesnikov, A., Sperl, G., Lampert, C.H.: icarl: Incremental classifier and representation learning. In: CVPR (2016)
4. Mazumder, P., Singh, P., Rai, P.: Few-shot lifelong learning. AAAI **35**(3), 2337–2345 (2021)
5. Tao, X., Hong, X., Chang, X., Dong, S., Wei, X., Gong, Y.: Few-shot class-incremental learning. In: CVPR, pp. 12 183–12 192 (2020)
6. Zhang, C., Song, N., Lin, G., Zheng, Y., Pan, P., Xu, Y.: Few-shot incremental learning with continually evolved classifiers. In: CVPR
7. Zhou, D.-W., Wang, F.-Y., Ye, H.-J., Ma, L., Pu, S., Zhan, D.-C.: Forward compatible few-shot class-incremental learning. In: CVPR (2022)
8. Chi, Z., Gu, L., Liu, H., Wang, Y., Yu, Y., Tang, J.: Metafscil: a meta-learning approach for few-shot class incremental learning. In: CVPR
9. Kalla, J., Biswas, S.: S3c: self-supervised stochastic classifiers for few-shot class-incremental learning. In: ECCV, pp. 432–448 (2022)
10. Zheng, L., Tse, D.N.C.: Communication on the grassmann manifold: a geometric approach to the noncoherent multiple-antenna channel. IEEE Trans. Inf. Theory **48**(2), 359–383 (2002)

11. Eichenbaum, H.: How does the brain organize memories? Science **277**(5324), 330–332 (1997)
12. C. Wah, S. Branson, P. Welinder, P. Perona, and S. Belongie, "The caltech-ucsd birds-200-2011 dataset," 2011
13. Krizhevsky, A.: Learning multiple layers of features from tiny images (2009)
14. Russakovsky, O., et al.: Imagenet large scale visual recognition challenge. Int. J. Comput. Vision **115**(3) (2015)
15. Zhu, K., Cao, Y., Zhai, W., Cheng, J., Zha, Z.-J.: Self-promoted prototype refinement for few-shot class-incremental learning. In: CVPR
16. Hersche, M., Karunaratne, G., Cherubini, G., Benini, L., Sebastian, A., Rahimi, A.: Constrained few-shot class-incremental learning (2022)
17. Yan, S., Xu, D., Zhang, B., Zhang, H.-J., Yang, Q., Lin, S.: Graph embedding and extensions: a general framework for dimensionality reduction. TPAMI **29**(1), 40–51 (2006)
18. Wang, R., Wu, X.-J., Liu, Z., Kittler, J.: Geometry-aware graph embedding projection metric learning for image set classification. TCDS **14**(3), 957–970 (2021)
19. Harandi, M.T., Sanderson, C., Shirazi, S., Lovell, B.C.: Graph embedding discriminant analysis on grassmannian manifolds for improved image set matching. In: CVPR. IEEE 2011, pp. 2705–2712 (2011)
20. Bansal, N., Chen, X., Wang, Z.: Can we gain more from orthogonality regularizations in training deep networks? NeurIPS, vol. 31 (2018)
21. Roweis, S.T., Saul, L.K.: Nonlinear dimensionality reduction by locally linear embedding. Science **290**(5500), 2323–2326 (2000)
22. Belkin, M., Niyogi, P.: Laplacian eigenmaps and spectral techniques for embedding and clustering. NeurIPS, vol. 14 (2001)
23. He, X., Niyogi, P.: Locality preserving projections. In: NeurIPS, vol. 16 (2003)
24. Hou, S., Pan, X., Loy, C.C., Wang, Z., Lin, D.: Learning a unified classifier incrementally via rebalancing. In: CVPR, pp. 831–839 (2019)
25. Rebuffi, S.-A., Kolesnikov, A., Sperl, G., Lampert, C.H.: icarl: incremental classifier and representation learning. In: CVPR, pp. 2001–2010 (2017)
26. Castro, F.M., Marín-Jiménez, M.J., Guil, N., Schmid, C., Alahari, K.: End-to-end incremental learning. In: ECCV, pp. 233–248 (2018)
27. Tao, X., Hong, X., Chang, X., Dong, S., Wei, X., Gong, Y.: Few-shot class-incremental learning. In: CVPR (2020)
28. Zhang, C., Cai, Y., Lin, G., Shen, C.: Deepemd: few-shot image classification with differentiable earth mover's distance and structured classifiers. In: CVPR, pp. 12 203–12 213 (2020)
29. Liu, B., Cao, Y., Lin, Y., Li, Q., Zhang, Z., Long, M., Hu, H.: Negative margin matters: understanding margin in few-shot classification. In: Vedaldi, A., Bischof, H., Brox, T., Frahm, J.-M. (eds.) ECCV 2020. LNCS, vol. 12349, pp. 438–455. Springer, Cham (2020). https://doi.org/10.1007/978-3-030-58548-8_26
30. Vinyals, O., Blundell, C., Lillicrap, T., Wierstra, D., et al.: Matching networks for one shot learning. NeurIPS

Global Variational Convolution Network for Semi-supervised Node Classification on Large-Scale Graphs

Yide Qiu, Tong Zhang$^{(\boxtimes)}$, Bo Huang, and Zhen Cui

PCA Lab, Key Lab of Intelligent Perception and Systems for High-Dimensional Information of Ministry of Education, School of Computer Science and Engineering, Nanjing University of Science and Technology, Nanjing, China
{ql15025886,tong.zhang,huangbo,zhen.cui}@njust.edu.cn

Abstract. Graph Neural Networks (GNNs) have received much attention in the graph deep learning. However, there are some issues in extending traditional aggregation-based GNNs to large-scale graphs. With the rapid increase of neighborhood width, we find that the direction of aggregation can be disrupted and quite unbalanced, which compromises graphic structure and feature representation. This phenomenon is referred to *Receptive Field Collapse*. In order to preserve more structural information on large-scale graphs, we propose a novel Global Variational Convolutional Networks (GVCNs) for Semi-Supervised Node Classifications, which consists of a variational aggregation mechanism and a guidance learning mechanism. Variational aggregation can moderately map the unbalanced neighborhood distribution to a prior distribution. And the guidance learning mechanism, based on positive pointwise mutual information (PPMI), encourages the model to concentrate on more prominent graphic structures, which increases information entropy and alleviates *Receptive Field Collapse*. In addition, we propose a variational convolutional kernel to achieve effective global aggregation. Finally, we evaluate GVCNs on the Open Graph Benchmark (OGB) Arxiv and Products datasets. Up to the submission date (Jan 20, 2023), GVCNs achieve significant performance improvements compared to other aggregation-based GNNs, even state-of-the-art decoupling-based methods, the performance of GVCNs remains competitive with moderate spatiotemporal complexity. Our code can be obtained from: https://github.com/Yide-Qiu/GVCN.

Keywords: Large-scale Graphs · Variational · Semi-Supervised Classification

1 Introduction

The Graph Neural Networks (GNNs) has benefits for various graph learning tasks in many domains, such as citation networks [11], social networks [15], biological

Q. Liu et al. (Eds.): PRCV 2023, LNCS 14432, pp. 192–204, 2024.
https://doi.org/10.1007/978-981-99-8543-2_16

graphs [8], and traffic networks [23]. In particular, with the exponential increase of data and the arrival of the big-model era, traditional graph learning methods are not scalable to be applied to large-scale graphs due to several issues such as *neighborhood explosion* and *over-smoothing*, there is an increasing need for specifically designed large-scale graph learning methods [26].

From the perspective of Message Passing Neural Networks (MPNNs) [1], the pipelines can be divided into two processes: *message passing* and *feature mapping*. Based on whether these two steps are alternated during the training process, large-scale graph methods can be classified into aggregation-based methods and decoupling-based methods. The paradigm of the aggregation-based GNNs can be formulated as:

$$H^{(l)} = \sigma(\widetilde{A}H^{(l-1)}W^{(l)}) \quad l \in [1,k] \tag{1}$$

Performing GCNs-like full-batch aggregation directly on large-scale graphs is often not plausible [19]. Sampling-based aggregation methods [5] repeatedly sample the graph using different sampling strategies and perform mini-batch learning on the sampled subgraphs. However, the frequent sampling process inevitably drops nodes or edges, which compromises the complete graphic structure and introduces additional spatiotemporal overhead [7].

To address the above issues, current mainstream decoupling-based methods [19,21] decouple the processes of *message passing* and *feature mapping*. By precomputing the results of the *message passing* process in CPU, the computationally expensive step is separated from the overall training process in GPU. The paradigm of decoupling-based GNNs can be expressed as:

$$\begin{cases} H^{(l)} = \widetilde{A}H^{(l-1)} & l \in [1,k] \\ H^{(l)} = H^{(l-1)}W^{(l)} & l \in [1,k] \end{cases} \tag{2}$$

In order to aggregate as much structural information as possible during *message passing*, several decoupling-based GNNs [18,25] precompute muti-hop propagations and utilise *label propagation* [13] as an additional input. Seemingly optimistically, decoupling-based GNNs can bypass the sampling and aggregation operations and significantly reduce the time cost by precomputation. As an efficient approximation to GCNs-like full-batch aggregation, decoupling-based GNNs have gradually emerged as general solutions for large-scale graphs.

However, decoupling-based GNNs have several inherent issues. Precomputing the propagation feature leads to a solidification of the neighborhood aggregation, which inevitably leads to inflexible neighborhood learning during *feature mapping*. Several methods [19,25] modestly alleviate the issue by incorporating node-wised, hop-wised or channel-wised attention mechanisms, or by introducing several operations to relearn or supplement the blurred neighborhood information, but they have not been solved spontaneously.

Due to the decoupling-based GNNs inevitably blur the aggregation direction, we attempt to perform global aggregation without decoupling, which encounters several issues such as *Receptive Field Collapse*(Sect. 3.2), which can potentially lead to the degradation of the model. Therefore, we propose a novel global variational aggregation method and a guided learning mechanism based on positive pointwise mutual information (PPMI). The former adaptively maps the unbalanced neighborhood distribution to a prior distribution moderately to increase the diversity of aggregation direction and the latter encourages the model to focus on more prominent graphic structures. Furthermore, the above two modules are integrated and form a novel framework Global Variational Convolutional Network (GVCN). By appropriately employing variational aggregation and guided learning, GVCNs achieve efficient global aggregation on large-scale graphs.

We summarize our contributions in the following four aspects:

1) We propose a novel framework named Global Variational Convolutional Networks (GVCNs) to preserve more structural information on large-scale graphs.
2) In the framework, a guided learning mechanism for measuring co-occurrence relationships through PPMI was introduced to moderately guide the model to concentrate on more significant structure.
3) To address the issue of *Receptive Field Collapse*, we design a novel global variational aggregation mechanism to adaptively increase the diversity of aggregation direction.
4) The experimental result demonstrates the significant performance of GVCNs on various large-scale graph datasets with moderate spatiotemporal complexity.

2   Related Work

Existing specially designed large-scale graph methods can be mainly divided into sampling-based aggregation GNNs [11] and decoupling-based GNNs [18]. Sampling-based aggregation methods separate the original graph into multiple subgraphs, seeking the optimal way to perform batch training. Instead, decoupling-based methods [18] precompute the feature propagation matrix once on CPU and store the computation results as inputs to the model on GPU.

Sampling-Based Aggregation Methods. For a large graph, a conventional naive approach is to decompose the graph through some reasonable methods to reduce the complexity. There are three common sampling strategies: node sampling [11], neighborhood sampling [4], and subgraph sampling [5,24]. Node sampling samples the same number of neighbors for each node. Hamilton et al. [11] proposes GraphSAGE, which utilizes a uniform distribution as the sampling distribution to effectively mitigate the issue of *neighborhood explosion*. However, experiments in [7] show that the mitigation is moderate due to the sampling quantity is not significantly smaller than the adjacency matrix density. Neighborhood sampling algorithms, which sample the same number of nodes for each layer, was proposed to further address this issue, but leads to insufficient structure learning by reducing the number of edge coarsely, although addresses

the *neighborhood explosion*. To preserve structural information and avoid dupli-
cate sampling, Chiang et al. [5] propose ClusterGCN and Zeng et al. [24] propose
GraphSAINT, in which the same sampled subgraphs are shared by all layers.
Although subgraph sampling reduces spatiotemporal complexity, inappropriate
sampling strategies can degrade performance.

Decoupling-Based Methods. Another category of methods involves precomputing approximate results of aggregation to replace *message passing*. Wu et
al. [21] proposes SGC, which utilizes the last-hop propagation features as input
for downstream tasks, but leads to neglect of the original feature. To avoid
this neglect, SIGN [9] considers concatenating multi-hop propagation features,
and SAGN [18] further introduces a hop-wised attention mechanism. Similarly,
GAMLP [25] explores another way to compute attention. As a compensation
strategy, this attention mechanism alleviates the inflexibility of neighborhood
aggregation and achieves promising results. On the other hand, Duan et al. [7]
proposed ENGCN, which introduces a voting mechanism to train the k-hop
propagation features iteratively. Considering the adaptive fusion of aggregation
features at node-wised and hop-wised during the graph diffusion process, Sun
et al. [19] propose AGDN, which adopts the architecture of GAT and calculates
attention between propagation features, and achieves promising performance.

In summary, compared to sampling-based aggregation methods, our framework eliminates the need for sampling operations and achieves efficient global
aggregation. Compared to decoupling-based methods, our approach does not
involve decoupling and thus avoids fixing the aggregation direction.

3   Proposed Methods

In this section, we present the proposed GVCNs framework, as well as the guided
learning module and the variational aggregation module, which are shown in
Fig. 1.

3.1   Positive Pointwise Mutual Information on Large-Scale Graphs

PPMI [3], measures correlation and captures global information, but the online
random walks on large-scale graphs has high time complexity.

Considering the complexity of large-scale graph, in this work, we generalize the PPMI algorithm to large-scale graph and provide an efficient parallel implementation as prior knowledge utilizing our proposed positional encod-
ing(Sect. 3.3).

Recall the paradigm of random walks. Given a graph $\mathcal{G} = (\mathcal{V}, \mathcal{E})$, where $\mathcal{V}$ is
the node set and $\mathcal{E}$ is the edge set, with an adjacency matrix $\boldsymbol{A} \in \mathbb{R}^{N \times N}$ and a
feature matrix $\boldsymbol{X} \in \mathbb{R}^{N \times d}$, d is the dimension of hidden feature, we can perform
node-wise random walks of length k on the graph $\mathcal{G}$, which can be formulated as:

$$p\left(\mathrm{s}(t+1) = x_j \mid \mathrm{s}(t) = x_i\right) = \frac{\boldsymbol{A}_{i,j}}{\sum_j \boldsymbol{A}_{i,j}} \tag{3}$$

where t is the time step, $s(t)$ denotes the index of node of time t, $\sum_j \boldsymbol{A}_{i,j}$ represents degree of node j. Then we count the frequency matrix $\boldsymbol{F} = \{F_{i,j}\} \in \mathbb{R}^{N \times N}$ of any node-pairs $(i, j) \in \mathcal{E}$ and calculate a co-occurrence matrix $\boldsymbol{PMI} \in \mathbb{R}^{N \times N}$. Furthermore, positive pointwise mulual information matrix $\boldsymbol{P} \in \mathbb{R}^{N \times N}$ can be formulated as:

$$P_{i,j} = \max \left\{ pmi_{i,j} = \log \left(\frac{F_{i,j}}{F_{i,*} \cdot F_{*,j}} \right), 0 \right\} \tag{4}$$

To capture global structure information, we propose a simple and effective guidance learning mechanism by pre-walking the PPMI matrix $\boldsymbol{P}$ as prior knowledge, which can be incorporated into the symmetric normalized adjacency $\widetilde{\boldsymbol{A}}$:

$$\overline{\boldsymbol{A}} = \widetilde{\boldsymbol{A}} \odot \boldsymbol{P} \tag{5}$$

where the $\odot$ represents Hadamard product. Introducing the matrix $\overline{\boldsymbol{A}}$, we can integrate the guidance learning into the model training process, which leads to aggregation direction more sensitive to serveral node-pairs of highly co-occurring and moderately guides the model to pay attention to more significant structure.

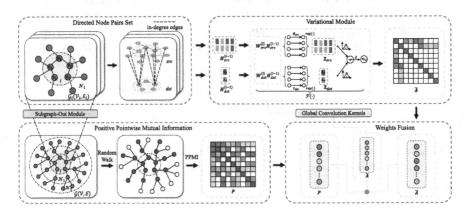

Fig. 1. Overall architecture of the GVCNs. To capture global structural information, we integrate offline PPMI and online variational learning. Where, $\mathcal{G}_i$ represents the selected subgraph, and $\mathcal{N}_1$ and $\mathcal{N}_2$ denote the 1-hop and 2-hop neighboring subgraphs, $\odot$ represents the element-wised matrix multiplication.

3.2 Global Variational Aggregation

Recall the paradigm of a GCNs layer based on MPNNs. Adopting the symbols introduced in Sect. 3.1, a single layer of GCNs can be formulated as follows:

$$\boldsymbol{H}^{(l)} = f \left(\boldsymbol{H}^{(l-1)}, \boldsymbol{A} \right) = \widetilde{\boldsymbol{A}}^{(l)} \boldsymbol{H}^{(l-1)} \boldsymbol{W}^{(l)} + \boldsymbol{H}^{(l-1)} \boldsymbol{W}_r^{(l)} \tag{6}$$

where, l represents the aggregation layer of l-th, $\boldsymbol{H}$ denotes the node representation matrix. Spatial-based GCNs stack multiple MPNN layers to perform multi-layer graph convolution. In [2], it was shown that the use of symmetric normalized adjacency is effective. GCNs utilize $\widehat{A}^{(l)} = D^{-1/2}AD^{-1/2}$ with self-loops, where D is the degree matrix with $D_{ii} = \sum_{j=1}^{N} A_{ij}$. GATs focus on the out-degree of the target nodes and the in-degree of the source nodes. To capture the mutual correlation, GATs consider the $A^{(l)}$ as a learnable adjacency matrix. Any weight of a node pair (i, j) can be formulated as:

$$
\begin{aligned}
e_{ij} &= \text{LeakyReLU}\left([h_i W \| h_j W] \cdot a\right) \\
&= \text{LeakyReLU}\left([h_i W] \cdot a_{dst} + [h_j W] \cdot a_{src}\right)
\end{aligned}
\tag{7}
$$

where, h_i denotes the hidden embedding of node i, a is the edge-wised attention. The src and dst represent the source and destination nodes, respectively. Furthermore, we utilize δ to denote the normalization function, the normalized adjacency matrix of GATs can be expressed as follows:

$$
\widetilde{A}_{ij} = \delta(e_{ij}) = \frac{\exp(e_{ij})}{\sum_{k \in \mathcal{N}(i)} \exp(e_{ik})}
\tag{8}
$$

However, we observe that $\widetilde{A}$ may be *extremely sparse* in the context of large-scale graphs. Recall that the computing of $\widetilde{A}_{ij}$ with a *softmax*, most elements of $\widetilde{A}_{i*}$ will have few non-zero values, just like a classifier, leading to an implicit collapse of the receptive field. Thus, a global aggregation GNN may degrade to a sampling-based aggregation GNN. We call this issue *Receptive Field Collapse*. In fact, as the average degree of the graph increases, the *collapse ratio* increases and the model degradation can be more severe, providing an alternative explanation for the poor performance of aggregation-based networks on large-scale graphs. Therefore, we introduce a soft variational constraint on aggregation weight $\bar{e}_{ij}$:

$$
\bar{e}_{ij} = \sigma\left(\text{V}\left(\left[h_i W_{src}^{(l)} \| h_j W_{dst}^{(l)}\right]\right)\right)
\tag{9}
$$

where σ denotes a nonlinear activation function, $V(\cdot)$ represents a generalized distribution constraint function that can smooth the directional distribution of the neighborhood aggregation. To simplify the notation, we use $\mathcal{F}(\cdot)$ to represent multiple feature mappings and distribution transform functions, and W_{drc} to integrate directed and undirected graphs. The adjacency matrix can be further expressed as follows:

$$
\overline{A}_{ij} = \delta(\bar{e}_{ij}) = \delta\left(\sigma\left(\text{V}\left(\left[h_i W_{src}^{(l)} \| h_j W_{dst}^{(l)}\right]\right)\right)\right) = \mathcal{F}(H^{(l-1)}, W_{drc}^{(l)})
\tag{10}
$$

Note that, due to nonlinear activation functions δ, σ can alter the distribution of features, $\overline{A}$ actually follows an approximate distribution by the constraint function $V(\cdot)$, which can be viewed as an acceptable tradeoff between the direction and strength of the neighborhood aggregation distribution. In Sect. 3.3, we provide details of the variational constraint of a standard normal distribution. We can denote the *message passing* as MP and formulate GVCNs layers as:

$$
\begin{aligned}
H_l &= (P \odot \overline{A}^{(l)})H^{(l-1)}W^{(l)} + H^{(l-1)}W_r^{(l)} \\
&= \text{MP}\{P \odot \mathcal{F}(H^{(l-1)}, W_{drc}^{(l)})\}W^{(l)} + H^{(l-1)}W_r^{(l)}
\end{aligned}
\tag{11}
$$

3.3 Variational Convolution Kernels

In standard Variational AutoEncoders (VAEs) [6,17], the encoder and decoder are tightly coupled. The encoder transforms input features into probability distribution parameters in the latent space, while the decoder samples output data from the latent variables by reparameterization. However, in our work, we find it is unnecessary to consider the generation effect of latent variables. We make the target probability distribution of the encoder obey a standard Gaussian distribution to moderately constrain the node features, which can be formulated as:

$$\overline{e}_{ij} = \sigma \left(rep([\boldsymbol{h}_i \boldsymbol{W}_{src}^{(l)}] \cdot \boldsymbol{\theta}_{src}) \oplus rep([\boldsymbol{h}_j \boldsymbol{W}_{dst}^{(l)}] \cdot \boldsymbol{\tau}_{dst}) \right) \tag{12}$$

where, $\oplus$ denotes various methods of merging two tensors such as taking the former, the latter, or their sum, $\overline{e}_{ij}$ represents the edge weight of a directed edge from node i to j, $\boldsymbol{\theta}_{src}$ and $\boldsymbol{\tau}_{dst}$ represent two pieces of parameter matrices, which can learn the mean μ and variance var of the features of source and target nodes, respectively, $rep(\cdot) = mu + var \cdot \epsilon$ denotes the reparameterization sampling process, where the ϵ is a random variable sampled from a standard Gaussian distribution. Then, we can couple the variation learning with the whole training process. Naturally, for undirected graphs, we can set $\boldsymbol{\theta}^{src} = \boldsymbol{\tau}^{dst}$. Therefore, we can construct $\overline{\boldsymbol{A}}$ from $\overline{e}_{ij}$, which approximately follows a Gaussian-like distribution. Furthermore, to constrain the distribution of latent variables of node features to follow the standard Gaussian distribution, we introduce the Kullback-Leibler divergence (kl-loss) [14] for μ and var:

Table 1. Dataset statistics

Datasets	#Nodes	#Features	#Edges	Classes	Metrics
ogbn-arxiv	169,343	128	1,166,243	40	Accuracy
ogbn-products	2,449,029	100	61,859,140	47	Accuracy

$$\begin{aligned} \mathcal{L} &= \mathcal{L}_{CE} + \mathcal{L}_{KL} \\ &= \mathcal{L}_{CE} + \frac{1}{2} \left(-\log var^2 + \mu^2 + var^2 - 1 \right) \end{aligned} \tag{13}$$

where the $\mathcal{L}_{CE}$ denotes the classical cross-entropy loss. Thus, we adaptively couple the distribution constraint with the *feature mapping* process. In addition, we propose a positional encoding that allows parallel computation of the PPMI prior based on random walk and global aggregation on large-scale graphs.

Existing aggregation methods for large-scale graphs require sampling [11, 24]. In this work, we propose a novel mechanism to perform layer-wise *message passing* without sampling by precomputing neighborhood position encodings $p(u) = \text{position}(u)$. The positional encoding matrixes can be represented as $\mathbf{NE}_{\text{ind}} \in \mathbb{R}^{1 \times 2E}$ and $\mathbf{SE}_{\text{ind}} \in \mathbb{R}^{2 \times N}$:

$$\mathbf{NE}_{\text{ind}} = \cup_{u \in N} \{p(v) \mid v \sim \mathcal{N}(u)\} \tag{14}$$

$$\mathbf{SE}_{\text{ind}} = \cup_{u \in N}\{(p(q), p(r)) \mid q, r = \mathcal{N}(u)(0), \mathcal{N}(u)(-1)\} \quad (15)$$

where $p(u)$ represents the position of node u in the aggregation matrix. We take the aggregation matrix to be $\boldsymbol{EI} \in \mathbb{R}^{2 \times 2E}$. If the position encoding matrices $\mathbf{NE}_{\text{ind}}$ and $\mathbf{SE}_{\text{ind}}$ are computed, we can organize a bipartite *src-dst* subgraph $\mathcal{G}_i$ in parallel for any subset $\forall \mathcal{V}_i \subseteq \mathcal{V}$. In this subgraph, $\mathcal{V}_i$ represents the set of destination nodes, while the neighborhoods of $\mathcal{V}_i$ form the set of source nodes. The size of the node subset $\mathcal{V}_i$ can be adaptively adjusted, allowing effective global aggregation with graphs of any scale.

4   Experiments

In this section, we conduct experiments on two OGB node classification datasets. Compared to the sampling-based aggregation GNNs, GVCNs achieves significant performance improvement. Compared to the decoupling-based GNNs, GVCNs performs competitively with less complexity and moderate runtime. We train all GVCN models on a single Nvidia RTX A6000 with 48Gb memory.

Table 2. The comparision experimental results on the ogbn-arxiv dataset. Except for GVCN, other results are from their papers or the OGB leaderboard.

Category	Baselines	Valid	Test
Sampling	GraphSAGE	72.77 ± 0.16	71.49 ± 0.27
	GCN	73.00 ± 0.17	71.74 ± 0.29
	DeeperGCN	72.62 ± 0.14	71.92 ± 0.16
	JKNet	73.35 ± 0.07	72.19 ± 0.21
	GAT	$\mathbf{75.04 \pm 0.06}$	73.52 ± 0.11
Decoupling	SIGN	73.23 ± 0.06	71.95 ± 0.11
	AGDN	74.23 ± 0.13	73.41 ± 0.25
Others	MLP	57.65 ± 0.12	55.50 ± 0.23
	LP	70.14 ± 0.00	68.32 ± 0.00
	Node2Vec	71.29 ± 0.13	70.07 ± 0.13
Ours	GVCN	74.19 ± 0.06	$\mathbf{73.78 \pm 0.12}$

Datasets. To fairly evaluate performance on datasets of different scales or types, We utilize two transductive OGB semi-supervised node classification datasets (ogbn-arxiv, ogbn-products) [12]. We summarize the detailed statistics of these datasets in Table 1. The ogbn-arxiv is a citation graph about Computer Science (CS) arXiv papers, whose data split is based on the publication dates of the papers. The ogbn-products dataset is an undirected and unweighted graph, representing an Amazon product co-purchasing network, whose data split is based on the sales ranking.

Baselines. In order to make a comprehensive comparison between GVCNs and other methods, we conducted extensive experiments and evaluations. Several representative GNNs and SOTA GNNs are selected as baselines. For sampling-based aggregatation GNNs, we utilize GCN [12], DeeperGCN [16], GAT [20], JKNet [22], GraphSAGE [11], FastGCN [4], ClusterGCN [5], GraphSAINT [24]. For decoupling-based GNNs, we utilize SGC [21], SIGN [9], SAGN [18], AGDN [19], GAMLP [25]. For other common methods, we utilize MLP [12], Node2vec [10], LP [13].

Global Settings. We evaluate our proposed models with 10 runs, fixing random seed 0–9, and report means and standard deviations. For OGB leaderboard, all final test scores are from the best model selected based on validation scores or losses. In the experiments tables of this paper, we highlight the results of GVCNs with underlined fonts and the best results with bold fonts.

4.1   Comparison Experiments

ogbn-arxiv Dataset. For ogbn-arxiv, we utilize 3 GVCN layers, 3 heads and 256 hidden dimensions, 0.002 learning rates, 0.65 embedding dropout rates, 0.1 input dropout rates, 0.2 edge dropout rates, Cross Entropy loss function and a residual linear connections without any tricks. Table 2 shows that the comparision experimental results on the ogbn-arxiv dataset. Compare to the best sampling-based aggregation GNN, GVCN achieves a performance improvement of 0.26%. Compare to the best decoupling-based GNN, GVCN gains a performance improvement of 0.37%. With a novel global variational convolution,

Table 3. The comparision experimental results on the ogbn-products dataset. Except for GVCN and GVCN+SLE, other results are from their papers or the OGB leaderboard.

Category	Baselines	Valid	Test
Sampling	GCN	92.00 ± 0.03	75.64 ± 0.21
	GraphSAGE	–	78.29 ± 0.16
	ClusterGCN	92.12 ± 0.09	78.97 ± 0.33
	GraphSAINT	91.62 ± 0.08	79.08 ± 0.24
	DeeperGCN	92.38 ± 0.09	80.98 ± 0.20
Decoupling	SIGN	92.99 ± 0.04	80.52 ± 0.16
	SAGN	93.09 ± 0.04	81.20 ± 0.07
	GAMLP+SLE	$\mathbf{93.12 \pm 0.03}$	83.54 ± 0.09
Others	MLP	75.54 ± 0.14	61.06 ± 0.08
	Node2Vec	90.32 ± 0.06	72.49 ± 0.10
	LP	90.91 ± 0.00	74.34 ± 0.00
Ours	GVCN	$\underline{92.47 \pm 0.11}$	$\underline{82.31 \pm 0.14}$
	GVCN+SLE	$\underline{92.81 \pm 0.09}$	$\mathbf{83.76 \pm 0.21}$

Table 4. The ablation experimental results of modules on the ogbn-products dataset, where the GA denotes the global aggregation, the GLM represents the guided learning mechanism and the VA denotes the variational aggregation.

GA	VA	GLM	Test Acc(%)
			55.50
✓			72.33
✓	✓		73.72
✓	✓	✓	**73.78**

Table 5. The ablation exprimental results of tricks on the ogbn-arxiv dataset. Except for GVCN, other results are from their papers or the OGB leaderboard. ①= BoT, ②= self-KD, ③= GIANT-XRT embedding.

Models	Valid	Test
UniMP	74.50 ± 0.15	73.11 ± 0.20
GAT+ ①	75.16 ± 0.08	73.91 ± 0.12
RevGAT+①	75.01 ± 0.10	74.02 ± 0.18
AGDN+①	75.25 ± 0.05	74.11 ± 0.12
Ours(GVCN+①)	$\underline{74.86 \pm 0.12}$	$\underline{74.22 \pm 0.06}$
GAT+①+②	75.14 ± 0.04	74.16 ± 0.08
RevGAT+①+②	74.97 ± 0.08	74.26 ± 0.17
AGDN+①+②	75.22 ± 0.09	74.31 ± 0.12
Ours(GVCN+①+②)	$\underline{74.73 \pm 0.11}$	$\underline{74.30 \pm 0.08}$
RevGAT+①+③	77.01 ± 0.09	75.90 ± 0.19
AGDN+①+③	77.24 ± 0.06	76.18 ± 0.16
Ours(GVCN+①+③)	$\mathbf{77.27 \pm 0.10}$	76.26 ± 0.14
RevGAT+①+②+③	77.16 ± 0.09	76.15 ± 0.10
AGDN+①+②+③	77.19 ± 0.08	76.37 ± 0.11
Ours(GVCN+①+②+③)	$\underline{77.06 \pm 0.09}$	$\mathbf{76.55 \pm 0.12}$

GVCN outperforms all other baselines and achieve a new SOTA performance of 73.78%. Furthermore, we only utilize 3 GVCN layers with hidden 256 with 1.43M parameters and RevGAT-Wide includes 5 layers with hidden dimension 1068 with 3.88M parameters but we achieve similar accuracy.

ogbn-products Dataset. Moreover, we evaluate GVCN on a larger dataset ogbn-products with our proposed position embedding and global aggregation kernel (Table 3). Without SLE, we utilize 2 GVCN layers, 512 hidden dimensions, 0.001 learning rates, 0.5 embedding dropout rates, 0.2 input dropout rates, 0.3 edge dropout rates, focal loss function with $\gamma = 1$. Compare to the best sampling-based aggregation GNN, GVCN achieves a performance improvement of 1.33%. Compare to best decoupling-based GNN, GVCN gains a performance improvement of 2.56%. For tricks of BoT [12], we adopt the same configuration as GAMLP to utilize SLE, and we make 0.9 temperature, 0.9 soft label threshold, 5 stages of 150 epochs of each stage, which can achieve a performance improvement of 0.20%. The experiment demonstrates the superiority of appropriately aggregating global structural information on larger graphs.

4.2   Ablation Study

In this section, to better identify the improvements from our proposed modules and other tricks, we conduct ablation studies with the following two aspects: In the first part, we view our three proposed contributions as three scalable modules and sequentially applied the modules to a base to investigate effectiveness of our methods: the variational aggregation (VA), the guided learning mechanism (GLM), and the global aggregation (GA). And the results are summarized in Table 4. Then, in order to evaluate the adaptability and scalability of GVCN, we adapt several proven effective tricks such as BoT, self-KD, and GIANT-XRT for GVCN from OGB leaderboard. And the results are summarized in Table 5. However, due to space limitations, we will provide a detailed description of the above complete ablation studies in Appendix 1.1.

4.3   Runtime Study

We compare the training time and inference time with other methods on the ogbn-proteins dataset under the same hardware devices, as shown in the Table 6.

Table 6. Runtime comparison on the ogbn-proteins dataset with similar Tesla V100 cards

Model	Training	Inference
RevGNN-Deep	13.5d/2000epochs	–
RevGNN-Wide	17.1d/2000epochs	–
AGDN	0.14d/2000epochs	12 s
Ours(GVCN)	0.03d/2000epochs	3.4 s

Compared to RevGNNs, GVCN greatly accelerates and is on par with the best decoupled network currently available, which is mainly due to the less parameters and lower computational redundancy.

5   Conclusion

In conclusion, we propose a novel GVCNs, an effective and flexible solution, to appropriately preserve more structural information during the aggregation process on large-scale graphs. In order to address the issue of *Receptive Field*

Collapse, we first employ the proposed positional encoding kernel function to achieve efficient parallel global aggregation. Furthermore, we introduce a guided learning mechanism based on PPMI and a variational learning mechanism, which encourage the model to incorporate all neighborhood features. Experimental results demonstrate that our proposed method achieves improvements of 0.26% and 2.78% on the ogbn-arxiv and ogbn-products datasets, respectively, compared to other similar aggregation-based GNNs, reaching the current state-of-the-art performance. In future work, we will explore the parallel acceleration, local directionality and weight distribution of aggregation on large-scale graphs.

References

1. Battaglia, P.W., et al.: Relational inductive biases, deep learning, and graph networks. arXiv preprint arXiv:1806.01261 (2018)
2. van den Berg, R., Kipf, T.N., Welling, M.: Graph convolutional matrix completion. arXiv preprint arXiv:1706.02263 (2017)
3. Bouma, G.: Normalized (pointwise) mutual information in collocation extraction. Proc. GSCL **30**, 31–40 (2009)
4. Chen, J., Ma, T., Xiao, C.: FastGCN: fast learning with graph convolutional networks via importance sampling. arXiv preprint arXiv:1801.10247 (2018)
5. Chiang, W.-L., Liu, X., Si, S., Li, Y., Bengio, S., Hsieh, C.-J.: Cluster-GCN: an efficient algorithm for training deep and large graph convolutional networks. In: Proceedings of the 25th ACM SIGKDD International Conference on Knowledge Discovery & Data Mining, pp. 257–266 (2019)
6. Carl Doersch. Tutorial on variational autoencoders. arXiv preprint arXiv:1606.05908 (2016)
7. Duan, K., et al.: A comprehensive study on large-scale graph training: benchmarking and rethinking. arXiv preprint arXiv:2210.07494 (2022)
8. Fout, A., Byrd, J., Shariat, B., Ben-Hur, A.: Protein interface prediction using graph convolutional networks. In: Advances in Neural Information Processing Systems, vol. 30 (2017)
9. Frasca, F., Rossi, E., Eynard, D., Chamberlain, B., Bronstein, M., Monti, F.: Sign: scalable inception graph neural networks. arXiv preprint arXiv:2004.11198 (2020)
10. Grover, A., Leskovec, J.: node2vec: scalable feature learning for networks. In: Proceedings of the 22nd ACM SIGKDD International Conference on Knowledge Discovery and Data Mining, pp. 855–864 (2016)
11. Hamilton, W., Ying, Z., Leskovec, J.: Inductive representation learning on large graphs. In: Advances in Neural Information Processing Systems, vol. 30 (2017)
12. Weihua, H., et al.: Open graph benchmark: datasets for machine learning on graphs. In: Advances in Neural Information Processing Systems, vol. 33, pp. 22118–22133 (2020)
13. Huang, Q., He, H., Singh, A., Lim, S.-N., Benson, A.R.: Combining label propagation and simple models out-performs graph neural networks. arXiv preprint arXiv:2010.13993 (2020)
14. Joyce, J.M.: Kullback-leibler divergence. In: Lovric, M. (ed.) International Encyclopedia of Statistical Science, pp. 720–722. Springer, Heidelberg (2011). https://doi.org/10.1007/978-3-642-04898-2_327
15. Kipf, T.N., Welling, M.: Variational graph auto-encoders. arXiv preprint arXiv:1611.07308 (2016)

16. Li, G., Xiong, C., Thabet, A., Ghanem, B.: DeeperGCN: all you need to train deeper GCNs. arXiv preprint arXiv:2006.07739 (2020)
17. Liang, D., Krishnan, R.G., Hoffman, M.D., Jebara, T.: Variational autoencoders for collaborative filtering. In: Proceedings of the 2018 World Wide Web Conference, pp. 689–698 (2018)
18. Sun, C., Gu, H., Hu, J.: Scalable and adaptive graph neural networks with self-label-enhanced training. arXiv preprint arXiv:2104.09376 (2021)
19. Sun, C., Wu, G.: Adaptive graph diffusion networks with hop-wise attention. arXiv e-prints, p. arXiv–2012 (2020)
20. Veličković, P., Cucurull, G., Casanova, A., Romero, A., Lio, P., Bengio, Y.: Graph attention networks. arXiv preprint arXiv:1710.10903 (2017)
21. Wu, F., Souza, A., Zhang, T., Fifty, C., Yu, T., Weinberger, K.: Simplifying graph convolutional networks. In: International Conference on Machine Learning, pp. 6861–6871. PMLR (2019)
22. Xu, K., Li, C., Tian, Y., Sonobe, T., Kawarabayashi, K., Jegelka, S.: Representation learning on graphs with jumping knowledge networks. In: International Conference on Machine Learning, pp. 5453–5462. PMLR (2018)
23. Le, Yu., Bowen, D., Xiao, H., Sun, L., Han, L., Lv, W.: Deep spatio-temporal graph convolutional network for traffic accident prediction. Neurocomputing **423**, 135–147 (2021)
24. Zeng, H., Zhou, H., Srivastava, A., Kannan, R., Prasanna, V.: Graphsaint: graph sampling based inductive learning method. arXiv preprint arXiv:1907.04931 (2019)
25. Zhang, W., et al.: Graph attention multi-layer perceptron. In: Proceedings of the 28th ACM SIGKDD Conference on Knowledge Discovery and Data Mining, pp. 4560–4570 (2022)
26. Zheng, C., et al.: ByteGNN: efficient graph neural network training at large scale. Proc. VLDB Endowment **15**(6), 1228–1242 (2022)

Frequency Domain Distillation for Data-Free Quantization of Vision Transformer

Gongrui Nan and Fei Chao$^{(\boxtimes)}$

School of Informatics, Xiamen University, Fujian 361005, People's Republic of China
nangongrui@stu.xmu.edu.cn, fchao@xmu.edu.cn

Abstract. The increasing size of deep learning models has made model compression techniques increasingly important. Neural network quantization is a technique that can significantly compress models while preserving their original precision. However, conventional quantization methods relies on real training data, making it unsuitable for scenarios where data is unavailable. Data-Free quantization methods address this issue by synthesizing pseudo data to calibrate or fine tune the quantized model. However, these methods overlook an important problem, i.e., the mismatch between the low-frequency and high-frequency components of the synthesized pseudo data. This is due to the simultaneous optimization of low-frequency and high-frequency information, which can interfere with each other. We analyze the reasons behind this phenomenon and propose a frequency domain distillation (FDD) method to address this issue. Specifically, we first optimize the low-frequency component, followed by the high-frequency component, and employ distillation to make the high-frequency component more consistent with the low-frequency component. Additionally, we apply a progressive optimization strategy by gradually increasing the optimized region of pseudo data. We achieved state-of-the-art results on all the Vit models involved in our experiments, and complete ablation study also demonstrated the effectiveness of our method. Our code can be found at here.

Keywords: Model compression · Data-free quantization · Vision transformer

1 Introduction

Quantization [7] is a key technique in model compression. It involves representing high-bit parameters or activation values using lower-bit representations, effectively compressing the model. However, obtaining a well-performing quantized network typically requires calibration or fine-tuning with real training data. When real data is unavailable, recovering the precision of a quantized network becomes challenging. Hence, the appearance of Data-Free quantization methods aims to address this issue.

© The Author(s), under exclusive license to Springer Nature Singapore Pte Ltd. 2024
Q. Liu et al. (Eds.): PRCV 2023, LNCS 14432, pp. 205–216, 2024.
https://doi.org/10.1007/978-981-99-8543-2_17

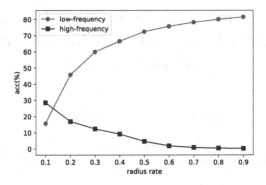

Fig. 1. Changes in accuracy corresponding to different frequency domain parts. The larger the "radius rate", the more low-frequency information and the less high-frequency information. The more low-frequency information, the higher the accuracy rate (red line), indicating that low-frequency information is very important, and the less high-frequency information, the lower the accuracy rate (blue line), indicating that high-frequency information is also important. (Color figure online)

Nagel et al. [12] achieves the recovery of quantized network precision through parameter calibration, without additional data. Methods like [1,2,6,8,16,17] utilize optimization techniques to optimize noise and generate pseudo data that follows the real data distribution, resulting in improved performance. However, these optimization methods have mainly been explored in the context of Convolutional Neural Networks (CNNs) and rely on the specific Batch Normalization (BN) layers of CNNs. They encounter challenges when dealing with the popular Vision Transformer (ViT) [4] architecture. PSAQ [9] made an initial attempt to generate good pseudo data on ViT, but it also has some limitations or issues.

Inspired by [15]'s analysis of high-frequency and low-frequency information in neural networks, we attempted to decompose the high-frequency and low-frequency components of the pseudo data and individually feed them into the neural network for testing. The results showed that both low-frequency and high-frequency information were helpful for classification. However, previous Data-Free methods suffered from significant inconsistency between the high-frequency and low-frequency components of the generated pseudo data. This inconsistency was manifested in a large KL divergence between the outputs when using the low-frequency and high-frequency components as inputs respectively, while the corresponding KL divergence for real data was comparably small. Therefore, we pondered whether it would be possible to enforce consistency between the low-frequency and high-frequency components during the optimization process to enhance the quality of the generated data.

Regarding the above issues, we propose a frequency domain distillation strategy. Specifically, we utilize fast Fourier transform to decompose the optimization of Gaussian noise into two stages. In the first stage, we use the low-frequency component of the noise as the input for the network and optimize it. In the second stage, we use both the low-frequency and high-frequency components of the noise as inputs for the network. We distill the high-frequency component by using the output obtained with the low-frequency component as the input. Additionally, the entire optimization process is designed as a progressive procedure. To be more specific, we repeat the aforementioned two-stage optimization process several times, each time expanding the region of low-frequency or high-frequency involvement in the optimization until the entire noisy image is optimized.

We validated the effectiveness of our method on DeiT [13]and Swin Transformer [10]. Compared to the current state-of-the-art method PSAQ, our approach outperforms it in terms of performance across models of different sizes. We also conducted analysis on the generated pseudo data, both visualization and KL divergence further supports the effectiveness of our method.

2   Related Work

2.1   Vision Transformer (ViT)

The Transformer [14] is a neural network architecture based on the self-attention mechanism, initially widely used in various tasks in the field of Natural Language Processing (NLP). Its introduction in the NLP domain has had a tremendous impact and has provided a foundation and inspiration for the development of other areas of artificial intelligence. The field of computer vision drew inspiration from the Transformer architecture and proposed the Vision Transformer [4], which, at one point, surpassed the monopolistic position held by CNN and became a new powerhouse in computer vision. The Vision Transformer utilizes the encoder part of the original Transformer and processes input images by dividing them into patches, converting them into token embeddings and positional embeddings, which are then combined as the network's input. It has demonstrated remarkable performance in tasks such as classification, object detection [5], and image segmentation [11].

2.2   Network Quantization

Data-Free Quantization (DFQ). Quantization Aware Training and Post-Training Quantization for recovering the accuracy of quantized networks both require the use of real training data. However, when real training data is not available, one can only consider using some Data-Free quantization methods. Methods such as [12] do not rely on data and achieve data-free recovery of quantized network performance through parameter calibration. Methods like [1, 2, 6, 8, 16, 17], on the other hand, keep the full-precision network fixed and optimize the noise input or use a generator to transform the noise into images that following the distribution of real data.

The Difficulty for Data-Free Quantization of Vision Transformer. The aforementioned Data-Free quantization methods are primarily based on CNNs.

However, when the network is switched to Vision Transformer (ViT), these methods cannot be seamlessly applied due to the absence of CNN-specific components such as the Batch Normalization (BN) layer and its associated statistical information in ViT. While PASQ [9] (Patch Similarity Aware Quantization) leverages the unique attention mechanism in transformers to generate pseudo-data that conforms to the distribution of real data. But it does not exploit another characteristic of transformers-its global receptive field. This can lead to excessive focus on low-frequency features in the data, unlike CNNs, which tend to prioritize high-frequency features. In this paper, we will delve deeper into the analysis and propose a novel method that leverages this characteristic of transformers to generate better pseudo data and aid in the more accurate recovery of quantized networks.

3   Preliminaries

3.1   Quantizer

Unless otherwise specified, this paper adopts post-training quantization and employs the basic min-max quantization strategy, similar to PSAQ [9].

For quantizing the weights of neural networks, we utilize symmetric quantization.

$$s = \frac{2max(|f_{max}|, |f_{min}|)}{2^b - 1} \tag{1}$$

$$q = s \cdot round(\frac{f}{s}) \tag{2}$$

In the formula above, s represents the scale factor used to scale a 32-bit full-precision value f to an integer value of b bits. f_{min} represents the minimum value of the full-precision value to be quantized, and f_{max} represents the maximum value. The $round$ denotes the rounding operation, and q represents the quantized value after quantization.

For the quantization of activation, we employ asymmetric quantization.

$$s = \frac{f_{max} - f_{min}}{2^b - 1} \tag{3}$$

$$ZeroPoint = -round(\frac{f_{min}}{s}) \tag{4}$$

$$q = s \cdot (round(\frac{f}{s} + ZeroPoint) - ZeroPoint) \tag{5}$$

In the formula above, $ZeroPoint$ is used to shift the integer range to achieve asymmetric quantization.

3.2   Fast Fourier Transform (FFT) and Frequency Domain

The Fourier Transform [3] is a mathematical technique used to convert a function from the time domain to the frequency domain. It can also be applied to 2D images to transform them from the spatial domain to the frequency domain.

$$F = FFT(I) \tag{6}$$

In the formula above, FFT represents Fast Fourier Transform, I represents the image in the spatial domain, and F represents the image in the frequency domain.

In the frequency domain, it is easy to decompose an image into its high-frequency and low-frequency components by simply multiplying it with the corresponding masks. Information near the center of the frequency domain image corresponds to lower frequency components, while information farther away from the center corresponds to higher frequencies.

$$LF = F \cdot mask_{lf} \tag{7}$$

$$HF = F \cdot mask_{hf} \tag{8}$$

In the formula above, LF represents the low-frequency components of the image, HF represents the high-frequency components of the image, $mask_{lf}$ refers to the low-frequency mask, and $mask_{hf}$ refers to the high-frequency mask.

The image processed in the frequency domain can be transformed back to the spatial domain.

$$I_{lf} = IFFT(LF) \tag{9}$$

$$I_{hf} = IFFT(HF) \tag{10}$$

In the formula above, $IFFT$ represents the inverse Fast Fourier Transform, which is used to transform the frequency domain image back to the spatial domain image.

4   Method

4.1   Our Insights

We found that testing with only the high-frequency or low-frequency components of an image separately leads to a decrease in accuracy. This is not surprising since the image loses its original information, which naturally affects the final outcome. Specifically, for the high-frequency components, gradually removing the low-frequency information from the image results in a gradual decline in accuracy, reaching significantly low values as shown by the blue line in Fig. 1. This indicates that the low-frequency information carries a significant amount of essential data for task completion. On the other hand, for the low-frequency components, gradually adding high-frequency information to the image leads to a gradual increase in accuracy (restoring it to the original image's precision level),

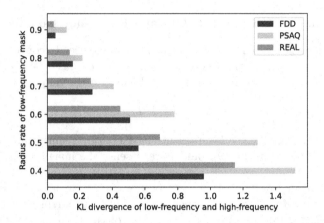

Fig. 2. Pseudo data generated by PSAQ will have serious inconsistencies in high frequency and low frequency information in swin_small model. Our method can solve this problem. The larger "Radius rate of low-frequency mask", the more low-frequency information and the less high-frequency information.

as shown by the red line in Fig. 1. This demonstrates that the high-frequency information in the image contributes to the task completion.

However, [15] pointed out that the high-frequency information not only contains helpful information for task completion but also includes various noises that could negatively impact accuracy. This is manifested in inconsistent classification results between high and low-frequency information. In other words, if we can minimize the noise within the high-frequency information during the generation of synthetic data, ensuring consistency between the low-frequency and high-frequency parts of the synthetic data, it will significantly improve the quality of the generated data and consequently affect the calibration results of the quantization network.

Previous data-free quantization approaches did not differentiate between high-frequency and low-frequency components in the images. Instead, they treated them equally and uniformly optimized all regions of Gaussian noise. These approaches lead to a significant problem because the distinctiveness of high-frequency and low-frequency information, as well as their different roles in neural networks, make the high-frequency and low-frequency information of optimized images not match effectively. This can be observed in the KL divergence between the results of inference using high-frequency information and low-frequency information separately, as shown in Fig. 2, which is not the desired outcome we expect to see.

We have been contemplating whether it is possible to establish a connection between high-frequency and low-frequency information during the noise optimization process, aiming to achieve a unified high-frequency and low-frequency representation in the generated pseudo data, thereby significantly improving the quality of the generated samples.

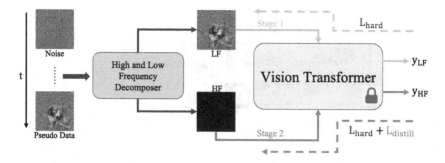

Fig. 3. The working process of our proposed method (FDD-DFQ-ViT).

4.2 Frequency Domain Distillation

To achieve consistency between low-frequency and high-frequency information, we propose a frequency domain distillation method (FDD-DFQ-ViT). Specifically, we divide the optimization process into two stages. In the first stage, we obtain the low-frequency version of the noise through fast Fourier transform and input it into the network for optimization. By only optimizing the low-frequency information in this stage, we quickly determine the direction of generation while excluding high-frequency information from the optimization process. This approach avoids mutual interference between low-frequency and high-frequency information during optimization. In this stage, we employ L_{class} as the loss function.

$$L_{class} = L_{hard} = CE(\hat{y}, y) \tag{11}$$

In the equation above, $\hat{y}$ represents the output of the network, while y represents the randomly generated label.

In the second stage, we obtain the high-frequency version of the noise through fast Fourier transform and input it into the network for optimization. Simultaneously, we also feed the low-frequency version into the network to obtain the network's output. We then utilize the network's output, with the low-frequency version as the input, to distill the high-frequency version of the noise.

$$L_{distill} = KL(\hat{y}, y_{lf}) \tag{12}$$

$$L_{class} = \alpha_{hard}L_{hard} + \alpha_{distill}L_{distill} \tag{13}$$

In the equation above, α_{hard} and $\alpha_{distill}$ represent the coefficients for the two losses. In our experiments, we found that setting both coefficients to 0.5 yielded the best results.

The entire process of our proposed method is illustrated in Fig. 3. Specifically, regarding the High and Low Frequency Decomposer, please refer to Fig. 4. For our proposed frequency domain distillation strategy, during the entire optimization process, our method iterates through the two aforementioned stages five times. Within these five iterations, the radius of the low-frequency mask gradually expands, allowing the low-frequency information to progressively evolve from

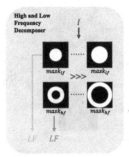

Fig. 4. The detail of the High and Low Frequency Decomposer. The *mask* here represents the selected part in the frequency domain, and the closer to the center, the lower the frequency of the information. As the optimization proceeds, the low-frequency area selected to participate in the optimization gradually expands, and the high-frequency area selected to participate in the optimization is also gradually extended outward.

low to mid and even high frequencies during the optimization. Not only does the low-frequency component undergo gradual changes, but the high-frequency component also experiences gradual modifications. We gradually increase the inner and outer radius of the high-frequency mask, enabling it to transition from mid-frequency to high-frequency gradually.

4.3   The Overall Pipeline

Our method, similar to most existing approaches, consists of two stages, as shown in Fig. 3. The first stage is dedicated to generating pseudo data, while the second stage involves using this pseudo data to calibrate the quantized network and recover its accuracy.

Fake Sample Generation Stage. In this stage, we first generate a batch of Gaussian noise images along with their corresponding random class labels. Then, while keeping the full-precision model fixed, we optimize these Gaussian noise images through backpropagation to approximate the distribution of real data, as described in Sect. 3.3. In addition to using the L_{class} and $L_{distill}$, we also continue to employ the TV loss and the PSE loss proposed in PSAQ.

The TV loss is utilized to smooth the generated pseudo data.

$$L_{TV} = \iint \bigtriangledown I(i,j)d_i d_j \tag{14}$$

$\bigtriangledown$ represents the operation of computing the gradient of an image, where i and j denote the positions of pixels.

The total loss is defined as follows:

$$L = L_{class} + \alpha_{TV}L_{TV} + L_{PSE} \tag{15}$$

Table 1. The comparison results.

Model	Method	Bit	Top1-Acc (%)	Bit	Top1-Acc (%)
swin_tiny	noise	W4A8	0.474	W8A8	0.510
	real	W4A8	70.980	W8A8	74.712
	PSAQ	W4A8	71.480	W8A8	75.004
	ours	W4A8	**72.420**	W8A8	**75.938**
swin_small	noise	W4A8	0.736	W8A8	0.796
	real	W4A8	71.258	W8A8	73.574
	PSAQ	W4A8	74.892	W8A8	76.538
	ours	W4A8	**75.126**	W8A8	**76.698**
deit_tiny	noise	W4A8	7.828	W8A8	10.306
	real	W4A8	65.248	W8A8	71.210
	PSAQ	W4A8	65.682	W8A8	71.618
	ours	W4A8	**65.708**	W8A8	**71.692**
deit_small	noise	W4A8	7.570	W8A8	11.858
	real	W4A8	72.338	W8A8	76.132
	PSAQ	W4A8	72.324	W8A8	76.172
	ours	W4A8	**72.442**	W8A8	**76.284**
deit_base	noise	W4A8	72.142	W8A8	15.732
	real	W4A8	76.280	W8A8	78.684
	PSAQ	W4A8	76.526	W8A8	78.964
	ours	W4A8	**76.730**	W8A8	**79.294**

Quantization Calibrating Stage. By passing the generated pseudo data through the network in a forward pass, we can obtain the maximum and minimum values of the activations. This information is then used to calibrate the activations of the network, enabling the recovery of the quantized network's accuracy.

5   Experimentation

5.1   Comparison Experiments

We compared our proposed method to the existing sole data-free post-training quantization method, PSAQ [9], and used both real data and Gaussian noise calibration as control groups to test the effectiveness of our approach. The comparative results are presented in Table 1, which indicate that our proposed method consistently outperforms PSAQ in terms of recovering quantized network accuracy, even surpassing the results achieved by calibrating the quantized network with real data. On the contrary, solely using Gaussian noise to calibrate the quantized network leads to significant performance degradation, highlighting the significance of data-free quantization methods.

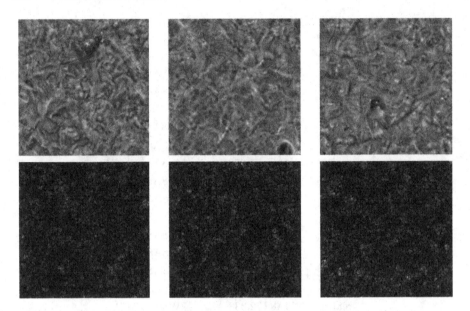

Fig. 5. The visualization of the data generated by PSAQ. The first row is the generated pseudo data, and the second row is the result of edge detection corresponding to the high-frequency part of the generated data.

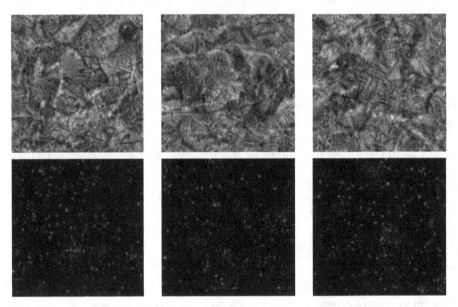

Fig. 6. The visualization of the data generated by ours (FDD-DFQ-ViT).

We visualized a comparison between the synthetic data generated using our method (FDD-DFQ-ViT) and the PSAQ method. Figure 5 displays the synthetic data generated by PSAQ, while Fig. 6 shows the synthetic data generated by ours. From the figures, it is evident that the high-frequency noise in the synthetic data generated by our method is significantly reduced compared to PSAQ.

Table 2. The Ablation Stduy. "Only LF" means to optimize only low-frequency information, "Only HF" means to optimize only high-frequency information, and "HF distill LF" means to optimize high-frequency information first and use the result to distill low-frequency information.

Method	Bit	Top1-Acc (%)	Bit	Top1-Acc (%)
PSAQ	W4A8	71.480	W8A8	75.004
Only LF	W4A8	72.086 (+0.606)	W8A8	75.404 (+0.4)
Only HF	W4A8	71.800 (+0.32)	W8A8	75.132 (+0.128)
HF distill LF	W4A8	72.004 (+0.524)	W8A8	74.210 (-0.794)
LF distill HF (ours)	W4A8	**72.420 (+0.94)**	W8A8	**75.938 (+0.934)**

5.2 Ablation Study

We investigated the impact of the proposed frequency domain distillation (FDD-DFQ-ViT) strategy on the performance, as shown in Table 2. It can be observed that optimizing only the low-frequency part or only the high-frequency part of the image both outperform the baseline (PSAQ), indicating that jointly optimizing the low-frequency and high-frequency components does indeed influence the quality of the generated data. Additionally, it is not reasonable to distill the low-frequency part with the high-frequency part for the vit architecture, as it may even perform worse than PSAQ, considering that vit focuses more on the low-frequency components of the image. By using our method, specifically distilling the high-frequency part with the low-frequency part, we achieve the best results.

6 Conclusions

We have analyzed the mismatch issue between low-frequency and high-frequency information in the generation process of existing data-free quantization methods for ViT. Consequently, we propose a frequency domain distillation strategy for pseudo data generation in data-free quantization for ViT. A comprehensive set of experiments demonstrates the effectiveness of our method. However, it should be noted that our method did not explore further enhancements in the CNN architecture, which is a significant aspect for future research.

References

1. Cai, Y., Yao, Z., Dong, Z., Gholami, A., Mahoney, M.W., Keutzer, K.: ZeroQ: a novel zero shot quantization framework. In: Proceedings of the IEEE/CVF Conference on Computer Vision and Pattern Recognition (CVPR), pp. 13169–13178 (2020)

2. Choi, K., et al.: It's all in the teacher: zero-shot quantization brought closer to the teacher. In: Proceedings of the IEEE/CVF Conference on Computer Vision and Pattern Recognition (CVPR), pp. 8311–8321 (2022)
3. Cooley, J.W., Tukey, J.W.: An algorithm for the machine calculation of complex Fourier series. Math. Comput. **19**(90), 297–301 (1965)
4. Dosovitskiy, A., et al.: An image is worth 16x16 words: transformers for image recognition at scale. arXiv preprint arXiv:2010.11929 (2020)
5. Girshick, R., Donahue, J., Darrell, T., Malik, J.: Rich feature hierarchies for accurate object detection and semantic segmentation. In: Proceedings of the IEEE Conference on Computer Vision and Pattern Recognition (CVPR), pp. 580–587 (2014)
6. Jeon, Y., Lee, C., Kim, H.Y.: Genie: show me the data for quantization. In: Proceedings of the IEEE/CVF Conference on Computer Vision and Pattern Recognition (CVPR), pp. 12064–12073 (2023)
7. Krishnamoorthi, R.: Quantizing deep convolutional networks for efficient inference: a whitepaper. arXiv preprint arXiv:1806.08342 (2018)
8. Li, H., et al.: Hard sample matters a lot in zero-shot quantization. In: Proceedings of the IEEE/CVF Conference on Computer Vision and Pattern Recognition (CVPR), pp. 24417–24426 (2023)
9. Li, Z., Ma, L., Chen, M., Xiao, J., Gu, Q.: Patch similarity aware data-free quantization for vision transformers. In: Proceedings of the IEEE/CVF conference on European Conference on Computer Vision (ECCV), pp. 154–170 (2022)
10. Liu, Z., et al.: Swin transformer: Hierarchical vision transformer using shifted windows. In: Proceedings of the IEEE/CVF International Conference on Computer Vision (ICCV), pp. 10012–10022 (2021)
11. Long, J., Shelhamer, E., Darrell, T.: Fully convolutional networks for semantic segmentation. In: Proceedings of the IEEE Conference on Computer Vision and Pattern Recognition (CVPR), pp. 3431–3440 (2015)
12. Nagel, M., Baalen, M.V., Blankevoort, T., Welling, M.: Data-free quantization through weight equalization and bias correction. In: Proceedings of the IEEE/CVF International Conference on Computer Vision (ICCV), pp. 1325–1334 (2019)
13. Touvron, H., Cord, M., Douze, M., Massa, F., Sablayrolles, A., Jégou, H.: Training data-efficient image transformers & distillation through attention. In: Proceedings of the IEEE/CVF International Conference on Machine Learning (ICML), pp. 10347–10357. PMLR (2021)
14. Vaswani, A., et al.: Attention is all you need. In: Advances in Neural Information Processing Systems, vol. 30 (2017)
15. Wang, H., Wu, X., Huang, Z., Xing, E.P.: High-frequency component helps explain the generalization of convolutional neural networks. In: Proceedings of the IEEE/CVF Conference on Computer Vision and Pattern Recognition (CVPR), pp. 8684–8694 (2020)
16. Zhang, X., et al.: Diversifying sample generation for accurate data-free quantization. In: Proceedings of the IEEE/CVF Conference on Computer Vision and Pattern Recognition (CVPR), pp. 15658–15667 (2021)
17. Zhong, Y., et al.: IntraQ: learning synthetic images with intra-class heterogeneity for zero-shot network quantization. In: Proceedings of the IEEE/CVF Conference on Computer Vision and Pattern Recognition (CVPR), pp. 12339–12348 (2022)

An ANN-Guided Approach to Task-Free Continual Learning with Spiking Neural Networks

Jie Zhang[1], Wentao Fan[2]([✉]) [iD], and Xin Liu[1]

[1] Department of Computer Science and Technology, Huaqiao University, Xiamen, China
zhangjie@stu.hqu.edu.cn, xliu@hqu.edu.cn
[2] Guangdong Provincial Key Laboratory IRADS and Department of Computer Science, Beijing Normal University-Hong Kong Baptist University (BNU-HKBU) United International College, Zhuhai, China
wentaofan@uic.edu.cn

Abstract. Task-Free Continual Learning (TFCL) poses a formidable challenge in lifelong learning, as it operates without task-specific information. Leveraging spiking neural networks (SNNs) for TFCL is particularly intriguing due to their promising results in low-energy applications. However, existing research has predominantly focused on employing SNNs for solving single-task classification problems. In this work, our goal is to utilize ANN to guide SNN in addressing catastrophic forgetting and model compression issues, while treating SNNs as the basic network of the model. We introduce AGT-SNN (ANN-Guided TFCL for Spiking Neural Networks), a novel framework that empowers SNNs to engage in lifelong learning without relying on task-specific information. We conceptualize the learning process of the model as a multiplayer game, involving participants in the roles of players and referees. Our model's fundamental components comprise player-referee pairs, where the player module adopts a SNN-based Variational Autoencoder (VAE) and the referee module employs a ANN-based Generative Adversarial Network (GAN). To dynamically expand the number of components, we propose an innovative method called Adversarial Similarity Expansion (ASE). ASE evaluates the performance of the current player against previously learned players without accessing any task-specific information. Additionally, we propose a innovative pruning strategy that selectively removes redundant components while preserving the diversity of knowledge, thereby reducing the model's complexity. Through comprehensive experimental validation, we demonstrate that our proposed framework enables SNNs to achieve exceptional performance while maintaining an appropriate network size.

Supported in part by the National Natural Science Foundation of China (62276106), the Guangdong Provincial Key Laboratory IRADS (2022B1212010006, R0400001-22) and the UIC Start-up Research Fund (UICR0700056-23) and the Artificial Intelligence and Data Science Research Hub (AIRH) of BNU-HKBU United International College (UIC).

Q. Liu et al. (Eds.): PRCV 2023, LNCS 14432, pp. 217–228, 2024.
https://doi.org/10.1007/978-981-99-8543-2_18

Keywords: Task-free continual learning · Spiking nerual networks · Model expansion and compression · Image classification

1 Introduction

In recent years, Artificial Neural Networks (ANNs) have undergone significant advancements and have emerged as key technologies in the fields of computer science and artificial intelligence. ANNs are renowned for their exceptional representation capabilities and find wide-ranging applications, including tasks such as image recognition, natural language processing and data mining [6,7,17,20]. These networks have demonstrated remarkable progress in tackling complex tasks. However, the scalability and growing complexity of ANNs come at the cost of increased computational resources and energy consumption.

To address the challenges of computational resources and energy consumption in ANNs, there has been growing interest in Spiking Neural Networks (SNNs) [19] as a promising neural network model. Unlike ANNs, SNNs closely mimic the functioning of biological neural systems by simulating the transmission and processing of information through spike signals between neurons. This spike-based encoding method offers high spatiotemporal efficiency, making SNNs well-suited for low-power applications and brain-inspired tasks. However, existing research has primarily focused on utilizing SNNs for single-task classification problems, with limited exploration of their application in Task-Free Continual Learning (TFCL) [2]. TFCL involves lifelong learning without any task-specific information, and it plays a crucial role in achieving more intelligent systems.

Given the significant advancements in ANNs and the pressing need to address computational resource requirements, the use of ANN to guide SNN in TFCL has emerged as an important research direction. Our objective is to harness the power of ANN to direct SNNs in the TFCL setting, effectively addressing the issues of catastrophic forgetting [8] and model compression. To achieve this, we propose AGT-SNN, a framework designed to facilitate lifelong learning in SNNs without task-specific information. AGT-SNN conceptualizes the learning process as a multiplayer game involving players and referees. The player module employs a Variational Autoencoder (VAE) based on SNNs [13], while the referee module utilizes a Generative Adversarial Network (GAN) [10]. This modular design enables efficient knowledge acquisition in the absence of tasks. To enhance knowledge diversity and dynamically expand the number of components, we introduce Adversarial Similarity Expansion (ASE), a method that evaluates the disparities between previously learned players and the current player to determine whether new components should be added. This expansion process occurs without access to task-specific information. Additionally, we propose a novel pruning strategy that selectively removes redundant components without compromising knowledge diversity, thereby reducing model complexity and improving task execution efficiency. By combining ASE with the pruning strategy, AGT-SNN achieves effective model compression while preserving performance. Our contributions in this work can be summarized as follows:

- We propose AGT-SNN, a framework that enables efficient knowledge acquisition without relying on task-specific information. It consists of player module implemented using SNN-based VAE and referee module employing ANN-based GAN.
- We introduce ASE as a method within AGT-SNN, which dynamically expands the model's components based on the disparities between previously learned players and the current player. Additionally, we propose a novel model compression strategy that selectively removes redundant components, effectively reducing model complexity while preserving knowledge diversity.
- We demonstrate the effectiveness of AGT-SNN in lifelong learning without task-specific information through image classification experiments. The results show that the proposed framework achieves state-of-the-art performance on the TFCL image classification task.
- We explore the use of ANNs to guide SNNs to facilitate continual learning of SNNs without task information. The proposed AGT-SNN has the potential to provide insights for developing energy-efficient lifelong learning systems for edge AI applications.

2   Related Works

2.1   Image Generation in SNNs

In recent years, SNNs have witnessed notable advancements and applications [24]. Kamata et al. and Kotariya et al. have demonstrated the feasibility of generating images using SNNs through the introduction of FSVAE [13] and Spiking-GAN [15], respectively. FSVAE, which leverages SNNs within a Variational Autoencoder (VAE) framework, has been shown to outperform ANNs in various aspects. Thus, in this work, we adopt FSVAE as the fundamental SNN component for capturing knowledge flow, ensuring its compatibility with neuromorphic devices [4].

2.2   Continual Learning

Continual learning, also known as lifelong learning, has garnered extensive attention in the field of ANNs. Various approaches, such as GANs [9] and VAEs [14], have been explored to mitigate catastrophic forgetting through generative replay [21,26]. However, these methods rely on task-specific information and are not directly applicable to TFCL. TFCL, which represents a more realistic scenario, has gained significant attention in recent years. Some existing TFCL solutions [5,16] incorporate a memory buffer to selectively store samples for model training. Nevertheless, these methods suffer from fixed memory capacity, limiting their scalability for learning from an infinite data stream, and the majority of the work is based on ANNs.

3   Preliminary

3.1   The Referee Module: WGAN

In this work, we propose the integration of an ANN-based WGAN (Wasserstein GAN with Gradient Penalty) [10] as the referee module. The ANN-based WGAN has demonstrated exceptional performance in generating high-quality images and effectively learning complex image distributions. It replaces the traditional Jensen-Shannon divergence with the Wasserstein distance, which improves training stability and mitigates the issue of vanishing gradients. Additionally, the WGAN incorporates a gradient penalty term to enforce Lipschitz continuity in the discriminator, thereby effectively preventing the problem of exploding gradients during training. Consequently, the ANN-based WGAN is an ideal choice for the referee module. The objective function of the WGAN is defined as follows:

$$
\begin{aligned}
\min_{G} \max_{D} V(D, G) = & \mathbb{E}_{x \sim p_{data}(x)}[D(x)] \\
& - \mathbb{E}_{z \sim p_z(z)}[D(G(z))] \\
& + \lambda \mathbb{E}_{\hat{x} \sim p_{\hat{x}}(\hat{x})}[(\|\nabla_{\hat{x}} D(\hat{x})\|_2 - 1)^2],
\end{aligned}
\tag{1}
$$

where $D(x)$ represents the discriminator's output for a real sample x, $G(z)$ represents the generator's output for a noise input z, λ is a hyperparameter that controls the gradient penalty term, $\hat{x}$ denotes a random interpolation between real and generated samples.

3.2   The Player Module: FSVAE

The design of the player module takes into consideration two crucial factors: 1) the use of SNNs in the network, and 2) the incorporation of inference mechanisms. To address these considerations, we employ the SNN-based VAE model called FSVAE [13], which integrates the concept of VRNN into SNNs. FSVAE tackles the limitation of floating-point computation in SNNs by utilizing an autoregressive SNN [11] to construct the prior and posterior. The latent variables are sampled from the autoregressive SNN's output using a Bernoulli process, eliminating the need for floating-point computation and enabling implementation on neuromorphic devices. Moreover, instead of using the traditional KL divergence as a metric for distribution distance, FSVAE employs MMD (Maximum Mean Discrepancy), which is more suitable in the context of SNNs [3]. The Evidence Lower Bound (ELBO) of the SNN-based VAE is defined as follows:

$$
\begin{aligned}
\mathcal{L}_{ELBO}(\theta, \phi; x) = & - E_{z_{1:t} \sim q_\phi(z_{1:t}|x_{1:t})}[\log(p_\theta(x_{1:t}|z_{1:t}))] \\
& + MMD[q_\phi(z_{1:t}|x_{1:t})\|p(z_{1:t})],
\end{aligned}
\tag{2}
$$

where t denotes the time step, and we set $t = 10$ in our experiments. The prior, denoted as p, is a composition of SNN and can generate binary time series outputs. The encoder, $q_\phi(z_{1:t}|x_{1:t})$, is based on SNNs and responsible for encoding the input data. On the other hand, the decoder, $p_\theta(x_{1:t}|z_{1:t})$, is also SNN-based and responsible for decoding the latent variables to reconstruct the original data.

Additionally, we introduce a SNN-based classifier and use the SNN training method in [11] to implement it :

$$\mathcal{L}_c = ce(E(x_{1:t}), y_{1:t}), \tag{3}$$

where E represents the SNN classifier, and ce denotes the cross-entropy loss.

4 Methodology

4.1 Problem Setting

Our work focuses on the sequential learning of diverse data domains using SNNs without the need for task/domain information access. We define $D_i^T = \left\{x_j^T\right\}_{j=1}^{N_i^T}$ and $D_i^S = \left\{x_j^S\right\}_{j=1}^{N_i^S}$ to represent the testing set and training set of the i-th data domain, respectively. D_i^T consists of a series of samples $\left\{x_j^T\right\}_{j=1}^{N_i^T}$, where N_i^T is the number of samples in the testing set. Similarly, D_i^S consists of a series of samples $\left\{x_j^S\right\}_{j=1}^{N_i^S}$, where N_i^S denotes the number of samples in the training set. During the i-th training step (T_i), the model only has access to a data sample $b_{T_i}^n \in D_i^S$. In our experiments, we set the sample size to be 10. After completing all training steps, the model's performance is evaluated using the testing set D_i^T to assess its effectiveness.

Definition 1: Memory Buffer. We define $\mathcal{M}(mc, mn)$ as a memory buffer, where samples from a data stream are sequentially stored. mc represents the current number of samples in the buffer, and mn denotes the maximum capacity of the buffer. The distribution of the memory buffer is denoted as $P_{\mathcal{M}}$.

Definition 2: Dynamic Expansion Model. We define $C = c_1, ..., c_n$ as a dynamically expanding model, where each component c consists of a WGAN and a SNN module, including a classifier and a VAE based on SNNs. At time T_i, c_n represents the component being learned, while c_1 to c_{n-1} represent the frozen components in the model, which have already acquired knowledge from the memory buffer.

4.2 Overview of Our Model

Figure 1 presents the architecture diagram of our model. It comprises two main components: a referee module and a player module. The referee module is implemented using WGAN, while the player module consists of a SNN-based VAE and a SNN-based classifier. At a given training step T_i, assuming the AGT-SNN has already learned n components, the referee and player modules are trained on the memory buffer, while the remaining components are kept frozen. As the

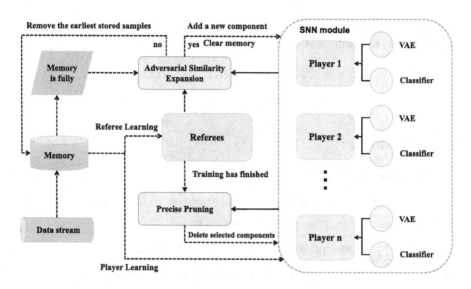

Fig. 1. The overview of the proposed AGT-SNN.

memory buffer fills up $(mc > mn)$, we employ the 'Adversarial Similarity Expansion' (ASE) method to determine whether component expansion is necessary. If expansion is not performed, we remove the earliest stored samples and continue training with the current component. Upon completion of training, we apply 'Precise Pruning' (PP) to compress the model while preserving its performance. In the following section, we provide detailed explanations of the proposed methods.

4.3 Adversarial Similarity Expansion

To ensure knowledge diversity among the components and promote learning from the data stream, we employ dynamic network expansion. Let C denote the current model that has learned n components. We define P as the knowledge distribution of the model C. The optimization function for enhancing knowledge diversity among the components can be formulated as follows:

$$P_C^\star = \arg\max_{i=1,\dots,t}\{\sum_{j=1}^{n-1}\{D(P_j, P_n^i)\}\}, \tag{4}$$

where i represents the current time step, n denotes the number of components, and D is a function that measures distribution distance. The objective of Eq. (4) is to find the optimal distribution $P_C^\star$ of the model, such that the distance between each previously learned distribution and itself is maximized, indicating that the model achieves an optimal state with maximum knowledge diversity. However, directly using Equation (4) is impractical in practice due to the unavailability of previous samples and the risk of catastrophic forgetting. To address this issue, we propose a novel expansion mechanism called Adversarial Similarity Expansion (ASE), which can be incorporated during the training process.

In the previous section, we described how the 'player' SNN model and the 'referee' WGAN model are trained simultaneously, both utilizing the memory buffer. When the memory buffer reaches its maximum capacity, we execute ASE. ASE leverages the knowledge acquired by the discriminator in the current WGAN model, which contains valuable information. The discriminator is employed to assess the disparities between the previously learned components and the current classification. Thus, we define ASE as follows:

$$\min or \max\{D(F_n(x_1), F_n(x_n)), \cdots, D(F_n(x_{n-1}), F_n(x_n))\}, \tag{5}$$

where F represents the feature embedding obtained from the current discriminator, and x is sampled from the VAE within the component. The choice of D depends on the desired measure, such as cosine similarity, mean squared error, or Manhattan distance. In this work, we adopt the cosine similarity as the measure D in Equation (5), resulting in the following reformulation:

$$CS = \max\{D_{cos}(F_n(x_1), F_n(x_n)), \cdots, D_{cos}(F_n(x_{n-1}), F_n(x_n))\}, \tag{6}$$

where CS represents the Component Similarity. Since it is evaluated on the low-dimensional latent space, the computation is efficient. And the D_{cos} can be defined as follows

$$D_{cos}(a, b) = \frac{\sum_i^{dim} a^i \cdot b^i}{\sqrt{\sum_i^{dim}(a^i)^2} \cdot \sqrt{\sum_i^{dim}(b^i)^2}}, \tag{7}$$

where dim is the dimension of the latent space and i represents the i-th dimensional value of a, the larger the cosine similarity, the higher the similarity between two features, so a larger output value from D_{cos} indicates a higher similarity between the two components, while a smaller value indicates greater dissimilarity. Thus, ASE can be redefined as

$$ASE = \begin{cases} \mathbb{S}, & CS \geq threshold \\ \\ \mathbb{A}, & CS < threshold \end{cases} \tag{8}$$

where $\mathbb{S}$ and $\mathbb{A}$ represent the two states obtained from ASE, indicating whether to continue the learning task with the current component or create a new component, respectively. The hyperparameter $threshold$ controls the expansion of the network, it can be set from -1 to 1, a larger threshold encourages the model to create more components during training, leading to performance improvement. Conversely, a smaller threshold results in fewer components, which may lead to performance degradation. The selection of the threshold will be demonstrated in the experiments.

4.4   Precise Pruning

While expanding components to ensure knowledge diversity can improve model performance, it also introduces a large number of parameters, leading to a heavier network. To address this, we propose a strategy called Precision Pruning (PP) to optimize the network. PP is executed after completing the model training, resulting in n components. Previously, Ye Fei et al. proposed a network compression method in TFCL [25], but it utilizes the computationally expensive Fréchet Inception Distance (FID) [12] as the basic algorithm, requiring a significant amount of computation during the training process, making it impractical for real-world applications. In PP, we leverage referee module to compare the similarity between each component. The referee module learns rich knowledge during the training process, making it effective for component comparison and more robust than existing methods. The knowledge difference between components is represented as a relationship matrix, and the loss function for PP is defined as follows:

$$\mathcal{L}_{PP} = D_{cos}(F_i(x_i), F_i(x_j)) \quad _{i,j=1,\cdots n} \quad _{i \neq j}, \tag{9}$$

where i and j represent the indices of the components, and we extract the feature embeddings of the discriminator in a similar manner to ASE.

By utilizing Equation (9), we can construct an asymmetric relationship matrix. From this matrix, we select $\max\{D_{cos}(F_i(x_i), F_i(x_j)) + D_{cos}(F_j(x_j), F_j(x_i))\}$ to determine i^* and j^*. Subsequently, we decide whether to remove the component i^* or j^* based on the following selection criteria:

$$\max\{\sum_{j=1}^{n} D_{cos}(F_{i^*}(x_{i^*}), F_{i^*}(x_j)), \sum_{i=1}^{n} D_{cos}(F_{j^*}(x_{j^*}), F_{j^*}(x_i))\}. \tag{10}$$

By comparing components i^* and j^* with all other components, we remove the one that exhibits a larger value, indicating a higher similarity. In our experiments, we set the number of components after PP to be 15.

We summarize the training of AGT-SNN into four steps as follows:

(1) **Buffer Update**: At time T_i, if the memory buffer is not full ($mc \leq mn$), we add new data samples $b_{T_i}^n$ from the data stream to the buffer and proceed to the second step. If the buffer is already full, we move to the third step.

(2) **Component Training**: Assuming that n components have been trained, at time T_i, we freeze the previously learned $n-1$ components and solely train c_n using the memory buffer. The referee module's learning is conducted using Equation (1), while the player module's learning is performed using Equations (2) and (3).

(3) **Expansion Check**: If the memory buffer is full ($mc > mn$), we consider executing the ASE algorithm to guide the network expansion of the model. This step yields two possible outcomes: 1) adding new components, clearing the memory area $\mathcal{M}(mc, mn)$, and then returning to Step 1; 2) removing the earliest stored samples and returning to Step 2.

(4) **Model Compression**: Once the training is completed, we employ the proposed PP algorithm to remove unnecessary components from the model until the number of components matches the predetermined number n.

5   Experimental Results

5.1   Dataset Setup

We conducted validation using three datasets to evaluate the performance of our model: Split MNIST, Split CIFAR10, and Split CIFAR100. In the Split MNIST dataset, we partitioned the original MNIST dataset, consisting of 70,000 images, into ten parts based on their respective categories. We used 60,000 images for training and 10,000 images for testing. Similarly, for the Split CIFAR10 dataset, we divided 6,000 images per category into different parts following the CIFAR-10 dataset division. As for the Split CIFAR100 dataset, we applied the same approach using the CIFAR-100 dataset and divided 600 images per category into different parts.

5.2   Classification Tasks Under TFCL

To facilitate fair comparisons, we adopted the same network architecture as [5] and implemented our VAE and classifier using SNN. For each of the three datasets (Split MNIST, Split CIFAR10, and Split CIFAR100), we set different buffer sizes and threshold values, specifically ([0.08, 1000], [0.09, 2000], [0.11, 3000]). During the testing phase, we use Equation (2) to evaluate the ELBO of each VAE. We identify the index of the component with the minimum ELBO and use it as the classifier for the test data. In the PP algorithm, we constrained the component size to be 15, and the results are presented in Table 1. Our model,

Table 1. The classification accuracy of various models on three datasets.

Methods	Split MNIST	Split CIFAR10	Split CIFAR100
finetune	19.75 ± 0.05	18.55 ± 0.34	3.53 ± 0.04
GEM [18]	93.25 ± 0.36	24.13 ± 2.46	11.12 ± 2.48
iCARL [22]	83.95 ± 0.21	37.32 ± 2.66	10.80 ± 0.37
reservoir [23]	92.16 ± 0.75	42.48 ± 3.04	19.57 ± 1.79
MIR [1]	93.20 ± 0.36	42.80 ± 2.22	20.00 ± 0.57
CoPE-CE [5]	91.77 ± 0.87	39.73 ± 2.26	18.33 ± 1.52
CoPE [5]	93.94 ± 0.20	48.92 ± 1.32	21.62 ± 0.69
CNDPM [16]	93.23 ± 0.09	45.21 ± 0.18	20.10 ± 0.12
DivSS [25]	96.95 ± 0.13	53.71 ± 0.19	26.03 ± 0.16
AGT-SNN	$\mathbf{98.26 \pm 0.24}$	$\mathbf{55.27 \pm 0.15}$	$\mathbf{27.42 \pm 0.18}$
AGT-SNN-PP	98.02 ± 0.34	54.10 ± 0.13	26.24 ± 0.22

AGT-SNN, exhibits remarkable competitiveness when compared to state-of-the-art models. Notably, even after applying the proposed PP method to compress the model, the observed performance degradation is minimal. This outcome can be attributed to the integration of a referee model, WGAN, with each player, enabling effective extraction of image features and accurate assessment of component differences.

5.3   The Impact of Different Thresholds and Buffer Sizes

In this experiment, we investigated the influence of varying thresholds and buffer sizes on the model's performance. We first fixed the buffer size to 1000 on the Split MNIST dataset and investigated the impact of different thresholds on the AGT-SNN model. The experimental results, depicted in Fig. 2, demonstrate the impact of not using PP (solid line) compared to using PP (dashed line), with the model compressed to 15 components, and it is evident that larger thresholds yield a greater number of generated components, resulting in improved model performance. Conversely, smaller thresholds lead to fewer components and a decline in performance. Additionally, we observed that the PP algorithm has only a small impact on performance when a large number of components are present. This indicates that we can achieve significant network size reduction while still maintaining model performance. From the experiment, it can be observed that using larger thresholds and applying the PP algorithm is a reasonable approach compared to using smaller thresholds directly. Figure 3 illustrates the results obtained by training Split MNIST with a fixed threshold of 0.08 and different buffer sizes. It shows that larger buffer sizes generally lead to better classification results.

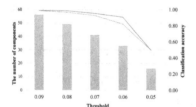

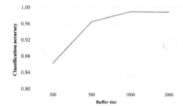

Fig. 2. The effect of varying threshold.   **Fig. 3.** The effect of varying buffer size.

5.4   ANN and SNN Under TFCL

We conducted an experiment to assess the efficiency and energy consumption advantages of event-driven SNNs. We performed a comparative analysis of the computational requirements between DivSS and AGT-SNN in classifying a single MNIST image. Both models were configured with the same network settings and subjected to the classification task. Table 2 presents the results obtained by traversing all components of the models. The findings revealed that AGT-SNN required approximately 6.2 times more addition operations than DivSS, but it required approximately 13.5 times fewer multiplication operations. It is noteworthy to acknowledge that in edge device, multiplication operations tend to incur greater time consumption compared to addition operations. Therefore, AGT-SNN exhibits a notable speed advantage over the ANN-based TFCL model, highlighting its efficiency in terms of computational demands.

Table 2. Comparison of the amount of computation required to classify a single image in MNIST

Model	Network	Addition	Multiplication
DivSS [25]	ANN	$\mathbf{1.5 \times 10^6}$	1.5×10^6
AGT-SNN	SNN	9.3×10^6	$\mathbf{1.1 \times 10^5}$

6   Conclusion

In this work, we introduce a framework called AGT-SNN, designed for TFCL using SNNs. Our framework leverages ANNs to facilitate network expansion and model compression in SNNs. By employing innovative network expansion and pruning strategies, we present a practical solution for dynamically scaling SNN-based models. Experimental results validate the effectiveness of our model in enabling SNNs to successfully carry out TFCL, thereby holding great potential for the future implementation of TFCL on edge devices. In future work, we will combine the latest research advances in SNN to explore the full use of SNN for TFCL.

References

1. Aljundi, R., et al.: Online continual learning with maximal interfered retrieval. In: Advances in Neural Information Processing Systems, vol. 32 (2019)
2. Aljundi, R., Kelchtermans, K., Tuytelaars, T.: Task-free continual learning. In: Proceedings of the IEEE/CVF Conference on Computer Vision and Pattern Recognition, pp. 11254–11263 (2019)
3. Arribas, D., Zhao, Y., Park, I.M.: Rescuing neural spike train models from bad MLE. In: Advances in Neural Information Processing Systems, pp. 2293–2303 (2020)
4. Davies, M., et al.: Loihi: a neuromorphic manycore processor with on-chip learning. IEEE Micro **38**(1), 82–99 (2018)
5. De Lange, M., Tuytelaars, T.: Continual prototype evolution: learning online from non-stationary data streams. In: Proceedings of the IEEE/CVF International Conference on Computer Vision, pp. 8250–8259 (2021)
6. Fan, W., Bouguila, N., Du, J.X., Liu, X.: Axially symmetric data clustering through Dirichlet process mixture models of Watson distributions. IEEE Trans. Neural Networks Learn. Syst. **30**(6), 1683–1694 (2019)

7. Fan, W., Yang, L., Bouguila, N.: Unsupervised grouped axial data modeling via hierarchical Bayesian nonparametric models with Watson distributions. IEEE Trans. Pattern Anal. Mach. Intell. **44**(12), 9654–9668 (2022)
8. French, R.M.: Catastrophic forgetting in connectionist networks. Trends Cogn. Sci. **3**(4), 128–135 (1999)
9. Goodfellow, I., et al.: Generative adversarial nets. In: Advances in Neural Information Processing Systems, pp. 2672–2680 (2014)
10. Gulrajani, I., Ahmed, F., Arjovsky, M., Dumoulin, V., Courville, A.C.: Improved training of Wasserstein GANs. In: Advances in Neural Information Processing Systems, pp. 5767–5777 (2017)
11. Hanle, Z., Yujie, W., Lei, D., Yifan, H., Guoqi, L.: Going deeper with directly-trained larger spiking neural networks. In: Proceedings of the AAAI Conference on Artificial Intelligence, pp. 11062–11070 (2021)
12. Heusel, M., Ramsauer, H., Unterthiner, T., Nessler, B., Hochreiter, S.: GANs trained by a two time-scale update rule converge to a local Nash equilibrium. In: Advances in Neural Information Processing Systems, pp. 6626–6637 (2017)
13. Kamata, H., Mukuta, Y., Harada, T.: Fully spiking variational autoencoder. In: Proceedings of the AAAI Conference on Artificial Intelligence, pp. 7059–7067 (2022)
14. Kingma, D.P., Welling, M.: Auto-encoding variational Bayes. In: International Conference on Learning Representations (2014)
15. Kotariya, V., Ganguly, U.: Spiking-GAN: a spiking generative adversarial network using time-to-first-spike coding. In: 2022 International Joint Conference on Neural Networks (IJCNN), pp. 1–7 (2022)
16. Lee, S., Ha, J., Zhang, D., Kim, G.: A neural Dirichlet process mixture model for task-free continual learning. arXiv preprint arXiv:2001.00689 (2020)
17. Li, Y.: Research and application of deep learning in image recognition. In: 2022 IEEE 2nd International Conference on Power, Electronics and Computer Applications (ICPECA), pp. 994–999. IEEE (2022)
18. Lopez-Paz, D., Ranzato, M.A.: Gradient episodic memory for continual learning. In: Advances in Neural Information Processing Systems, vol. 30 (2017)
19. Maass, W.: Networks of spiking neurons: the third generation of neural network models. Neural Netw. **10**(9), 1659–1671 (1997)
20. Otter, D.W., Medina, J.R., Kalita, J.K.: A survey of the usages of deep learning for natural language processing. IEEE Trans. Neural Networks Learn. Syst. **32**(2), 604–624 (2020)
21. Ramapuram, J., Gregorova, M., Kalousis, A.: Lifelong generative modeling. Neurocomputing **404**, 381–400 (2020)
22. Rebuffi, S.A., Kolesnikov, A., Sperl, G., Lampert, C.H.: iCaRL: incremental classifier and representation learning. In: Proceedings of the IEEE Conference on Computer Vision and Pattern Recognition, pp. 2001–2010 (2017)
23. Vitter, J.S.: Random sampling with a reservoir. ACM Trans. Math. Software (TOMS) **11**(1), 37–57 (1985)
24. Yamazaki, K., Vo-Ho, V.K., Bulsara, D., Le, N.: Spiking neural networks and their applications: a review. Brain Sci. **12**(7), 863 (2022)
25. Ye, F., Bors, A.G.: Lifelong compression mixture model via knowledge relationship graph. In: AAAI Conference on Artificial Intelligence. AAAI Press (2023)
26. Zhang, J., Fan, W., Liu, X.: Spiking generative networks in lifelong learning environment. In: Fujita, H., Wang, Y., Xiao, Y., Moonis, A. (eds.) IEA/AIE 2023. LNCS, vol. 13925, pp. 353–364. Springer, Cham (2023). https://doi.org/10.1007/978-3-031-36819-6_31

Multi-adversarial Adaptive Transformers for Joint Multi-agent Trajectory Prediction

Qihuang Chen[✉], Zhongwen Xiao, Zhen Zhang, and Yaonong Wang

Zhejiang Leapmotor Technology Co., LTD, HangZhou 310051, China
qihuang.chen@gmail.com

Abstract. Multi-agent trajectory prediction is of vital importance for autonomous driving and robotic systems, particularly in situations where frequent interaction happens. Existing methods essentially suppose that training and testing data are drawn from the identical distribution, while ignoring the potential domain discrepancy. This is not hold in practice and results in inevitable performance degradation. To alleviate the problem of domain discrepancy, we propose a novel multi-adversarial adaptive transformers framework, which jointly conducts multi-agent trajectory prediction and domain adaptation in a unified framework. Specifically, the framework consists of a simple but effective transformer-based encoder-decoder architecture and three domain adaptation components. The former generates multi-modal trajectories of multi-agents simultaneously, and the latter reduces the domain disparity from different aspects: the temporal aspect, the social aspect, and the contextual aspect. The three domain adaptation components are implemented by learning a domain classifier in an adversarial training manner, respectively. By this way, domain-invariant feature representations are learned and domain discrepancies will be better alleviated. Practical and challenging experiments are conducted cross multiple domains, and the results demonstrate the effectiveness of our proposed framework for robust multi-agent trajectory prediction.

Keywords: multi-agent trajectory prediction · domain adaptation · autonomous driving

1 Introduction

Multi-agent trajectory prediction aims to predict simultaneously the future trajectories of all agents from the given historical trajectories and the environmental information. It plays a major role in larger number of real world applications

---

Supplementary Information The online version contains supplementary material available at https://doi.org/10.1007/978-981-99-8543-2_19.

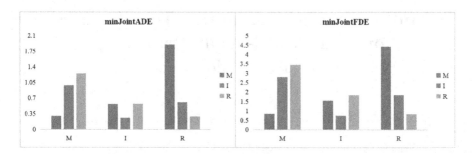

Fig. 1. The performance drops when it encounters a domain discrepancy. The evaluation metric is minJointADE and minJointFDE. Lower is better.

such as autonomous driving, robotic system, visual surveillance, and so on. Predicting the accurate future trajectory is still a challenging task, because of the uncertainty of the driver's behavior and the changes in the environment around it. Existing successful methods always take into account the history trajectory of agents and the intersections between these agents in the same scene. To consider such intersections, a social pooling layer is designed in [1] to encode the intersection of the history trajectories of all agent in the same scene. Following it, many methods [3,4,25] have been proposed for model the intersections via different mechanisms. Furthermore, the environmental information is also introduced into trajectory predictor [10,14], which is viewed as the prior knowledge to guide the behavior of the agent.

However, these existing methods typically assume that the training and testing data collected from the same domain, and ignore the potential domain discrepancies. We argue that this strategy leads to inevitable performance degradation and limits the application scope of the model, due to domain-bias and discrepancies. To illustrate the challenge of the problem, we apply a representative and state-of the-art method, AutoBots [7], which achieves the best results of multi-agent trajectory prediction task in INTERACTION challenge [24], to demonstrate the performance drop when it comes to different trajectory domains. As for three domains: merging (M), intersection (I) and roundabout (R), each domain consists of training dataset and validation dataset. We illustrate this challenge by training the model on the training dataset from one domain and evaluating the model on validation dataset from three others. Figure 1 show that the performance is decreased significantly when the training and validation dataset are from different domain.

In this work, we delve into the challenging domain discrepancy problem in multi-agent trajectory prediction, and propose a multi-adversarial adaptive transformers method, which jointly conducts multi-agent trajectory prediction and domain adaptation in a unified framework. Specifically, the proposed method reduces the domain discrepancy from three aspects: the temporal aspect, the social aspect, and the contextual aspect. Accordingly, these three components are designed and implemented by learning a domain classifier in adversarial training manner, respectively. The contributions of this paper can be summarized as follows:

(1) We delve into a more general and practical problem of domain discrepancy in multi-agent trajectory prediction, and propose a multi-adversarial adaptive transformers framework for jointly predicting future trajectories and adaptively learning domain-invariant features.

(2) We design three domain adaptation components to alleviate the domain discrepancy from temporal, social and contextual aspect, respectively, each of which is learned by a domain classifier in adversarial training manner.

(3) We conduct exhaustive experiments on cross diverse domains for joint multi-agent trajectory prediction to verify the effectiveness of our proposed method.

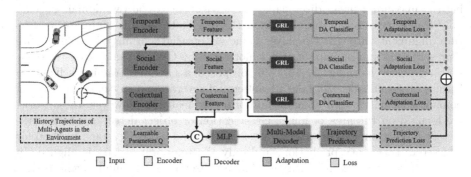

Fig. 2. Architecture Overview. Our model consists of three main components: 1) transformer based encoders, which extract the temporal, social and contextual features of multi-agents in a scene. 2) a transformer based decoder, which generates jointly multi-modal future trajectories of multi-agents. 3) multi-aspect domain adaptation modules, which explore domain-invariant feature from different aspects.

2   Related Works

2.1   Multi-agent Trajectory Prediction

Multi-agent trajectory prediction aims to predict the trajectories of all agents in the same scene simultaneously, which have flourished in recent years, due to the rise in interest in autonomous driving and robotic applications and the release of related datasets and benchmarks [24]. Generally, there are two main difficulties are considered in this task: 1) how to model the intersection among agents in a scene. 2) how to relieve the uncertainty of the results of trajectory prediction. Therefore, successful trajectory prediction models must learn the relationships between these agents in the environment. Further, the multimodal trajectories are predicted for relieving the uncertainty. [7] use the set transformer [11] to model the intersection between agents. [6] uses raster graph to model all the agents in the scene. Recent work has employed variations of self-attention and Transformers for modeling different axes: temporal trajectory encoding and decoding [8,22,23], encoding relationships between agents [2,13,18,19,23], and encoding relationships with road elements. However, these methods always assume the training data and testing data are from the identical domain, which is not hold in real application. In this work, we take inspiration from [7], and design a simpler model to deal with the domain discrepancy.

2.2 Domain Adaptation

Domain adaptation aims to bridge different domains or tasks by reducing the distribution discrepancy, which has a large number of approaches [5,15] to resolve the domain shift problem. [17] implemented the domain adaptation by minimizing the maximum mean discrepancy (MMD) between the two domain-specific fully connected branches of the CNN. Besides that, domain confusion for feature alignment through two-player game adversarial learning between feature representation and domain classifier motivated by GAN [9] was extensive studied in transfer learning [12,16]. [26] proposed a two-step training scheme to learn a target encode. [20] take advantages of several domain classifiers to learn domain informative and domain uninformative features. However, these works are focused on the area of image classification or object detection. As for trajectory prediction, [21] proposes a attention based adaptive learning method for pedestrian trajectory prediction, and only the temporal feature is aligned. In our work, we are dealing with the domain discrepancy in multi-agent trajectory prediction, and the feature alignment is perform from temporal, social and contextual aspect, respectively.

3   Proposed Method

The design principle of each module in our proposed method is as simple and effective as possible with a high degree of separation, which helps us to delve into the impact of domain discrepancies on model performance. The overall framework of the proposed method is illustrated in Fig. 2.

3.1   Encoder: Processing Multi-aspect Data

Robust trajectory prediction model must take into account the history trajectories of agents, and their environmental information. Moreover, the relationship between agents in the environment is also crucial to the performance of the model.

Temporal Encoder. The goal of temporal encoder is to process the current and prior state from all agents in a scene, which takes as an input a set of history state from all agents $H = \{H^1, \cdots, H^M\}$, where $H^i = \{h^i_{t-T_h}, \cdots, h^i_t\}$ denotes the temporal information from an individual agent i, M denotes the number of agents in a scene. To process the temporal information, a transformer encoder is applied to each agent, respectively.

Social Encoder. The goal of social encoder is to capture the interaction information among all agents in a scene, which takes as an input the result $\mathcal{H}$ output from temporal encoder. The social encoder processes information at each time step across agents.

Contextual Encoder. In this work, the static environmental information is used, that is, road structure information. The road is usually represented as sets of points, which can be denoted as $E' \in \mathbb{R}^{pn \times ra}$, where pn is the number of point, and ra is the number of attributes in each point (e.g. position, lane type). A multi-head self attention (MHSA) is used to encode the road information in the scene. And we perform the operation by casting E' to query, key and value matrices and adding a residual connection:

$$E = e_c(E') + MHSA\left(e_c(E'), e_c(E'), e_c(E')\right) \tag{1}$$

where $e_t(\cdot)$ is an embedding layer projecting E' to the size of the hidden state d, $E \in \mathbb{R}^{pn \times d}$ represents the environmental feature. In this aspect, the global environmental features are encoded.

3.2   Decoder: Generating Multi-modal Trajectories

The goal of decoder is to generate multi-modal trajectories of all agents simultaneously in a scene, which takes inputs as the social feature $\mathcal{S}$ and contextual feature $\mathcal{E}$. To generate $c \in \mathbb{N}$ different future trajectories for all agents in the same scene, inspired by AutoBots, the encoder employs c matrix of learnable parameters $Q_j \in \mathbb{R}^{T_f \times d}$s, where T_f is the prediction horizon and $j \in \{1, 2, \cdots, c\}$. Each learnable parameter matrix Q_j is then repeated M times to produce the input tensor $Q'_j \in \mathbb{R}^{T_f \times d \times M}$. To make the contextual information can be seen in the future time steps in c modes of all agents in a scene, we first reshape the contextual feature E to a one-dimensional vector $E'' \in \mathbb{R}^l$, where $l = pn \times d$. And then a embedding layer is applied to project E'' to $\mathcal{E}' \in \mathbb{R}^d$. We copy the vector along the M and T_f dimensions, to produce the tensor $\mathcal{E}_j \in \mathbb{R}^{T_f \times d \times M}$. Each Q'_j is concatenated with $\mathcal{E}_j$ along d dimension, and then processed by a MLP layer to produce a tensor $G_j \in \mathbb{R}^{T_f \times d \times M}$.

A multi-head attention block decoders (MABD), which were also introduced in, is used to process G_j:

$$O_j = MABD(\mathcal{H}'_j, G_j) \tag{2}$$

where $\mathcal{H}'_j$ is the output of social tensor. The final output of the decoder is a tensor $O \in \mathbb{R}^{c, M, T_f, d}$, which can then be processed a trajectory predictor $\varphi(\cdot)$, which is a MLP in this work, to generate c trajectories for each agents in a scene. In this work, $\varphi(\cdot)$ produces the parameters of a bivariate Gaussian distribution. The assumption is that a trajectory of a agent i coordinates $p_t^i = (x_t^i, y_t^i) \in \mathbb{R}^{\mathcal{K}}$ in time step t follows a bi-variate Gaussian distribution as $(x_t^i, y_t^i) \sim \mathcal{N}\left(\mu_t^i, \alpha_t^i, \rho_t^i\right)$, where $\mu_t^i = (\mu_x, \mu_y)_t^i$ is the mean, $\sigma_t^i = (\sigma_x, \sigma_y)_t^i$ is the standard deviation, and ρ_t^i is the correlation coefficient,

$$(\mu_t^i, \sigma_t^i, \rho_t^i)_j = \varphi(O) \tag{3}$$

where, $i \in \{1, 2, \cdots, M\}$, $t \in \{1, 2, \cdots, T_f\}$ and $j \in \{1, 2, \cdots, c\}$.

3.3 Adaptation: Learning Doamin Invaint Feature

To minimize the domain discrepancies, we propose the multi-adversarial domain adaptation module which is formulated by multi-adversarial domain classifier sub-modules from temporal, social and contextual aspect, respectively. The adversarial domain classifier aims to confuse the domain features, with min-max optimization between the domain classifiers and features from encoders. We consider multi-adversarial domain classifiers from several aspects instead of a single adversarial domain classifier, because the temporal, social and contextual features are all helpful to the performance of trajectory prediction.

Temporal Domain Adaptation Module. Given a sample $H \in \mathbb{R}^{(T_h+1) \times k \times M}$, from suorce domain or target domain, the temporal encoder outputs the temporal domain feature $\mathcal{H} = \{\mathcal{H}_1, \cdots, \mathcal{H}_M\} \in \mathbb{R}^{(T_h+1) \times d \times M}$. And the temporal adversarial classifier, which is denoted as D_t, is learned to predict the domain label of H. Following the the adversarial learning strategy, the minimax learning of the adversarial classifier can be written as:

$$\min_{\theta_t} \max_{W_t} L_t \tag{4}$$

where $\min_{\theta_t} \max_{W_t} L_t (D_t (T_e (H, W_t), \phi_t), g_t)$, in which L_t is the cross entropy loss, ϕ_t is parameter of temporal domain classifier, W_t is parameter of temporal encoder. g_t is the domain label of sample H, which is labeled as 1 for the source domain and 0 for the target domain. By adversarial learning of the domain classifier with back-propagated gradient reverse (i.e. GRL [5]), the temporal feature representations are characterized to be domain-invariant.

The goal of the temporal encoder is to model the historical states of a single agent in a scene. To efficiently train the temporal domain feature alignment module, we select a representative historical states in a scene using Max Pooling Layer along the number of agents M axis, denoted as $\mathcal{H}_T \in \mathbb{R}^{(T_h, d)}$. And then two fully connected layers with an ReLU activation function in between them is applied as the domain classifier.

Social Domain Adaptation Module. Given the temporal domain feature $\mathcal{H} = \{\mathcal{H}_1, \cdots, \mathcal{H}_M\} \in \mathbb{R}^{(T_h+1) \times d \times M}$, the social encoder outputs the social domain feature $\mathcal{S} = \{S_1, \cdots, S_{T_h+1}\} \in \mathbb{R}^{(T_h+1) \times d \times M}$. The social adversarial classifier, which is denoted as D_s, is learned to predict the domain label of the sample H. Following the the adversarial learning strategy, the minimax learning of the adversarial classifier can be written as:

$$\min_{\theta_s} \max_{W_s} L_s \tag{5}$$

where $\min_{\theta_s} \max_{W_s} L_t \left(D_s \left(S_e \left(\mathcal{H}, W_s \right), \phi_s \right), g_s \right)$, in which L_s is the cross entropy loss, ϕ_s is parameter of temporal domain classifier, W_s is parameter of temporal encoder. g_s is the domain label of sample H, which is labeled as 1 for the source domain and 0 for the target domain. By adversarial learning of the domain classifier with back-propagated gradient reverse (i.e. GRL), the social feature representations are characterized to be domain-invariant.

Different from the temporal encoder, which model information for a single agent, the social models the global interaction information among all agents in a scene. To efficiently train the social encoder, a 1×1 convolution layer is used to aggregate the features of all agents along M axis. And then two fully connected layers with an ReLU activation function in between them is applied as the domain classifier.

Contextual Domain Adaptation Module. Given the environmental information $E^{'} \in \mathbb{R}^{pn \times ra}$ in a scene, from suorce domain or target domain, the contextual encoder outputs the contextual domain feature $E \in \mathbb{R}^{pn \times d}$. The contextual adversarial classifier, which is denoted as D_c, is learned to predict the domain label of the sample $E^{'}$. Following the adversarial learning strategy, the minimax learning of the adversarial classifier can be written as:

$$\min_{\theta_c} \max_{W_c} L_c \tag{6}$$

where $\min_{\theta_c} \max_{W_c} L_t \left(D_c \left(C_e \left(E^{'}, W_c \right), \phi_c \right), g_c \right)$, in which L_c is the cross entropy loss, ϕ_c is parameter of temporal domain classifier, W_c is parameter of temporal encoder. g_c is the domain label of sample $E^{'}$, which is labeled as 1 for the source domain and 0 for the target domain. By adversarial learning of the domain classifier with back-propagated gradient reverse (i.e. GRL), the contextual feature representations are characterized to be domain-invariant. A 1×1 convolution layer and two fully connected layers with an ReLU activation function in between them are applied as the contextual domain classifier.

3.4   Loss Function

The overall objective function consists of three terms, the multi-agent trajectory prediction loss L_{pred} for predicting future trajectory, the adversarial losses L_{src} from source domain and the adversarial losses L_{tgt} from target domains. The prediction loss is the negative log-likelihood as,

$$L_{pred} = -\sum_{t=1}^{T_f} log \left(\mathbb{P}((x_t^i, y_t^i)|(\mu_t^i, \sigma_t^i, \rho_t^i)) \right) \tag{7}$$

Note that only samples from source trajectory domain participate in the prediction phase. The whole model is trained by jointly optimizing the prediction loss L_{pred} and the alignment loss,

$$L = L_{pred} + \gamma_1 L_{src} + \gamma_2 L_{tgt} \tag{8}$$

where $L_{src} = (L_t + L_s + L_c)_{src}$, which is computed from source domain samples. $L_{tgt} = (L_t + L_s + L_c)_{tgt}$, which is computed from target domain samples. γ_1 and γ_2 are hyper-parameters for balancing the loss functions.

Table 1. minJointADE results of our model in comparison representative state-of-the-art method, AutoBots. "2" represents from source domain to target domain. M, I and R denote merging, intersection and roundabout scenario, respectively.

Method	M2R	M2I	R2M	R2I	I2M	I2R	Ave
AutoBots	1.254	0.982	1.901	0.612	0.569	0.577	0.983
ours	1.040	0.763	1.712	0.461	0.453	0.526	0.825

Table 2. minJointFDE results of our model in comparison representative state-of-the-art method, AutoBots. "2" represents from source domain to target domain. M, I and R denote merging, intersection and roundabout scenario, respectively.

Method	M2R	M2I	R2M	R2I	I2M	I2R	Ave
AutoBots	3.449	2.816	4.418	1.832	1.539	1.850	2.651
ours	3.133	2.677	4.397	1.611	1.332	1.717	2.477

4 Experiments

4.1 Dataset

Experiments are conducted on Interaction dataset [24]. It is a large-scale real-world dataset which consists of diverse scenarios. The dataset is challenging as it includes interactions between vehicles, different environments, and potentially multiple plausible predictions. The dataset can be divided into three trajectory domains based on scenario differences, merging (M), intersection (I), roundabout (R). Each of scenarios has a training dataset and a validate dataset.

4.2 Problem Setting

In this work, we explore the influence of domain disparity on multi-agent trajectory prediction. For this, a more challenging and practical problem setting is introduced that a model is trained on one domain and tested on other domains, respectively. Given three challenging trajectory domains, we have total 6 multi-agent trajectory prediction tasks: M2R, M2I, I2M, I2R, R2M, R2I, where M, I, R represents merging domain, intersection domain and roundabout domain, respectively. Each domain has a training dataset and a validation dataset, and there is no overlap sample between them. To ensure the fair comparison under the new setting, the representative existing method compared with our proposed model is trained with training dataset from source domain as well as the validation dataset of the target domain. And our proposed model only has access to the observed trajectory from the validation set.

4.3   Evaluation Metrics

The following two multi-agent metrics are used to for performance evaluation. Minimum Joint Average Displacement Error (minJointADE), which represents the minimum value of the euclidean distance averaged by time and all agents between the ground truth and modality with the lowest value. The minJointADE of a single case is calculated as:

Table 3. The minJointADE of different variants of our proposed model.

contextual	temporal	social	M2R	M2I	R2M	R2I	I2M	I2R	Ave
×	×	×	1.239	0.951	1.922	0.611	0.593	0.629	0.991
×	×	√	1.341	1.212	2.335	0.891	0.693	0.755	1.205
×	√	×	1.299	1.117	2.109	0.798	0.652	0.703	1.113
×	√	√	1.443	1.512	2.401	0.915	0.773	0.806	1.308
√	×	×	1.145	0.833	1.801	0.571	0.601	0.589	0.923
√	×	√	1.097	0.779	1.754	0.512	0.554	0.536	0.872
√	√	×	1.096	0.791	1.736	0.497	0.491	0.530	0.857
√	√	√	1.040	0.763	1.712	0.461	0.453	0.526	0.826

Table 4. The minJointFDE of different variants of our proposed model.

contextual	temporal	social	M2R	M2I	R2M	R2I	I2M	I2R	Ave
×	×	×	3.631	3.002	4.701	2.081	1.655	2.050	2.853
×	×	√	3.901	3.450	5.011	2.335	1.776	2.434	3.151
×	√	×	3.777	3.339	4.803	2.403	1.709	2.576	3.101
×	√	√	4.054	3.786	5.324	2.786	1.899	2.778	3.438
√	×	×	3.553	2.749	4.507	1.871	1.593	1.893	2.694
√	×	√	3.301	2.708	4.443	1.739	1.398	1.776	2.567
√	√	×	3.199	2.698	4.404	1.684	1.465	1.805	2.543
√	√	√	3.133	2.677	4.397	1.611	1.332	1.717	2.477

$$minJointADE = \min_{c \in \{1,\cdots,C\}} \frac{1}{MT} \sum_{m,t} \sqrt{\left(x'_{m,t} - x^c_{m,t}\right)^2 + \left(y'_{m,t} - y^c_{m,t}\right)^2} \qquad (9)$$

where M is the number of agents to be predicted in this case, T is the number of predicted timestamps, C is the number of modalities in this challenge, x' and y' means the ground truth. The final value is averaged over all cases. Minimum Joint Final Displacement Error (minJointFDE), which represents the minimum value of the euclidean distance at the last predicted timestamps averaged by all agents between the ground truth and modality with the lowest value. The minJointFDE of a single case is calculated as:

$$minJointFDE = \min_{c \in \{1, \cdots, C\}} \frac{1}{M} \sum_m \sqrt{\left(x'_{m,T} - x^c_{m,T}\right)^2 + \left(y'_{m,T} - y^c_{m,T}\right)^2} \quad (10)$$

where M is the number of agents to be predicted in this case, T is the number of predicted timestamps, C is the number of modalities in this challenge, x' and y' means the ground truth. The final value is averaged over all cases.

4.4 Implementation Details

We use a 1 s of historical trajectory and a 3 s prediction horizon. We down-sample the sampling rate of 10 Hz to 5 Hz to reduce the model complexity, that is $T_h = 5$ and $T_f = 15$. The hidden state d in the model is always set to 128. We use Adam with learning rate 0.001 for 250 epochs. The batch size is 64. γ_1 and γ_2 is all set to 1. The number of predicted modes is 6. The model is implemented using PyTorch. Note that the number of agents M is not fixed in a scene. For the convenience of training and testing, we will set a maximum number of agents and eliminate the interference of invalid agent information by means of mask.

4.5 Quantitative Analysis

Compared with the Representative Method. AutoBots achieves the best results in multi-agent trajectory prediction task of INTERACTION challenge, and is also far better than the method in the second place. Table 1 and Table 2 show that our proposed model consistently outperforms AutoBots. It validates that our model has the ability to learn transferable knowledge from source to target domain and alleviate the domain discrepancy. In addition, for tasks M2R and R2M, all the models have relatively higher minJointADE and minJointFDE. One possible reason is that domain M (merging) and R (roundabout) have a bigger difference environmental information, in which agents may have bigger different moving pattern. In contrast, for tasks I2R and R2I, all the models have relatively smaller minJointADE and minJointFDE. One possible reason is that domain I (intersection) and R (roundabout) have the similar environmental information, in which agents may have similar moving pattern. This phenomenon further illustrates the importance of considering the domain discrepancy problem in trajectory prediction task.

4.6 Ablation Study

Effectiveness of Multi-aspect Domain Adaptation Modules. Temporal, social and contextual domain adaptation modules are designed in the proposed method. And we verify the effectiveness of each component by removing it in the proposed model. As shown in Table 3 and Table 4, when the contextual feature is not aligned, the minJointADE and minJointFDE drops sharply even the temporal and social features are aligned, which indicates that contextual information restrains the behavior of agents in the environment, and the effectiveness of contextual domain adaptation component. When the temporal, social and contextual features are all aligned, the performance of our model exceeds AutoBots, even though our model architecture is simpler, which further reveals the importance of the domain adaptation components we designed.

5 Conclusion

In this work, we delved into the domain discrepancies in the multi-agent trajectory prediction task. Specifically, a more general, practical yet challenging trajectory prediction problem setting was proposed. Then we proposed a unified model which performed multi-agent trajectory prediction and domain adaptation simultaneously. We relieved the domain discrepancy from three aspects: temporal, social and contextual aspects. Extensive experiments proved the superiority of our model in both multi-agent trajectory prediction and trajectory domain discrepancy alleviation.

References

1. Alahi, A., Goel, K., Ramanathan, V., Robicquet, A., Fei-Fei, L., Savarese, S.: Social LSTM: human trajectory prediction in crowded spaces. In: Proceedings of the IEEE Conference on Computer Vision and Pattern Recognition, pp. 961–971 (2016)
2. Bhat, M., Francis, J., Oh, J.: Trajformer: Trajectory prediction with local self-attentive contexts for autonomous driving. arXiv preprint arXiv:2011.14910 (2020)
3. Chen, Q., Li, B., Xiao, Z., Zhang, Z., Wen, S., Wang, Y.: Strip and spatial social pooling for trajectory prediction. In: Kountchev, R., Nakamatsu, K., Wang, W., Kountcheva, R. (eds.) WCI3DT 2022. LNCS, vol. 323, pp. 183–195. Springer, Cham (2023). https://doi.org/10.1007/978-981-19-7184-6_15
4. Deo, N., Trivedi, M.M.: Convolutional social pooling for vehicle trajectory prediction. In: Proceedings of the IEEE Conference on Computer Vision and Pattern Recognition Workshops, pp. 1468–1476 (2018)
5. Ganin, Y., Lempitsky, V.: Unsupervised domain adaptation by backpropagation. In: International Conference on Machine Learning, pp. 1180–1189. PMLR (2015)
6. Gilles, T., Sabatini, S., Tsishkou, D., Stanciulescu, B., Moutarde, F.: Thomas: trajectory heatmap output with learned multi-agent sampling. arXiv preprint arXiv:2110.06607 (2021)
7. Girgis, R., et al.: Autobots: latent variable sequential set transformers. arXiv preprint arXiv:2104.00563 (2021)
8. Giuliari, F., Hasan, I., Cristani, M., Galasso, F.: Transformer networks for trajectory forecasting. In: 2020 25th International Conference on Pattern Recognition (ICPR), pp. 10335–10342. IEEE (2021)
9. Goodfellow, I., et al.: Generative adversarial networks. Commun. ACM **63**(11), 139–144 (2020)
10. Gu, J., Sun, C., Zhao, H.: DensetNT: end-to-end trajectory prediction from dense goal sets. In: Proceedings of the IEEE/CVF International Conference on Computer Vision, pp. 15303–15312 (2021)

11. Lee, J., Lee, Y., Kim, J., Kosiorek, A., Choi, S., Teh, Y.W.: Set transformer: a framework for attention-based permutation-invariant neural networks. In: International Conference on Machine Learning, pp. 3744–3753. PMLR (2019)

12. Li, C.L., Chang, W.C., Cheng, Y., Yang, Y., Póczos, B.: MMD GAN: towards deeper understanding of moment matching network. In: Advances in Neural Information Processing Systems, vol. 30 (2017)

13. Li, L.L., et al.: End-to-end contextual perception and prediction with interaction transformer. In: 2020 IEEE/RSJ International Conference on Intelligent Robots and Systems (IROS), pp. 5784–5791. IEEE (2020)

14. Liang, M., et al.: Learning lane graph representations for motion forecasting. In: Vedaldi, A., Bischof, H., Brox, T., Frahm, J.-M. (eds.) ECCV 2020. LNCS, vol. 12347, pp. 541–556. Springer, Cham (2020). https://doi.org/10.1007/978-3-030-58536-5_32

15. Long, M., Cao, Y., Wang, J., Jordan, M.: Learning transferable features with deep adaptation networks. In: International Conference on Machine Learning, pp. 97–105. PMLR (2015)

16. Long, M., Cao, Z., Wang, J., Jordan, M.I.: Conditional adversarial domain adaptation. In: Advances in Neural Information Processing Systems, vol. 31 (2018)

17. Long, M., Zhu, H., Wang, J., Jordan, M.I.: Unsupervised domain adaptation with residual transfer networks. In: Advances in Neural Information Processing Systems, vol. 29 (2016)

18. Mercat, J., Gilles, T., El Zoghby, N., Sandou, G., Beauvois, D., Gil, G.P.: Multi-head attention for multi-modal joint vehicle motion forecasting. In: 2020 IEEE International Conference on Robotics and Automation (ICRA), pp. 9638–9644. IEEE (2020)

19. Park, S.H., et al.: Diverse and admissible trajectory forecasting through multimodal context understanding. In: Vedaldi, A., Bischof, H., Brox, T., Frahm, J.-M. (eds.) ECCV 2020. LNCS, vol. 12356, pp. 282–298. Springer, Cham (2020). https://doi.org/10.1007/978-3-030-58621-8_17

20. Tzeng, E., Hoffman, J., Saenko, K., Darrell, T.: Adversarial discriminative domain adaptation. In: Proceedings of the IEEE Conference on Computer Vision and Pattern Recognition, pp. 7167–7176 (2017)

21. Xu, Y., Wang, L., Wang, Y., Fu, Y.: Adaptive trajectory prediction via transferable GNN. In: Proceedings of the IEEE/CVF Conference on Computer Vision and Pattern Recognition, pp. 6520–6531 (2022)

22. Yu, C., Ma, X., Ren, J., Zhao, H., Yi, S.: Spatio-temporal graph transformer networks for pedestrian trajectory prediction. In: Vedaldi, A., Bischof, H., Brox, T., Frahm, J.-M. (eds.) ECCV 2020. LNCS, vol. 12357, pp. 507–523. Springer, Cham (2020). https://doi.org/10.1007/978-3-030-58610-2_30

23. Yuan, Y., Weng, X., Ou, Y., Kitani, K.M.: AgentFormer: agent-aware transformers for socio-temporal multi-agent forecasting. In: Proceedings of the IEEE/CVF International Conference on Computer Vision, pp. 9813–9823 (2021)

24. Zhan, W., et al.: Interaction dataset: an international, adversarial and cooperative motion dataset in interactive driving scenarios with semantic maps. arXiv preprint arXiv:1910.03088 (2019)

25. Zhang, P., Ouyang, W., Zhang, P., Xue, J., Zheng, N.: SR-LSTM: state refinement for LSTM towards pedestrian trajectory prediction. In: Proceedings of the IEEE/CVF Conference on Computer Vision and Pattern Recognition, pp. 12085–12094 (2019)
26. Zhang, W., Ouyang, W., Li, W., Xu, D.: Collaborative and adversarial network for unsupervised domain adaptation. In: Proceedings of the IEEE Conference on Computer Vision and Pattern Recognition, pp. 3801–3809 (2018)

Enhancing Open-Set Object Detection via Uncertainty-Boxes Identification

Wei Ji[1], Dongqin Wu[1], Rui-Wei Zhao[2], Weijia Fu[3], Yingwen Wang[3], Yuejie Zhang[1(✉)], Rui Feng[1,2,3], and Xiaobo Zhang[3(✉)]

[1] School of Computer Science, Shanghai Key Laboratory of Intelligent Information Processing, Fudan University, Shanghai, China
{jiw20,yjzhang,fengrui}@fudan.edu.cn
[2] Academy for Engineering and Technology, Fudan University, Shanghai, China
rwzhao@fudan.edu.cn
[3] Children's Hospital, Fudan University, Shanghai, China
xiaobozhang@fudan.edu.cn

Abstract. Open-set object detection is a challenging task in computer vision, which aims to detect known object categories while simultaneously identifying unknown objects. Inspired by how humans naturally distinguish unseen objects by comparing their similarities and dissimilarities with known objects, we propose a novel network called UNBDet (Uncertainty-Boxes Detection) for enhancing open-set object detection. Our approach, UNBDet, utilizes a Pseudo Proposal Advisor (PPA) to generate a wide range of unknown object candidates to improve learning and make the distribution of pseudo unknowns more consistent with the actual unknown object distribution. Furthermore, we employ an Unknown Probability Estimator (UPE) and Uncertainty-NMS modules to reason the predicted overlapping and uncertainty-boxes from multiple known classes, thus enabling easy identification of unknown objects with high uncertainty. Our experimental results demonstrate that UNBDet significantly outperforms state-of-the-art models in open-set object detection.

Keywords: Open-set object detection · Pseudo Proposal Advisor · Unknown Probability Estimator · Uncertainty-NMS

1 Introduction

As deep learning techniques have advanced, close-set object detection methods have seen remarkable performance in recent years [14,16,17]. However, these methods rely on the assumption that the object categories present in the testing phase must also be included in the training phase. This leads to a drop in

Supplementary Information The online version contains supplementary material available at https://doi.org/10.1007/978-981-99-8543-2_20.

detection performance when attempting to identify and locate unknown objects. In contrast, the open-set object detection (OSOD) paradigm requires the model to detect both a set of known objects and find unknown objects in each training episode.

Recognizing unknown objects is a major challenge for conventional detectors [1,3,7,18,21]. As shown in Fig. 1, a detector trained on PASCAL VOC [6] such as Faster R-CNN [17], can accurately locate and classify known objects using a few candidate boxes with the same predicted class labels. However, it struggles to identify unknown objects (such as a zebra or a kite) and instead classifies them as known objects (like a cat, bird or sheep) with multiple overlapping boxes from various classes. We refer to these overlapping boxes with high-IoU neighboring boxes as *uncertainty-boxes*. This phenomenon reflects the uncertainty of the close-set training model when predicting unknown objects. This uncertainty is similar to the one mentioned by Zheng et al. [20], which they attribute to the different sizes of receptive fields of Feature Pyramid Networks (FPN) [13]. With this inspiration, our approach aims to identify unknown objects by re-estimating the unknown probabilities of the uncertainty-boxes.

Fig. 1. An illustration of prediction results for a conventional object detector (Faster R-CNN) trained on VOC dataset. It always incorrectly detects unknown objects (*giraffe*, *kite*) into multiple known classes boxes (*cat*, *bird*, *train*), due to their predicted categories are uncertain, we called it **uncertainty-boxes**. The uncertainty-boxes tend to have known probabilities of low confidence and adjacent boxes of high IoU, and can be seen as unknown instances.

OSOD models are tasked with identifying and locating objects that were not seen during training. This problem can be considered as an extension of open-set recognition (OSR) [9,18] to the object detection domain. While OSR has been widely studied in recent years [2,11], only a few works have attempted to address the challenging problem of OSOD. For example, Miller et al. [15] proposed using dropout samples to extract better label uncertainty information. Joseph et al. [10] employed auto-labeling of unknown objects and an energy-based unknown identifier for OSOD. Han et al. [9] utilized an Unknown Probability Learner to divide low-density latent regions around the clusters of known classes. Overall, these works employ unique unknown identifiers to separate objects, but the relationship between known and unknown objects is rarely considered.

Conventional detectors tend to predict multiple different categories of uncertainty-boxes for unknown objects, as shown in Fig. 1. This is because the model is too confident in predicting known classes and uncertain in identifying unknown objects. In order to address this issue, we propose a novel UNBDet network. Our network includes a Comprehensive Pseudo Proposal Advisor (PPA), Unknown Probability Estimator (UPE), and Uncertainty-NMS module. The PPA generates many pseudo labels that align with the natural distribution of unknown objects for the unknown objects learning. The UPE re-calculates the unknown probability for each proposal through a superficial confidence estimation branch in order to distinguish knowns and unknowns. Finally, the Uncertainty-NMS processes the network output by considering the known probability, unknown probability, and IoU of different boxes in order to detect more unknown boxes.

Our proposed UNBDet network provides a new approach for identifying unknown objects in OSOD. To demonstrate the effectiveness of UNBDet, we conducted experiments using the PASCAL VOC [6] dataset for close-set training, and several open-set settings that involve both VOC and COCO [13]. The main contributions of our work are:

- We propose a novel UNBDet network for OSOD. To the best of our knowledge, this is the first time mining unknown objects using uncertainty-boxes has been proposed. The proposed Uncertainty-NMS module comprehensively considers known probability, unknown probability and IoU for predicted boxes.
- We introduce a comprehensive pseudo proposal advisor that generates pseudo labels for unknown object learning. Compared to previous pseudo methods, the distribution of pseudo labels generated by PPA is more consistent with the distribution of unknown objects in the real world.
- We design an additional unknown-aware probability branch to estimate the unknown confidence. This is not only beneficial for decision boundaries between known and unknown objects in high-dimensional space, but also provides new unknown probabilities for uncertainty-boxes, avoiding the detection of known-boxes as unknowns.

2   Related Work

Open-Set Object Detection. Open-set object detection (OSOD) is an extension of open-set recognition. Dhamija et al. [5] first formalized object detection as an open-set problem and benchmarked several representative detectors by their classifiers. The state-of-the-art two-stage multi-class detector, Faster R-CNN [17], performs better than one-vs-rest [14] and objectness-based classifiers [16] in handling unknown objects. Miller et al. [15] extended the concept of Dropout Sampling to OSOD for the first time. They estimated uncertainty in object detection, thus reducing open-set errors. Joseph et al. [10] proposed an unknown-aware proposals network and energy-based unknown identification to address the challenges of open-world detection. Han et al. [9] presented OpenDet with a Contrastive Feature Learner and an Unknown Probability Learner, which expands low-density regions from two folds. Zhao et al. [19] proposed a simple yet effective open-world object detection framework, including an auxiliary proposal advisor and a class-specific expelling classifier, which could assist RPN in identifying unknown proposals and expelling confusing predictions for each known class. Unlike these approaches, our UNBDet employs PPA to generate more quality pseudo labels. Based on the uncertainty prediction results predicted by close-set training, we re-estimate the unknown probabilities for uncertainty-boxes. These boxes can be seen as an expression of the uncertainty of an object on different FPN features.

Pseudo Label Learning. Pseudo labels are a simple and efficient semi-supervised learning method for deep neural networks [12]. In the context of OSOD, the OWOD approach of ORE [10] selects proposals with high objectness scores and those that do not overlap with ground-truth known instances as pseudo-unknowns. OW-DETR [8] introduces a bottom-up attention-driven pseudo-labeling scheme that is better generalized and can be applied to single-stage object detectors. Zhao et al. [19] deploys an auxiliary nonparametric Proposal ADvisor module (PAD) to assist the RPN in identifying ambiguous proposals. However, these pseudo-label generation methods will inevitably introduce a lot of non-object noise. Many background proposals are regarded as unknown, which hurts the precision of known and unknown objects. Our method, as shown in Fig. 2, generates pseudo labels of unknown objects from known instances. These pseudo labels naturally belong to the object rather than the background, which will not cause confusion between the semantics of the object and the background.

Uncertainty Estimation. A neural network typically comprises many parameters and non-linear activation functions, making it challenging to estimate the posterior distribution of network predictions. These networks tend to produce over-confident predictions on known categories, and often lead to uncertain predictions on unknown objects. Therefore, it is crucial to estimate the uncertainty of model predictions for real-world applications. Currently, uncertainty estimation can be divided into sampling-based and sampling-free methods. Sampling-based methods involve integrating multiple runs or multiple models, but these methods are not suitable for fast object detection. Sampling-free methods, on the other hand, learn additional confidence values to estimate uncertainty [9]. Our model belongs to the sampling-free method and, unlike other sampling-free methods, we use the uncertainty of prediction results of uncertainty-boxes that belong to the same object, which expands the scope of uncertainty estimation.

3   Methodology

This paper proposes a model called UNBDet for OSOD, where images are divided into *known* and *unknown* categories. The model is trained on the *known* categories and is able to classify and localize objects from both *known* and *unknown* categories. The UNBDet model contains a base version of Faster R-CNN, as well as additional components such as a pseudo proposal advisor, a multi-task detection head, and an Uncertainty-NMS post-processing module. Additionally, a known-aware contrastive head is used during training to improve the class boundary of the latent space (see Fig. 3).

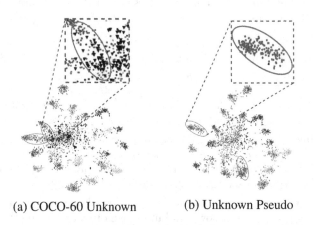

(a) COCO-60 Unknown   (b) Unknown Pseudo

Fig. 2. t-SNE visualization of instance features generated by ROIAlign (pre-trained on 20 VOC categories). In (a), we consider 20 VOC categories as known classes (colored dots) and 60 non-VOC categories in COCO as unknown classes (black triangles). The distribution of unknown objects is divergent and mixed with some known objects. In (b), we visualize the distribution of pseudo labels (red triangles) and VOC categories. The pseudo unknown objects generated by our PPA have a coincident distribution with the actual unknown objects. This indicates that our PPA is able to generate pseudo labels that are similar to the true distribution of unknown objects. (Color figure online)

3.1 Preliminary

Our OSOD problem can be formally described as follows. Let $D = \{(x, y), x \in X, y \in Y\}$ denote an object detection dataset, where x is input image and $y = \{(c_i, b_i)\}_{i=1}^{N}$ denotes a set of objects with corresponding class label c and bounding box b. Let $y = \{(c_u, b_i)\}$ specially denote a set of objects with unknown labels c_u and bounding boxes. OSOD requires training the detectors on a training set D_{train} with K known classes $C_K = \{1, ..., K\}$, and testing on a set D_{test} with objects from both known classes C_k and unknown classes C_u. The goal of OSOD is not only to detect all known objects (objects $\in C_u$), but also to identify the unknown objects (objects $\in C_K$), preventing them from being misclassified into incorrect known classes. In the real world, the number of unknown categories is countless. Since it is impossible to list unlimited unknown classes, so we denote all unknown classes with $C_u = K + 1$.

3.2 Baseline Setup

Faster R-CNN [17] is a classic two-stage object detection network consisting of an FPN-based backbone, RPN, and R-CNN. R-CNN includes a share fully connected (FC) layer and two separate FC layers for object classification and box regression. Without loss of generality, we set Faster R-CNN as our backbone and augment it to achieve OSOD in the following three folds. (a) The shared FC layer in R-CNN is replaced with two parallel FC layers, so that the module applied to the classification branch will not affect the regression task. (b) A cosine similarity-based classifier is used to alleviate the over-confidence issue mentioned by [3]. Specifically, we adopt scaled cosine similarity scores $s_{i,j} = \frac{\alpha \mathcal{F}(x)_i^T w_j}{\|\mathcal{F}(x)_i\|\|w_j\|}$ as

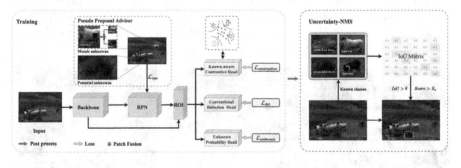

Fig. 3. Overview of our proposed UNBDet. UNBDet is a two-stage detector that consists of three main components: (1) a pseudo proposal advisor (PPA) that generates pseudo labels that align with the natural distribution of unknown objects, (2) an unknown probability estimator branch (UPE) that provides extra unknown probabilities for uncertainty-boxes, and (3) an Uncertainty-NMS module that identifies unknown objects by combining known scores, unknown scores, and Intersection over Union (IoU) of different boxes. These three modules work together to enhance the ability of OSOD to identify and locate both known and unknown objects in an image.

output logits, where $s_{i,j}$ denotes the similarity score between the i-th proposal feature $\mathcal{F}(x)_i$ and weight vector of class j, α is the scaling factor. (c) The box regression branch is set to class-agnostic and outputs a vector of length 4.

3.3   Pseudo Proposal Advisor

Our UNBDet model takes a different approach than other methods for OSOD that only mine unknown objects from the background [10] or use GAN [7] to generate unknown objects. Instead, UNBDet uses a pseudo proposal advisor to synthesize informative unknown object patterns from known objects. As shown in Fig. 2(a), the feature distributions of known class objects are grouped in colored dots. The unknown instances (in black triangles) are distributed in the low-density areas or are somehow overlapped with the known objects, which indicates that some unknown objects are similar to known ones, and some are not. Thus, the pseudo labels are divided into known-similar and known-dissimilar ones. Let P_{sim} and P_{dis} denote the known-similar unknown objects and the known-dissimilar unknowns, respectively. For P_{sim}, the approach of Auto-labeling Unknowns of ORE [10] is implemented. For P_{dis}, new unknown pseudo labels are generated by sampling two instances of different categories.

Inspired by these image-level augmentation methods, Fig. 4(a) Grid mask [4] and (b) mosaic augmentation of images, we propose instance-level mosaic augmentation to generate pseudo labels. Specially, an instance is fixed and another instance from a different category is cropped out. Multiple grid masks are randomly cropped from the second instance and pasted onto the first instance, and the resulting new instance is regarded as an unknown object. The unknown pseudo label is then pasted onto the original image. The distribution of the generated P_{dis} pseudo labels, as shown in Fig. 2(b), is found to be relatively similar to that of the real unknown instances. During training, the PPA module generates unknown pseudo labels with a probability of ρ, and too large ρ can decrease the precision of known objects as it will interfere with the recognition of known classes.

(a)   (b)   (c)

Fig. 4. An illustration of the evolution process of mosaic instance in our pseudo proposal advisor. (a) Grid mask. (b) Mosaic augmentation. (c) Mosaic instance, pseudo label generated using sheep instance and dog instance.

3.4 Uncertainty-Box Detection

The PPA module generates pseudo labels to help identify unknown objects, but a significant number of unknown boxes are still classified as known. Some of these misclassified boxes are clustered together and are referred to as uncertainty-boxes. The paper proposed an uncertainty-NMS post-processing module to address this issue by suppressing these uncertainty-boxes. We reasoned that the definition of uncertainty-box is as follows. Let C_{ij} as class of j-th box of $object_i$, and $P(C_{ij_1} = C_a) = P_a$ ($C_a \in C_{known}$) as the probability of box_{j_1} attributed to class a. According to the winner takes all principle of *softmax*, for any box_{j_2}, $P(C_{ij_2} = C_b) = P_b$ ($C_b \in C_{known}$), $P_b \ll P_a$. Conversely, if $|P_b - P_a| < \tau$ (τ is a small number), then the class probability of $object_i$ is uncertain, and P_a, P_b are uncertainty-boxes of $object_i$.

Unknown Probability Estimator. We detect uncertainty-boxes by calculating the IoU matrix of all boxes, and convert some unknown boxes into unknown boxes. Unknown instances will be detected as uncertainty-boxes, in addition, in a crowded scene, many highly overlapping boxes of different categories are also detected as uncertainty-boxes. In order to distinguish the real unknown objects, our UNBDet estimate the unknown probabilities for uncertainty-boxes, we employ the Unknown Probability Estimator (UPE) to predict the unknown confidences of uncertainty-boxes. Since uncertainty-boxes are special known boxes, we design UPE as a superficial binary classification branch, which takes unknown proposals as positive samples and known proposals as negative samples, and abandons background proposals for simplifying training. Binary Cross Entropy Loss is used to optimize the UPE model.

Uncertainty-NMS. The Uncertainty-NMS module for UNBDet is used to transform uncertainty-boxes into unknown boxes by taking known and unknown probabilities as inputs to refine detection results. In general, we apply NMS to all boxes to eliminate redundant prediction boxes, and then calculate the IoU matrix M for known boxes. M_{ij} is the IoU of i-th box_i and j-th box_j, if M_{ij} is greater than T_{IoU} (T_{IoU} represents the overlap of the two boxes), box_i and box_j are neighbor boxes. Uncertainty-box is the box which has n neighbor boxes ($n > 1$). We select boxes which unknown score $> S_{\mathcal{U}_t}$ and known score $< S_{\mathcal{K}_t}$ as unknown boxes, then convert these boxes to unknown. Finally, we apply NMS for unknown boxes to eliminate redundant unknown boxes. The steps of the uncertainty NMS algorithm are detailed in the supplementary materials.

4 Experiment

4.1 Experimental Setup

Datasets. We evaluate our UNBDet on PASCAL VOC [6] and MS COCO [13]. We take the trainval set of VOC for close-set training. Meanwhile, following [9], we use two settings, i.e., VOC-COCO-T_1 and VOC-COCO-T_2, to evaluate our method. VOC-COCO-T_1 contains VOC testing images and 20, 40, 60 non-VOC classes in COCO, respectively. VOC-COCO-T_2 constructs four joint datasets with VOC testing images and $\{0.5n, n, 2n, 4n\}$ ($n = 5000$) COCO images disjointing with VOC classes for increasing the Wilderness Ratio (WR) [5].

Evaluation Metrics. Wilderness Impact (WI) [5] is employed to measure the degree of unknown objects misclassified to known classes:

$$WI = (\frac{P_\mathcal{K}}{P_{\mathcal{K} \cup \mathcal{U}}} - 1) \times 100 \tag{1}$$

where $P_\mathcal{K}$ and $P_{\mathcal{K} \cup \mathcal{U}}$ are the precision of close-set and open-set classes. We report WI under a recall level of 0.8. Besides, Absolute Open-Set Error (AOSE) [15] is used to count the number of misclassified unknown objects. Furthermore, we report the mean Average Precision (mAP) of known classes (mAP$_\mathcal{K}$) and AP of unknown classes (AP$_u$) for unknown class discovery ability. WI, AOSE, and AP$_u$ are measured for open-set evaluation, and mAP$_\mathcal{K}$ is for close-set evaluation.

Implementation Details. We implement UNBDet on Detectron2, and ResNet-50 with Feature Pyramid Network [13] is the backbone for all methods. SGD optimizer is adopted with an initial learning rate of 0.02, momentum of 0.9, and weight decay of 0.0001. Our model is trained on 8 GPU$_s$ with a batch size of 16. For Pseudo Proposal Advisor, we use a probability of 0.2 to generate pseudo labels randomly, mainly considering that excessive pseudo labels will affect the AP$_k$. For contrastive learning, following OpenDet [9], we set memory size 256 and sampling size 16. For Uncertainty-NMS, we use the IoU threshold of uncertainty-boxes > 0.95, and the number of uncertainty-boxes > 2 to judge whether a box is unknown, and boxes with an unknown score > 0.9 will be considered as unknown.

Table 1. Comparisons with other methods on VOC and VOC-COCO-T_1.

Method	VOC	VOC-COCO-20				VOC-COCO-40				VOC-COCO-60			
	mAP$_{\mathcal{K}\uparrow}$	WI$_\downarrow$	AOSE$_\downarrow$	mAP$_{\mathcal{K}\uparrow}$	AP$_{u\uparrow}$	WI$_\downarrow$	AOSE$_\downarrow$	mAP$_{\mathcal{K}\uparrow}$	AP$_{u\uparrow}$	WI$_\downarrow$	AOSE$_\downarrow$	mAP$_{\mathcal{K}\uparrow}$	AP$_{u\uparrow}$
FR-CNN [17]	80.10	18.39	15118	58.45	0	22.74	23391	55.26	0	18.49	25472	55.83	0
FR-CNN * [17]	80.01	18.83	11941	57.91	0	23.24	18257	54.77	0	18.72	19566	55.34	0
PROSER [21]	79.68	19.16	13035	57.66	10.92	24.15	19831	54.66	7.62	19.64	21322	55.20	3.25
ORE [10]	79.80	18.18	12811	58.25	2.60	22.40	19752	55.30	1.70	18.35	21415	55.47	0.53
DS [15]	80.04	16.98	12868	58.35	5.13	20.86	19775	55.31	3.39	17.22	21921	55.77	1.25
OpenDet [9]	80.02	14.95	11286	58.75	14.93	18.23	16800	55.83	10.58	14.24	18250	56.37	4.36
UNBDet(Ours)	**80.17**	**14.81**	**10428**	**58.77**	**15.53**	**18.11**	**15801**	**55.87**	**11.29**	**14.12**	**18058**	**56.41**	**5.04**

Table 2. Comparisons with other methods on VOC and VOC-COCO-T_2.

Method	VOC-COCO-0.5n				VOC-COCO-n				VOC-COCO-4n			
	WI↓	AOSE↓	mAP$_{K↑}$	AP$_{U↑}$	WI↓	AOSE↓	mAP$_{K↑}$	AP$_{U↑}$	WI↓	AOSE↓	mAP$_{K↑}$	AP$_{U↑}$
FR-CNN [17]	9.25	6015	77.97	0	16.14	12409	74.52	0	32.89	48618	63.92	0
FR-CNN * [17]	9.01	4299	77.66	0	16.00	9477	74.17	0	33.11	37012	63.80	0
PROSER [21]	9.32	5105	77.35	7.48	16.65	10601	73.55	8.88	34.60	41569	63.09	11.15
ORE [10]	8.39	4945	77.84	1.75	15.36	10568	74.34	1.81	32.40	40865	64.59	2.14
DS [15]	8.30	4862	77.78	2.89	15.43	10136	73.67	4.11	31.79	39388	63.12	5.64
OpenDet [9]	6.44	3944	78.61	9.05	11.70	8282	75.56	12.30	26.69	32419	65.55	16.76
UNBDet(Ours)	**6.10**	**3815**	**78.66**	**9.45**	**10.55**	**8077**	**76.31**	**13.87**	**24.79**	**31401**	**65.62**	**17.14**

4.2   Comparison with Other Methods

We compare our UNBDet with other methods on VOC-COCO-$\{T_1, T_2\}$ with FR-CNN [17], FR-CNN* [17] (* means a higher score threshold (i.e., 0.1) for testing), PROSER [21], DS [15] and OpenDet [9]. Table 1 shows comparison results on VOC-COCO-T_1. For close-set evaluation (mAP_K) on VOC, same as the previous methods, UNBDet achieves a consistent performance with Faster R-CNN, which shows that our method does not affect the performance of the model in close-set detection. For OSOD, our model outperforms all the other existing methods on all the metrics, which shows the effectiveness of our model. Specially, we have obtained 1.1%–7.6% reduction on AOSE, respectively, compared with other methods. This shows that our model misclassifies fewer unknown objects. For AP$_U$, UNBDet achieves an increase of 2.2%–16%, and along with increasing known classes, our model has better performance. Table 2 shows comparison results on VOC-COCO-T_2. Comparing UNBDet with other methods by increasing the WR, we get a similar conclusion with Table 1. For example, the AP$_U$ gains are $\{0.4, 1.57, 0.54\}$ on VOC-COCO-$\{0.5n, n, 4n\}$, which indicates our method has a strong unknown discovery ability.

4.3   Ablation Studies

To further evaluate the effectiveness of each component in our proposed model, we conduct ablation experiments on VOC-COCO-20 to verify the effectiveness of our components. We implement a basic Faster R-CNN following the setting in subsection *Baseline Setup.*

Pseudo Proposal Advisor. To study the effectiveness of PPA in UNBDet, we design a comparative experiment with the Unknown-aware-RPN [10]. As shown in Table 3, compared with **Faster R-CNN,** due to **Unknown-aware-RPN** that can provide *Potential unknowns* during training, **Faster R-CNN + Unknown-aware-RPN** has an outstanding ability to discover unknown objects, but our PPA module has 26.2% improvement on AP$_U$ by providing more *Mosaic unknowns*. It proves the effectiveness of the PPA module.

Table 3. Ablation study on VOC-COCO-T_1.

Method	VOC-COCO-20			
	WI$_\downarrow$	AOSE$_\downarrow$	mAP$_{\mathcal{K}\uparrow}$	AP$_{\mathcal{U}\uparrow}$
FR-CNN	19.50	16518	58.70	0
FR-CNN + Unknown-aware-RPN	16.86	16862	58.65	10.28
FR-CNN + PPA	17.91	16646	58.72	12.97
FR-CNN + PPA + U-NMS*	15.68	10814	58.77	14.59
FR-CNN + PPA + U-NMS(UPE)	**14.81**	**10428**	**58.77**	**15.53**

Uncertainty-NMS. In order to show the superiority of Uncertainty-NMS, we design a comparative experiment on Uncertainty-NMS* (without unknown probability) and Uncertainty-NMS, respectively, as shown in Table 3. When using Uncertainty-NMS* to process detection results, we get 2.23, 35% reduction on (WI, AOSE) and 12.5% improvement on AP$_{\mathcal{U}}$, while when using Uncertainty-NMS we still obtain noticeable improvements on WI ($15.58 \rightarrow 14.81$), AOSE ($10814 \rightarrow 10428$) and AP$_{\mathcal{U}}$ ($14.59 \rightarrow 15.53$). Above all, Uncertainty-NMS has

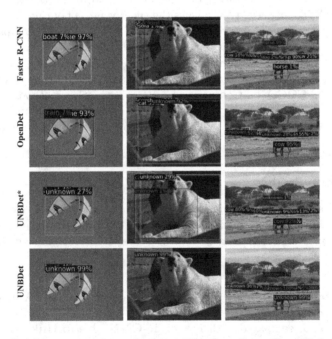

Fig. 5. Qualitative comparisons between the baseline, OpenDet, UNBDet* **and UNBDet.** We do not use score thresholds to guide visualization. UNBDet* means testing without Uncertainty-NMS. Our UNBDet has a more strong performance in discovering unknown objects.

a solid ability to reduce AOSE and find more unknown objects. UPE further improves the performance of the model in an open-set setting.

4.4 Visualization and Qualitative Analysis

Figure 5 shows the qualitative results of baseline, OpenDet, and UNBDet. Compared with baseline, OpenDet has detected some unknown objects, but our UNBDet* has further improved the performance of unknown detection. For example, in the first column, Faster R-CNN and OpenDet have no unknown prediction on image *kite*, but UNBDet* finds out unknown boxes and some redundant known boxes for it. Furthermore, UNBDet (equipped with Uncertainty-NMS) reduces the number of redundant boxes and optimizes the unknown probability.

5   Conclusions

OSOD is a challenging task in computer vision. We introduce the concept of uncertainty-boxes and propose a novel approach, UNBDet, to improve OSOD performance by identifying these boxes. UNBDet comprises three key components: a Pseudo Proposal Advisor (PPA) that generates unknown pseudo labels, an Unknown Probability Estimator (UPE) that reassesses the probabilities of uncertainty-boxes, and a Uncertainty-NMS post-processing module determines the Uncertainty-boxes by calculating the IoU between the predicted known objects. Our proposed approach, UNBDet, demonstrates a significant improvement in performance metrics as compared to existing methods, by using uncertainty-boxes to refine open-set detection results.

Acknowledgement. This work was supported by the National Natural Science Foundation of China (No.62172101); and the Science and Technology Commission of Shanghai Municipality(No. 21511100500.No. 22DZ1100101); and Auxiliary Diagnosis and Rare Disease Screening System for Children's Pneumonia(No.yg2022-7).

References

1. Bendale, A., Boult, T.E.: Towards open set deep networks. In: Proceedings of the IEEE Conference on Computer Vision and Pattern Recognition, pp. 1563–1572 (2016)
2. Bevandić, P., Krešo, I., Oršić, M., Šegvić, S.: Dense open-set recognition based on training with noisy negative images. Image Vis. Comput. **124**, 104490 (2022)
3. Chen, G., et al.: Learning open set network with discriminative reciprocal points. In: Vedaldi, A., Bischof, H., Brox, T., Frahm, J.-M. (eds.) ECCV 2020. LNCS, vol. 12348, pp. 507–522. Springer, Cham (2020). https://doi.org/10.1007/978-3-030-58580-8_30
4. Chen, P., Liu, S., Zhao, H., Jia, J.: GridMask data augmentation. arXiv preprint: arXiv:2001.04086 (2020)

5. Dhamija, A., Gunther, M., Ventura, J., Boult, T.: The overlooked elephant of object detection: open set. In: Proceedings of the IEEE/CVF Winter Conference on Applications of Computer Vision, pp. 1021–1030 (2020)
6. Everingham, M., Van Gool, L., Williams, C.K., Winn, J., Zisserman, A.: The pascal visual object classes (VOC) challenge. Int. J. Comput. Vision **88**(2), 303–338 (2010)
7. Ge, Z., Demyanov, S., Chen, Z., Garnavi, R.: Generative openmax for multi-class open set classification. arXiv preprint: arXiv:1707.07418 (2017)
8. Gupta, A., Narayan, S., Joseph, K., Khan, S., Khan, F.S., Shah, M.: OW-DETR: open-world detection transformer. arXiv preprint: arXiv:2112.01513 (2021)
9. Han, J., Ren, Y., Ding, J., Pan, X., Yan, K., Xia, G.S.: Expanding low-density latent regions for open-set object detection. In: Proceedings of the IEEE/CVF Conference on Computer Vision and Pattern Recognition, pp. 9591–9600 (2022)
10. Joseph, K., Khan, S., Khan, F.S., Balasubramanian, V.N.: Towards open world object detection. In: Proceedings of the IEEE/CVF Conference on Computer Vision and Pattern Recognition, pp. 5830–5840 (2021)
11. Lavín, Á., et al.: On the determination of uncertainty and limit of detection in label-free biosensors. Sensors **18**(7), 2038 (2018)
12. Lee, D.H., et al.: Pseudo-label: the simple and efficient semi-supervised learning method for deep neural networks. In: Workshop on Challenges in Representation Learning, ICML, vol. 3, p. 896 (2013)
13. Lin, T.Y., Dollár, P., Girshick, R., He, K., Hariharan, B., Belongie, S.: Feature pyramid networks for object detection. In: Proceedings of the IEEE Conference on Computer Vision and Pattern Recognition, pp. 2117–2125 (2017)
14. Lin, T.Y., Goyal, P., Girshick, R., He, K., Dollár, P.: Focal loss for dense object detection. In: Proceedings of the IEEE International Conference on Computer Vision, pp. 2980–2988 (2017)
15. Miller, D., Nicholson, L., Dayoub, F., Sünderhauf, N.: Dropout sampling for robust object detection in open-set conditions. In: 2018 IEEE International Conference on Robotics and Automation (ICRA), pp. 3243–3249. IEEE (2018)
16. Redmon, J., Divvala, S., Girshick, R., Farhadi, A.: You only look once: unified, real-time object detection. In: Proceedings of the IEEE Conference on Computer Vision and Pattern Recognition, pp. 779–788 (2016)
17. Ren, S., He, K., Girshick, R., Sun, J.: Faster R-CNN: towards real-time object detection with region proposal networks. In: Advances in Neural Information Processing Systems, vol. 28 (2015)
18. Scheirer, W.J., de Rezende Rocha, A., Sapkota, A., Boult, T.E.: Toward open set recognition. IEEE Trans. Pattern Anal. Mach. Intell. **35**(7), 1757–1772 (2012)
19. Zhao, X., Liu, X., Shen, Y., Ma, Y., Qiao, Y., Wang, D.: Revisiting open world object detection. arXiv preprint: arXiv:2201.00471 (2022)
20. Zheng, Z., Yang, Y.: Rectifying pseudo label learning via uncertainty estimation for domain adaptive semantic segmentation. Int. J. Comput. Vision **129**(4), 1106–1120 (2021)
21. Zhou, D.W., Ye, H.J., Zhan, D.C.: Learning placeholders for open-set recognition. In: Proceedings of the IEEE/CVF Conference on Computer Vision and Pattern Recognition, pp. 4401–4410 (2021)

Interventional Supervised Learning for Person Re-identification

Qingwang Zeng[1] and Jing Xiao[2(✉)]

[1] School of Artificial Intelligence, South China Normal University,
Foshan, Guangdong, China
`2021023264@m.scnu.edu.cn`
[2] School of Computer Science, South China Normal University,
Guangzhou, Guangdong, China
`xiaojing@scnu.edu.cn`

Abstract. Person re-identification (ReID) is a problem of retrieving pedestrian images. The complexity of ReID is attributed to image resolution, shooting angles, and object occlusion. With the evolution of computer vision, researchers have proposed effective solutions to tackle ReID. Causal inference has demonstrated its utility across various computer vision tasks. This paper investigates the potential of causal inference technology to enhance ReID performance. We construct a structural causal model for ReID and determine that dataset knowledge can constrain models from grasping genuine causality by affecting image features. Consequently, we propose intervention-based methods for supervised learning ReID (ISReID), introducing three effective backdoor adjustment schemes to intervene in supervised learning ReID models. To evaluate our approach's effectiveness, we conduct experiments on four widely-used ReID datasets: DukeMTMC-reID, Market-1501, CUHK03, and MSMT17. The experimental results confirm the efficacy of our backdoor adjustment schemes in enhancing supervised learning ReID models' performance. The source code can be found at https://github.com/ SCNU203/ISReID.

Keywords: Supervised learning · Person re-identification · Structural Causal Model · Backdoor adjustment

1 Introduction

Person re-identification [3,27,36] refers to the process of using computer vision techniques to retrieve target pedestrians in image sequences. Unlike face recognition technology, person re-identification data has high complexity and low pixelation [9,25]. The images contained in a single dataset come from multiple shooting angles, multiple backgrounds [19], different walking postures [4,18,33], and light intensity [7]. In the real world, we often prefer simple solutions that take into account factors like implementation cost, computational speed, and

© The Author(s), under exclusive license to Springer Nature Singapore Pte Ltd. 2024
Q. Liu et al. (Eds.): PRCV 2023, LNCS 14432, pp. 255–267, 2024.
https://doi.org/10.1007/978-981-99-8543-2_21

interpretability. Therefore, the study of easy ReID is of significant value. Typically, supervised learning easy ReID models (SReID) can only learn correlations between the features. These correlations have the potential to restrict the performance of both straightforward and intricate SReID models.

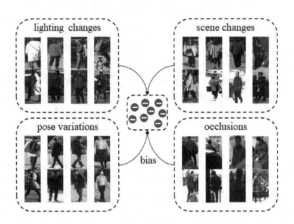

Fig. 1. Four common bias factors in MSMT17: lighting changes, scene changes, pose variations, and occlusions. Each column of images represents the same pedestrian.

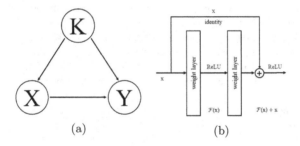

Fig. 2. (a) The structural causal model we constructed. Nodes are abstract data and directed edges are causal relationships. Node X represents features; node Y represents labels, and node K represents knowledge. Among them, node K is a confounding factor. (b) Structure of a residual block. Skip connections in the residual block are used to alleviate the gradient vanishing in deep learning.

The dataset is the most direct cause of the correlation. Figure 1 illustrates bias among images of the same class. For example, consider a scenario where the subject we are trying to retrieve walks in an area with intense lighting. This can result in captured images exhibiting varying colors of clothing, thereby elevating the potential for inaccurate outcomes. The absence of effective generalization within SReID exacerbates the bias issue, particularly since the images encompass numerous extraneous factors.

This paper aims to employ causal inference techniques to resolve correlation challenges within the context of SReID. We utilize Judea Pearl's Structural Causal Model (SCM) [14,15] for causal analysis of SReID. The structural causal model we constructed is shown in Fig. 2(a). Here, X signifies the features of person images; Y denotes the labels of these images, and K encapsulates the dataset's inherent knowledge. We introduce the concept of "knowledge," encompassing various elements such as backgrounds, light intensity, and poses from within the same dataset. $X \to Y$ means that feature X can be classified to get label Y, which is the real causal relationship we want to get. $K \to X$ means that knowledge K can generate feature X. $K \to Y$ means that knowledge K corresponds to label Y through the knowledge subset. In our SCM, three nodes constitute the collision structure. Thus, our findings indicate that knowledge K can mislead SReID into learning spurious correlations.

We employ backdoor adjustment techniques [14,15] to intervene during the training of the SReID model, utilizing our SCM as the foundation. We propose and implement three efficient intervention-based methods for supervised learning ReID, known as ISReID. We validate the adjusted model on four widely-used ReID datasets. The experimental results indicate that our adjusted model effectively improves accuracy and outperforms several state-of-the-art SReID methods. In summary, our contributions are as follows:

- We perform causal analysis on SReID and construct SCM. Based on SCM, we find that knowledge K is a confounding factor for SReID.
- We propose and implement three backdoor adjustment schemes to intervene in SReID training.
- Experiments show that our model has obvious intervention effect and is superior to several advanced SReID models.

2   Related Work

2.1   Supervised Person Re-identification

Research focused on supervised learning can contribute to unsupervised learning endeavors, including areas like unsupervised domain adaptation. Consequently, several researchers continue to delve into the study of the SReID model. We summarize the current mainstream SReID models into three types, namely global feature representation learning [1,6,8,24,30,37], local feature representation learning [5,10,13,21,22], and auxiliary feature representation learning [12,20,23,34]. Global feature representation learning is derived from image classification. For example, Wang et al. [24] proposed a learning framework for joint single-image feature matching and cross-image feature binary classification (SIR&CIR). SIR&CIR is a joint optimization model based on convolutional neural network, and achieved good accuracy. Local feature representation learning is a segmentation retrieval strategy for person images. Sun et al. [22] adopted a consistent localization method to learn the part information of each part for person re-identification (PCB+RPP). The author proposed a horizontal average slicing strategy and a partially refined pooling strategy to make the sharding more consistent. Auxiliary feature representation learning is a multi-annotation training method. For example, Tay et al. [23] proposed a classification framework (AANet) integrating person and attribute information to solve the person re-identification problem. The authors think that attribute attention maps formed by class combinations of individual attributes are beneficial for discriminative representations.

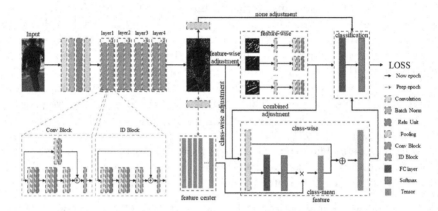

Fig. 3. The proposed model framework. The black arrow indicates the data flow of the current epoch; the gray arrow indicates the data flow before the current epoch; × denotes matrix multiplication, and ⊕ denotes vector concatenation. The input image is obtained through a feature extraction network to obtain 3D Tensor. Before training, we calculate the current feature centroid. If there is no adjustment, the data flow is directly input into the classification component after pooling. The feature-wise dataflow is first directed to the feature-wise component, then to the classification component. The class-wise dataflow is first directed to the class-wise component, then to the classification component. The combined adjustment dataflow is first directed to the feature-wise component, then to the class-wise component, and finally to the classification component.

2.2   Causal Intervention via Backdoor Adjustment

As shown in Fig. 2(a), in our SCM, K, X and Y constitute a collision structure, namely $K \rightarrow X \rightarrow Y$ and $K \rightarrow Y$. If we want to get the causal relationship of $X \rightarrow Y$, we must block the influence of knowledge K. In the real world, knowledge K often cannot be instantiated. The definition of knowledge K, in this context meaning knowledge representation, is analogous to a channel in a neural network.

We derive the adjustment formula using the backdoor criterion in causal inference techniques. The backdoor criterion ensures that the observed causal effect remains consistent with the trend when we eliminate the confounding factor, knowledge K. We assume that knowledge can be instantiated, specifically,

$K = k$, where k represents an element in knowledge K. Therefore, our backdoor adjustment formula is as follows:

$$P(Y \mid do(X = x)) = \sum_k P(Y \mid X = x, K = k)P(K = k) \qquad (1)$$

The key to backdoor adjustment lies in the instantiation of knowledge K. Although the formulas for various instantiation methods differ, the underlying principle remains consistent. In this paper, knowledge K is instantiated in the feature and class dimensions. Our specific backdoor adjustment is inspired by the literature [22,31]. Different from them, we apply an easy SReID model and the knowledge K comes from the dataset.

3   The Proposed Method

In this section, we will outline the SReID model used in our experiment, as well as the three methods we used to make backdoor adjustments. The SReID model serves as our baseline model and will be included in the following comparative experiments. In the context of SReID, our dataset can be represented as X, Y, where X_i has a corresponding Y_i. The length of X and Y is defined as N, which is the number of images.

3.1   Overview of Framework

Our model is based on three backdoor adjustments intervening supervised learning (ISReID), and the structure is shown in Fig. 3. The SReID model uses a feature extraction network based on ResNet-50 that has been pre-trained on ImageNet. We fix the training parameters of the modules before layer3. The input image first enters a 7×7 convolution layer, followed by batch normalization and max-pooling. The resulting output is then sequentially input into 4 additional layers. Among them, 4 layers have 3, 4, 6, and 3 Bottlenecks respectively. Bottleneck has two kinds of convolutional layer combinations, the structure is similar to Fig. 2(b), which are Conv Block and ID Block. The output of layer4 is input into the average pooling to obtain the features of the image sample. The size of the feature map is the standard output of ResNet-50, which has a size of $12 \times 4 \times 2048$. In the absence of backdoor adjustment, the features are directly input into the classification component for supervised classification learning. In feature-wise adjustment, features are first split horizontally and globally pooled. Then, the features are input into a 1×1 convolutional layer to generate new features. In class-wise adjustment, we prioritize generating class-mean features. The class-mean feature is the product of the current predicted value and the center of the class feature. Then, the feature is concatenated with the class-mean feature to obtain a new feature. In combined adjustment, features are sequentially input into feature-wise and class-wise, and then new features are generated. The new features generated by all adjustment methods are finally input into the classification component for supervised classification learning. The classification component consists of FC layer and softmax function. We optimize the model using the cross-entropy loss.

We will introduce three backdoor adjustment methods in the following sections.

3.2   Feature-Wise Adjustment

We assume that the output of the last layer of the feature extraction network is x, then $\mathcal{F}$ is the index set of elements of each dimension on x. We think that the features at each level contain elements of knowledge K. Therefore, knowledge K is also split into n subsets. Knowledge K can be defined as $K := \{k_1, k_2, \ldots, k_n\}$, and prior knowledge $P(k_i) = \frac{1}{n}$. Classification tasks [11,22,36] are also transformed into n classification tasks and are independent of each other. We set $k_i = \mathcal{F}_i$, the formula is as follows:

$$I(x, k_i) := \{idx \mid idx \in \mathcal{F}_i \cap \mathcal{I}_t\} \tag{2}$$

where $\mathcal{I}_t$ is the index to ensure that the feature is the active state; t is the feature threshold of the active state; idx is the selected index; $I(x, k_i)$ is the index subset of x, and $[x]_{I(x,k_i)}$ is the feature subset. In our experiments, $t = 1e - 3$. We substitute into formula 1, we can get:

$$P(Y \mid do(X = x)) = \frac{1}{n} \sum_{i=1}^{n} P(Y \mid [x]_{I(x,k_i)}) \tag{3}$$

We optimize the network using the cross-entropy loss, namely:

$$\mathcal{L} = -\frac{1}{n \cdot N} \sum_{i=1}^{N} \sum_{j=1}^{n} \log p(y_i \mid [x_i]_{I(x_i,k_j)}) \tag{4}$$

3.3   Class-Wise Adjustment

We assume that there are m classes in the dataset, denoted as $C = \{c_1, c_2, \ldots, c_m\}$. Each class can be defined as a subset of knowledge K, i.e. $K := \{k_1, k_2, \ldots, k_m\}$, and prior knowledge $P(k_i) = \frac{1}{m}$. We formulate k_i as:

$$k_i = P(c_i \mid x)\overline{x_i} \tag{5}$$

where $\overline{x_i}$ is the feature center of the ith class. Different from [31], k_i and $\overline{x_i}$ vary with training. We substitute into formula 1, we can get:

$$P(Y \mid do(X = x)) = \frac{1}{m} \sum_{i=1}^{m} P(Y \mid x \oplus P(c_i \mid x)\overline{x_i})$$

$$\approx P(Y \mid x \oplus \frac{1}{m} \sum_{i=1}^{m} P(c_i \mid x)\overline{x_i}) \tag{6}$$

where $\oplus$ denotes vector concatenation. Because $P(Y \mid do(X = x))$ needs to consume a lot of forward propagation resources. We optimize $\sum P$ to $P(\sum)$ using normalized weighted geometric mean (NWGM) [10,13]. Finally, $P(Y \mid do(X = x))$ is an estimate. We optimize the network using the cross-entropy loss, namely:

$$\mathcal{L} = -\frac{1}{N} \sum_{i=1}^{N} \log p(y_i \mid x_i \oplus \frac{1}{m} \sum_{j=1}^{m} P(c_j \mid x_i)\overline{x_j}) \tag{7}$$

Compared with our SReID, the length of the features is enlarged by two times.

3.4 Combined Adjustment

Combined adjustment is to instantiate knowledge K from the dimension of the feature and class. We think that instantiating knowledge K in multiple dimensions can make the adjustment more fine-grained. Specifically, we perform class-wise adjustment after feature-wise adjustment. Therefore, we substitute into formula 1 to get:

$$P(Y \mid do(X = x)) \approx \frac{1}{n} \sum_{i=1}^{n} P(Y \mid [x]_{I(x,k_i)} \oplus [\frac{1}{m} \sum_{j=1}^{m} P(c_j \mid x)\overline{x_j}]_{I(x,k_i)}) \tag{8}$$

We optimize the network using the cross-entropy loss, namely:

$$\mathcal{L} = -\frac{1}{n \cdot N} \sum_{i=1}^{N} \sum_{j=1}^{n} \log p(y_i \mid [x_i]_{I(x_i,k_j)} \oplus [\frac{1}{m} \sum_{l=1}^{m} P(c_l \mid x_i)\overline{x_l}]_{I(x_i,k_j)}) \tag{9}$$

4 Experiment

4.1 Datasets and Settings

Datasets. Our model is evaluated on four widely-used ReID datasets, DukeMTMC-reID [17], Market-1501 [35], CUHK03 [17], and MSMT17 [28]. DukeMTMC-reID collects a total of 36411 images. These images are derived from 1812 pedestrians, of which 1404 pedestrians are captured by multiple cameras. Market-1501 collects a total of 32668 images of 1501 pedestrians. Every pedestrian appears on at least two cameras. CUHK03 collects a total of 14097 images. These images are derived from 1467 pedestrians, each captured by two cameras. MSMT17 collects a total of 126,411 images of 4,101 pedestrians. But unlike other datasets, the ratio of the training set to the test set is 1 : 3, in order to encourage efficient training strategy.

Evaluation Metrics. We will evaluate the model using Cumulative Matching Characteristics (CMC) and mean Average Precision (mAP), which are the most popular performance evaluation methods in ReID. CMC tends to favor single-image retrieval scenarios. In contrast, its performance in multi-image retrieval is suboptimal, making mAP a more robust and convincing metric in such cases.

Baseline. We use the SReID model shown in Fig. 3 as the baseline model.

Experiment Settings. In all experiments, we randomly flip, crop, and erase the dataset. Our split horizon factor n is set to 12 in both feature-wise and combined adjustment. Our training has a total of 60 epochs. In the first 40 epochs, the learning rate of the baseline model and the three backdoor adjustment methods is 0.1. After that, the learning rate is reduced to 0.01.

Table 1. Comparison of our model tested with the baseline model on DukeMTMC-reID and Market-1501. **Ft**: feature-wise adjustment, the segmentation coefficient $n = 12$. **Cl**: class-wise adjustment. **Ft+cl**: combined adjustment.

Methods	DukeMTMC-reID			Market-1501		
	R-1	R-5	mAP	R-1	R-5	mAP
SReID Only	76.3	86.9	56.4	87.4	95.3	67.9
ISReID(ft)	**78.9**	87.4	60.0	89.6	**96.1**	**73.2**
ISReID(cl)	76.3	86.8	56.1	87.0	95.6	68.9
ISReID(ft+cl)	78.1	**87.9**	**61.5**	**89.9**	**96.1**	**73.2**

Table 2. Comparison of our model tested with the baseline model on CUHK03 and MSMT17. **Ft**: feature-wise adjustment, the segmentation coefficient $n = 12$. **Cl**: class-wise adjustment. **Ft+cl**: combined adjustment.

Methods	CUHK03			MSMT17		
	R-1	R-5	mAP	R-1	R-5	mAP
SReID Only	30.4	50.1	28.4	54.4	69.7	24.9
ISReID(ft)	**46.4**	**67.7**	**44.2**	50.7	67.5	25.9
ISReID(cl)	32.8	51.6	29.9	**55.6**	**70.6**	25.2
ISReID(ft+cl)	45.3	66.4	43.4	52.1	69.0	**26.9**

4.2   Evaluation

Performance on DukeMTMC-reID. Table 1 reports the evaluation results of the baseline model and three adjustment methods on DukeMTMC-reID. Our baseline model already achieves high accuracy. ISReID(cl) did not show an advantage, and even mAP and rank-5 showed slight negative growth. However, ISReID significantly improves performance in feature-wise and combined adjustments. Especially, ISReID(ft+cl) achieves 5.1% mAP increase and 1.8% rank-1 increase. Combined adjustment is also significantly improved compared to feature-wise adjustment because combined adjustment is more fine-grained. From the data of Table 1, the images collected by DukeMTMC-reID are more suitable for segmentation, indicating that the knowledge of DukeMTMC-reID is more suitable for instantiation with part information.

Performance on Market-1501. Table 1 reports the evaluation results of the baseline model and three adjustment methods on Market-1501. On Market-1501, the baseline model performs better than DukeMTMC-reID. ISReID(cl) achieves 68.9% mAP. Compared with class-wise adjustment, feature-wise and combined adjustment work better. Specifically, ISReID(ft+cl) achieves a 5.3% increase in mAP, and rank-1 has different degrees of growth. From the data of Table 1, the three backdoor adjustment methods are effective, indicating that the knowledge of Market-1501 can be instantiated from the dimension of the feature and class.

Table 3. Performance comparison of our model with state-of-the-art methods for ReID on Market-1501 and DukeMTMC-reID. **Ft**: feature-wise adjustment, the segmentation coefficient $n = 12$. **Cl**: class-wise adjustment. **Ft+cl**: combined adjustment.

Methods	DukeMTMC-reID			Market-1501		
	R-1	R-5	mAP	R-1	R-5	mAP
PAN [38]	71.6	83.9	51.5	82.8	93.5	63.3
PN-GAN [16]	73.6	-	53.2	89.4	-	72.6
BraidNet [26]	76.4	-	59.5	83.7	-	69.5
BCL [32]	78.2	-	58.6	86.4	-	67.6
DPFL [2]	**79.2**	-	60.6	88.6	-	72.0
AACN [29]	76.8	-	59.2	85.9	-	66.9
APR [12]	73.9	-	55.6	87.0	95.1	66.9
SReID Only	76.3	86.9	56.4	87.4	95.3	67.9
ISReID(ft)	78.9	87.4	60.0	89.6	**96.1**	**73.2**
ISReID(cl)	76.3	86.8	56.1	87.0	95.6	68.9
ISReID(ft+cl)	78.1	**87.9**	**61.5**	**89.9**	**96.1**	**73.2**

Performance on CUHK03. Table 2 reports the evaluation results of the baseline model and three adjustment methods on CUHK03. On CUHK03, the performance of the baseline model is poor. ISReID(ft) achieves 44.2% mAP and 46.4% rank-1, an increase of 15.8% and 16.0% compared to the baseline model, respectively. ISReID(cl) improves mAP and rank-1 by 1.5% and 2.4%, respectively. ISReID(ft+cl) also has a significant improvement, achieving 43.4% mAP and 45.3% rank-1 respectively. From the data of Table 2, ISReID(ft) achieves the best performance. All three backdoor adjustment methods are effective, but CUHK03's knowledge is more suitable for instantiation in the dimension of the feature.

Performance on MSMT17. Table 2 reports the evaluation results of the baseline model and three adjustment methods on MSMT17. MSMT17 is generally widely used in unsupervised ReID because it chooses a training strategy that encourages fewer labels. On MSMT17, the baseline model and the three adjustment methods perform poorly. Compared with the baseline model, all three adjustment methods have different degrees of growth. Especially, ISReID(ft+cl) achieves the best mAP. ISReID(cl) achieves the best rank-1 and rank-5. From the data of Table 2, the knowledge of MSMT17 is effective in instantiating from the dimension of the feature and class, which is due to the complex background and lighting characteristics of MSMT17.

Comparison of Three Adjustments. It can be seen from Table 1 and Table 2 that the baseline model has different degrees of improvement in the three adjustments. This is because three adjustments are more fine-grained in training than the baseline model. While all three backdoor adjustments assume knowledge instantiation, they differ in their dimensions. In our experiments, feature-wise adjustment is more pronounced. Because the pedestrians in the ReID dataset are easy to identify with information such as clothes and walking postures, that can be processed hierarchically. In our SCM, the label Y can be obtained by the knowledge K, from which the idea of class-wise adjustment. Estimated causal effects and class-mean feature make the effect of class-wise adjustment insignificant. Combined adjustment is not a superposition of the two adjustments but instantiates the knowledge K in two dimensions. The advantage of combined adjustment is only revealed when the effects of feature-wise adjustment and class-wise adjustment are similar, such as MSMT17. In other cases, the combined adjustment will float on the adjustment method with the best effect.

4.3 Comparison with State-of-the-Art Methods

Table 3 reports the results of our model compared with state-of-the-art models in recent years. Compared with global feature representation learning includes PAN [38], PN-GAN [16], BraidNet [26], and BCL [32]. Compared with local feature representation learning includes DPFL [2] and AACN [29]. Compared with auxiliary feature representation learning is only APR [12]. Compared with state-of-the-art global feature representation learning, our feature-wise and combined adjustment have significant advantages. Specifically, on DukeMTMC-reID, combined adjustment achieves 61.5% mAP and feature-wise adjustment achieves 78.9% rank-1. On Market-1501, combined adjustment outperforms PN-GAN by a small margin. Compared with advanced local feature representation learning, we achieve the best mAP and rank-5 accuracy on DukeMTMC-reID, and we achieve mAP improvement of 1.2% on Market-1501. Compared with advanced auxiliary feature representation learning, all three backdoor adjustment methods outperform APR.

5   Conclusion

In this paper, we utilize structural causal models to perform causal analysis of the easy SReID Model, investigating the impact of dataset knowledge on the training process. Due to the influence of dataset knowledge, different adjustment methods may have varying effects. To address this issue, we introduce ISReID, intervention-based methods for supervised learning ReID. We propose three backdoor adjustment methods: feature-wise adjustment, class-wise adjustment, and combined adjustment, which effectively eliminate the correlation caused by dataset knowledge. We conduct experiments on four widely-used datasets and observe that our backdoor adjustment methods successfully reduce the impact of misleading knowledge. Moving forward, we aim to investigate more dimensions of dataset knowledge and apply our approach beyond SReID.

Acknowledgements. This paper is supported by the National Natural Science Foundation of China No. 62177015.

References

1. Chen, W., Chen, X., Zhang, J., Huang, K.: Beyond triplet loss: a deep quadruplet network for person re-identification. In: CVPR, pp. 403–412 (2017)
2. Chen, Y., Zhu, X., Gong, S.: Person re-identification by deep learning multi-scale representations. In: ICCV, pp. 2590–2600 (2017)
3. Chen, Y.C., Zhu, X., Zheng, W.S., Lai, J.H.: Person re-identification by camera correlation aware feature augmentation. IEEE Trans. Pattern Anal. Mach. Intell. **40**(2), 392–408 (2017)
4. Cho, Y.J., Yoon, K.J.: Improving person re-identification via pose-aware multi-shot matching. In: CVPR, pp. 1354–1362 (2016)
5. Guo, J., Yuan, Y., Huang, L., Zhang, C., Yao, J.G., Han, K.: Beyond human parts: dual part-aligned representations for person re-identification. In: ICCV, pp. 3642–3651 (2019)
6. Hermans, A., Beyer, L., Leibe, B.: In defense of the triplet loss for person re-identification. In: CVPR (2017)
7. Huang, Y., Zha, Z.J., Fu, X., Zhang, W.: Illumination-invariant person re-identification. In: ACM MM, pp. 365–373 (2019)
8. Li, H., Wu, G., Zheng, W.S.: Combined depth space based architecture search for person re-identification. In: CVPR, pp. 6729–6738 (2021)
9. Li, X., Zheng, W.S., Wang, X., Xiang, T., Gong, S.: Multi-scale learning for low-resolution person re-identification. In: ICCV, pp. 3765–3773 (2015)
10. Li, Y.J., Chen, Y.C., Lin, Y.Y., Du, X., Wang, Y.C.F.: Recover and identify: a generative dual model for cross-resolution person re-identification. In: ICCV, pp. 8090–8099 (2019)
11. Li, Y., He, J., Zhang, T., Liu, X., Zhang, Y., Wu, F.: Diverse part discovery: occluded person re-identification with part-aware transformer. In: CVPR, pp. 2898–2907 (2021)
12. Lin, Y., et al.: Improving person re-identification by attribute and identity learning. Pattern Recognit. **95**, 151–161 (2019)

13. Liu, F., Zhang, L.: View confusion feature learning for person re-identification. In: CVPR, pp. 6639–6648 (2019)
14. Pearl, J., Glymour, M., Jewell, N.P.: Causal inference in statistics: a primer. Internet Res. (2016)
15. Pearl, J., et al.: Models, reasoning and inference, vol. 19, no. 2. Cambridge University Press, Cambridge (2000)
16. Qian, X., et al.: Pose-normalized image generation for person re-identification. In: ECCV, pp. 650–667 (2018)
17. Ristani, E., Solera, F., Zou, R., Cucchiara, R., Tomasi, C.: Performance measures and a data set for multi-target, multi-camera tracking. In: Hua, G., Jégou, H. (eds.) ECCV 2016. LNCS, vol. 9914, pp. 17–35. Springer, Cham (2016). https://doi.org/10.1007/978-3-319-48881-3_2
18. Sarfraz, M.S., Schumann, A., Eberle, A., Stiefelhagen, R.: A pose-sensitive embedding for person re-identification with expanded cross neighborhood re-ranking. In: CVPR, pp. 420–429 (2018)
19. Song, C., Huang, Y., Ouyang, W., Wang, L.: Mask-guided contrastive attention model for person re-identification. In: CVPR, pp. 1179–1188 (2018)
20. Su, C., Zhang, S., Xing, J., Gao, W., Tian, Q.: Deep attributes driven multi-camera person re-identification. In: Leibe, B., Matas, J., Sebe, N., Welling, M. (eds.) ECCV 2016. LNCS, vol. 9906, pp. 475–491. Springer, Cham (2016). https://doi.org/10.1007/978-3-319-46475-6_30
21. Suh, Y., Wang, J., Tang, S., Mei, T., Lee, K.M.: Part-aligned bilinear representations for person re-identification. In: ECCV, pp. 402–419 (2018)
22. Sun, Y., Zheng, L., Yang, Y., Tian, Q., Wang, S.: Beyond part models: person retrieval with refined part pooling (and a strong convolutional baseline). In: ECCV, pp. 480–496 (2018)
23. Tay, C.P., Roy, S., Yap, K.H.: Aanet: attribute attention network for person re-identifications. In: CVPR, pp. 7134–7143 (2019)
24. Wang, F., Zuo, W., Lin, L., Zhang, D., Zhang, L.: Joint learning of single-image and cross-image representations for person re-identification. In: CVPR, pp. 1288–1296 (2016)
25. Wang, Y., et al.: Resource aware person re-identification across multiple resolutions. In: CVPR, pp. 8042–8051 (2018)
26. Wang, Y., Chen, Z., Wu, F., Wang, G.: Person re-identification with cascaded pairwise convolutions. In: CVPR, pp. 1470–1478 (2018)
27. Wang, Y., Shen, J., Petridis, S., Pantic, M.: A real-time and unsupervised face re-identification system for human-robot interaction. Pattern Recogn. Lett. **128**, 559–568 (2019)
28. Wei, L., Zhang, S., Gao, W., Tian, Q.: Person transfer GAN to bridge domain gap for person re-identification. In: ICCV, pp. 79–88 (2018)
29. Xu, J., Zhao, R., Zhu, F., Wang, H., Ouyang, W.: Attention-aware compositional network for person re-identification. In: CVPR, pp. 2119–2128 (2018)
30. Yuan, Y., Chen, W., Yang, Y., Wang, Z.: In defense of the triplet loss again: learning robust person re-identification with fast approximated triplet loss and label distillation. In: CVPR, pp. 354–355 (2020)
31. Yue, Z., Zhang, H., Sun, Q., Hua, X.S.: Interventional few-shot learning. In: NIPS, vol. 33, pp. 2734–2746 (2020)
32. Zhang, G., Xu, J.: Discriminative feature representation for person re-identification by batch-contrastive loss. In: ACML, pp. 208–219. PMLR (2018)
33. Zhao, H., et al.: Spindle net: person re-identification with human body region guided feature decomposition and fusion. In: CVPR, pp. 1077–1085 (2017)

34. Zhao, Y., Shen, X., Jin, Z., Lu, H., Hua, X.S.: Attribute-driven feature disentangling and temporal aggregation for video person re-identification. In: CVPR, pp. 4913–4922 (2019)
35. Zheng, L., Shen, L., Tian, L., Wang, S., Wang, J., Tian, Q.: Scalable person re-identification: a benchmark. In: ICCV, pp. 1116–1124 (2015)
36. Zheng, L., Yang, Y., Hauptmann, A.: Person re-identification: past, present and future. In: CVPR (2016)
37. Zheng, L., Zhang, H., Sun, S., Chandraker, M., Yang, Y., Tian, Q.: Person re-identification in the wild. In: CVPR, pp. 1367–1376 (2017)
38. Zheng, Z., Zheng, L., Yang, Y.: Pedestrian alignment network for large-scale person re-identification. IEEE Trans. Circuits Syst. Video Technol. **29**(10), 3037–3045 (2018)

DP-INNet: Dual-Path Implicit Neural Network for Spatial and Spectral Features Fusion in Pan-Sharpening

Jingjia Huang[1], Ge Meng[1], Yingying Wang[2], Yunlong Lin[1], Yue Huang[1(✉)], and Xinghao Ding[1,2]

[1] School of Informatics, Xiamen University, Xiamen 361001, China
yhuang2010@xmu.edu.cn
[2] Institute of Artificial Intelligence, Xiamen University, Xiamen 361001, China

Abstract. Pan-sharpening is a technique that fuses a high-resolution panchromatic (PAN) image with its corresponding low-resolution multispectral (MS) image to create a high-resolution multispectral image. Due to the powerful representation ability of Convolutional Neural Networks (CNNs), deep learning-based pan-sharpening methods have rapidly developed in recent years. However, existing methods often ignore the representation of multimodal information from the perspective of continuous physical signals, which inevitably leads to the loss of detailed information during the fusion process. Therefore, this paper proposes a novel pan-sharpening method that integrates spectral information with structural information in a continuous domain by using implicit neural representation (INR). Specifically, an implicit representation function is used to align the spatial information of multimodal images in the continuous domain, which preserves structural details. Additionally, a gated convolutional network is utilized to achieve interaction between different order spectral information in multispectral images. Finally, an MLP is used to fuse the spatial and spectral information in the continuous space to generate the expected high-resolution multispectral image. Extensive experiments on different datasets show that our method outperforms existing methods in terms of both quantitative and qualitative metrics.

Keywords: Pan-sharpening · Continuous domain · Implicit neural representation · Multimodal images

1 Introduction

Multispectral images contain more information compared to regular RGB images. Different objects usually have different spectral reflectivities, which are particularly evident at specified wavelengths. Therefore, multispectral imaging has been widely used in environmental monitoring, agriculture, mapping services and so on. However, due to physical limitations, satellites are unable to

© The Author(s), under exclusive license to Springer Nature Singapore Pte Ltd. 2024
Q. Liu et al. (Eds.): PRCV 2023, LNCS 14432, pp. 268–279, 2024.
https://doi.org/10.1007/978-981-99-8543-2_22

simultaneously capture high-resolution multispectral images from a single sensor. Instead, they can only acquire high-resolution panchromatic images and corresponding low-resolution multispectral images separately. Traditional pansharpening methods have been extensively explored in the past, such as component substitution [3,8,14], multi-resolution analysis [16,24], and variation-based methods [1,12]. These methods have already achieved good results, but most of them only rely on prior knowledge and manual feature extraction, which limits the representation of high-resolution and spectral information. Thanks to the powerful feature representation ability of deep neural networks, numerous models based on CNN [5–7,9,17,26–28,30] have been developed to tackle pansharpening task in recent years, and these models have shown better performance than previous ones. However, all of them overlook the continuity preservation of spatial signals and spectral signals in the feature representation process, which is crucial for pan-sharpening. INR has benefited from its powerful ability to represent signals in the continuous domain, and has achieved good results in tasks such as image super-resolution [4], texture synthesis [10,20], scene rendering [18,19] and 3D reconstruction [11,21,22] in recent years. The idea of INR is to use an implicit function to map continuous spatial coordinates to signals in a specified domain, such as RGB values in an image. This implicit function is commonly parameterized by a neural network. Inspired by this, we aim to address the issue of preserving spatial and spectral information in pan-sharpening by using INR.

Based on the insights mentioned above, this paper proposes a new pansharpening framework from the perspective of INR. Specifically, we first use an encoder to extract high-frequency features from both PAN and MS images at different scales simultaneously, and treat features as latent codes uniformly distributed in the continuous space. We align the multi-level high-frequency features in the continuous spatial domain. Secondly, we extract high-order spectral information of MS images using a gate convolution network and achieve interaction between spectral features of various orders. Finally, we aggregate the aligned high-frequency features and spectral features in the latent space, and generate a high-resolution MS image in the continuous domain through an MLP. Extensive experiments on different datasets demonstrate that our method outperforms existing methods in both quantitative and qualitative metrics.

In summary, the contributions of this work are as follows:

- This paper proposes a new pan-sharpening method that fuses the high-frequency information from the PAN image and the spectral information from the MS image from the perspective of INR.
- We propose an implicit multimodal image representation function, which aligns multi-scale high-frequency information in the continuous domain by self-learning latent coding, and fuses multimodal information in the continuous 3D spatial-spectral domain.
- Extensive experiments on different datasets show that the proposed method significantly outperforms existing methods in both quantitative and qualitative metrics.

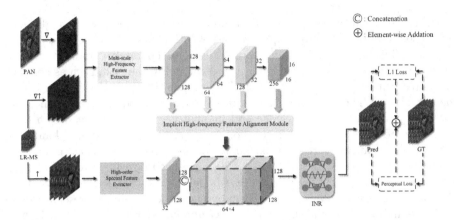

Fig. 1. The overall framework of our proposed pan-sharpening method.

2 Method

Figure 1 shows the overall framework of our method, which mainly consists of the following three modules: 1) Implicit high-frequency feature alignment module; 2) High-order spectral feature extraction module; 3) Implicit information fusion module. In this section, we further provide a detailed introduction to each module mentioned above.

2.1 Implicit High-Frequency Feature Alignment Module

Implicit Feature Interpolation. When generating PAN and MS images, different wavelengths are captured by sensors. This can result in a phase shift between multimodal images, leading to edge misalignment in objects. Therefore, it is essential to properly align cross-modal multi-level high-frequency features at a uniformed resolution.

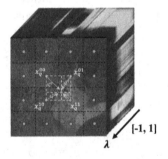

Fig. 2. Implicit Feature Interpolation.

Inspired by the implicit image representation function [4], we align high-frequency features of different multimodal images in the continuous domain. Specifically, given a PAN image $P \in R^{H \times W \times 1}$ and an MS image $M \in R^{\frac{H}{r} \times \frac{W}{r} \times C}$, where r represents the upsampling factor required to upscale the resolution of the M to match that of P. We first extract their high-frequency information respectively:

$$P^H = \nabla(P)$$
$$M^H = \text{UP}(\nabla(M)) \tag{1}$$

where UP $(\cdot)$ represents upsampling the edge image of the MS image to match the resolution of the PAN image, and ∇ denotes gradient operation. Next, we extract high-frequency features from the PAN and MS images at different scales using gated convolution:

$$F^* = E_\phi(P^H, M^H) \tag{2}$$

where E_ϕ represents an encoder that extracts multi-level features. As shown in Fig. 2, the high-frequency features at different scales are aligned through an implicit feature interpolation operation:

$$F^{HP}(x_q) = \sum_{i \in N_q} w_q^i v_q^i$$
$$w_q^i = \frac{S_q^i}{S} \tag{3}$$

where x_q is the coordinate of the pixel q in the aligned feature map F^{HP}, N_q is the set of neighbor pixels for q in the multi-level feature maps, Fig. 2 shows the relationship between the query coordinate q and N_q, N_q contains four adjacent pixels to pixel q, which are the top-left, top-right, bottom-left, and bottom-right pixels. To create a coordinate mapping between the feature maps before and after alignment, we normalize the pixel coordinates of F^{HP} to the range of $[-1,1]$. w_q^i represents the weight of the latent code v_q^i, which satisfies $\sum_{i \in N_q} w_q^i = 1$. S_q^i represents the rectangular area formed by pixels i and q, which satisfies $S = \sum_{i \in N_q} S_q^i$. v_q^i represents the potential encoding at pixel i:

$$v_q^i = \Phi(F_{x_q^i}^*) \tag{4}$$

where x_q^i represents the adjacent pixel i of query coordinate x_q in the multi-level feature map and $F_{x_q^i}^*$ represents its corresponding feature value. Φ is an INR network that maps 3D coordinates in the continuous domain to RGB values.

Implicit Multispectral Image Representation Function. The multiscale high-frequency feature maps can be seen as latent codes distributed in continuous 3D space, with each feature vector assigned a 2D spatial coordinate and a 1D spectral coordinate. The value at any coordinate x in band λ_b can be represented as:

$$I(x) = f_\theta(q_x^*, x - x^*, \lambda_b) \tag{5}$$

where x^* represents the coordinate nearest to the query coordinate x in the discrete multi-level feature maps, x - x^* represents the offset between x^* and x, q_x^* represents the latent code at x^*, λ_b represents the spectral to which the feature map belongs, which is normalized to $[-1, 1]$, and f_θ represents an MLP network with parameters θ. Figure 2 illustrates the approach for calculating the feature value at query coordinate x in an aligned feature map. We find neighboring nodes of x_q in each feature map, and calculate the RGB value:

$$F^{HP}(x_q) = \sum_{i \in \{00,01,10,11\}} \frac{S_q^i}{S_q} \cdot f_\theta(q_i, x_q - x_q^i, \lambda_b) \tag{6}$$

where $i \in \{00, 01, 10, 11\}$ represents the four neighboring nodes of x_q (top-left, top-right, bottom-left, and bottom-right), q_i represents the latent code at x_q^i, S_q^i represents the rectangular area formed by x_q and x_q^i, and $S_q = \sum_{i \in \{00,01,10,11\}} S_q^i$.

2.2   High-Order Spectral Feature Extraction Module

To achieve interaction of contrast information across the different bands of MS image M, we employ a gated convolutional network to extract high-order spectral features:

$$[u_k^{HW \times C}, v_k^{HW \times C}] = \psi_{in}^k(\text{UP}(M)) \in \mathbb{R}^{HW \times 2C}$$
$$v_{k+1} = f(v_k) \odot u_k \in \mathbb{R}^{HW \times C}, \ k = 0, 1, \cdots, n-1 \tag{7}$$
$$F^{SPEC} = \psi_{out}(u_n)$$

where ψ_{in}^k represents an arbitrary k-order linear projection that projects the MS image (after bilinear interpolation) into neighboring features u_k and v_k. $f(\cdot)$ is a depth-wise convolution, the interaction among u_k and v_k through the element-wise multiplication. Finally, spectral features are generated through a linear projection ψ_{out}. Figure 3 shows the details of high-order feature extraction process, and we also use a similar module to extract multi-scale high-frequency features.

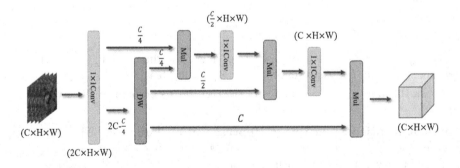

Fig. 3. High-order Spectral Feature Extractor.

2.3   Implicit Information Fusion Module

We fuse the high-frequency feature F^{HP} and spectral feature F^{SPEC} in the implicit space by using an MLP network with sine activation function [23]:

$$\widehat{X} = \text{SIREN}(\text{Cat}(F^{HP}, F^{SPEC})) \tag{8}$$

where $\widehat{X}$ is the resulting high-resolution multispectral image.

2.4   Loss Function

In this paper, we use L1 loss to supervise the fusion results at the pixel level:

$$\mathcal{L}1(\widehat{X}, X) = \frac{1}{n} \sum_{i}^{n} |\widehat{X}(i) - X(i)|. \tag{9}$$

where n is the total number of sampled pixels, i represents any pixel, $\widehat{X}(i)$ represents the predicted value for pixel i, and $X(i)$ represents its corresponding ground truth. We also use perceptual loss to measure the stylistic difference between the predicted and ground truth:

$$\mathcal{L}_{feat}^{\phi,j}(\widehat{X}, X) = \frac{1}{C_j H_j W_j} \parallel \phi_j(\widehat{X}) - \phi_j(X) \parallel_2^2 \tag{10}$$

ϕ represents a VGG19 network, where C_j, H_j, and W_j respectively denote the number of channels, height, and width of the feature maps at different level j. Finally, we optimize the network using a combination of the two loss functions:

$$\mathcal{L} = \alpha \cdot \mathcal{L}1(\widehat{X}, X) + \beta \cdot \mathcal{L}_{feat}^{\phi,j}(\widehat{X}, X) \tag{11}$$

Here, α and β are the weight factors, which both set to 0.5 empirically.

3   Experiments

3.1   Baseline Methods

We compare the performance of our method with three traditional pan-sharpening methods, including Brovey [8], IHS [3], and GFPCA [15], and five deep learning-based pan-sharpening methods, including PANNET [26], MSD-CNN [29], SRPPNN [2], GPPNN [25], and BAM [13].

3.2   Implementation Details

We implement our network on the PC with a single NVIDIA TITAN RTX 3090 GPU, we build our network in Pytorch framework. In the training phase, the Adam optimizer is used to optimize the model parameters over 1000 epochs. The learning rate is set to 8×10^{-4} and the batch size is set to 4. When reaching 200 epochs, the learning rate is decayed by multiplying 0.5.

3.3   Datasets and Evaluation Metrics

Datasets. We use three satellite image datasets in our experiments, including WorldView-II, WorldView-III, and GaoFen2, each containing several hundred PAN-MS image pairs. The PAN images are cropped into patches with the size of 128×128, and the corresponding MS patches are with the size of 32×32. We also test models on 200 full-resolution real datasets without down-sampling to compare their generalization ability.

Evaluation Metrics. We use PSNR, SSIM, SAM, and the relative dimensionless global error in synthesis(ERGAS) as quantitative metrics to evaluate the image quality on three datasets. Since the full-resolution real dataset does not contain ground truth, we use three no-reference image quality evaluation metrics to assess model performance, including the spectral distortion index D_λ, the spatial distortion index D_S, and the quality without reference QNR.

3.4   Comparison with SOTA Methods

Table 1 and Table 2 present the performance of our proposed method and the baseline methods on three datasets. Compared to the second-best results, our method achieves a PSNR improvement of 0.01, 0.009, and 0.008 dB. Moreover, our method has shown significant improvements across other metrics as well. These comparisons demonstrate the effectiveness of our proposed method. Additionally, we show the fusion results of representative samples from the WV2, WV3, and GF2 datasets, respectively (as shown in Figs. 4, 5, and 6). Table 3 shows the performance of our method on full-resolution real datasets, which indicates that our method has better generalization ability.

Table 1. Quantitative comparison of reference metrics on WV2 and WV3 datasets. The best and the second-best values are highlighted by the red and blue respectively. The up or down arrow indicates higher or lower metric corresponding to better images.

Method	WorldView-II				WorldView-III			
	PSNR ↑	SSIM ↑	SAM ↓	ERGAS ↓	PSNR ↑	SSIM ↑	SAM ↓	ERGAS ↓
Brovey	35.8646	0.9216	0.0403	1.8238	22.5060	0.5466	0.1159	8.2331
IHS	35.2962	0.9027	0.0461	2.0278	22.5579	0.5354	0.1266	8.3616
GFPCA	34.5581	0.9038	0.0488	2.1411	22.3344	0.4826	0.1294	8.3964
PANNET	40.8176	0.9626	0.0257	1.0557	29.6840	0.9072	0.0851	3.4263
MSDCNN	41.3355	0.9664	0.0242	0.9940	30.3038	0.9184	0.0782	3.1884
SRPPNN	41.4538	0.9679	0.0233	0.9899	30.4346	0.9202	0.0770	3.1553
GPPNN	41.1622	0.9684	0.0244	1.0315	30.1785	0.9175	0.0776	3.2593
BAM	41.3527	0.9671	0.0239	0.9932	30.3845	0.9188	0.0773	3.1679
Ours	41.4665	0.9689	0.0231	0.9897	30.4443	0.9222	0.0772	3.1547

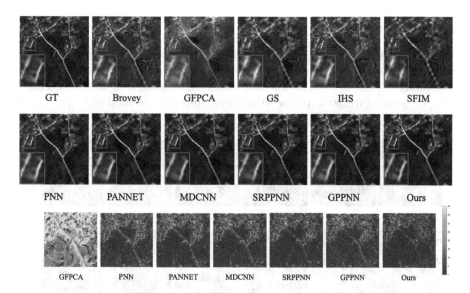

Fig. 4. The visual comparisons between other pan-sharpening methods and our method on WorldView-II satellite.

Table 2. Quantitative comparison of reference metrics on GF2 dataset.

Method	GaoFen2			
	PSNR ↑	SSIM ↑	SAM ↓	ERGAS ↓
Brovey	37.7974	0.9026	0.0218	1.372
IHS	38.1754	0.9100	0.0243	1.5336
GFPCA	37.9443	0.9204	0.0314	1.5604
PANNET	43.0659	0.9685	0.0178	0.8577
MSDCNN	45.6874	0.9827	0.0135	0.6389
SRPPNN	47.1998	0.9877	0.0106	0.5586
GPPNN	44.2145	0.9815	0.0137	0.7361
BAM	45.7419	0.9836	0.0134	0.6267
Ours	47.2079	0.9881	0.0102	0.6194

Table 3. Evaluation on the real-world full-resolution dataset. The best and the second-best values are highlighted by the red and blue respectively.

Metrics	Brovey	IHS	GFPCA	PANNET	MSDCNN	SRPPNN	GPPNN	BAM	Ours
$D_\lambda\downarrow$	0.1378	0.0770	0.0914	0.0737	0.0734	0.0767	0.0782	0.0755	0.0347
$D_S\downarrow$	0.2605	0.2985	0.1635	0.1224	0.1151	0.1162	0.1253	0.1159	0.1110
QNR↑	0.6390	0.6485	0.7615	0.8143	0.8251	0.8173	0.8073	0.8211	0.8527

The top two rows compare the fusion results with SOTA methods, and the bottom row compares the MSE residual between the fusion results and the ground truth. Upon zooming in on local regions of the fused image, it becomes evident that our proposed method preserves both spatial and spectral details better than existing methods. Moreover, the MSE residual comparison also indicates powerful information preservation capabilities of our method.

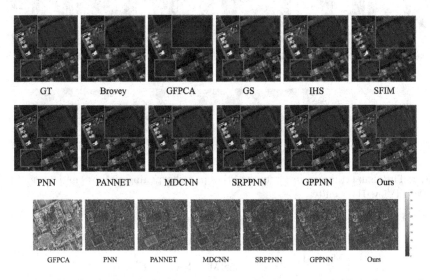

Fig. 5. The visual comparisons between other pan-sharpening methods and our method on WorldView-III satellite.

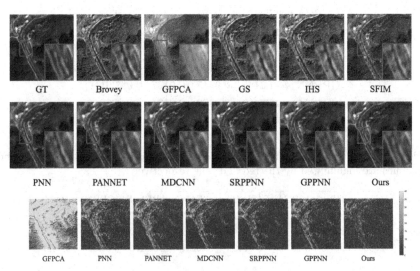

Fig. 6. The visual comparisons between other pan-sharpening methods and our method on GaoFen2 satellite.

3.5   Ablation Experiments

To validate the effectiveness of our proposed Implicit High-frequency Feature Alignment Module (IHFA), we further conduct ablation experiments. Specifically, we replace IHFA with bilinear interpolation and compare the fusion results of the two approaches. Table 4 shows the performance comparison. The results demonstrate that IHFA can significantly improve the fusion performance.

Table 4. Ablation studies about IHFA on WV3 dataset.

Method	WorldView-III			
	PSNR ↑	SSIM ↑	SAM ↓	ERGAS ↓
MSDCNN	30.3038	0.9184	0.0782	3.1884
SRPPNN	30.4346	0.9202	0.0770	3.1553
GPPNN	30.1785	0.9175	0.0776	3.2593
Ours w/o IHFA	29.0897	0.9000	0.0862	3.6729
Ours w/ IHFA	30.4443	0.9222	0.0772	3.1547

4   Conclusion

In this paper, we propose a novel pan-sharpening method from the perspective of implicit neural representation. The core idea is to fuse the multi-modal information containing spatial and spectral information in a continuous implicit space. To the best of our knowledge, this is the first attempt to introduce the concept of implicit space into the pan-sharpening field. In the future, we will further explore how we can better represent spectral and high-frequency information in an implicit space to improve the performance of our approach.

Acknowledgements. The work was supported in part by the National Natural Science Foundation of China under Grant 82172033, U19B2031, 61971369, 52105126, 82272071, 62271430, and the Fundamental Research Funds for the Central Universities 20720230104.

References

1. Ballester, C., Caselles, V., Igual, L., Verdera, J., Rougé, B.: A variational model for P+XS image fusion. Int. J. Comput. Vision **69**(1), 43 (2006)
2. Cai, J., Huang, B.: Super-resolution-guided progressive pansharpening based on a deep convolutional neural network. IEEE Trans. Geosci. Remote Sens. **59**(6), 5206–5220 (2020)
3. Carper, W., Lillesand, T., Kiefer, R., et al.: The use of intensity-hue-saturation transformations for merging spot panchromatic and multispectral image data. Photogramm. Eng. Remote. Sens. **56**(4), 459–467 (1990)

4. Chen, Y., Liu, S., Wang, X.: Learning continuous image representation with local implicit image function. In: Proceedings of the IEEE/CVF Conference on Computer Vision and Pattern Recognition, pp. 8628–8638 (2021)

5. Deng, L.J., Vivone, G., Jin, C., Chanussot, J.: Detail injection-based deep convolutional neural networks for pansharpening. IEEE Trans. Geosci. Remote Sens. **59**(8), 6995–7010 (2020)

6. Dong, C., Loy, C.C., He, K., Tang, X.: Image super-resolution using deep convolutional networks. IEEE Trans. Pattern Anal. Mach. Intell. **38**(2), 295–307 (2015)

7. Fu, X., Wang, W., Huang, Y., Ding, X., Paisley, J.: Deep multiscale detail networks for multiband spectral image sharpening. IEEE Trans. Neural Netw. Learn. Syst. **32**(5), 2090–2104 (2020)

8. Gillespie, A.R., Kahle, A.B., Walker, R.E.: Color enhancement of highly correlated images. ii. channel ratio and "chromaticity" transformation techniques. Remote Sens. Environ. **22**(3), 343–365 (1987)

9. He, L., et al.: Pansharpening via detail injection based convolutional neural networks. IEEE J. Sel. Top. Appl. Earth Obs. Remote Sens. **12**(4), 1188–1204 (2019)

10. Henzler, P., Mitra, N.J., Ritschel, T.: Learning a neural 3D texture space from 2D exemplars. In: Proceedings of the IEEE/CVF Conference on Computer Vision and Pattern Recognition, pp. 8356–8364 (2020)

11. Jiang, C., Sud, A., Makadia, A., Huang, J., Nießner, M., Funkhouser, T., et al.: Local implicit grid representations for 3D scenes. In: Proceedings of the IEEE/CVF Conference on Computer Vision and Pattern Recognition, pp. 6001–6010 (2020)

12. Jiang, Y., Ding, X., Zeng, D., Huang, Y., Paisley, J.: Pan-sharpening with a hyper-laplacian penalty. In: Proceedings of the IEEE International Conference on Computer Vision, pp. 540–548 (2015)

13. Jin, Z.R., Deng, L.J., Zhang, T.J., Jin, X.X.: Bam: bilateral activation mechanism for image fusion. In: Proceedings of the 29th ACM International Conference on Multimedia, pp. 4315–4323 (2021)

14. Kwarteng, P., Chavez, A.: Extracting spectral contrast in landsat thematic mapper image data using selective principal component analysis. Photogramm. Eng. Remote. Sens. **55**(1), 339–348 (1989)

15. Liao, W., et al.: Two-stage fusion of thermal hyperspectral and visible RGB image by PCA and guided filter. In: 2015 7th Workshop on Hyperspectral Image and Signal Processing: Evolution in Remote Sensing (WHISPERS), pp. 1–4. IEEE (2015)

16. Mallat, S.G.: A theory for multiresolution signal decomposition: the wavelet representation. IEEE Trans. Pattern Anal. Mach. Intell. **11**(7), 674–693 (1989)

17. Masi, G., Cozzolino, D., Verdoliva, L., Scarpa, G.: Pansharpening by convolutional neural networks. Remote Sens. **8**(7), 594 (2016)

18. Mildenhall, B., Srinivasan, P.P., Tancik, M., Barron, J.T., Ramamoorthi, R., Ng, R.: Nerf: representing scenes as neural radiance fields for view synthesis. Commun. ACM **65**(1), 99–106 (2021)

19. Niemeyer, M., Mescheder, L., Oechsle, M., Geiger, A.: Differentiable volumetric rendering: Learning implicit 3D representations without 3D supervision. In: Proceedings of the IEEE/CVF Conference on Computer Vision and Pattern Recognition, pp. 3504–3515 (2020)

20. Oechsle, M., Mescheder, L., Niemeyer, M., Strauss, T., Geiger, A.: Texture fields: learning texture representations in function space. In: Proceedings of the IEEE/CVF International Conference on Computer Vision, pp. 4531–4540 (2019)

21. Park, J.J., Florence, P., Straub, J., Newcombe, R., Lovegrove, S.: Deepsdf: learning continuous signed distance functions for shape representation. In: Proceedings of the IEEE/CVF Conference on Computer Vision and Pattern Recognition, pp. 165–174 (2019)

22. Saito, S., Huang, Z., Natsume, R., Morishima, S., Kanazawa, A., Li, H.: PIFu: pixel-aligned implicit function for high-resolution clothed human digitization. In: Proceedings of the IEEE/CVF International Conference on Computer Vision, pp. 2304–2314 (2019)

23. Sitzmann, V., Martel, J., Bergman, A., Lindell, D., Wetzstein, G.: Implicit neural representations with periodic activation functions. Adv. Neural. Inf. Process. Syst. **33**, 7462–7473 (2020)

24. Vivone, G., et al.: A critical comparison among pansharpening algorithms. IEEE Trans. Geosci. Remote Sens. **53**(5), 2565–2586 (2014)

25. Xu, S., Zhang, J., Zhao, Z., Sun, K., Liu, J., Zhang, C.: Deep gradient projection networks for pan-sharpening. In: Proceedings of the IEEE/CVF Conference on Computer Vision and Pattern Recognition, pp. 1366–1375 (2021)

26. Yang, J., Fu, X., Hu, Y., Huang, Y., Ding, X., Paisley, J.: Pannet: a deep network architecture for pan-sharpening. In: Proceedings of the IEEE International Conference on Computer Vision, pp. 5449–5457 (2017)

27. Yang, Y., Tu, W., Huang, S., Lu, H., Wan, W., Gan, L.: Dual-stream convolutional neural network with residual information enhancement for pansharpening. IEEE Trans. Geosci. Remote Sens. **60**, 1–16 (2021)

28. Yuan, Q., Wei, Y., Meng, X., Shen, H., Zhang, L.: A multiscale and multidepth convolutional neural network for remote sensing imagery pan-sharpening. IEEE J. Sel. Top. Appl. Earth Obs. Remote Sens. **11**(3), 978–989 (2018)

29. Yuan, Q., Wei, Y., Zhang, Z., Shen, H., Zhang, L.: A multi-scale and multi-depth convolutional neural network for remote sensing imagery pan-sharpening (2017)

30. Zhang, H., Ma, J.: GTP-PNet: a residual learning network based on gradient transformation prior for pansharpening. ISPRS J. Photogramm. Remote. Sens. **172**, 223–239 (2021)

Stable Visual Pattern Mining via Pattern Probability Distribution

Yifan Ma[2], Xuefeng Liang[1,2(✉)], Xiaoyu Lin[2], and Guanghui Shi[1]

[1] School of Artificial Intelligence, Xidian University, Xi'an, China
xliang@xidian.edu.cn
[2] Guangzhou Institute of Technology, Xidian University, Guangzhou, China

Abstract. Visual patterns are the fundamental elements that compose an image and often convey higher-level semantics. Visual pattern mining can be widely applied to real-world applications and various downstream tasks, such as tourist destination recommendation, scenic area promotion, etc. Therefore, it is a fundamental issue in computer vision. Visual pattern has two properties: *frequency* and *discrimination*. Most emerging studies first perform self-supervised representation learning and a supervised classification task to extract discriminative feature, and then mine the visual patterns through clustering algorithms to ensure the frequency. However, these methods require the manual setting of the hyperparameters (e.g. clustering radius *eps* in DBSCAN, etc.) to which the mining results are sensitive. Improper hyperparameters will result in obtaining redundant visual patterns or missing visual patterns, and the optimal hyperparameters differ among categories. To address these problems, we propose a new stable mining method which directly obtains visual patterns in each category without using clustering algorithms. First, we propose a Pattern Distribution Extractor (PDE) to ensure the discrimination and extract the pattern probability distribution, which represents the probability of the image belonging to each candidate pattern cluster. Experiments demonstrate pattern probability distribution provides more stability in mining stage than the previous methods. Further, we propose a Semantic Density Mining (SDM) strategy which first adaptively removes the non-frequent images to ensure the frequency and then directly obtains the visual patterns in each category according to the pattern probability distribution. Extensive experiments demonstrate the effectiveness of our proposed method.

Keywords: visual pattern mining · self-supervised learning · joint optimization

Supplementary Information The online version contains supplementary material available at https://doi.org/10.1007/978-981-99-8543-2_23.

Q. Liu et al. (Eds.): PRCV 2023, LNCS 14432, pp. 280–292, 2024.
https://doi.org/10.1007/978-981-99-8543-2_23

1   Introduction

Visual patterns are the basic elements that compose an image and tend to convey higher-level semantics than raw pixels. Therefore, visual patterns can be considered as combinations of low-level visual features that exhibit some high-level semantics. Visual patterns are widely available in the real world. For example, when recommending tourist destinations [21] or promoting scenic areas [24] on a travel website, we expect to mine scenic hot spots that tourists are interested in from massive amounts of photos taken by tourists, and then recommend them to other people. Figure 1 shows the mined hot spots in Toore de Belem and Taipei 101 from the Travel dataset [19] by methods PaclMap [19], which can be considered as visual patterns. The task of hot spot mining (visual pattern mining) is to mine all the hot spots (visual patterns) without prior knowledge of these places (number and label of visual pattern) but only knowing the scenic area (category label) of photographs where they were taken. Visual pattern mining can be also applied to various downstream tasks, such as image retrieval [14], co-localization [23] and co-segmentation [15,28]. Therefore, visual pattern mining has become an important computer vision task. Visual patterns have two

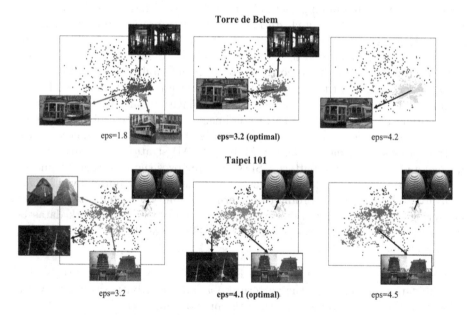

Fig. 1. The mining results of PaclMap [19] attained by DBSCAN with different clustering radius (*eps*) on category Toore de Belem and Taipei 101 from the Travel dataset [19]. It shows that (1) the mining results are sensitive to the hyperparameters of clustering algorithm. Improper settings will result in redundant visual patterns (see images pointed by the red arrows in the left subfigure) or missing visual patterns (see images in the right subfigure). (2) the optimal clustering radius (*eps*) varies among different categories, it is difficult to find the best hyperparameter for each category.

properties: (1) *discrimination*, which means it represents only one particular category rather than the other categories, and (2) *frequency*, which means it frequently appears in images of the category. The aim of the visual pattern mining task is to mine all visual patterns in each category when neither the label nor the number of patterns is known (we assume that one image contains one visual pattern at most).

Emerging visual pattern mining methods [19,25] usually design a deep model to extract discriminative features by jointly optimizing supervised classification loss and self-supervised loss functions, and then apply a clustering algorithm (e.g., DBSCAN [8]) to obtain pattern clusters, which ensures the frequency. However, the clustering algorithm requires the manual setting of hyperparameters (e.g. clustering radius *eps* in DBSCAN), to which the mining results are sensitive. Figure 1 shows the mining result of PaclMap with different *eps*. Improper hyperparameters will result in obtaining redundant visual patterns or missing visual patterns. In addition, since the semantic of patterns vary greatly among different categories, the inter-pattern features are more compact (see Torre de Belem in Fig. 1) if visual patterns are semantically similar to each other, and sparse (see Taipei 101 in Fig. 1) if they are not. Therefore, it is difficult to set a hyperparameter that can be universally applied to all categories. To address these issues, we propose a new stable mining method which directly obtains visual patterns in each category without using clustering algorithms. First, we propose a Pattern Distribution Extractor (PDE) which extracts the pattern probability distribution through self-supervised learning and performs a supervised classification task to ensure the discrimination. The pattern probability distribution represents the probability of the image belonging to each candidate pattern cluster. And images containing the same visual pattern have similar pattern probability distributions after optimizing PDE. Experiments demonstrate probability distributions provides more stability in mining stage than the previous methods. We further propose a Semantic Density Mining (SDM) strategy to ensure the frequency and obtain visual patterns. SDM first removes the non-frequent images adaptively according to semantic density, and then directly obtains visual patterns for each category according to pattern probability distributions instead of clustering algorithms. Extensive experiments on multiple benchmark datasets demonstrate the effectiveness of our method. The contributions of this paper can be summarized as follows:

1. We propose a Pattern Distribution Extractor (PDE) which ensures the discrimination and extracts the pattern probability distribution which provides more stability in mining stage than the previous methods.
2. We propose a Semantic Density Mining (SDM) strategy which ensures the frequency and obtains visual patterns in each category directly according to pattern probability distribution instead of using clustering algorithms.
3. Extensive experiments show that our method achieves the best results on multiple benchmark datasets.

2   Related Work

2.1   Visual Pattern Mining

Visual pattern mining is a fundamental issue in computer vision. The earliest methods applied the association rules [1] on HOG [7] or SIFT [20] to mine the visual patterns. However, such handcrafted features have little semantic information. Recently, many methods inclined to choose the middle-level features of the convolutional network because of their high-level semantic information. These methods can be grouped into three streams. The first group [18,29] focused on the frequency. Most of them extracted middle-level features from CNN, then used association rules [18] or mean shift [29] to obtain visual patterns. However, the mined patterns usually have worse discrimination. The second group [16,27] focused on the discrimination. Such methods generally defined a classification task and mined visual patterns according to the combinations of the activations of convolutional kernels [16] or the activations to different patches in the image [27]. However, these methods can not guarantee the frequency. The third group [19,25] is able to mine both frequent and discriminative patterns. Wang et al. [25] used triplet loss and cross-entropy loss to ensure the frequency and discrimination simultaneously. Liang et al. [19] jointly optimized Pacl and Map to ensure the frequency and discrimination. However, these methods obtain visual patterns through clustering algorithms with manual setting of hyperparameters, to which the mining results are sensitive. In contrast, we propose a new method which obtains visual patterns directly without using clustering algorithms.

2.2   Self-supervised Learning

There are large numbers of work [2,4,11,26] on self-supervised learning which focus on instance classification. These methods considered each image as a different class and the model needs to distinguish them up to the data augmentations. Alexey et al. [2] learned a classifier to distinguish all images which can not scale well with the number of images. Wu et al. [26] used contrastive loss [10] instead of classification, which requires large batchsize [4] or memory bank [11,26]. However, these methods may mistake the samples with the same pattern for negatives which leads the network to overly focus on instance features. Other works [3,5,9] have shown we can learn unsupervised features without discriminating between images. Grill et al. [9] learned the feature by matching them to representations obtained with a momentum encoder. Chen et al. [5] learned the feature through constraining cosine similarity between different views of image. However, these methods mainly focused on learning semantic features rather than clustering. Caron et al. [3] proposed a method called DINO, which learns the same distribution between different views of image and encourages clustering of similar images in the feature space. Motivated by DINO, we extract the pattern probability distribution, which can obtain visual patterns directly.

3   Proposed Method

In this section, we first formulate the visual pattern mining task (Sect. 3.1). Then we propose a Pattern Distribution Extractor (PDE) (Sect. 3.2) to ensures the discrimination and extract the pattern probability distribution. Finally, we propose a Semantic Density Mining (SDM) strategy (Sect. 3.3) to ensure the frequency and obtain visual patterns in each category directly.

3.1   The Definition of Visual Pattern Mining

Given an image dataset $X = \{x_i\}_{i=1}^N$ and the given label $Y = \{y_i\}_{i=1}^N$, where $y_i = \{0,1\}^{|C|}$, $\sum_{c=1}^C y_{i,c} = 1$, C is the number of categories. The visual pattern set of the whole dataset can be represented as $R = \{R_c\}_{c=1}^C$ (unknown for mining methods), where $R_c = \{r_c^i\}_{i=1}^{T_c}$ is the visual pattern set of category c, r_c^i is the ith pattern and T_c is the pattern number of category c. The visual pattern mining task aims to train a model M, which can unsupervisedly mine all visual patterns R_c in each category c from the dataset X according to the given label Y, and it can be formulated as Eq. 1.

$$R = \text{Mining}(M(X,Y)) \tag{1}$$

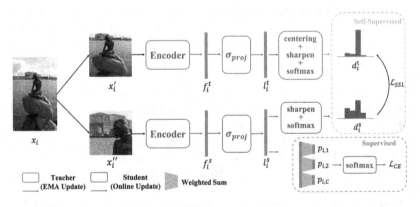

Fig. 2. The structure of Pattern Distribution Extractor (PDE). The PDE mainly contains encoder E which extracts feature f, and MLP σ_{proj} generates l, which can be normalized to pattern probability distribution d through $softmax$. For the self-supervised learning, we constrain the KL divergence between d_i^t and d_i^s. For the supervised classification, we apply the weighted sum on l_i^s to obtain classification logit $p_{i,c}$ for each class c (c=1,...,C in the figure), which can be used in cross-entropy loss.

3.2   Pattern Distribution Extractor (PDE)

The proposed Pattern Distribution Extractor (PDE) is shown in Fig. 2. It contains a self-supervised learning which aims to extract the pattern probability distribution, and a supervised classification task which aims to ensure the discrimination through the given label.

For the self-supervised learning, we first use an encoder E to extract the feature $f = E(x)$ of the image x. Emerging pattern mining methods [19,25] usually focused on feature f and performed representation learning through contrastive loss. Even if they were able to extract the optimal features, they would still need additional clustering algorithms to obtain the visual patterns. Different from these studies, we expect to extract the pattern probability distribution d of each image x so that visual patterns can be obtained directly. Specifically, we assume that each category has $K/C(K \gg C)$ visual patterns at most, and build a MLP σ_{proj} with $l2$ normalization and a weight normalized FC layer to further map the feature f into a K-dimensional vector $l = \sigma_{proj}(f) \in \mathbb{R}^K$, where each weight vector of the last layer of σ_{proj} and each value of l can be considered as a candidate pattern cluster center and the similarity between feature f and candidate pattern cluster center, respectively. Then, we normalize l through a *softmax* layer to obtain pattern probability distribution d. Each value d^k can be regarded as the probability of image x belonging to the kth candidate pattern cluster. We expect that images containing the same visual pattern have similar pattern probability distributions. To achieve this, motivated by DINO [3], we additionally build a teacher model with the same structure updated with an exponential moving average (EMA) of student model (mentioned above, updated online). Specifically, for image x_i, we generate two views of x_i by multi-crop strategy and augmentation, and then obtain pattern probability distributions d_i^s and d_i^t as shown in Eq. 2, where l_i^s and l_i^t are two vectors from student model and teacher model, τ_s and τ_t are sharpen temperature. Finally, we minimize the KL divergence between d_i^t and d_i^s as shown in Eq. 3. In addition, we also add a bias *center* to teacher model to avoid model collapse [3].

$$
\begin{cases}
d_i^s = \text{softmax}(l_i^s/\tau_s) \\
d_i^t = \text{softmax}\left(\left(l_i^t - center\right)/\tau_t\right)
\end{cases}
\tag{2}
$$

$$
\mathcal{L}_{SSL} = \frac{1}{N} \sum_i^N KL\left(d_i^t \parallel d_i^s\right)
\tag{3}
$$

For the supervised classification task, we first divide l_i^s into C parts along the channel dimension, and each part $l_{i,c}^s \in \mathbb{R}^{K/C}$ represents the similarities between f and all candidate pattern cluster centers in category c. In order to obtain the classification logit $p_{i,c}$ of each category c, motivated by SoftTriple [22], we perform weighted sum as shown in Eq. 4, where $l_{i,c}^{s,k}$ denotes the kth value of the $l_{i,c}^s$. Finally, the cross-entropy loss is formulated as Eq. 5, where $\hat{p}_i = \text{softmax}(p_i)$. The total objective function is shown in Eq. 6. After optimizing the PDE, we just maintain the teacher model [19] for further processing.

$$
p_{i,c} = \sum_k^{K/C} \frac{\exp\left(\frac{1}{\tau_s}l_{i,c}^{s,k}\right)}{\sum_k \exp\left(\frac{1}{\tau_s}l_{i,c}^{s,k}\right)} l_{i,c}^{s,k}
\tag{4}
$$

$$\mathcal{L}_{CE} = -\frac{1}{N} \sum_i^N y_i \log \hat{p}_i \tag{5}$$

$$\mathcal{L}_{\text{total}} = \mathcal{L}_{CE} + \mathcal{L}_{SSL} \tag{6}$$

3.3   Semantic Density Mining (SDM) Strategy

There are two types of features after optimizing PDE as shown in Fig. 3 (a): (1) features with low density of non-frequent images which may damage the mining performance. (2) discriminative and frequent features with high density. In order to adaptively remove these non-frequent images (ensure the frequency) and then directly obtain the visual patterns in each category, we propose a Semantic Density Mining (SDM) strategy, which contains a preprocessing stage and a mining stage. For preprocessing stage, we first define ε_i, the average similarity between f_i^t and its N_m nearest features, as the semantic density of image x_i. Then, we construct the histogram of semantic density $\{\varepsilon_i\}_{y_i=c}$ and fit it through a GMM with two Gaussian components for each category c (see Fig. 3). Further, we define a threshold β which indicates the posterior probability of non-frequent image and simply set $\beta = 0.9$ for all categories. Finally, we directly remove the images with the posterior probability greater than β. As shown in Fig. 3 (b), the preprocessing stage can adaptively remove the non-frequent images without damaging the visual pattern clusters. For the mining stage, since the pattern probability distribution d is approximately a one-hot vector after training PDE and images containing the same visual patterns share similar pattern probability distribution [3], we can directly obtain the visual pattern clusters in each category. Specifically, we consider the pattern cluster which the image x_i belongs to is $argmax \ d_i$. Thus, x_i and x_j belong to the same pattern cluster when $argmax \ d_i = argmax \ d_j$. Finally, we can obtain all effective pattern clusters and each effective pattern cluster is considered as a visual pattern.

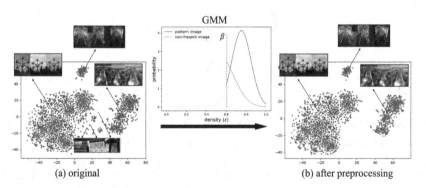

(a) original   (b) after preprocessing

Fig. 3. The T-SNE visualization result of f^t from Atomium in Travel. (a) Original visualization result. There exist non-frequent images which may damage the mining performance. (b) After preprocessing stage in SDM. The preprocessing stage can adaptively remove the non-frequent images without damaging the visual pattern clusters.

4 Experiment

To demonstrate the effectiveness of our proposed method, we compare it with the six SOTA methods: MDPM [17], CBMS [29], P-CNN [27], PatternNet [16], JDFR [25] and PaclMap [19] on the following benchmark datasets: Place-20 [30], ILSVRC-20 [6], Travel [19] and CIFAR100 [13]. For CIFAR100, we consider the five subclasses in a superclass as different visual patterns in one category (more dataset details please see the Supplementary). We use the same backbone for all methods above for a fair comparison.

4.1 Evaluation Metrics

We evaluate the discrimination and frequency of our method through the following two metrics, respectively. Following [19], we select 20 images for each pattern which are nearest to the pattern cluster center for evaluation. For the discrimination metric, we follow the previous studies and use the image classification task as a proxy to evaluate the discrimination. We first train a Resnet50 on each dataset [19], and then evaluate the average accuracy (ACC) of those 20 images. For the frequency metric, we follow [19] to use the Frequency Rate (FR) to measure the frequency of mined patterns, which computes the percentage of the images that are similar to the mined visual patterns in the feature space.

Table 1. ACC and FR of seven methods on four datasets.

ACC	Single pattern ($T_c = 1$)				Multiple patterns ($T_c > 1$)				All patterns				
	ILSVRC20	CIFAR100	Place20	Travel	ILSVRC20	CIFAR100	Place20	Travel	ILSVRC20	CIFAR100	Place20	Travel	Average
MDPM	0.930	–	0.835	0.950	0.932	0.935	0.925	0.947	0.932	0.935	0.898	0.948	0.9283
CBMS	0.965	–	0.910	0.956	0.965	0.962	0.927	0.968	0.965	0.962	0.922	0.964	0.9530
PCNN	0.980	–	0.910	0.972	0.973	0.965	0.946	0.978	0.974	0.965	0.935	0.976	0.9625
PatternNet	0.970	–	0.930	0.967	0.968	0.960	0.956	0.980	0.969	0.960	0.948	0.976	0.9633
JDFR	0.995	–	**0.980**	0.956	0.973	0.968	0.951	0.953	0.975	0.968	0.960	0.964	0.9668
PaclMap	0.995	–	**0.980**	**1.000**	0.988	0.982	0.989	**0.989**	0.989	0.982	0.986	**0.993**	0.9875
Ours	**1.000**	–	**0.980**	**1.000**	**0.992**	**0.987**	0.989	**0.989**	**0.993**	**0.987**	**0.988**	**0.993**	**0.9903**

FR	Single pattern ($T_c = 1$)				Multiple patterns ($T_c > 1$)				All patterns				
	ILSVRC20	CIFAR100	Place20	Travel	ILSVRC20	CIFAR100	Place20	Travel	ILSVRC20	CIFAR100	Place20	Travel	Average
MDPM	0.716	–	0.173	0.515	0.943	0.246	0.316	0.492	0.867	0.246	0.291	0.500	0.4760
CBMS	0.705	–	0.162	0.623	0.960	0.286	0.301	0.594	0.876	0.286	0.305	0.604	0.5178
PCNN	0.696	–	0.160	0.507	0.933	0.237	0.269	0.472	0.861	0.237	0.265	0.484	0.4618
PatternNet	0.694	–	0.160	0.506	0.972	0.228	0.277	0.484	0.877	0.228	0.268	0.492	0.4663
JDFR	0.782	–	0.196	0.495	0.978	0.287	0.271	0.379	**0.909**	0.287	0.270	0.453	0.4798
PaclMap	0.786	–	**0.197**	**0.639**	0.987	0.294	0.392	**0.677**	0.922	0.294	0.372	**0.664**	0.5630
Ours	**0.789**	–	**0.197**	**0.639**	**0.994**	**0.301**	**0.402**	**0.677**	**0.930**	**0.301**	**0.381**	**0.664**	**0.5690**

4.2 Implementation Details

The model is optimized by Adam optimizer [12]. We set $\tau_s = 0.1$ and $\tau_t = 0.04$ as the same as DINO [3]. The batchsize is set to 128 and the learning rate is 5e-4 with a cosine decay schedule for 300 epochs. We set $N_m = 20$ for the SDM.

4.3 Comparison with the State-of-the-Art Methods

Quantitative Comparison. We first evaluate the ACC and FR of our method and six SOTA models. As the categories in these datasets may have a single pattern or multiple patterns, the results are organized in terms of the number of patterns in a category as shown in Table 1. Our proposed method can ensure the discrimination and frequency through the PDE and SDM, respectively, achieving the best performances on all four datasets. For discrimination metric ACC, MDPM [17] and CBMS [29] are of the lowest performances, because they both mainly concentrate on the frequency other than the discrimination. JDFR [25] uses cross-entropy to ensure the discrimination. PatternNet [16] and P-CNN [27] use the mid-level features of networks to mine discriminative patterns, but are inferior to JDFR. PaclMap [19] can ensure discrimination because of using the multiple linear combinations mechanism and achieve a good result. For frequency metric FR, CBMS achieves a good result because of using mean shift to guarantee the frequency. JDFR utilizes the triplet loss to ensure the frequency and also achieves a good result. MDPM divides the image into patches, which may lose the semantic information, resulting in inferior to JDFR. P-CNN and PatternNet focus on the discrimination, their results are the worst. PaclMap can ensure high frequency because of using DBSCAN with the optimal hyperparameters.

Table 2. The number of coarse categories with different number of visual patterns mined by our method and PaclMap on CIFAR100 (T': the number of mined visual patterns in each category).

CIFAR100	$T' = 1$	$T' = 2$	$T' = 3$	$T' = 4$	$T' = 5$
PaclMap ($eps = 3.0$)	0	1	2	7	10
PaclMap ($eps = 4.5$)	3	4	10	2	1
PaclMap ($eps = 6.0$)	8	7	4	1	0
Ours ($\beta = 0.5$, $N_m = 20$)	0	1	3	7	9
Ours ($\beta = 0.7$, $N_m = 20$)	0	1	2	8	9
Ours ($\beta = 0.9$, $N_m = 20$)	0	1	1	8	10
Ours ($\beta = 0.999$, $N_m = 20$)	0	1	1	8	10
Ours ($\beta = 0.9$, $N_m = 5$)	0	1	1	8	10
Ours ($\beta = 0.9$, $N_m = 20$)	0	1	1	8	10
Ours ($\beta = 0.9$, $N_m = 50$)	0	1	3	7	9
Ours ($\beta = 0.9$, $N_m = 100$)	0	2	4	6	8

Additionally, We evaluate the stability of our method and PaclMap which achieves the best performance in previous six SOTA methods mentioned above. Table 2 shows the number of coarse categories with different number of visual patterns mined by our method and PaclMap on CIFAR100. The total number of coarse categories is 20 and each coarse category contains five visual patterns (the sum of each line in Table 2 is 20). The results are organized by the number of mined visual patterns. For PaclMap, we evaluate the clustering radius (*eps*) in DBSCAN. When the clustering radius (*eps*) is small (such as $eps = 3.0$), PaclMap can obtain most of the visual patterns in most coarse categories. However, the number of mined patterns significantly decreases as the clustering radius gradually increases. For our proposed method, we evaluate all two hyperparameters in the mining stage of our method, the posterior probability β and the N_m, respectively. When the β is small (such as $\beta = 0.5$), SDM tends to remove more images in each category, which may lead to incorrect removal of images containing visual pattern and thus resulting in the omission of visual patterns. However, it also achieves a similar result to the optimal performance of our proposed method, which show the stability of our method. We also evaluate our method with different N_m ($N_m = 5, 20, 50, 100$), and the results show that a large N_m will slightly degrade the mining performance but still can mine most visual patterns in the most categories, which demonstrates the stability of our proposed method (more quantitative comparisons please see the Supplementary).

Qualitative Comparison. Figure 4 shows the visual patterns of Taipei 101 in Travel and Arena-Rodeo in Place20 mined by all methods. MDPM and P-CNN can mine more than one pattern, but are not separated and include a few off-target errors. CBMS, PatternNet and JDFR can mine only one visual pattern. PaclMap can mine all visual patterns when setting the optimal hyperparameter in the clustering algorithm. However, our proposed method can also mine all visual patterns directly without using clustering algorithms.

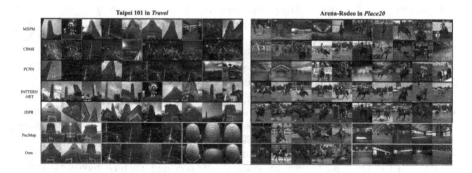

Fig. 4. The results discovered by seven methods from the category Taipei 101 in Travel and Arena-Rodeo in Place20. The visual patterns are marked with different color boxes. In particular, the wrong results are marked with red boxes. (Color figure online)

Table 3. Quantitative results of the ablation study on SDM.

Backbone	SDM (W/O preprocessing)	SDM	DBSCAN	CIFAR100		Travel		ILSVRC20		Place20	
				ACC	*FR*	*ACC*	*FR*	*ACC*	*FR*	*ACC*	*FR*
✓			✓	0.983	0.295	**0.993**	**0.664**	0.992	0.920	0.986	0.377
✓	✓			0.975	0.251	0.989	0.596	0.989	0.898	0.980	0.350
✓		✓		**0.987**	**0.301**	**0.993**	**0.664**	**0.993**	**0.930**	**0.988**	**0.381**

4.4   Ablation Study

We use all four datasets for the ablation study on SDM, and the quantitative results are listed in Table 3. First, we use DBSCAN with optimal hyperparameters to mine visual patterns on feature f^t, which achieves a similar result to SDM. We also evaluate the effectiveness of the preprocessing stage in SDM, the result shows that preprocessing stage can ensure the frequency by removing the non-frequent images achieving large FR improvements on all datasets.

5   Conclusion

Visual pattern mining is a fundamental problem in computer vision, which aims at discovering all visual patterns from a category when neither the label nor the number of patterns is known. However, Emerging studies mine visual patterns through clustering algorithms with manual setting of hyperparameters, to which the mining results are sensitive. We propose a stable mining method without using clustering algorithms, which contains a Pattern Distribution Extractor (PDE) and a Semantic Density Mining (SDM) strategy. PDE extracts pattern probability distribution and ensures the discrimination. SDM strategy ensures the frequency according to semantic density and directly obtains visual patterns instead of clustering algorithms. Extensive experiments demonstrate the effectiveness and stability of our proposed method.

References

1. Agarwal, R., Srikant, R., et al.: Fast algorithms for mining association rules. In: Proceedings of the 20th VLDB Conference, vol. 487, p. 499 (1994)
2. Alexey, D., Fischer, P., Tobias, J., Springenberg, M.R., Brox, T.: Discriminative unsupervised feature learning with exemplar convolutional neural networks. IEEE Trans. Pattern Anal. Mach. Intell. **38**(9), 1734–1747 (2016)
3. Caron, M., et al.: Emerging properties in self-supervised vision transformers. In: Proceedings of the IEEE/CVF International Conference on Computer Vision, pp. 9650–9660 (2021)
4. Chen, T., Kornblith, S., Norouzi, M., Hinton, G.: A simple framework for contrastive learning of visual representations. In: International Conference on Machine Learning, pp. 1597–1607. PMLR (2020)
5. Chen, X., He, K.: Exploring simple siamese representation learning. In: Proceedings of the IEEE/CVF conference on computer vision and pattern recognition, pp. 15750–15758 (2021)
6. Deng, J., Dong, W., Socher, R., Li, L.J., Li, K., Fei-Fei, L.: ImageNet: a large-scale hierarchical image database. In: 2009 IEEE Conference on Computer Vision and Pattern Recognition, pp. 248–255. IEEE (2009)
7. Doersch, C., Singh, S., Gupta, A., Sivic, J., Efros, A.: What makes Paris look like Paris? ACM Trans. Graph. **31**(4), 1–9 (2012)
8. Ester, M., Kriegel, H.P., Sander, J., Xu, X.: A density-based algorithm for discovering clusters in large spatial databases with noise. In: Proceedings of the Second International Conference on Knowledge Discovery and Data Mining, p. 226–231. KDD 1996, AAAI Press (1996)

9. Grill, J.B., et al.: Bootstrap your own latent: a new approach to self-supervised learning. Adv. Neural. Inf. Process. Syst. **33**, 21271–21284 (2020)

10. Gutmann, M., Hyvärinen, A.: Noise-contrastive estimation: a new estimation principle for unnormalized statistical models. In: Proceedings of the Thirteenth International Conference on Artificial Intelligence and Statistics, pp. 297–304. JMLR Workshop and Conference Proceedings (2010)

11. He, K., Fan, H., Wu, Y., Xie, S., Girshick, R.: Momentum contrast for unsupervised visual representation learning. In: Proceedings of the IEEE/CVF Conference on Computer Vision and Pattern Recognition, pp. 9729–9738 (2020)

12. Kingma, D.P., Ba, J.: Adam: a method for stochastic optimization. arXiv preprint arXiv:1412.6980 (2014)

13. Krizhevsky, A., Hinton, G., et al.: Learning multiple layers of features from tiny images (2009)

14. Lei, Z.: Large-scale web image search and pattern mining. Sci. Sin. Inform. **43**(12), 1641–1653 (2013)

15. Li, B., Tang, L., Kuang, S., Song, M., Ding, S.: Toward stable co-saliency detection and object co-segmentation. IEEE Trans. Image Process. **31**, 6532–6547 (2022)

16. Li, H., Ellis, J.G., Zhang, L., Chang, S.F.: PatternNet: visual pattern mining with deep neural network. In: Proceedings of the 2018 ACM on International Conference on Multimedia Retrieval, pp. 291–299 (2018)

17. Li, Y., Liu, L., Shen, C., van den Hengel, A.: Mid-level deep pattern mining. In: Proceedings of the IEEE Conference on Computer Vision and Pattern Recognition, pp. 971–980 (2015)

18. Li, Y., Liu, L., Shen, C., Hengel, A.v.d.: Mining mid-level visual patterns with deep CNN activations. Int. J. Comput. Vision **121**, 344–364 (2017)

19. Liang, X., Liang, Z., Shi, H., Zhang, X., Zhou, Y., Ma, Y.: Multipattern mining using pattern-level contrastive learning and multipattern activation map. IEEE Trans. Neural Networks Learn. Syst. 1–15 (2022)

20. Lowe, D.G.: Object recognition from local scale-invariant features. In: Proceedings of the Seventh IEEE International Conference on Computer Vision, vol. 2, pp. 1150–1157. IEEE (1999)

21. Memon, I., Chen, L., Majid, A., Lv, M., Hussain, I., Chen, G.: Travel recommendation using geo-tagged photos in social media for tourist. Wireless Pers. Commun. **80**, 1347–1362 (2015)

22. Qian, Q., Shang, L., Sun, B., Hu, J., Li, H., Jin, R.: SoftTriple Loss: deep metric learning without triplet sampling. In: IEEE International Conference on Computer Vision, ICCV 2019 (2019)

23. Tang, K., Joulin, A., Li, L.J., Fei-Fei, L.: Co-localization in real-world images. In: Proceedings of the IEEE Conference on Computer Vision and Pattern Recognition, pp. 1464–1471 (2014)

24. Vu, H.Q., Li, G., Law, R., Ye, B.H.: Exploring the travel behaviors of inbound tourists to Hong Kong using geotagged photos. Tour. Manag. **46**, 222–232 (2015)

25. Wang, Q., Zhou, Y., Zhu, Z., Liang, X., Gu, Y.: Jointly discriminating and frequent visual representation mining. In: Proceedings of the Asian Conference on Computer Vision (2020)

26. Wu, Z., Xiong, Y., Yu, S.X., Lin, D.: Unsupervised feature learning via non-parametric instance discrimination. In: Proceedings of the IEEE Conference on Computer Vision and Pattern Recognition, pp. 3733–3742 (2018)

27. Yang, L., Xie, X., Lai, J.: Learning discriminative visual elements using part-based convolutional neural network. Neurocomputing **316**, 135–143 (2018)

28. Yuan, Z.H., Lu, T., Wu, Y., et al.: Deep-dense conditional random fields for object co-segmentation. In: IJCAI, vol. 1, p. 2 (2017)
29. Zhang, W., Cao, X., Wang, R., Guo, Y., Chen, Z.: Binarized mode seeking for scalable visual pattern discovery. In: Proceedings of the IEEE Conference on Computer Vision and Pattern Recognition, pp. 3864–3872 (2017)
30. Zhou, B., Lapedriza, A., Khosla, A., Oliva, A., Torralba, A.: Places: a 10 million image database for scene recognition. IEEE Trans. Pattern Anal. Mach. Intell. **40**(6), 1452–1464 (2017)

Dynamic Visual Prompt Tuning for Parameter Efficient Transfer Learning

Chunqing Ruan[1] and Hongjian Wang[2(✉)]

[1] Beijing University of Posts and Telecommunications, Beijing, China
[2] Jilin University, Jilin, China
forever020200@gmail.com

Abstract. Parameter efficient transfer learning (PETL) is an emerging research spot that aims to adapt large-scale pre-trained models to downstream tasks. Recent advances have achieved great success in saving storage and computation costs. However, these methods do not take into account instance-specific visual clues for visual tasks. In this paper, we propose a Dynamic Visual Prompt Tuning framework (DVPT), which can generate a dynamic instance-wise token for each image. In this way, it can capture the unique visual feature of each image, which can be more suitable for downstream visual tasks. We designed a Meta-Net module that can generate learnable prompts based on each image, thereby capturing dynamic instance-wise visual features. Extensive experiments on a wide range of downstream recognition tasks show that DVPT achieves superior performance than other PETL methods. More importantly, DVPT even outperforms full fine-tuning on 17 out of 19 downstream tasks while maintaining high parameter efficiency. Our code will be released soon.

Keywords: parameter efficient transfer learning (PETL) · instance-specific visual clues · Meta-Net · Dynamic Visual Prompt Tuning (DVPT)

1 Introduction

Recently, the Transformer [1] has shown significant potential in achieving various objectives, including natural language processing(NLP) [2], visual recognition [3,4], dense prediction [5], Generative Adversarial Network (GAN) [7,8], reinforcement learning [9], robotics [10], etc. Especially, the large models based on transformer exhibit extremely strong generalization performance. Currently, there is a growing interest in adapting large pre-trained models to various downstream tasks, since this approach offers the advantage of reducing the need for designing and training task-specific models.

Existing literature related to adapting pre-trained model [11,12] in computer vision tend to focus on two prevalent approaches, fine-tuning and linear probing. Fine-tuning involves updating all the model parameters based on

Q. Liu et al. (Eds.): PRCV 2023, LNCS 14432, pp. 293–303, 2024.
https://doi.org/10.1007/978-981-99-8543-2_24

new datasets, which typically leads to better performance than linear probing on various tasks [13]. However, fine-tuning can be computationally expensive and parameter inefficient, especially for larger models like Transformers, whose parameters grow exponentially. On the other hand, linear probing only updates and stores new prediction heads while keeping the backbone frozen, making it more computational and parameter efficient. However, linear probing often results in inferior performance compared to fine-tuning.

Recently, there have been some new techniques to overcome such a dilemma, such as adapter [14,15], LoRA [17], and NOAH [20]. These PETL methods add lightweight modules to pre-trained models, freeze the pre-trained weights, and fine-tune the model to adapt to downstream tasks. Another approach is based on Visual Prompt Tuning (VPT) [18]. It introduces a small number of learnable prompts into the embeddings of image patches while freezing the whole backbone during downstream training. However, the learnable prompts are fixed and do not consider instance-specific visual clues. It is updated entirely through the network's training process, so it is difficult for the network to capture visual features. Therefore, a well-designed vision-oriented PETL method is expected to introduce additional inductive bias and visual instance-wise features.

To overcome the issue, we proposed a dynamic visual prompt tuning framework (DVPT), which can adaptively learn a unique prompt for each image instead of a fixed prompt. To ensure that the model remains parameter efficient, we implement DVPT through a lightweight Meta-Net module, which generates learnable prompts for each image. The learnable prompts are designed to accurately represent each image, rather than only serving certain classes. Therefore, DVPT can fully leverage each input image's visual features, making it more suitable for downstream visual tasks. We validate DVPT on 19 downstream recognition tasks using a pre-trained ViT as the backbone, and the results show that DVPT achieves higher accuracy than other PETL methods.

Overall, we summarize our contributions as follows. (1) We point out that existing PETL methods ignore the visual instance-wise feature, which limits their transferability across different tasks. (2) We propose DVPT, a simple yet effective PETL method that generates dynamic prompts using instance-specific visual clues to adapt pre-trained models to downstream visual tasks. (3) Extensive results on VTAB show that DVPT outperforms full fine-tuning and previous PETL methods, validating the effectiveness of DVPT.

2   Related Works

The DVPT approach incorporates a plug-and-play Meta-Net module to fine-tune the current vision Transformer models. To provide context for this approach, we perform a literature review focused on two key areas: the vision Transformers and efficient transfer learning for vision Transformers.

2.1   Transformer in Vision

The Transformer architecture was introduced in [1] and then revolutionized the field of natural language processing [2]. The remarkable achievements of

Transformers in natural language processing have also inspired computer vision researchers to adopt this architecture since Vision Transformer (ViTs) [3]. The Transformer's ability to effectively model long-range relationships has proven useful in a variety of computer vision tasks, including image classification [3,4], object detection [5,22], semantic/instance segmentation [6], video understanding [23,24], point cloud modeling [25,26], 3D object recognition [27] and even low-level image processing [28,29]. Transformer models have further improved performance in vision recognition by large-scale pretraining [30,31]. With the advent of pre-trained Transformer models significantly larger than previous CNN backbones, a key challenge is how to fine-tune these models for downstream vision tasks effectively.

2.2 Efficient Transfer Learning for Transformers

Researchers are increasingly interested in PETL to optimize large-scale pre-trained models. PETL was first used in natural language processing [14,17], showing that fine-tuning lightweight modules in a large pre-trained model can achieve almost optimal performance. Based on this success, PETL principles have been applied to large pre-trained vision models for various vision tasks [14,16,19,20]. Two main approaches include adapter-based methods [14,19], such as Adaptformer, and prompt tuning-based methods [16,32], such as VPT. Adapter-based methods insert small MLP networks into the vision model to adapt to downstream tasks, while prompt tuning-based methods add trainable tokens to the input sequence of the vision Transformer to reduce the gap between pre-training and downstream data distributions. LoRA [17] obtains a low-rank representation of multi-head attention [1] while keeping the parameters frozen. Zhang [20] et al. propose a prompt search algorithm that combines adapter, prompt tuning, and LoRA automatically.

3 Approach

We propose the DVPT for adapting large, pre-trained vision transformer models to downstream tasks. The overall framework is presented in Fig. 1.

3.1 Preliminaries

Vision Transformer. In a Vision Transformer (ViT) [3] with N layers, the input image $\mathbf{x}$ is partitioned into fixed-sized patches $\{I_j \in \mathbb{R}^{3 \times h \times w} | j \in \mathbb{N}, 1 \leq j \leq m\}$, where h, w are the height and width of image patches and m is the number of image patches. Each patch is embedded into a d-dimensional latent space and undergoes positional encoding:

$$\mathbf{e}_0^j = Embed(I_j) \qquad \mathbf{e}_0^j \in \mathbb{R}^d, j = 1, 2, ...m. \tag{1}$$

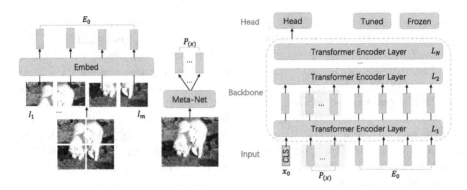

Fig. 1. Our approach, Dynamic Visual Prompt Tuning (DVPT), consists of three learnable components: a set of prompts, linear head and a lightweight Meta-Net that generates dynamic prompts for each input image. During training on downstream tasks, only the learnable components are updated while the whole Transformer encoder is frozen.

We use image patch embeddings, $\mathbf{E}_i = \{\mathbf{e}_i^j \in \mathbb{R}^d | j \in \mathbb{N}, 1 \le j \le m\}$ as inputs for the $i+1$-th Transformer layer (L_{i+1}). Together with an additional learnable classification token [CLS], the entire ViT can be expressed as follows,

$$[\mathbf{x}_i, \mathbf{E}_i] = L_i([\mathbf{x}_{i-1}, \mathbf{E}_0]) \qquad i = 1, 2, ..., N \tag{2}$$

$$\mathbf{y} = Head(\mathbf{x}_N), \tag{3}$$

where $\mathbf{x}_i \in \mathbb{R}^d$ is the embedding of [CLS] at L_{i+1}'s input space. The notation $[\cdot, \cdot]$ denotes stacking and concatenating along the dimension of sequence length, i.e., $[\mathbf{x}_i, \mathbf{E}_i] \in \mathbb{R}^{(1+m)\times d}$. Each Transformer encoder has two sub-layers: a multihead self-attention layer (MHSA) and an MLP layer. A neural classification head translates the [CLS] embedding, $\mathbf{x}_N$, from the final layer into a distribution of predicted class probabilities represented by $\mathbf{y}$. Additional information on this topic can be found in [3].

Visual Prompt Tuning(VPT). Given a pre-trained Transformer model, VPT adds a set of p continuous embeddings of dimension d, i.e., *prompts*, to the input space after the embedding layer. Only the task-specific prompts are updated during fine-tuning while the Transformer backbone remains frozen.

Each prompt token can be represented as a learnable vector in a d-dimensional space. $\mathbf{P} = \{\mathbf{p}^k \in \mathbb{R}^d | k \in \mathbb{N}, 1 \le k \le p\}$ denotes a collection of p prompts. The prompted ViT is:

$$[\mathbf{x}_1, \mathbf{Z}_1, \mathbf{E}_1] = L_1([\mathbf{x}_0, \mathbf{P}, \mathbf{E}_0]) \tag{4}$$

$$[\mathbf{x}_i, \mathbf{Z}_i, \mathbf{E}_i] = L_i([\mathbf{x}_{i-1}, \mathbf{Z}_{i-1}, \mathbf{E}_{i-1}]) \qquad i = 2, 3, ..., N \tag{5}$$

$$\mathbf{y} = Head(\mathbf{x}_N), \tag{6}$$

where $\mathbf{Z}_i \in \mathbb{R}^{p \times d}$ refers to the features computed by the i-th Transformer layer, and $[\mathbf{x}_i, \mathbf{Z}_i, \mathbf{E}_i] \in \mathbb{R}^{(1+p+m)\times d}$.

3.2   Dynamic Visual Prompt Tuning

As discussed in Sect. 3.1, VPT uses a fixed prompt for different images. We argue that visual instance-wise feature can achieve better performance because it is centered on analyzing each input, rather than relying on a fixed set of classes. Therefore, we propose an efficient approach DVPT. It extends VPT by further introducing a Meta-Net module, which generates instance-wise visual features for each input.

The input image $\mathbf{x}$ is fed into the Meta-Net module to get dynamic prompts. Inspired by the VPT [18], the dynamic prompts are the combination of learnable prompts and dynamic instance-wise visual clues, $\mathbf{P}(\mathbf{x}) = \mathbf{P} + \boldsymbol{\pi}$. The $\mathbf{P} = \{\mathbf{p}^k \in \mathbb{R}^d | k \in \mathbb{N}, 1 \leq k \leq p\}$ is a collection of p prompts following VPT. Finally, the DVPT is formulated as:

$$[\mathbf{x}_1, \mathbf{Z}_1, \mathbf{E}_1] = L_1([\mathbf{x}_0, \mathrm{P}(\mathrm{x}), \mathbf{E}_0]) \tag{7}$$

$$[\mathbf{x}_i, \mathbf{Z}_i, \mathbf{E}_i] = L_i([\mathbf{x}_{i-1}, \mathbf{P}_{i-1}, \mathbf{E}_{i-1}]) \qquad i = 2, 3, ..., N \tag{8}$$

$$\mathbf{y} = Head(\mathbf{x}_N), \tag{9}$$

During training, we update both the Meta-Net parameters and the prompt vectors $\mathbf{P}(\mathbf{x})$. In our experiments, 4 linear layers are used to implement the Meta-Net module for efficiency. More complex networks can lead to better performance. The detailed analysis can be found in Sect. 4.3.

4   Experiments

We utilized pre-trained Transformer backbones to evaluate the effectiveness of DVPT in various downstream recognition tasks. We provide a detailed account

Table 1. ViT-B/16 pre-trained on supervised ImageNet-21k. For each method and each downstream task, we report the test accuracy score and the number of wins in (·) compared to Full. The best results among all methods except FULL are **bolded**.

	CIFAR-100	Caltech101	DTD	Flowers102	Pets	SVHN	Sun397	Mean	Patch Camelyon	EuroSAT	Resisc45	Retinopathy	Mean	Clevr/count	Clevr/distance	DMLab	KITTI/distance	dSprites/location	dSprites/orientation	SmallNORB/azimuth	SmallNORB/elevation	Mean
FULL	68.9	87.7	64.3	97.2	86.9	87.4	38.8	75.88	79.7	95.7	84.2	73.9	83.36	56.3	58.6	41.7	65.5	57.5	46.7	25.7	29.1	47.64
LINEAR	63.4	85.0	63.2	97.0	86.3	36.6	51.0	68.93 (1)	78.5	87.5	68.6	74.0	77.16 (1)	34.3	30.6	33.2	55.4	12.5	20.0	9.6	19.2	26.84 (0)
PARTIAL-1	66.8	85.9	62.5	97.3	85.5	37.6	50.6	69.44 (2)	78.6	89.8	72.5	73.3	78.53 (0)	41.5	34.3	33.9	61.0	31.3	32.8	16.3	22.4	34.17 (0)
MLP-2	63.2	84.8	60.5	97.6	85.9	34.1	47.8	67.70 (2)	74.3	88.8	67.1	73.2	75.86 (0)	45.2	31.6	31.8	55.7	30.9	24.6	16.6	23.3	32.47 (0)
MLP-5	59.3	84.4	59.9	96.1	84.4	30.9	46.8	65.98 (1)	73.7	87.2	64.8	71.5	74.31 (0)	50.8	32.3	31.5	56.4	7.5	20.8	14.4	20.4	29.23 (0)
SIDETUNE	60.7	60.8	53.6	95.5	66.7	34.9	35.3	58.21 (0)	58.5	87.7	65.2	61.0	68.12 (0)	27.6	22.6	31.3	51.7	8.2	14.4	9.8	21.8	23.41 (0)
BIAS	72.8	87.0	59.2	97.5	85.3	59.9	51.4	73.30 (3)	78.7	91.6	72.9	69.8	78.25 (0)	61.5	55.6	32.4	55.9	66.6	40.0	15.7	25.1	44.09 (2)
ADAPTER-64	74.2	85.8	62.7	97.6	87.2	36.3	50.9	70.65 (4)	76.3	87.5	73.7	70.9	77.10 (0)	42.9	39.9	30.4	54.5	31.9	25.6	13.5	21.4	32.51 (0)
ADAPTER-8	74.2	85.7	62.7	97.8	87.2	36.4	50.7	70.67 (4)	76.9	89.2	73.5	71.6	77.80 (0)	45.2	41.8	31.1	56.4	30.4	24.6	13.2	22.0	33.09 (0)
VPT	78.8	90.8	65.8	**98.0**	88.3	78.1	49.6	78.48 (6)	81.8	96.1	83.4	68.4	82.43 (2)	68.5	60.0	46.5	72.8	73.6	47.9	32.9	37.8	54.98 (8)
DVPT	**80.3**	**91.9**	**67.6**	97.8	**89.5**	**80.3**	**51.9**	**79.9**(6)	**83.8**	**96.9**	**85.3**	70.5	**84.13**(3)	**70.6**	**63.9**	**48.9**	**76.8**	**76.5**	**50.3**	**35.6**	**39.9**	**57.81**(8)

of our experimental setup in Sect. 4.1, covering the pre-trained backbone, the downstream tasks, and an overview of alternative transfer learning methods. In Sect. 4.2, we demonstrate our approach's effectiveness and practical utility. Furthermore, we conducted a thorough analysis of how different design choices can influence performance in Sect. 4.3, contributing to a better understanding of our method.

4.1   Experiment Setup

Pre-trained Backbones. In the experiment, we utilize the Vision Transformer(ViT) [3] as the Transformer architecture in vision. The backbone in this section is pre-trained on ImageNet-21k [33]. We have adhered to the original configurations, such as the number of image patches and the existence of [CLS] token.

Baselines. We are evaluating how DVPT performs in comparison to other fine-tuning methods that are commonly used:

(a) FULL: update all the parameters of both the backbone and classification head.
(b) Methods concentrate on fine-tuning only the classification head while keeping the weights of the pre-trained backbone frozen.
  - LINEAR: utilize a single linear layer as the trainable classification head.
  - PARTIAL-k: fine-tune the last k layers of backbone while freezing the others. It sets a new boundary between the backbone and the classification head.
  - MLP-k: use a multilayer perceptron (MLP) with k layers as the classification head instead of a linear layer.
(c) Methods that add new learnable parameters to the backbone during fine-tuning process:
  - SIDETUNE [34]: develop a "side" network and perform linear interpolation between pre-trained features and side-tuned features before being fed into the head.
  - BIAS [35]: fine-tune only the bias terms of a pre-trained backbone.
  - ADAPTER [14,19]: introduce new MLP modules with the residual connection into Transformer layers.
  - VPT [18]: add a set of prompts in the input space after the Embedding layer.

Table 2. Ablation on the number of linear layers in the Meta-Net. We vary the number of linear layers in the DVPT and show the top-1 accuracy on 5 VTAB datasets. The best results among all numbers of linear layers are **bolded**.

Number of layers	DTD	EuroSAT	KITTI/distance	Pets	Resisc45
2	63.4	92.59	75.95	87.05	83.92
4	67.61	96.91	76.79	89.52	85.29
6	**68.73**	**97.02**	**78.62**	**91.24**	**87.05**

Downstream Tasks. The VTAB-1K benchmark consists of 19 image classification tasks from various domains, which can be broadly classified into three
groups: Natural, Specialized, and Structured. Each task has only 1,000 training
samples. We perform an 80/20 split on the 1000 training images in each task for
hyperparameters searching. We run the final evaluation using the entire training
data. The average accuracy score on the test set is reported based on three runs.

FGVC includes 5 established Fine-Grained Visual Classification tasks,
namely CUB-200-2011, NABirds, Oxford Flowers, Stanford Dogs, and Stanford
Cars. If a certain dataset only provides train and test sets, we randomly split the
training set into 90% for training and 10% for validation. We use the validation
set to select the best hyperparameters for our models.

4.2   Main Results

Table 1 displays the results of fine-tuning a pre-trained ViT-B/16 across three
different downstream task groups. Our DVPT is compared with 8 other tuning
methods. We can see that:

1. DVPT achieves better results than FULL (as shown in Table 1) on 17 out
   of 19 tasks, while using a significantly small number of total model parameters. The underline reason is the FULL updates all parameters, destroying the
   knowledge learned by the pre-trained model, while DVPT does not change
   the weights of the pre-trained model and only learns some tokens, improving performance based on the pre-trained model. Thus, the DVPT approach
   shows great potential for adapting larger Transformers in vision tasks.
2. As shown in Table 1, DVPT surpasses all other parameter-efficient tuning methods across all task groups. Our DVPT outperforms the current
   SOTA PETL method, VPT, on 18 out of 19 VTAB-1K datasets, specifically, DTD(3.1%), Retinopathy(2.12%), and Clevr/distance(3.86%). It validates the effectiveness of our DVPT when working with limited storage
   resources. Moreover, we discover that instance-wise visual clues play a key
   role in visual tasks.

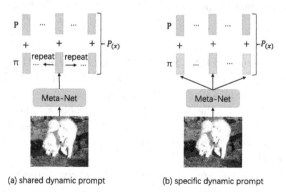

(a) shared dynamic prompt   (b) specific dynamic prompt

Fig. 2. Two implementation approaches for dynamic prompts: (a) shared dynamic
prompt (b) specific dynamic prompt.

4.3   Ablation on Model Design Variants

We ablate our DVPT to study what properties make for a good DVPT. The ablation studies conducted in this work are all performed on the VTAB-1K, with the same setup in Table 1.

Different Number of Linear Layers in the Meta-Net module. The linear layers number of a Meta-Net controls the number of introduced parameters in DVPT. Using fewer linear layers can lead to a lower parameter count, with a possible performance cost. We ablate DVPT on the linear layers number to study these effects. Task results on different linear layers number are presented in Table 2. The result shows a consistent improvement in top-1 accuracy as the number of linear layers increases, e.g., for Resisc45, DVPT with 6 linear layers surpasses 4 linear layers and 2 linear layers by 1.76%, 3.13% on top-1 accuracy, respectively. In other experiments, we set the number of layers to 4 to achieve the trade-off between performance and parameter count.

Varying the Dimension of Instance-Wise Visual clues. As mentioned in section Sect. 3.2, the output of the Meta-Net module is a *shared dynamic prompt* (Fig. 2 (a)). In this section, we let the Meta-Net module outputs a *specific dynamic prompt*, as in Fig. 2 (b). We study which case can achieve better performance. As shown in Table 3, we observe that the specific dynamic prompt method outperforms the shared dynamic prompt method, e.g., the specific method achieves 2.74% top-1 accuracy gains over the shared method on KITTI/distance dataset.

Table 3. Ablation on the dimension of instance-wise visual clues in the Meta-Net. We show the top-1 accuracy on 3 VTAB datasets. The best results are **bolded**.

Method	KITTI/distance	Pets	Resisc45
shared dynamic prompt	76.79	89.52	85.29
specific dynamic prompt	**79.53**	**91.05**	**87.37**

Different Downstream Data Sizes. We further study the performance of DVPT under varying training data sizes on the FGVC dataset. Following VPT, we reduce the training data sizes to 10%, 20%, 30%, 40%, 50%, 60%, and 70% and compare DVPT with the Full and Linear methods. The results for each method on different training data scales are presented in Fig. 3. It shows that DVPT consistently yields better performance than the Full and Linear across data scales. When there is limited data available, DVPT and Linear perform better than Full. However, as more training data becomes available, Full begins to outperform Linear. In contrast, DVPT still consistently outperforms Full across training data sizes.

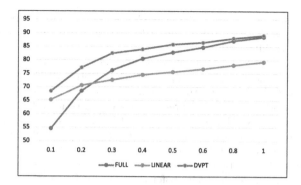

Fig. 3. Performance comparison on different downstream data scales averaged across 5 FGVC tasks. DVPT is compared with Full and Linear.

5 Conclusions

In this paper, we point out that the current PETL methods ignore the visual instance-wise feature, which leads to suboptimal performance on downstream tasks. Based on the above findings, we propose a DVPT framework, a vision-oriented PETL method that employs Meta-Net to adapt pre-trained ViT to downstream tasks. In this way, DVPT can capture the unique visual features of each image, which can be more suitable for downstream visual tasks. The experimental results on the VTAB-1K benchmark indicate that DVPT achieves better performance than other PETL methods. Our approach highlights the importance of the visual instance-wise feature for visual tasks during the development of PETL techniques. Our work will encourage further research into developing more effective methods for fine-tuning large vision models.

References

1. Vaswani, A., et al.: Attention is all you need. In: Advances in Neural Information Processing Systems (2017)
2. Brown, T., et al.: Language models are few-shot learners. In: Advances in Neural Information Processing Systems (2020)
3. Dosovitskiy, A., Gelly, et al. An image is worth 16×16 words: transformers for image recognition at scale. arXiv preprint arXiv:2010.11929 (2020)
4. Liu, Z., et al:. Swin transformer: hierarchical vision transformer using shifted windows. In Proceedings of the IEEE/CVF International Conference on Computer Vision (2021)
5. Carion, N., Massa, F., Synnaeve, G., Usunier, N., Kirillov, A., Zagoruyko, S.: End-to-end object detection with transformers. In: European Conference on Computer Vision (2020)
6. Xie, E., Wang, W., Yu, Z., Anandkumar, A., Alvarez, J.M., Luo, P.: SegFormer: simple and efficient design for semantic segmentation with transformers. In: Advances in Neural Information Processing Systems (2021)

7. Jiang, Y., Chang, S., Wang, Z.: TransGAN: two pure transformers can make one strong GAN, and that can scale up. In: Advances in Neural Information Processing Systems, vol. 34 (2021)
8. Hudson, D.A., Zitnick, L.: Generative adversarial transformers. In: International Conference on Machine Learning, pp. 4487–4499. PMLR (2021)
9. Chen, C., Wu, Y.-F., Yoon, J., Ahn, S.: TransDreamer: reinforcement learning with transformer world models. arXiv preprint arXiv:2202.09481 (2022)
10. Jangir, R., Hansen, N., Ghosal, S., Jain, M., Wang, X.: Look closer: bridging egocentric and third-person views with transformers for robotic manipulation. IEEE Robot. Autom. Lett. **7**, 3046–3053 (2022)
11. Evci, U., Dumoulin, V., Larochelle, H., Mozer, M.C.: Head2Toe: utilizing intermediate representations for better transfer learning. In: International Conference on Machine Learning, pp. 6009–6033. PMLR (2022)
12. Kornblith, S., Shlens, J., Le, Q.V.: Do better ImageNet models transfer better? In: Proceedings of the IEEE/CVF Conference on Computer Vision and Pattern Recognition, pp. 2661–2671 (2019)
13. Zhai, X., et al.: A large-scale study of representation learning with the visual task adaptation benchmark. arXiv preprint arXiv:1910.04867 (2019)
14. Houlsby, N., et al.: Parameter-efficient transfer learning for NLP. In: International Conference on Machine Learning, pp. 2790–2799. PMLR (2019)
15. Mahabadi, R.K., Henderson, J., Ruder, S.: Compacter: efficient low-rank hypercomplex adapter layers. In: Advances in Neural Information Processing Systems, vol. 34 (2021)
16. Zhou, K., Yang, J., Loy, C.C., Liu, Z.: Learning to prompt for vision-language models. arXiv preprint arXiv:2109.01134 (2021)
17. Hu, E.J., et al.: Lora: low-rank adaptation of large language models. arXiv preprint arXiv:2106.09685 (2021)
18. Jia, M., et al.: Visual prompt tuning. arXiv preprint arXiv:2203.12119 (2022)
19. Chen, S., et al.: AdaptFormer: adapting vision transformers for scalable visual recognition. In: Advances in Neural Information Processing Systems (2022)
20. Zhang, Y., Zhou, K., Liu, Z.: Neural prompt search. arXiv preprint arXiv:2206.04673 (2022)
21. Lian, D., Zhou, D., Feng, J., Wang, X.: Scaling shifting your features: a new baseline for efficient model tuning. In: Advances in Neural Information Processing Systems (2022)
22. Zhu, X., Su, W., Lu, L., Li, B., Wang, X., Dai, J.: Deformable DETR: deformable transformers for end-to-end object detection. arXiv preprint arXiv:2010.04159 (2020)
23. Bertasius, G., Wang, H., Torresani, L.: Is space-time attention all you need for video understanding. arXiv preprint arXiv:2102.05095 (2021)
24. Arnab, A., Dehghani, M., Heigold, G., Sun, C., Lučić, M., Schmid, C.: ViViT: a video vision transformer. In Proceedings of the IEEE/CVF International Conference on Computer Vision, pp. 6836–6846 (2021)
25. Zhao, H., Jiang, L., Jia, J., Torr, P.H.S., Koltun, V.: Point transformer. In: Proceedings of the IEEE/CVF International Conference on Computer Vision, pp. 16259–16268 (2021)
26. Guo, M.-H., Cai, J.-X., Liu, Z.-N., Tai-Jiang, M., Martin, R.R., Hu, S.-M.: PCT: point cloud transformer. Comput. Visual Media **7**, 187–199 (2021)
27. Chen, S., Yu, T., Li, P.: MVT: multi-view vision transformer for 3D object recognition. arXiv preprint arXiv:2110.13083 (2021)

28. Chen, H., et al.: Pre-trained image processing transformer. In: IEEE/CVF Conference on Computer Vision and Pattern Recognition (2021)
29. Liang, J., Cao, J., Sun, G., Zhang, K., Van Gool, L., Timofte, R.: SwinIR: image restoration using swin transformer. In: IEEE/CVF International Conference on Computer Vision (2021)
30. Chen, X., Xie, S., He, K.: An empirical study of training self-supervised vision transformers. In: IEEE/CVF International Conference on Computer Vision (2021)
31. Caron, M., et al.: Emerging properties in self-supervised vision transformers. In: IEEE/CVF International Conference on Computer Vision (2021)
32. Zhou, K., Yang, J., Loy, C.C., Liu, Z.-w.: Conditional prompt learning for vision-language models. In: CVPR (2022)
33. Deng, J., Dong, W., Socher, R., Li, L.J., Li, K., Fei-Fei, L.: ImageNet: a large-scale hierarchical image database. In: CVPR (2009)
34. Zhang, J.O., Sax, A., Zamir, A., Guibas, L., Malik, J.: Side-tuning: a baseline for network adaptation via additive side networks. In: Vedaldi, A., Bischof, H., Brox, T., Frahm, J.-M. (eds.) ECCV 2020. LNCS, vol. 12348, pp. 698–714. Springer, Cham (2020). https://doi.org/10.1007/978-3-030-58580-8_41
35. Cai, H., Gan, C., Zhu, L., Han, S.: TinyTL: reduce memory, not parameters for efficient on-device learning. NeurIPS **33**, 11285–11297 (2020)

C-volution: A Hybrid Operator for Visual Recognition

Feng He[1] and Fuchuan Ni[1,2,3](✉)

[1] College of informatics, Huazhong Agricultural University, Wuhan 430070, China
[2] Engineering Research Center of Intelligent Technology for Agriculture,
Ministry of Education, Wuhan, China
[3] Key Laboratory of Smart Farming for Agricultural Animals,
Ministry of Agriculture and Rural Affairs, Wuhan 430070, China
fcni_cn@mail.hzau.edu.cn

Abstract. Convolution is a fundamental building block of modern neural networks, playing a critical role in the success of deep learning for vision tasks. However, convolutional neural networks exhibit limited spatial context due to their local receptive field, which also neglects global/long-term dependent relations. To this end, we propose a lightweight hybrid structure operator, called C-volution. The operator utilizes a multi-branch architecture to extract spatial and channel information from input data separately, enabling the network to capture abstract features while preserving important spatial information. In addition, summarize context information in a larger spatial range by generating dynamic kernels to strengthen the spatial contextual aggregation capability, overcoming the difficulty of long-term interactions in convolutions. This paper validates the efficacy of our operator through extensive experiments on ImageNet classification, COCO detection and segmentation, and the results have demonstrated the proposed C-volution when paired with ResNet50 achieves an outstanding boost in performance on visual tasks(+2.0% top-1 accuracy, +3.1% box mAP, and +2.0% mask mAP) while having low parameters (i.e., CedNet50@16.3M Params).

Keywords: Convolution · Local receptive field · Spatial information · Long-term interactions

1 Introduction

Convolution is an essential primitive in deep learning, with convolutional kernels exhibiting spatial-agnostic and channel-specific characteristics(i.e., in terms of spatial extent, ensuring that convolution kernels are efficient by allowing them to be reused across different locations and achieving translation invariance, and the use of multiple convolutional kernels in the channel dimension to collect diverse information encoded in different channels.). The aforementioned characteristics endow the network with an exceptional ability to perceive local features, which has led to the widespread application of convolutional neural networks

© The Author(s), under exclusive license to Springer Nature Singapore Pte Ltd. 2024
Q. Liu et al. (Eds.): PRCV 2023, LNCS 14432, pp. 304–315, 2024.
https://doi.org/10.1007/978-981-99-8543-2_25

(CNNs) [2,5,7,13] in diverse computer vision (CV) tasks, including image classification [10], object detection [4], and instance segmentation [6].

As the core ingredient of CNNs, discrete convolution operators achieve translation-equivariant properties by reusing convolutional kernels across different positions in space and collecting a variety of information encoded in different channels with a set of convolutional kernels in the channel domain, thereby enhancing perceptual ability [7,14]. However, the limitations of the receptive field of convolution hinder the acquiring of global/long-term dependencies which promotes performance in many CV tasks [4,18,21]. Recently, the popularity of transformer models has prompted the consideration of convolution. Li et al. [15] rethink the convolution structure in view of the advantages of the transformer, and propose an involution operation, which has inverse symmetry compared with convolution i.e., strengthens the spatial modeling and weakens the channel modeling. Dynamically generating convolution kernels through the mapping of input features enables the network to prioritize information-rich regions, thus overcoming the difficulty of long-term interactions in convolutions. However, some studies [8,27] show that the sharing of channel dimension parameters fixes the abstract features of data and weakens the interaction of channel information.

In this work, we design a lightweight hybrid operator called C-volution, which incorporates the channel-specificity character while retaining the spatial information interaction capability. The proposed operator aims to address the limitations of traditional convolution in terms of the receptive field and the weakness of the involution operator in channel information interaction. We extract spatial information and channel information separately through a multi-branch structure, and at the same time refer to the residual idea to combine the extracted information features with the original input. This enables the output features to capture the abstract features of the input data while retaining more spatial information. Our key contributions are as follows:

- We propose a simple and effective hybrid operator, c-volution, which can be embedded into any modern network as an atomic operation.
- We verify the effect of different kernel sizes on operator performance through extensive ablation experiments.
- We explore the performance of the operator and find it achieves greatly improved in multiple benchmarks(ImageNet-1K and MS COCO).

The experiments show that our proposed hybrid structure operator improves the top1 accuracy of image classification on ImageNet dataset [24] by 2.0%, the box mAP of object detection on the COCO dataset [17] by 3.1%, and the mask mAP of instance segmentation on the COCO dataset by 2.0%.

2   Related Work

Convolution is a crucial component of deep neural networks, and there has been a significant increase in CNN architectures that use various aggregation techniques in recent years, and these models have demonstrated outstanding performance in various tasks [1, 9, 25]. Jaderberg. et al. [12] introduce a differentiable spatial transformation module into a neural network, which learns to geometrically transform input data, thereby enhancing the performance of the network. Hu et al. [8] design embeddable network substructures to perform channel feature recalibration by capturing the interdependencies between channels to achieve superior accuracy and efficiency. Wang et al. [26] dissect the SENet [8] and separately demonstrate that avoiding dimensionality reduction and proper cross-channel interactions are important. These methods [8, 12, 20, 26] supplement the information of the convolution operator by adding modules, rather than optimizing the operators themselves. Moreover, the convolution operator is limited to the local receptive field, and the fixed convolution kernel of this static model during training largely restricts its feature extraction and expression capabilities [5, 10].

Recently, Li et al. [15] purpose an involution kernel that is conditioned on the original input, ensuring that the resulting output kernel is appropriately aligned with the input kernel. The dynamic parameterized involution kernel is shared in the channel dimension and has wide coverage in the spatial dimension, effectively aggregates the semantic information of the spatial context, and thus overcomes the difficulty of pre-remote interaction modeling. But parameter sharing in the channel dimension can lead to poor interaction of channel information [8, 27]. A novel interactive attention module that combines spatial and channel information is proposed in [19, 27], and the experimental results are presented to validate the significance of these two dimensions as well as the advantages of their interplay. In summary, it is necessary to design an operator that encompasses both spatial and channel information.

3   Methodology

We design a lightweight mixed structure operator, named C-volution, which unifies the advantages of convolution and involution operators in a single module. By effectively integrating channel and spatial information, this module enables the network to achieve superior feature extraction and performance capabilities. In this section, we first review the mathematical formulas of convolution and involution and then introduce our design philosophy of a mixed structure. Finally, we demonstrate the rationality of our design through mathematical expressions.

3.1   Convolution Module

The convolution operator uses linear weighting to output information from local regions in the network. Given an input feature tensor $\mathbf{X} \in \mathbb{R}^{W \times H \times C_i}$, where C_i represents the number of input channels, W represents the width, and H represents the height, the mathematical formula for the estimated result $\mathbf{Y} \in \mathbb{R}^{W \times H \times C_o}$ (C_o represents the number of output channels) after convolution is defined as follows:

$$\mathbf{Y}_{i,j,k} = \sum_{c=1}^{C_i} \sum_{(u,v) \in \Omega_{i,j}} \mathcal{W}_{k,c_i,u+\lfloor K/2 \rfloor, v+\lfloor K/2 \rfloor} \mathbf{X}_{i+u,j+v,c} \tag{1}$$

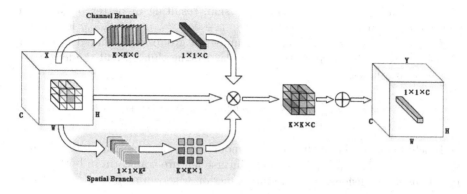

Fig. 1. Our designed C-volution model consists of two distinct branches that process the input feature, as illustrated in the schematic diagram: one for enhancing channel information and the other for mapping spatial information. Then, the processed information is aggregated with the input features. The symbol ⊗ represents the multiplication operation that is applied to all C channels, while ⊕ indicates the summation performed within the spatial neighborhood of size $K \times K$.

A convolution filter with a fixed kernel size of $K \times K$ is denoted as $\mathcal{W}$, where each filter contains C_i convolution kernels. (u, v) is the center pixel coordinate corresponding to the convolution kernel. k represents the index of the corresponding convolution kernel. $X_{i,j}$ are pixels in the feature tensor X. Standard convolution estimates result through weighted averaging of the window information, which is essentially a first-order linear weighting operation.

3.2 Group Convolution

The feature maps obtained by group convolution [11] through different convolution paths have low coupling and complementary characteristics. In conventional convolution, each kernel operates over all input channels, but group convolution groups the channels corresponding to the convolution kernel and performs convolution on each group separately, which can be mapped separately to improve performance. This allows the network to reduce the number of parameters while increasing computational efficiency. The basic formula is as follows:

$$\mathbf{Y}_{i,j,k} = \sum_{c=1}^{C_i/g} \sum_{(u,v) \in \Omega i,j} \mathcal{G}_{k,\mathrm{u}+\lfloor K/2 \rfloor, \mathrm{v}+\lfloor K/2 \rfloor} \mathbf{X}_{i+\mathrm{u},j+\mathrm{v},k} \tag{2}$$

Different from the standard convolution, the convolution kernel $\mathcal{G}_k$ represents the slice x corresponding to the kth channel, and the spatial information in the slice is shared.

3.3 Involution

Involution uses a generative kernel to dynamically form convolution kernels by mapping the input features X. It enhances spatial modeling and weakens modeling relationships in the channel dimension, resulting in better performance and fewer parameters. Specifically, $\mathbf{W}_{i,j...g} \in \mathbb{R}^{K \times K}$, $g = 1, 2, 3..g$ is adapted to $\mathbf{X}_{i,j} \in \mathbb{R}^C$, where g represents the number of groups with shared involution kernels. However, its computation is similar to convolution, and the formula is expressed as follows:

$$Y_{i,j,k} = \sum_{(u,v) \in \Delta_K} \mathcal{W}_{i,j,u+[K/2],v+[K/2],[kg/C]} \mathbf{X}_{i+u,j+v,k} \tag{3}$$

It achieves spatial and channel-wise modeling by designing involution kernels $\mathbf{W} \in \mathbb{R}^{H \times W \times k \times k \times g}$ that performs the inverse feature transform in both spatial and channel dimensions. The mapping function is shown as f :

$$\mathcal{W}_{i,j} = f\left(\mathbf{X}_{\psi_{i,j}}\right) \tag{4}$$

where $X_{\Psi_{i,j}}$ is based on the pixel set of $W_{(i,j)}$, and to reduce computational complexity, the mapping function f is composed of a 1×1 convolution.

$$\mathcal{W}_{i,j} = f\left(\mathbf{x}_{\Psi_{i,j}}\right) = \mathbf{W}_1 \sigma\left(\mathbf{W}_0 \mathbf{X}_{i,j}\right) \tag{5}$$

where $X_{i,j}$ denotes the input feature, $\mathbf{W}_0 \in \mathbb{R}^{C^2/r}$ and $\mathbf{W}_1 \in \mathbb{R}^{C/r \times (K \times K \times g)}$ are the matrices for feature mappings, and $W_{(i,j)}$ represents the parameters of the involution kernel at (i,j). and σ denotes a combination of operations(Batch Normalization and interleaving non-linear activation functions).

3.4   C-volution

We aim to combine the advantages of these two operators to achieve weight distribution of channel dimensions and adaptive variation of convolution kernels. In brief, we designed a multi-branch structure. The channel branch is designed to capture channel-wise information, while the spatial branch enhances the ability to represent diverse features across different spatial locations. Finally, it is fused with the input features, so that the output strengthens the channel weight ratio and the problem of long-distance pixel dependence has been improved. As shown in Fig. 1, we provide an input feature map X with size $H \times W \times C$ and the kernel generation function $f(\cdot)$, the overall process can be summarized as:

$$Y'_{i,j,c} = \sum_{(u,v) \in \Omega_{i,j}} X_{i+u,j+v,c} f\left(X_{i,j,c}\right) W_c \tag{6}$$

Channel Branch. We obtain a local matrix $\mathbf{G} \in \mathbb{R}^{W \times H \times k \times k \times C}$ and divide the channel into g groups each of which includes C/g channels. The purpose is to independently learn each set of channel information. The channel information extraction process is as follows:

$$R = G \circledast X \tag{7}$$

where ⊛ denotes the local matrix multiplication operation. Then, we employ a matrix $\mathcal{W} = [w_1, w_2, w_3, ...w_C]$ to learn channel information.

$$\mathcal{W}_c = \frac{1}{H_i + W_j}(\sum_{h=0}^{H_{i-1}}\sum_{w=0}^{W_{j-1}}\mathcal{R}_c(h, w)), \mathcal{R}_c \in \Omega_c^k \tag{8}$$

where Ω^k indicates the set of $k(k = 3)$ adjacent channels of $\mathcal{R}$, and $\mathcal{R}$ demonstrates the weight calculated by the interaction between $\mathcal{R}$ and its k neighbors.

Spatial Branch. We dynamically generate a corresponding kernel on a single pixel from the input feature map and then perform element-wise multiplication between the generated kernel and the feature pixels in the neighborhood of the corresponding coordinate point on the original input feature map. This enables the current position to gather contextual information from the neighboring area, thereby aggregating spatial context information over a larger range. The spatial information extraction process is as follows:

$$f(X) = \sum_{(u,v)\in\Delta_K} \mathcal{H}_{i,j,u+[K/2],v+[K/2],[kg/C]} \tag{9}$$

$$\mathcal{H}_{i,j} = \mathbf{W}_1\sigma\left(\mathbf{W}_0\mathbf{X}_{i,j}\right) \tag{10}$$

After acquiring the channel's information $\mathcal{W}$ and the spatial information $f(X)$ from two branches. To perform adaptive feature optimization, we will multiply the input feature map with the obtained information. The formulas such as Eq. 6.

To specify the Parameter of previous methods and our channel branch. Given the input feature $\mathbf{X} \in \mathbb{R}^{h\times w\times c}$ and output channel c_{out}, and applies c filters $\mathbf{w} \in \mathbb{R}^{k\times k}$ to compute parameter. The intput $\mathbf{X}$ through a 1×1 convolution takes $(k^2*c)c_{out}+c_{out}$ parameter. If paired with our method, it will introduce an additional parameter of $(k^2*c/g)c_{out}+c_{out}$ for the purpose of channel information aggregation. The increase of additional parameters is inversely proportional to g. When g takes a large value, the total parameters are $k^2(c + 1)c_{out} + 2c_{out}$, and only a modest number of parameters will be added. Here, we set g to 16.

4 Experiments

4.1 Implementation Details

To evaluate the effectiveness of our proposed algorithm, we adapted the C-volution operator into the ResNet backbone network, named CedNet, and verified it through three different experiments: (a) image classification based on the Imagenet dataset; (b) object detection and instance segmentation based on the COCO dataset; (c) ablation experiments on different kernel sizes based on the Imagenet and COCO datasets.

Table 1. Image Classification on ImageNet.

Architecture	Params	FLOPs	Top-1 Acc
ResNet-38(2016)	19.6M	3.2G	76.0
Stand-Alone ResNet-38 (2019) [22]	14.1M	3.0G	76.9
SAN15 (2020) [29]	14.1M	3.0G	77.1
RedNet-38 (2021)	12.4M	2.2G	77.3
BiNet-38 (dc)(2022) [9]	13.2M	2.2G	78.2
CedNet-38	**12.4M**	**2.2G**	**78.0**
ResNet-50(2016)	25.6M	4.1G	76.8
ResNeXt-50 (32×4d) (2017) [28]	25.0M	4.3G	77.8
SE-ResNet-50 (2018)	28.1M	4.1G	77.6
Res2Net-50 (14w-8 s) (2019) [3]	25.7M	4.2G	78.0
AA-ResNet-50 (2019) [1]	25.8M	4.2G	77.7
Stand-Alone ResNet-50 (2019)	18.0M	3.6G	77.6
SAN19 (2020)	17.6M	3.8G	77.4
ECA-ResNet-50 (2020c) [26]	25.6M	4.1G	77.5
Axial ResNet-S (2020b) [25]	12.5M	3.3G	78.1
Fca-ResNet-50 (2021) [20]	28.1M	4.1G	78.5
RedNet-50 (2021)	15.5M	2.7G	78.1
BiNet-50 (dc)(2022)	19.3M	2.8G	78.8
CedNet-50	**16.3M**	**2.9G**	**78.8**

The ImageNet dataset is widely regarded as one of the most complex and demanding datasets in computer vision. It comprises a total of 1.28 million images for training, along with an additional set of 50k validation images, which includes 1k different object classes. In this experiment, we utilized data augmentation techniques as part of our training optimization strategy, whereby the input images were subject to random cropping to dimensions of 224 × 224 and horizontal flipping, we employed the Stochastic Gradient Descent (SGD) optimization algorithm, with an initial learning rate of 0.8, the momentum of 0.9, and weight decay set to 1×10^{-4}. The network was trained for a total of 130 epochs, with a batch size of 2048.

For object detection, we used the representative framework Faster-RCNN [23] and FPN [16] as our baseline frameworks for the experiments. For instance segmentation, we employed the classic segmentation framework Mask-RCNN [6]. These experiments were conducted on the COCO dataset, which contains 115K training images and 5K validation images, and the mean average precision (mAP) was used as the evaluation metric [17]. Both experiments were performed with a batch of 16, the optimizing process was the Stochastic Gradient Descent (SGD) optimizer(initial learning rate: 0.02, weight decay: 0.0001, the momentum: 0.9.), and the learning rate underwent division by 10 during the 8th and 11th epochs(on Tesla V100 GPUs).

The code was implemented primarily using PyTorch, and we will be released the source code for reproducibility.

4.2   Main Results

Image Classification. As shown in Table 1, we compare CedNet with the classic ResNet-38, ResNet-50, and the latest versions of the ResNet series. Under the same standard, CedNet-50 improves accuracy by 2.0% and reduces Params by 36.3% compared to ResNet-50. Under an acceptable model computational cost, our proposed method shows a 0.7% effective improvement compared to the involution algorithm. We further conduct experiments using class activation maps, as shown in Fig. 2. The feature coverage obtained by CedNet-50 extends to more parts of the target object for identification, which indicates that our approach effectively facilitates the ability of the network to acquire pertaining to salient information.

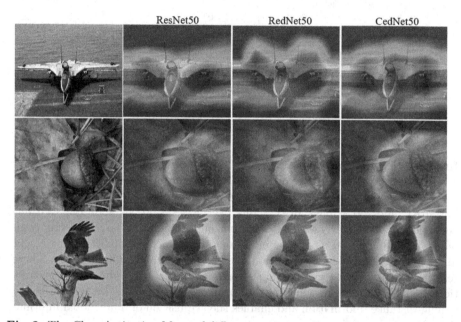

Fig. 2. The Class Activation Maps of different operators. From left to right, the first column shows the origin pictures and the last three columns demonstrate the results of convolution(ResNet), involution(RedNet), and our operator(CedNet).

Object Detection. In order to further validate the effectiveness and generalization of the proposed C-volution operator, we assess its performance on downstream tasks. We compared our model with ResNet-50 and RedNet-50, and it is observed that the backbone incorporating our proposed method showed significant improvements in accuracy. As shown in Table 2 for object detection, our proposed CedNet network improved by 2.5% compared to ResNet, increased by 0.4% on RedNet, and we further observe performance improvement after replacing the convolution module in FPN, with an increase of 3.1% compared to ResNet-50 and 0.3% compared to RedNet-50.

Table 2. Object Detection results on COCO val 2017.

Detector	Backbone	Neck	AP^{bbox}	AP^{bbox}_{50}	AP^{bbox}_{75}	AP^{bbox}_{S}	AP^{bbox}_{M}	AP^{bbox}_{L}
	ResNet-50	convolution	37.4	58.1	40.4	21.2	41.0	48.1
Faster-R-CNN	RedNet-50	convolution	39.5(+2.1)	60.9(+2.8)	42.8(+2.4)	**23.3(+2.1)**	42.9(+1.9)	52.2(+4.1)
	Cednet-50	convolution	**39.9(+2.5)**	**61.1(+3.0)**	**43.5(+3.1)**	22.7(+1.5)	**43.6(+2.6)**	**52.7(+4.6)**
Faster-R-CNN	RedNet-50	involution	40.2(+2.8)	62.1(+4.0)	43.4(+3.0)	24.2(+3.0)	43.3(+2.3)	52.7(+4.6)
	Cednet-50	C-volution	**40.5(+3.1)**	**62.6(+4.5)**	**43.6(+3.2)**	**24.4(+3.2)**	**43.6(+2.6)**	**53.8(+5.7)**

Table 3. Instance segmentation results on COCO val 2017.

Detector	Backbone	Neck	AP^{bbox} AP^{mask}	AP^{bbox}_{50} AP^{mask}_{50}	AP^{bbox}_{75} AP^{mask}_{75}	AP^{bbox}_{S} AP^{mask}_{S}	AP^{bbox}_{M} AP^{mask}_{M}	AP^{bbox}_{L} AP^{mask}_{L}
Mask-R-CNN	ResNet-50	convolution	38.2	59.0	41.8	22.3	41.6	48.9
			34.7	55.7	36.9	16.3	37.4	46.9
	RedNet-50	convolution	39.9(+1.7)	60.9(+1.9)	43.2(+1.4)	23.4(+1.1)	43.1(+1.5)	53.3(+4.4)
			35.7(+1.0)	57.5(+1.8)	37.7(+0.8)	**19.7(+3.4)**	**38.8(+1.4)**	49.0(+2.1)
	CedNet-50	convolution	40.5(+2.3)	60.9(+1.9)	**43.9(+2.1)**	**24.3(+2.0)**	43.1(+1.5)	**54.1(+5.2)**
			36.2(+1.5)	**57.8(+2.1)**	38.5(+1.6)	18.0(+1.7)	38.6(+1.2)	**52.0(+5.1)**
	RedNet-50	involution	40.8(+2.6)	62.3(+3.3)	44.3(+2.5)	24.2(+1.9)	44.0(+2.4)	53.0(+4.1)
			36.4(+1.7)	59.0(+3.3)	38.5(+1.6)	**19.9(+3.6)**	39.4(+2.0)	49.1(+2.2)
	CedNet-50	C-volution	41.0(+2.8)	62.3(+3.3)	44.5(+2.7)	24.8(+2.5)	**44.6(+3.0)**	54.9(+6.0)
			36.7(+2.0)	**59.2(+3.5)**	38.8(+1.9)	19.0(+2.7)	**40.0(+2.6)**	**53.2(+6.3)**

Instance Segmentation. Table 3 presents the performance comparison of our proposed operator with other operators for instance segmentation. Specifically, C-volution achieves 41.0% box mAP and 36.7% mask mAP, which brings 2.8% box mAP and 2.0% mask mAP gains over ResNet-50. In addition, CedNet-50 performs particularly well in detecting and segmenting large objects, compared with ResNet-50 as the baseline, our model achieves 6.3% improvement in AP^{mask}_{L}. We provide visualization of some prediction results in Fig. 4. These results demonstrate the aggregation of channel information enhances the extraction of local information, which further promotes the dynamic kernel's modeling of long-distance information and enables the acquisition of more global information.

Table 4. Ablation study for different kernel sizes on imagenet dataset(Image Classification).

Operator	Kernel Type	Params	FLOPs	Top-1 Acc
C-volution	1 × 1	15.5M	2.7G	78.2
	3 × 3	15.6M	2.7G	78.5
	5 × 5	15.8M	2.7G	78.6
	7 × 7	16.3M	2.9G	78.8

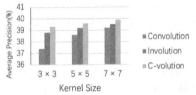

Fig. 3. Ablation study for different kernel sizes on COCO dataset(Object Detection).

Ablation Study. To investigate the impact of the hybrid structure on the receptive field, we conduct ablation experiments with different kernel sizes on the Imagenet and COCO datasets. As shown in Table 4 and Fig. 3, we find that the kernel size has different effects on model accuracy, increasing the kernel size will expand the receptive field of the network, which also enables the network to have greater modeling capability over larger regions. Compared with convolution and involution, our proposed hybrid structure operator exhibits excellent performance and stationarity. Moreover, within a certain range, as the spatial range increases, the degree of improvement also increases. Which seems to be related to the feature resolution in the network.

Fig. 4. Example of instance segmentation results on the COCO test set. The first column shows the original image. The last three columns in the data display the prediction outcomes of three different methods: baseline(ResNet), involution(RedNet), and our method(CedNet).

5   Conclusion

This paper introduces a lightweight hybrid operator to enhance the ability of the network to aggregate spatial information and capture the abstract feature through a multi-branch structure. Our hybrid branch structure achieves significant improvement in the accuracy of common computer vision tasks while maintaining low parameters, demonstrating the rational design of our operator. We will focus on refining the operator by strengthening the coupling degree of spatial and channel information in the future, additionally, we will conduct various experiments to evaluate the adaptability and generalization ability of the operator on diverse tasks.

Acknowledgments. This work was supported by the Key Special Project National Key R&D Program of China (grant number: 2018YFC1604000), partly by the National Natural Science Foundation of China (grant number: 62106081), partly Supported by "the Fundamental Research Funds for the Central Universities", Huazhong Agricultural University (grant number: 2662022JC005).

References

1. Bello, I., Zoph, B., Vaswani, A., Shlens, J., Le, Q.V.: Attention augmented convolutional networks. In: Proceedings of the IEEE/CVF International Conference on Computer Vision, pp. 3286–3295 (2019)
2. Dai, J., Qi, H., Xiong, Y., Li, Y., Zhang, G., Hu, H., Wei, Y.: Deformable convolutional networks. In: Proceedings of the IEEE International Conference on Computer Vision, pp. 764–773 (2017)
3. Gao, S.H., Cheng, M.M., Zhao, K., Zhang, X.Y., Yang, M.H., Torr, P.: Res2Net: a new multi-scale backbone architecture. IEEE Trans. Pattern Anal. Mach. Intell. **43**(2), 652–662 (2019)
4. Girshick, R., Donahue, J., Darrell, T., Malik, J.: Rich feature hierarchies for accurate object detection and semantic segmentation. In: Proceedings of the IEEE Conference on Computer Vision and Pattern Recognition, pp. 580–587 (2014)
5. Graves, A.: Adaptive computation time for recurrent neural networks. arXiv preprint arXiv:1603.08983 (2016)
6. He, K., Gkioxari, G., Dollár, P., Girshick, R.: Mask R-CNN. In: Proceedings of the IEEE International Conference on Computer Vision, pp. 2961–2969 (2017)
7. He, K., Zhang, X., Ren, S., Sun, J.: Deep residual learning for image recognition. In: Proceedings of the IEEE Conference on Computer Vision and Pattern Recognition, pp. 770–778 (2016)
8. Hu, J., Shen, L., Sun, G.: Squeeze-and-excitation networks. In: Proceedings of the IEEE Conference on Computer Vision and Pattern Recognition, pp. 7132–7141 (2018)
9. Hu, X., Chen, X., Ni, B., Li, T., Liu, Y.: Bi-volution: a static and dynamic coupled filter. In: Proceedings of the AAAI Conference on Artificial Intelligence, vol. 36, pp. 960–968 (2022)
10. Huang, G., Chen, D., Li, T., Wu, F., Van Der Maaten, L., Weinberger, K.Q.: Multi-scale dense networks for resource efficient image classification. arXiv preprint arXiv:1703.09844 (2017)
11. Ioannou, Y., Robertson, D., Cipolla, R., Criminisi, A.: Deep roots: improving CNN efficiency with hierarchical filter groups, pp. 1231–1240 (2017)
12. Jaderberg, M., Simonyan, K., Zisserman, A., et al.: Spatial transformer networks. In: Advances in Neural Information Processing Systems 28 (2015)
13. Krizhevsky, A., Sutskever, I., Hinton, G.E.: ImageNet classification with deep convolutional neural networks. Commun. ACM **60**(6), 84–90 (2017)
14. LeCun, Y., Bottou, L., Bengio, Y., Haffner, P.: Gradient-based learning applied to document recognition. Proc. IEEE **86**(11), 2278–2324 (1998)
15. Li, D., et al.: Involution: inverting the inherence of convolution for visual recognition. In: Proceedings of the IEEE/CVF Conference on Computer Vision and Pattern Recognition, pp. 12321–12330 (2021)
16. Lin, T.Y., Dollár, P., Girshick, R., He, K., Hariharan, B., Belongie, S.: Feature pyramid networks for object detection. In: Proceedings of the IEEE Conference on Computer Vision and Pattern Recognition, pp. 2117–2125 (2017)

17. Lin, T.-Y., et al.: Microsoft COCO: common objects in context. In: Fleet, D., Pajdla, T., Schiele, B., Tuytelaars, T. (eds.) ECCV 2014. LNCS, vol. 8693, pp. 740–755. Springer, Cham (2014). https://doi.org/10.1007/978-3-319-10602-1_48

18. Mottaghi, R., et al.: The role of context for object detection and semantic segmentation in the wild. In: Proceedings of the IEEE Conference on Computer Vision and Pattern Recognition, pp. 891–898 (2014)

19. Park, J., Woo, S., Lee, J.Y., Kweon, I.S.: BAM: bottleneck attention module. arXiv preprint arXiv:1807.06514 (2018)

20. Qin, Z., Zhang, P., Wu, F., Li, X.: FcaNet: frequency channel attention networks. In: Proceedings of the IEEE/CVF International Conference on Computer Vision, pp. 783–792 (2021)

21. Rabinovich, A., Vedaldi, A., Galleguillos, C., Wiewiora, E., Belongie, S.: Objects in context. In: 2007 IEEE 11th International Conference on Computer Vision, pp. 1–8. IEEE (2007)

22. Raghu, M., Unterthiner, T., Kornblith, S., Zhang, C., Dosovitskiy, A.: Do vision transformers see like convolutional neural networks? Adv. Neural. Inf. Process. Syst. **34**, 12116–12128 (2021)

23. Ren, S., He, K., Girshick, R., Sun, J.: Faster R-CNN: towards real-time object detection with region proposal networks. In: Advances in Neural Information Processing Systems 28 (2015)

24. Russakovsky, O., et al.: ImageNet large scale visual recognition challenge. Int. J. Comput. Vision **115**, 211–252 (2015)

25. Wang, H., Zhu, Y., Green, B., Adam, H., Yuille, A., Chen, L.-C.: Axial-DeepLab: stand-alone axial-attention for panoptic segmentation. In: Vedaldi, A., Bischof, H., Brox, T., Frahm, J.-M. (eds.) ECCV 2020. LNCS, vol. 12349, pp. 108–126. Springer, Cham (2020). https://doi.org/10.1007/978-3-030-58548-8_7

26. Wang, Q., Wu, B., Zhu, P., Li, P., Zuo, W., Hu, Q.: ECA-Net: efficient channel attention for deep convolutional neural networks. In: Proceedings of the IEEE/CVF Conference on Computer Vision and Pattern Recognition, pp. 11534–11542 (2020)

27. Woo, S., Park, J., Lee, J.-Y., Kweon, I.S.: CBAM: convolutional block attention module. In: Ferrari, V., Hebert, M., Sminchisescu, C., Weiss, Y. (eds.) ECCV 2018. LNCS, vol. 11211, pp. 3–19. Springer, Cham (2018). https://doi.org/10.1007/978-3-030-01234-2_1

28. Xie, S., Girshick, R., Dollár, P., Tu, Z., He, K.: Aggregated residual transformations for deep neural networks. In: Proceedings of the IEEE Conference on Computer Vision and Pattern Recognition, pp. 1492–1500 (2017)

29. Yang, Z., He, X., Gao, J., Deng, L., Smola, A.: Stacked attention networks for image question answering. In: Proceedings of the IEEE Conference on Computer Vision and Pattern Recognition, pp. 21–29 (2016)

Motor Imagery EEG Recognition Based on an Improved Convolutional Neural Network with Parallel Gate Recurrent Unit

Junbo Zhang, Wenhui Guo, Haoran Yu, and Yanjiang Wang$^{(\boxtimes)}$

China University of Petroleum (East China), Qingdao, China
yjwang@upc.edu.cn

Abstract. Motor imagery (MI) electroencephalogram (EEG) recognition is currently widely used in brain-computer interface (BCI) devices for people with motor disabilities to achieve various motor interaction functions with the outside world. Over 70% of recent researches use convolutional neural networks (CNN) for recognition of MI. However, using CNN is often difficult to fully utilize the temporal features of long time series in EEG. It may lead to part of the difference information among subjects not being learned by the CNN. In this paper, we introduce a multi-branch CNN that integrates a gate recurrent unit (GRU) module. In this model, the serial module extracts rough features in the temporal and spatial domains, and the parallel module uses different scale convolution blocks and GRU modules to extract time series information in different ranges to improve the precision of learning. This training strategy not only retains the advantages of CNN in extracting temporal and spatial features, but also makes full use of the long time series information extracted by GRU so as to improve the classification accuracy. Experimental results show that our proposed framework has higher performance compared to other typical models, such as DeepNet, EEGNet. The within-subject average classification accuracy reaches 74.4% on BCI competition IV-2a dataset, and the minimum accuracy among subjects increases by 5.6%. This indicates that the proposed model has good generalization ability among different subjects, thus promoting the implementation of personalized real-time BCI devices in the future.

Keywords: Electroencephalogram (EEG) · Motor Imagery (MI) · Convolutional neural network (CNN) · Gate Recurrent Unit (GRU)

1 Introduction

Electroencephalogram (EEG) signals contain important information about human conditions, and brain-computer interface (BCI) systems with EEG are

This work is supported by the National Natural Science Foundation of China under Grant No. 62072468.

currently widely used for normal interaction between people with related disabilities through external devices [3], as well as brainprint identification [7]. For these purposes, using computers to decode EEG into understandable language becomes one of the research focuses of BCI technology.

Motor imagery (MI) refers to imagining an action in the brain but not actually executing it [14], which is different from the EEG signals generated during actual actions [16], and has a wide range of application scenarios. In the field of healthcare, it can be used for the rehabilitation process of patients with paralysis and stroke [11], and can also help patients with motor disorders communicate with the outside world through BCI devices [10], such as mobile assisted robots [5].

EEG signals have relatively low signal to noise ratio and large individual differences [25], and the current research results are not sufficient to achieve a satisfactory level of application. The ultimate goal of EEG signal recognition should be robust, automatic, and with high-precision [1]. In traditional methods, the processing of MI-EEG signals is usually composed of three steps: signal preprocessing, feature extraction, and classification [2]. Traditional feature extraction methods such as common spatial pattern can be used to construct optimal spatial filters. And Principal Component Analysis can be used for data preprocessing which can extract more effective features [24]. However, they are often difficult to meet the actual needs, and often need manual debugging.

In recent years, the widely used deep learning model not only overcomes the limitations of specific tasks, but also further improves the classification accuracy based on data preprocessing. With the future progress in the fields of computer vision and natural language processing, the recognition of MI will be more accurate and rapid. The convolutional neural network (CNN) framework, which is widely used in EEG processing, is successfully applied in MI recognition since its inception. The deep CNN model can use parallel multiscale filter banks to process data, and the resulting inter subject individual network has a good effect on cross subject classification [18]. Using batch normalization (BN) [8], exponential linear units (ELU), and reasonably designing training strategies, the decoding performance of deep convolutional networks [17] are effectively improved.

Recent research on recognizing MI mainly concentrates on various CNN models, as presented in Fig. 1 [1]. This figure also affirms that CNN has an excellent capability to classify EEG-based MI. Nonetheless, the employment of CNN models in MI classification encounters some limitations as the research progressed. For example, although CNN has strong ability to extract local features, its performance in extracting long time series features is poor. However, most studies in improving CNN do not consider the long term series features of the data itself.

In this study, we propose an improved CNN model for EEG-based MI classification tasks. Long-term features in MI signals are extracted using the gate recurrent unit (GRU), while temporal and spatial features are extracted by the convolutional blocks. The biggest difference between our method and other hybrid CNN methods is that GRU extracts features in parallel. This process does not forget the features obtained after convolution operations, and can fully leverage the advantages of CNN in extracting local features and GRU in extracting long time series features. This approach proves to be highly effective in compensating

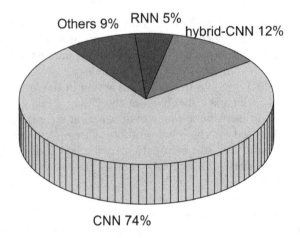

Fig. 1. Proportion of deep neural networks architecture used in recent years.

for the large individual difference present in EEG-based MI signals. The proposed model is evaluated on the BCI competition IV-2a, a highly challenging dataset, and achieves competitive accuracy level in within-subject classification. The manuscript is structured as follows: Sect. 2 introduces the related works. Section 3 describes the proposed model's architecture in detail. Section 4 covers the experimental setup including the dataset, performance metrics, and analysis of the experimental results. The limitations and potential of the proposed model are further discussed. Section 5 summarizes the paper.

2   Related Works

For the problem of less EEG-based MI data, literature [23] uses data augmentation to construct artificial EEG frames to expand the dataset and maintain the original multi-channel EEG input. In order to better address the problem of large individual differences in EEG, literature [5] conducts more in-depth research on feature extraction methods using mixed scale convolution blocks, and finds that the optimal convolution scale varies depending on the subject. And then they use different convolution kernels to extract temporal features. Further, a series parallel structure is used in the deep framework to extract deep features at different scales in the time domain, space domain, and frequency domain [25]. Distinguishing temporal convolution from spatial convolution to extract the feature representation of EEG signals in different spaces simultaneously is a design idea for general parallel structures [9].

A small number of researchers focus on developing and utilizing other networks to improve classification accuracy, such as using wavelet instead of convolutional layer [23]. Among them, recurrent neural networks (RNN) [19] have a good performance. And long short-term memory (LSTM) networks are feasible in analyzing and learning longer range data sequences [15], such as stacked LSTM using FFT preprocessing and downsampling [4] or hybrid LSTM models with CNN decoders [6]. Literature [22] designs a shared neural network independent of the subjects. It uses all data from different subjects to form a model.

3 Methods

3.1 The Framework of the Model Proposed

The main framework is shown in Fig. 2. We first extract the shallow features using a shallow feature extraction module (SFE) that can retain time series features [10]. In this process, a portion of the disordered noise can also be removed. Next, we use a deep feature extraction module (DFE) to extract temporal features from three different views, namely, features in a short time period, features in a medium time period, and features in a long time period. Then, we make corresponding feature concatenate according to the data format to get the classification results. The framework of the SFE and DFE will be described in the next section.

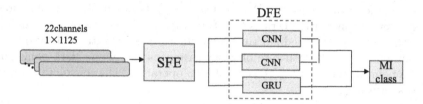

Fig. 2. The main framework for the proposed model.

3.2 The Framework of the Two Modules

The SFE is shown in Fig. 3. Firstly, the EEG signals from multiple channels are converted into 1D multi-channel data through channel projection. We used a total of 4.5 s of MI and rest process data, which is an input length of 1125. The advantage is that the collected 1D signals are directly utilized, without the need for additional transformations, such as converting to a designed fixed form of 2D feature map. Next we apply temporal and spatial convolutions after converting the channel dimensions to extract shallow features. The input is 1D time series information from 22 channels. Channel projection using 1×1 kernel convolution layer and converts the features into 35 channels. In order to better process the

Fig. 3. Shallow feature extraction module (SFE).

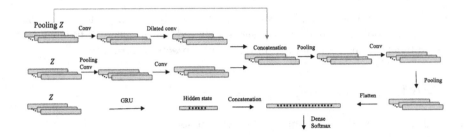

Fig. 4. Deep feature extraction module (DFE) and feature fusion.

feature, the number of channels is converted into the second dimension of the data, which is achieved through BN. Then 1×11 kernel is applied to extract spatial features and transforming the output into 25 channels, i.e. (25, 35, 1115). 35×1 time kernel is used to convert the second dimension to the temporal dimension, the common input for the DFE module is obtained. We record it as Z.

The DFE module consists of three branches. On the one hand, we continue to exploit the advantages of CNN in extracting EEG-based MI signal features, using different convolution kernels to extract high-level features on medium and short time scales. The medium time scale information is extracted by dilated convolution with a dilation rate of 2, which means skipping a data point and select a point as the object of the convolution operation. The convolution kernel size is shown in (1).

$$d = ker + (ker - 1) \cdot (p - 1) \tag{1}$$

where d is the size of the convolution kernel, ker is the original size of the kernel, and p is the dilation rate. This can expand the receptive field of the convolution, resulting in the desired medium length of time feature. The receptive field can be calculated by (2).

$$f = ker + p \tag{2}$$

where f is the receptive field, ker is the original size of the kernel, and p is the dilation rate. On the other hand, there are large individual differences in EEG-based MI signals. Using GRU to extract the long time series features of EEG data can better solve the problem without destroying the feature extracted by CNN. The DFE architecture is shown in Fig. 4. The first branch performs a max pooling and a convolution operation with a size of (100, 1×1). And then performing 11×1 filtered convolution to obtain a medium length time feature (100, 1, 371). The second branch performs a max pooling and convolution operation with 100 convolution kernels of size 1×1. And then performing a convolution operation with size 11×1 to obtain a short time feature, and then fuses the first two branches and the Z after max pooling to obtain a feature of (225, 1, 371). In order to highlight the role of long-term features, we perform continuous pooling, convolution, and pooling of the first two branch fused feature to reduce the size of it. Finally, we obtain a 1D feature by flatten, and the temporal dimension of the feature obtained after each pooling is one-third of the original size with convolution kernel size 1×11. In the third branch, Z is put into GRU, and the hidden size of the GRU is set to 128, that is, 128 hidden state neurons. We select the last hidden state of the hidden layer as the output of the third branch, and concatenate the flattened results obtained from the first two branches to obtain a 1D feature with a length of 8453. The hidden layer output is as follows:

$$y_t = (1 - z_t) \cdot y_{t-1} + z_t \cdot d_t \tag{3}$$

where y_t is the hidden state variable at time t, z_t is the state variable of the update gate at time t, y_{t-1} is the hidden state variable at time $t-1$, d_t is the candidate hidden state variable at time t, and y_t is affected by both the input at time t and the candidate hidden state variable at time $t-1$. Equation (3) means that each hidden state variable y_t is affected by the previous hidden state variable y_{t-1}. When long-term sequence information is used as input, GRU continuously forgets and remembers to enable the hidden state variable y_t to obtain important information in the sequence.

At last, we put the feature into fully connected layer to obtain an output with a length of 4, and use it as the input of softmax to finally obtain the classification result. ELU is used as the activation function for convolution and fully connected layers. Dropout and BN are used before and after convolution and fully connected layer to reduce the risk of overfitting.

The loss function uses cross entropy loss.

$$l = - \sum td_i(x) log pd_i(x), i = 1, 2, 3, ..., n \tag{4}$$

where l is the loss, $td_i(x)$ is the probability distribution of the i-th tag in the target domain, and $pd_i(x)$ is the probability distribution of the i-th tag in the prediction domain.

4   Experiment

4.1   Dataset

The BCI competition IV-2a dataset is a widely used MI dataset [1], which is a 4-class dataset that includes four types of MI: left hand, right hand, tongue, and foot. EEG data were obtained from 9 subjects, each of whom completed a session on two different dates. One session was used as the training and validation set for the classifier, and the other session was used as the test set. Each session is further divided into 6 runs, and each run contains 48 trials. These 48 trials have 12 of each type, including left hand, right hand, tongue, and foot, but the order is random. The process of a trial is as follows. When t = 0 s, the screen displays the cross and emits a prompt tone. When t = 2 s, an MI category prompt appears on the screen, indicating the MI category that the subject needs to perform next. The prompt lasts for 1.25 s. When t = 3 s, the subject begins to perform the corresponding MI, which lasts for 3 s. When t = 6 s, this trial ends, enters a short rest, and then proceeds to the next trial. The experimental content and corresponding time are shown in Fig. 5.

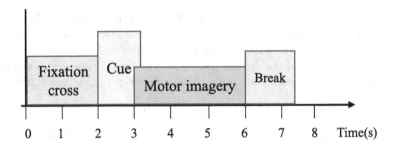

Fig. 5. The experimental paradigm of data collection.

Including the short break, each trial takes approximately 8 s. The collected data includes a total of 22 EEG channels and 3 electrooculogram channels, with a sampling signal frequency of 250 Hz. Bandpass filtering is performed between 0.5 Hz and 100 Hz, and a 50 Hz notch filter is used to remove line noise.

4.2   Train Approach

The experiment uses the Adam optimizer, which has been proven to be the most widely used and effective EEG-based MI classification optimization algorithm [2], with advantages such as simple implementation and low memory requirements. The initial learning rate is set to 0.0001, and the adaptive attenuation learning rate can ensure the generalization ability for multiple subjects to a certain extent. Because during the experiment, it can be observed that the convergence speed is slightly different when using data from different subjects for training. If a fixed learning rate is used, it may lead to significant differences in training speed among subjects. The batch size for training is 260, and the batch size for validation is 20.

The BCI competition IV-2a dataset is stored in gdf format during collection, which is a commonly used format for EEG signals. It needs to be decoded into mat format for training. The converted data size should be (2, 288, 22, 1125), where 2 is the data and its labels, 288 is the number of trials per session, and 1125 is the number of sampling points per trial.

This paper randomly divides the training data of each subject into 10 equal parts, with one part serving as the validation part and the other parts serving as the training parts. We repeat the experiment 10 times and calculate the average of the results. The above method is called the 10-fold cross method. After the experiment, 9 separate models were established for 9 subjects, and for each model, the training and testing data came from the same subject. This kind of within-subject classification does not require data from others for training, and has high application value in real-time BCI devices.

4.3   Experiment Results

We compared the proposed framework with deep learning methods such as Deep-Net, EEGNet, ShallowNet and SPCNN. The results of these methods are introduced from literature [25], and we used the same experimental setup and training methodologies as these methods. The average test accuracy is shown in Fig. 6. In our method, the average testing accuracy of within-subject classification reaches 74.4%.

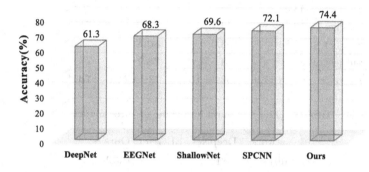

Fig. 6. The average testing accuracy of within-subject classification.

Due to the individual differences among subjects, the average test accuracy often fluctuates greatly in the within-subject classification [21]. The comparison for each subject is shown in Table 1. We focus on the maximum deviation from the average, as this directly affects the experience among BCI users. In our method, the lowest testing accuracy can reach 63.3%, which is 5.6% higher than the SPCNN [25] method. The maximum deviation from the average is 17.8% which means that the user experience of BCI will not be too bad.

The smaller the value of cross entropy loss, the better the prediction effect of the model. One reason why the recognition accuracy is not very high is that the data distribution of the test data and the training data in the BCI competition IV-2a dataset is different, and they are collected on different days. One usage scenario for BCI devices is to collect data from the same day for real-time training and use. In this case, the magnitude of cross entropy loss under the same training time for the same equipment has become an important indicator of practicality. Due to the different convergence rates of various models, we uniformly select the minimum loss comparison at 1000 epochs, as shown in Table 2. Our method is superior to the classic DeepNet [25] and EEGNet [25] methods.

Table 1. 10-fold within-subject testing classification performance and maximum deviation from the average.

Network	DeepNet(%)	EEGNet(%)	SPCNN(%)	Ours(%)
Subject 1	66.3	74.7	78.9	**83.0**
Subject 2	46.9	53.9	51.0	**63.3**
Subject 3	74.2	83.5	**89.2**	89.1
Subject 4	67.4	59.3	67.6	**77.2**
Subject 5	63.1	**71.4**	63.5	70.0
Subject 6	43.4	56.9	**57.4**	56.6
Subject 7	59.7	72.8	**82.0**	74.7
Subject 8	66.8	68.0	**80.8**	78.5
Subject 9	64.0	74.5	**78.7**	77.6
Average	61.3±17.9	68.3±**15.2**	72.1±21.1	**74.4**±17.8

Table 2. The minimum loss value of three models.

Network	DeepNet	EEGNet	Ours
Subject 1	0.81	0.49	**0.32**
Subject 2	0.54	0.99	**0.24**
Subject 3	0.89	0.41	**0.36**
Subject 4	0.68	0.79	**0.33**
Subject 5	**0.57**	0.61	0.66
Subject 6	0.84	**0.70**	0.97
Subject 7	0.83	**0.49**	0.93
Subject 8	0.81	**0.48**	0.85
Subject 9	0.71	0.35	**0.29**
Average	0.74	0.59	**0.55**

4.4   Ablation Study

The effectiveness of CNN in extracting features at different scales has been proven by a large number of studies [24]. For the effectiveness of the GRU branch, the accuracy comparison before and after removing the GRU branch is shown in Table 3. It can be seen that the extraction of deep features by GRU plays a certain role in improving the accuracy of the model.

The training set and the test set are data from different dates with a certain number of days interval, and their distribution may vary. Therefore, there are certain requirements for the generalization ability of the network. By using GRU to learn more deep features from EEG, the problem of low recognition accuracy caused by such differences can be alleviated. Compared with LSTM, GRU has a lower complexity and a corresponding reduction in the amount of parameters and computation, which makes it more advantageous when faced with multi-layer neural networks with less data.

Table 3. Model accuracy without and with GRU.

Network	No GRU(%)	With GRU(%)
Subject 1	82.3	**83.0**
Subject 2	60.6	**63.3**
Subject 3	87.9	**89.1**
Subject 4	76.2	**77.2**
Subject 5	68.2	**70.0**
Subject 6	**59.5**	56.6
Subject 7	74.3	**74.7**
Subject 8	78.1	**78.5**
Subject 9	77.2	**77.6**
Average	73.8	**74.4**

4.5 Limitation and Prospect

The network framework proposed in this paper is a little complex for the dataset used. Although methods such as dropout and BN are used, overfitting is still prone to occur on some subjects. In addition, It is obvious that significantly increasing the GRU output will amplify the noise in EEG data, which will cause some invalid features learned, thus discounting the classification accuracy [20]. In addition, an ideal real-time BCI device should use fewer channels rather than a large number of electrodes in the head [13]. However, in existing studies, using fewer channels is difficult to achieve satisfactory performance. Appropriate processing of δ (δ, 0.5-3 Hz), θ (θ, 4-7 Hz), α (α, 8-13 Hz), β (β, 14-30 Hz) and γ (γ, 30-80 Hz) frequency bands can also improve the recognition performance [12], which will further accelerate the development of the BCI field.

5 Conclusion

In this paper, we propose a parallel model combining CNN and GRU, which is capable of learning features from three views. The model uses convolution layer and dilated convolution layer to extract local features, and GRU to extract global time features. Our model can extract time and spatial features of different scales, overcoming individual differences in MI recognition to a certain extent, and has better recognition performance than mainstream networks in within-subject classification. The experimental results also show that the multi-branch parallel structure have good generalization ability among different subjects and the parallel GRU is proven to be effective.

References

1. Al-Saegh, A., Dawwd, S.A., Abdul-Jabbar, J.M.: Deep learning for motor imagery EEG-based classification: a review. Biomed. Signal Process. Control **63**, 102172 (2021)

2. Altaheri, H., et al.: Deep learning techniques for classification of electroencephalogram (EEG) motor imagery (MI) signals: a review. Neural Comput. Appl. **35**, 1–42 (2021)

3. Ang, K.K., Guan, C.: EEG-based strategies to detect motor imagery for control and rehabilitation. IEEE Trans. Neural Syst. Rehabil. Eng. **25**(4), 392–401 (2016)

4. Bai, Z., Yang, R., Liang, Y.: Mental task classification using electroencephalogram signal. arXiv preprint arXiv:1910.03023 (2019)

5. Dai, G., Zhou, J., Huang, J., Wang, N.: HS-CNN: a CNN with hybrid convolution scale for EEG motor imagery classification. J. Neural Eng. **17**(1), 016025 (2020)

6. Garcia-Moreno, F.M., Bermudez-Edo, M., Rodríguez-Fórtiz, M.J., Garrido, J.L.: A CNN-LSTM deep learning classifier for motor imagery EEG detection using a low-invasive and low-cost BCI headband. In: 2020 16th International Conference on Intelligent Environments (IE), pp. 84–91. IEEE (2020)

7. Jin, X., et al.: CTNN: a convolutional tensor-train neural network for multi-task brainprint recognition. IEEE Trans. Neural Syst. Rehabil. Eng. **29**, 103–112 (2020)

8. Li, G., et al.: An EEG data processing approach for emotion recognition. IEEE Sens. J. **22**(11), 10751–10763 (2022)

9. Li, X., et al.: EEG motor imagery classification based on multi-spatial convolutional neural network. In: 2022 5th International Conference on Artificial Intelligence and Big Data (ICAIBD), pp. 433–437. IEEE (2022)

10. Li, Y., Zhang, X.R., Zhang, B., Lei, M.Y., Cui, W.G., Guo, Y.Z.: A channel-projection mixed-scale convolutional neural network for motor imagery EEG decoding. IEEE Trans. Neural Syst. Rehabil. Eng. **27**(6), 1170–1180 (2019)

11. Lin, P.J., et al.: CNN-based prognosis of BCI rehabilitation using EEG from first session BCI training. IEEE Trans. Neural Syst. Rehabil. Eng. **29**, 1936–1943 (2021)

12. Mai, N.D., Long, N.M.H., Chung, W.Y.: 1D-CNN-based BCI system for detecting emotional states using a wireless and wearable 8-channel custom-designed EEG headset. In: 2021 IEEE International Conference on Flexible and Printable Sensors and Systems (FLEPS), pp. 1–4. IEEE (2021)

13. Mattioli, F., Porcaro, C., Baldassarre, G.: A 1D CNN for high accuracy classification and transfer learning in motor imagery EEG-based brain-computer interface. J. Neural Eng. **18**(6), 066053 (2022)

14. Musallam, Y.K., et al.: Electroencephalography-based motor imagery classification using temporal convolutional network fusion. Biomed. Signal Process. Control **69**, 102826 (2021)

15. Pathan, S.M.K., Rana, M.M.: Investigation on classification of motor imagery signal using bidirectional LSTM with effect of dropout layers. In: 2022 International Conference on Advancement in Electrical and Electronic Engineering (ICAEEE), pp. 1–5. IEEE (2022)

16. Petoku, E., Capi, G.: Object movement motor imagery for EEG based BCI system using convolutional neural networks. In: 2021 9th International Winter Conference on Brain-Computer Interface (BCI), pp. 1–5. IEEE (2021)

17. Schirrmeister, R.T., et al.: Deep learning with convolutional neural networks for EEG decoding and visualization. Hum. Brain Mapp. **38**(11), 5391–5420 (2017)

18. Wu, H., et al.: A parallel multiscale filter bank convolutional neural networks for motor imagery EEG classification. Front. Neurosci. **13**, 1275 (2019)

19. Yang, Y., Wu, Q., Qiu, M., Wang, Y., Chen, X.: Emotion recognition from multi-channel EEG through parallel convolutional recurrent neural network. In: 2018 International Joint Conference on Neural Networks (IJCNN), pp. 1–7. IEEE (2018)

20. Zhang, D., Yao, L., Chen, K., Wang, S., Chang, X., Liu, Y.: Making sense of spatio-temporal preserving representations for EEG-based human intention recognition. IEEE Trans. Cybern. **50**(7), 3033–3044 (2019)
21. Zhang, H.Y., Stevenson, C.E., Jung, T.P., Ko, L.W.: Stress-induced effects in resting EEG spectra predict the performance of SSVEP-based BCI. IEEE Trans. Neural Syst. Rehabil. Eng. **28**(8), 1771–1780 (2020)
22. Zhang, R., Zong, Q., Dou, L., Zhao, X.: A novel hybrid deep learning scheme for four-class motor imagery classification. J. Neural Eng. **16**(6), 066004 (2019)
23. Zhang, Z., et al.: A novel deep learning approach with data augmentation to classify motor imagery signals. IEEE Access **7**, 15945–15954 (2019)
24. Zhao, X., Zhang, H., Zhu, G., You, F., Kuang, S., Sun, L.: A multi-branch 3D convolutional neural network for EEG-based motor imagery classification. IEEE Trans. Neural Syst. Rehabil. Eng. **27**(10), 2164–2177 (2019)
25. Zhao, X., et al.: Deep CNN model based on serial-parallel structure optimization for four-class motor imagery EEG classification. Biomed. Signal Process. Control **72**, 103338 (2022)

A Stable Vision Transformer
for Out-of-Distribution Generalization

Haoran Yu[1], Baodi Liu[1], Yingjie Wang[1], Kai Zhang[1], Dapeng Tao[2],
and Weifeng Liu[1(✉)]

[1] China University of Petroleum (East China), Qingdao, China
`liuwf@upc.edu.cn`
[2] Yunnan University, Kunming, China

Abstract. Vision Transformer (ViT) has achieved amazing results in
many visual applications where training and testing instances are drawn
from the independent and identical distribution (I.I.D.). The performance will drop drastically when the distribution of testing instances is
different from that of training ones in real open environments. To tackle
this challenge, we propose a Stable Vision Transformer (SViT) for out-
of-distribution (OOD) generalization. In particular, the SViT weights
the samples to eliminate spurious correlations of token features in Vision
Transformer and finally boosts the performance for OOD generalization.
According to the structure and feature extraction characteristics of the
ViT models, we design two forms of learning sample weights: SViT(C)
and SViT(T). To demonstrate the effectiveness of two forms of SViT for
OOD generalization, we conduct extensive experiments on the popular
PACS and OfficeHome datasets and compare them with SOTA methods.
The experimental results demonstrate the effectiveness of SViT(C) and
SViT(T) for various OOD generalization tasks.

Keywords: Out-of-Distribution Generalization · Independence
Samples Weighting · Vision Transformer

1 Introduction

ViT achieves amazing performance in various visual applications, including
image classification [2], object detection [11], and semantic segmentation [29].
Most of these scenarios assume that training and testing instances follow the
independent and identical distribution (I.I.D.) hypothesis. However, it is hard to
hold in real open environments where the data is usually heterogeneous. To solve
the above problems, out-of-distribution (OOD) generalization aims to improve
the performance for unknown distribution shifts [6]. From the representation

This work was supported by the National Natural Science Foundation of China (Grant
No.62372468), in part by the National Natural Science Foundation of China (Grant
No.61671480), Shandong Natural Science Foundation (Grant No. ZR2023MF008) and
Qingdao Natural Science Foundation (Grant No. 23-2-1-161-zyyd-jch).

Q. Liu et al. (Eds.): PRCV 2023, LNCS 14432, pp. 328–339, 2024.
https://doi.org/10.1007/978-981-99-8543-2_27

learning point of view, the OOD generalization can be roughly divided into invariant-based, causal-based, and independence-based methods.

The invariant-based methods assume that there are invariant representations that have transferability in different distributions. These representations are beneficial to improve the generalization ability of the model under distribution shifts. Domain alignment [23] learns domain-invariant representations by minimizing the difference between source domains. Domain adversarial learning [10] learns them via adversarial learning. Most invariant-based methods require clear and significant heterogeneity between domains. It is usually necessary to manually divide and label the domains (domain labels). However, in practical applications, domain labels are difficult to obtain and are primarily influenced by human subjective consciousness [14]. Although recent partial works learn latent domains from data without domain labels, these works assume the latent domains are balanced. It is also difficult to establish in real open environments.

Causal-based methods aim to find causal factors to solve the OOD generalization. [15] directly uses causal inference to find causal factors. Causal representation learning [26] learns the variables of the causal graph in data through unsupervised learning. However, these methods have strict theoretical and conditional limitations. So their application is usually limited to large models.

Independence-based methods solve the OOD generalization problem by learning independent representations of features to eliminate spurious correlations in the feature space. For example, if the birds appears in the sky in the training samples, the models will correlate the sky with the bird and make predictions based on it. However, when the bird in the test samples appear in the cage (OOD), this spurious correlation is no longer established. [7] theoretically proves that spurious correlation is the core reason for model performance degradation in OOD generalization. Disentangled representation learning [21] aims to learn potential independent representations where distinct and informative factors of variations in data are separated. Independence sample weighting [1] learns a set of sample weights to make the features satisfy statistical independence so that the treatment effect can be estimated accurately. It makes the model pay more attention to real and stable relationships under OOD generalization.

However, the above research is few focusing on the OOD generalization performance of ViT. It raises doubts about the stability and robustness of ViT in real open environments. Based on the above analysis of OOD generalization, considering that spurious correlation is the core reason for performance degradation, and the limitation of domain labels. In this paper, we propose a Stable Vision Transformer (SViT) for OOD generalization. It eliminates the spurious correlations of features in the token by sample weighting to boost the OOD generalization performance of ViT. According to the structural characteristics of ViT, we propose two schemes of models to learn sample weights from tokens: SViT(C) and SViT(T). SViT(C) directly learns sample weights from class token, and SViT(T) learns sample weights by synthesizing tokens other than class token. We conduct extensive experimental verification on widely used datasets, including PACS and OfficeHome. The experimental results demonstrate the superiority of the SViT for OOD generalization tasks.

2   Related Work

Vision Transformer: Vision transformer (ViT) [2] is the first model to apply Transformer structure to image classification directly. It has achieved amazing performance in many scenarios. Recently, there have been various improved models based on ViT. [20] improves the training process of ViT. [24] proposes to integrate convolution into ViT. [19] proposes self-distillation can improve the model's generalization ability and robustness. However, there is few research focusing on the OOD generalization performance of ViT. Although some recent research focus on the distribution shift problems of ViT, these works require the acquisition of target distribution, it is not a strict OOD generalization problem. [27] analyzed the performance of ViT in different distribution shifts and allowed the model to access the unlabeled target data. [5] improves the model's OOD performance by the test time adaptation (TTA). TTA still updates parameters during the test. So it still obtains the target distribution, albeit with strictly limited access. OOD generalization requires the model to improve the generalization ability when the test distribution is strictly unseen. So the previous works have not proposed specific methods for the OOD generalization of ViT when the test distribution is strictly unseen. Our approach achieves OOD generalization without any target distribution and domain labels.

Independence Sample Weighting: Independence sample weighting is an emerging method to solve the OOD generalization. It weights the samples to make the features independent of each other, removing the spurious correlations between features to produce stable predictions. DWR [8] and SRDO [17] directly weight the samples to decorrelate all features. [25] theoretically proved the effectiveness of them. DVD [16] divides all features into two categories: stable features and unstable features. After that, it only decorrelates the features of different categories. Independence sample weighting has been widely used in various models, [7] applied it to the linear models, [28] demonstrated the feasibility of independence sample weighting in CNN, generalizing it to deep learning for OOD generalization.

3   Stable Vision Transformer

3.1   Overall Architecture of Stable Vision Transformer

The goal of OOD generalization is to train with multiple source domains $D = \{d\}_{k=1}^{K}$ to obtain a model that performs well on the unseen target domain, which is characterized by a different probability distribution $P(X, Y) \neq P^{d_k}(X, Y)$ for all $k \in \{1, ..., K\}$. We use ViT-B-16 as the backbone network for SViT. SViT uses independence sample weighting to eliminate spurious correlations of token features in ViT to improve the performance of OOD generalization. We noticed that ViT uses class token extracted by the encoder as classification information to make predictions, but some related improved models [12] do not use the class token when processing classification tasks. Therefore, we think both class token and token (without class token) can be used as the extracted feature information for independence sample weighting. So we design two schemes of SViT: class token decorrelation (SViT(C)) and token decorrelation (SViT(T)). SViT(C) and SViT(T) learn sample weights from the class token and token (except class token) output by the encoder respectively, and eliminate spurious correlations in feature space to improve the model's OOD generalization ability.

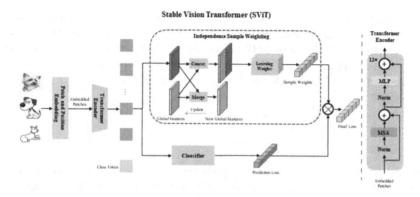

Fig. 1. The overall architecture of the proposed Stable Vision Transformer (SViT). SViT (C) learns sample weights from class tokens, while SViT (T) performs mean pooling on other tokens and learns sample weights.

3.2 Statistical Guidelines for Independence

To make the features independent of each other, SViT needs a suitable method to measure their independence and use it as an optimization criterion. Assuming that there are two variables: A and B, they are any pair of features in the feature space. According to the Hilbert-Schmidt Independence Criterion (HSIC) [3], when the square of the Hilbert-Schmidt norm of the cross-covariance matrix of A and B is zero, it can supervise independence sample weighting. However, the HSIC norm requires high computing costs. It is not suitable for the training of deep learning models. The Frobenius norm is equivalent to it in Euclidean space [18], so we choose the Frobenius norm of the cross-covariance matrix as the supervision for independence sample weighting.

In addition, when the cross-covariance matrix is zero, it can only guarantee linear independence between features. In the deep learning model, there will be nonlinear correlations between features. Random Fourier Feature (RFF) maps the original features to the high-dimension space. After that, we will eliminate the linear correlation between the features in the high-dimension space to ensure the independence between the original features.

$$H_{RFF} = \{h : x \to \sqrt{2}\cos(wx + \varphi) \mid w \sim N(0,1), \varphi \sim \text{Uniform}(0, 2\pi)\} \quad (1)$$

So the independence statistic between features is defined as:

$$I_{AB} = \|\mathrm{cov}(u(A), v(B))\|_F^2 \tag{2}$$

where u, v represent RFF. When I_{AB} tends to zero, features are independent of each other.

3.3   Learning Independence Sample Weights

The class token is the classification feature extracted by the transformer encoder. Therefore, SViT uses the method of independence sample weighting to eliminate the correlation of the features in the class token. Based on the statistical guidelines for independence, it uses Eq. 2 to define the independence statistics and learn the sample weights to decorrelate features by Eq. 3.

$$w^* = \left(\arg\min \sum_{1 \leq i < j \leq dim}^{w \in \Delta_n} \|\mathrm{cov}\left(Z_{:,i}, Z_{:,j}; w\right)\|_F^2 \right) \tag{3}$$

where $\Delta_n = \{\sum_{i=1}^n w_i = n\}$, dim represents the dimension of feature space, $Z_{:,i}, Z_{:,j}$ are features of samples in the RFF map generates by class token.

   We also considered that Transformer and some related improved models [12] could also be used to process classification tasks without class token. So it is possible to directly decorrelate the feature information contained in the token without class token. We take a similar approach to the related works above, averaging all tokens (except class token) and merging them into a synthetic token for learning. After that, we decorrelate the features of the synthetic token. This scheme is consistent with the decorrelation of class token in form. We use the same method to learn independent sample weights and still use the class token as input to the classifier.

$$\text{synthetic token} = \frac{1}{n} \sum_{i=1}^n \text{token}_i \tag{4}$$

As in the class token decorrelation, we learn sample weights in Eq. 3, RFF map is generated by the synthetic token.

3.4   Global Weight Learning and Joint Optimization

The goal of independence sample weighting is to ensure that the features between all samples are independent via sample weighting. But in the training phase of deep learning, the models only observe part of the samples in each batch. The sample weights learned from a batch can only make the part of samples' features independent of each other. For this problem, it needs to merge and save the global features and sample weights in the training phase. [28] proves the necessity of this mechanism for deep learning models to learn global sample weights. For each input batch (X_L, Y_L), we first generate features Z_O by concatenating the current batch features Z_L and global features information Z_G. After that, W_L is initialized as a learnable all-ones vector and generates W_O like Z_O. Then bring them into Eq. 3 to learn the sample weights.

$$Z_O = \text{Concat}(Z_G, Z_L)$$
$$W_O = \text{Concat}(W_G, W_L) \tag{5}$$

For joint optimization, the classifier uses class token directly to predict without sample weights. All parameters of the model are optimized by the weighted loss function as follows:

$$f^{(t+1)}, g^{(t+1)} = \arg\min_{f,g} \sum_{i=1}^{n} w_{L_i} L\left(g\left(f\left(X_{L_i}\right)\right), Y_{L_i}\right) \tag{6}$$

where f represents transformer encoder, g represents classification layer, and L represents the loss function. Finally, we update global features and weights.

$$Z'_G = \alpha_i Z_G + (1 - \alpha_i) Z_L$$
$$W'_G = \alpha_i W_G + (1 - \alpha_i) W_L \tag{7}$$

Algorithm 1. Training of Stable Vision Transformer(SViT)

Input: Samples
Output: Stable Vision Transformer(SViT)
1: **for** j in 1 to epoch number **do**
2:　　**for** k in 1 to batch number **do**
3:　　　　Load the global feature, weight vector by Equation 5
4:　　　　Learning sample weights W_L by Equation 3
5:　　　　Back propagate loss of model by Equation 6
6:　　　　Update Z_G and W_G by Equation 7
7:　　**end for**
8: **end for**

---

4 Experiments

4.1 Experimental Settings and Dataset

We set up two types of experiments: unbalanced setting, unbalanced and flexible setting. Unbalanced setting includes PACS and OfficeHome datasets, unbalanced and flexible setting includes PACS.

Dataset: We conduct extensive experiments on PACS [9] and OfficeHome [22] datasets. PACS is a widely used OOD generalization dataset containing 4 domains: Art-painting (A), Cartoon (C), Sketch (S), Photo (P), and 7 categories. OfficeHome includes 4 domains: Art (Ar), Clipart (Cl), Product (Pr), Real World (Rl), and 65 categories. For PACS, dividing the training set and validation set according to the standard in [28]. For OfficeHome, We follow convention of dividing 80% of the source domain as the training set and 20% as the validation set.

Unbalanced: Common OOD generalization settings usually consider that the capacity of the source domain is comparable, and the dataset is generated by equal sampling in the latent domain. But most datasets are formed by mixing multiple potential domains, so it can hardly assume that the amount of samples from these domains is consistent since these datasets are not generated by equally sampling from latent domains. We simulate this scenario with this setting. It divides all domains in the dataset into the source and target domains. After that, we follow the settings in [28]. The number of samples in each source domain is unbalanced, one of the domains is dominant, with a 5:1 sample ratio to other domains. The domain with the dominant number of source domain is called the dominant domain. The source domain, dominant domain, and target domain in different categories are the same.

Unbalanced and Flexible: Based on the unbalance setting, the domains of different categories might be different in real open environments. Making the source domain and target domain of all categories the same is not enough to verify the model's OOD generalization performance in real open environments. Therefore, in this setting, different categories contain different dominant domain, source domain, and target domain. It requires the model to have excellent OOD generalization ability.

4.2 Implementation Details and Comparison with Other Methods

SViT uses the ViT-B-16 pre-trained on ImageNet as the backbone network. We use SGD optimizer, setting the initial learning rate to 0.01 and using cosine decay scheduler. The batch size is 64, and the weight decay is 1e-4.

Since the SViT proposed in this paper does not need to use domain labels. For fairness, we compare SViT with StableNet [28], RSC [4], and MMLD [13], which do not require domain labels. These works are based on CNN. We use the ResNet which commonly used in the three papers as the backbone network, and report their performance with the officially published code and default parameter settings. In addition, there is few research focus on the OOD generalization ability of ViT. We have sorted out the current research situation in detail above, there is currently a lack of suitable ViT-based OOD generalization methods to compare with us. So in this paper, we compare SViT with the baseline (ViT-B-16) under the ViT framework. ViT-B-16 has the same parameters setting as SViT. The experimental results are run three times and averaged.

4.3 Specific Results of Unbalanced Setting

We follow the setting of [28], selecting 3 domains to form source domains, and the sample ratio of the domains is 5:1:1. The remaining domain is used as the target domain, and the domain with the dominant number of the source domains is called the dominant domain. For OfficeHome, the art domain contains relatively fewer samples. Using art as the dominant domain would result in many classes having only around 20 samples. That causes the source domain other than the art domain to contain few samples (only 1, even 0), resulting in a small number of samples. It brings difficulties and uncertainties to partitioning the validation set and model selection. We have provided detailed explanations of this issue in the supplementary materials. Therefore, we excluded 3 cases where art is the dominant domain. Apart from the mentioned special cases, we conducted extensive experiments on all possibilities for PACS and OfficeHome. The overall performance of SViT and other comparison methods is shown in Table 1, and the detailed performance of PACS and OfficeHome in each group is shown in Tables 2 and 3. We bold the best results and underline the second.

Table 1. The overall performance of PACS and OfficeHome under the unbalanced setting. The title of each column indicates the target domain name, and the result is the average of selecting different dominant domain in the source domain. For example, A represents the average performance when A is used as the target domain, C, S and P as source domains and are respectively used as the dominant domain.

		PACS					OfficeHome				
		A	C	S	P	Avg	Ar	Cl	Pr	Rl	Avg
MMLD	CNN	69.62	71.10	61.99	92.37	73.77	54.02	46.29	70.84	71.53	60.67
RSC	CNN	72.74	72.76	_66.12_	92.72	76.09	50.12	44.20	67.68	67.69	56.63
StableNet	CNN	72.75	73.07	**70.39**	89.34	76.39	52.33	46.89	69.28	69.41	59.47
ViT-B-16	ViT	_85.27_	78.04	39.54	**98.69**	75.39	75.17	59.14	_85.38_	86.77	76.62
SViT(C)	ViT	**88.14**	**78.83**	44.60	97.48	_77.26_	**75.73**	_59.91_	**85.72**	**87.12**	**77.12**
SViT(T)	ViT	84.05	_78.04_	52.68	_98.48_	**78.31**	_75.55_	**60.52**	85.11	_87.07_	_77.06_

From the above results, both SViT(C) and SViT(T) outperform the ViT and other SOTA methods on two datasets in unbalanced setting. In this setting, the generalization ability of ViT on most domains is better than SOTA methods based on CNN. However, the performance on the sketch domain in PACS, neither ViT nor SViT performance are worse than CNN-based methods. That is because ViT and CNN have different characteristics. The distribution shifts can be roughly summarized as background, texture, and structure. For example, when the target domain is art-painting in PACS, the distribution shift mainly consists of texture factors. When the target domain is photo, the distribution shifts are primarily caused by the background. The distribution shifts of the sketch primarily consist of structure shifts. Compared to CNN, ViT learns weaker biases on backgrounds and textures but is equipped with stronger inductive biases toward shapes and structures. The sketch domain exhibits significant structure shifts compared to other domains, which makes ViT-based models' overall poor performance when the sketch is the target domain [27]. The effectiveness of our proposed SViT has been demonstrated by the good results achieved in the vast majority of experiments. At the same time, SViT has significantly improved performance in the sketch domain compared to ViT.

Table 2. Detailed results of PACS. Taking an example, C-A means that A is the target domain. C, S and P are source domain, and C is the dominant source domain.

		C-A	P-A	S-A	A-C	P-C	S-C	A-S	C-S	P-S	A-P	C-P	S-P	Avg
MMLD	CNN	64.25	76.96	67.65	71.03	70.31	71.95	64.16	64.41	57.40	94.85	90.61	91.64	73.77
RSC	CNN	75.72	75.68	66.83	75.96	68.50	73.82	66.10	66.71	65.56	94.85	89.39	93.93	76.09
StableNet	CNN	80.16	78.38	59.71	69.31	74.15	75.75	**70.10**	**75.13**	65.94	95.15	78.63	94.24	76.39
ViT-B-16	ViT	79.39	91.40	85.01	77.47	79.19	77.47	50.77	25.77	42.09	**99.70**	96.67	**99.70**	75.39
SViT(C)	ViT	**89.22**	89.19	86.00	76.40	**79.19**	**80.90**	51.79	43.11	38.90	97.28	95.76	99.39	77.26
SViT(T)	ViT	73.77	**92.14**	**86.24**	**77.68**	77.25	79.19	55.36	55.49	47.19	99.39	**96.67**	99.39	**78.31**

Table 3. Detailed results of OfficeHome. The meaning of each column is similar to Table 2.

		Cl-Ar	Pr-Ar	Rl-Ar	Ar-Cl	Pr-Cl	Rl-Cl	Ar-Pr	Cl-Pr	Rl-Pr	Ar-Rl	Cl-Rl	Pr-Rl	Avg
MMLD	CNN	54.55	49.94	57.76	–	44.28	48.29	–	68.17	73.51	–	70.46	72.60	60.67
RSC	CNN	48.95	46.60	54.80	–	41.81	46.58	–	63.00	72.36	–	65.83	69.75	56.63
StableNet	CNN	51.96	47.92	57.11	–	44.51	49.26	–	65.83	72.72	–	68.17	70.64	59.47
ViT-B-16	ViT	72.85	74.08	78.57	–	58.26	60.02	–	83.46	87.29	–	85.43	88.11	76.62
SViT(C)	ViT	74.29	75.28	77.63	-	58.79	61.03	–	**83.87**	87.56	–	**86.11**	88.13	**77.12**
SViT(T)	ViT	**74.74**	73.38	78.53	–	**59.02**	**62.02**	–	83.40	86.82	–	86.00	88.13	77.06

4.4   Specific Results of Unbalanced and Flexible Setting

To simulate the situation in which different categories contain different domains in the real open environments. In this setting, we randomly choose different source domain, dominant domain, and target domain for each category. We randomly generate three groups of experiments on PACS. Experimental results in the unbalanced setting show that ViT and CNN perform differently on different

Table 4. Results of unbalanced and flexible setting on PACS.

	Group1	Group2	Group3	Avg
ViT-B-16	41.73	40.72	**30.64**	37.70
SViT(C)	52.98	46.12	27.29	42.29
SViT(T)	55.88	**48.62**	29.71	**44.74**

generalization tasks. To rule out this difference, we only conducted experiments within the framework of ViT in this setting. The experimental results of unbalanced and flexible setting are shown in Table 4. Both SViT (C) and SViT (T) have greatly improved the overall generalization performance of ViT.

4.5   Further Analysis and Discussion

SViT improves the performance of ViT by eliminating spurious correlations in the feature space. According to the characteristics of ViT, we designed two forms of SViT. We noticed a difference in the performance of them. In unbalanced setting, their performance is similar in OfficeHome. In PACS, SViT(T) performs slightly better than SViT(C). Compared with the unbalanced setting, unbalanced and flexible setting poses a higher challenge to the OOD generalization performance of the model. In this setting, the improvement effect of SViT is more obvious. SViT(T) is better than SViT(C). From experimental results, The performance of SViT(C) is more stable, only slightly worse in the photo domain of PACS, and has obvious improvement in all of the other domains. SViT(T) has greatly improved in some experiments, especially in the sketch domain and unbalanced and flexible setting, which are the most challenging OOD generalization task. But the stability is slightly worse than SViT(C). Both the average and worst performance is commonly used to measure the OOD generalization ability of the model. Their application scenarios are slightly different. Currently, no research gives a clear distinction between class token and token, especially in OOD generalization. Based on analysis of ViT and related models, we propose different SViTs from the experience. We will theoretically prove their difference.

In addition, SViT suffers a slight degradation in photo domain performance. It is common in previous work. The pre-training dataset is equivalent to a large photo domain, so it is not a strict OOD generalization task. SViT removes all the correlations in the feature space during the training process, causing the performance of the model to decline in scenarios similar to the pre-training distribution. Some recent works [16] have discussed this issue and proposed the idea of differentiated decorrelation. However, they are limited to linear models and lack theoretical proof.

At the same time, the ViT-based and the CNN-based models have different performances in different distribution shifts, and previous research [27] conclusions are also consistent with our experimental results. The performance of ViT and CNN in different tasks has been a research focus. Some ViT-based work [24] also introduces convolution to build models. Compared with CNN, ViT has more baselines to choose from, so analyzing the performance of different ViTs in different OOD generalization tasks is important.

5   Conclusion

In this paper, we proposed a Stable Vision Transformer (SViT) for the OOD generalization problem. SViT improves the OOD generalization performance of

the ViT by eliminating spurious correlations between features through independence sample weighting. According to the structural characteristics of ViT, we
designed two schemes: class token decorrelation and token decorrelation. It allows
our work to be more flexibly migrated to different ViT-based models. Extensive
experiments proved the effectiveness of our approach for OOD generalization.

References

1. Cui, P., Athey, S.: Stable learning establishes some common ground between causal
   inference and machine learning. Nat. Mach. Intell. **4**(2), 110–115 (2022)
2. Dosovitskiy, A., et al.: An image is worth 16x16 words: transformers for image
   recognition at scale. arXiv preprint arXiv:2010.11929 (2020)
3. Gretton, A., Fukumizu, K., Teo, C., Song, L., Schölkopf, B., Smola, A.: A kernel
   statistical test of independence. In: Advances in Neural Information Processing
   Systems 20 (2007)
4. Huang, Z., Wang, H., Xing, E.P., Huang, D.: Self-challenging improves cross-
   domain generalization. In: Vedaldi, A., Bischof, H., Brox, T., Frahm, J.-M. (eds.)
   ECCV 2020. LNCS, vol. 12347, pp. 124–140. Springer, Cham (2020). https://doi.
   org/10.1007/978-3-030-58536-5_8
5. Iwasawa, Y., Matsuo, Y.: Test-time classifier adjustment module for model-agnostic
   domain generalization. Adv. Neural. Inf. Process. Syst. **34**, 2427–2440 (2021)
6. Krueger, D., et al.: Out-of-distribution generalization via risk extrapolation (rex).
   In: International Conference on Machine Learning, pp. 5815–5826. PMLR (2021)
7. Kuang, K., Cui, P., Athey, S., Xiong, R., Li, B.: Stable prediction across unknown
   environments. In: proceedings of the 24th ACM SIGKDD International Conference
   on Knowledge Discovery & Data Mining, pp. 1617–1626 (2018)
8. Kuang, K., Xiong, R., Cui, P., Athey, S., Li, B.: Stable prediction with model
   misspecification and agnostic distribution shift. In: Proceedings of the AAAI Conference on Artificial Intelligence, vol. 34, pp. 4485–4492 (2020)
9. Li, D., Yang, Y., Song, Y.Z., Hospedales, T.M.: Deeper, broader and artier domain
   generalization. In: Proceedings of the IEEE International Conference on Computer
   Vision, pp. 5542–5550 (2017)
10. Li, H., Pan, S.J., Wang, S., Kot, A.C.: Domain generalization with adversarial
    feature learning. In: Proceedings of the IEEE Conference on Computer Vision and
    Pattern Recognition, pp. 5400–5409 (2018)
11. Li, Y., Mao, H., Girshick, R., He, K.: Exploring plain vision transformer backbones
    for object detection. In: Avidan, S., Brostow, G., Cissé, M., Farinella, G.M., Has-
    sner, T. (eds.) Computer Vision – ECCV 2022. ECCV 2022. LNCS, vol. 13669.
    Springer, Cham. https://doi.org/10.1007/978-3-031-20077-9_17
12. Liu, Z., et al.: Swin transformer: Hierarchical vision transformer using shifted win-
    dows. In: Proceedings of the IEEE/CVF International Conference on Computer
    Vision, pp. 10012–10022 (2021)
13. Matsuura, T., Harada, T.: Domain generalization using a mixture of multiple latent
    domains. In: Proceedings of the AAAI Conference on Artificial Intelligence, vol.
    34, pp. 11749–11756 (2020)
14. Niu, L., Li, W., Xu, D.: Visual recognition by learning from web data: a weakly
    supervised domain generalization approach. In: Proceedings of the IEEE Confer-
    ence on Computer Vision and Pattern Recognition, pp. 2774–2783 (2015)

15. Peters, J., Bühlmann, P., Meinshausen, N.: Causal inference by using invariant prediction: identification and confidence intervals. J. Royal Statist. Soc. Ser. B (Statist. Methodol.) **78**(5), 947–1012 (2016)
16. Shen, Z., Cui, P., Liu, J., Zhang, T., Li, B., Chen, Z.: Stable learning via differentiated variable decorrelation. In: Proceedings of the 26th ACM SIGKDD International Conference on Knowledge Discovery & Data Mining, pp. 2185–2193 (2020)
17. Shen, Z., Cui, P., Zhang, T., Kunag, K.: Stable learning via sample reweighting. In: Proceedings of the AAAI Conference on Artificial Intelligence, vol. 34, pp. 5692–5699 (2020)
18. Strobl, E.V., Zhang, K., Visweswaran, S.: Approximate Kernel-based conditional independence tests for fast non-parametric causal discovery. J. Causal Infer. **7**(1), 17 (2019)
19. Sultana, M., Naseer, M., Khan, M.H., Khan, S., Khan, F.S.: Self-distilled vision transformer for domain generalization. In: Proceedings of the Asian Conference on Computer Vision, pp. 3068–3085 (2022)
20. Touvron, H., Cord, M., Douze, M., Massa, F., Sablayrolles, A., Jégou, H.: Training data-efficient image transformers & distillation through attention. In: International Conference on Machine Learning, pp. 10347–10357. PMLR (2021)
21. Träuble, F., et al.: On disentangled representations learned from correlated data. In: International Conference on Machine Learning, pp. 10401–10412. PMLR (2021)
22. Venkateswara, H., Eusebio, J., Chakraborty, S., Panchanathan, S.: Deep hashing network for unsupervised domain adaptation. In: Proceedings of the IEEE Conference on Computer Vision and Pattern Recognition, pp. 5018–5027 (2017)
23. Wang, Z., Loog, M., Van Gemert, J.: Respecting domain relations: hypothesis invariance for domain generalization. In: 2020 25th International Conference on Pattern Recognition (ICPR), pp. 9756–9763. IEEE (2021)
24. Wu, H., et al.: CvT: introducing convolutions to vision transformers. In: Proceedings of the IEEE/CVF International Conference on Computer Vision, pp. 22–31 (2021)
25. Xu, R., Zhang, X., Shen, Z., Zhang, T., Cui, P.: A theoretical analysis on independence-driven importance weighting for covariate-shift generalization. In: International Conference on Machine Learning, pp. 24803–24829. PMLR (2022)
26. Yang, M., Liu, F., Chen, Z., Shen, X., Hao, J., Wang, J.: CausalVAE: disentangled representation learning via neural structural causal models. In: Proceedings of the IEEE/CVF Conference on Computer Vision and Pattern Recognition, pp. 9593–9602 (2021)
27. Zhang, C., et al.: Delving deep into the generalization of vision transformers under distribution shifts. In: Proceedings of the IEEE/CVF Conference on Computer Vision and Pattern Recognition, pp. 7277–7286 (2022)
28. Zhang, X., Cui, P., Xu, R., Zhou, L., He, Y., Shen, Z.: Deep stable learning for out-of-distribution generalization. In: Proceedings of the IEEE/CVF Conference on Computer Vision and Pattern Recognition, pp. 5372–5382 (2021)
29. Zheng, S., et al.: Rethinking semantic segmentation from a sequence-to-sequence perspective with transformers. In: Proceedings of the IEEE/CVF Conference on Computer Vision and Pattern Recognition, pp. 6881–6890 (2021)

Few-Shot Classification with Semantic Augmented Activators

Ruixuan Gao[1], Han Su[1,2,3(✉)], and Peisen Tang[1]

[1] School of Computer Science, Sichuan Normal University, Chengdu 610101, China
jkxy_sh@sicnu.edu.cn
[2] Visual Computing and Virtual Reality Key Laboratory of Sichuan Province,
Chengdu, China
[3] School of Computer Science, University of Hull, Hull, UK

Abstract. Metric-based methods predict class labels by measuring the distance between a few given samples, often failing to preserve more useful semantic details in their vectorial representations. In this paper, we propose Semantic Augmented Activators (SAA), which are generated based on the variance of the intra-set samples in an unsupervised manner, to enhance the discriminability of feature vectors with more class-related semantic information. This generation process does not rely on any learnable parameters. Meanwhile, to align the SAA preferred to operate in the intra-set and sufficiently leverage the finite samples, we treat the Self-Cross loss as an auxiliary loss, which bi-directionally complements the limitations of the traditional loss function. Additionally, we introduce Map-To-Cluster, a transductive module to map the SAA-enhanced features to a lower-dimensional embedding space. This encourages proximity among similar samples and separation among dissimilar samples. The resulting methods are lightweight and computationally efficient. Our methods demonstrate competitive performance on the *mini*-ImageNet and *tiered*-ImageNet benchmarks, and achieve outstanding results in Cross-Domain Few-Shot classification.

Keywords: Few-shot learning · Metric-based learning · Transductive inference

1 Introduction

Traditional convolutional neural classifiers rely on abundant annotated image sets, such as ImageNet [18], and COCO [12], to achieve brilliant performance. However, acquiring such extensive human-labeled datasets is extremely challenging in reality. Inspired by the remarkable ability of humans to classify objects

Supported by Natural Science Foundation of Sichuan Province, Grant/Award Number: 2023NSFSC1080, 2023NSFSC0210, 2023YFS0202; Chengdu Science and Technology Program, Grant/Award Number: 2022-YF09-00019-SN; Chengdu Research Base of Giant Panda Breeding, Grant/Award Number: 2021CPB-B06; China Scholarship Council.

Q. Liu et al. (Eds.): PRCV 2023, LNCS 14432, pp. 340–352, 2024.
https://doi.org/10.1007/978-981-99-8543-2_28

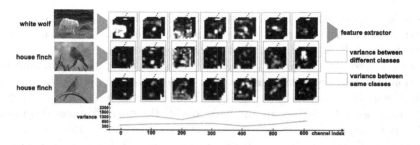

Fig. 1. Variance of channels: We calculate the mean variances for the same class and different classes per 100 channels. The variances of channels from the same class are obviously lower than those from different classes.

with a small amount of data and quickly adapt to new tasks, few-shot learning emerged and grew in popularity. Few-shot learning aims to predict image class labels with only a few annotated samples. Nevertheless, the limited support images fail to provide sufficient information for the model to learn the underlying semantic representation, and the model needs to recognize numerous novel classes that were not present in the training set.

Metric-based methods, as one of the solution branches, utilize distance functions to measure the similarity between vector representations of images in the embedding space for predicting class labels. Commonly used distance functions include cosine similarity [24], Euclidean distance [21], or learnable modules [22]. In general, performance heavily relies on the class-related semantic content encoded in the representations. Therefore, enhancing the semantic information within single vectorial representations becomes a crucial challenge for researchers. Previous approaches such as feature reconstruction [9,27] focus on capturing similar semantic information between support and query samples, but they often require additional attention modules or parameters. Alternatively, incorporating prior knowledge [7] as supplementary semantic information to feature vectors is also explored, but it asks a weight generator for help. In this paper, we aim to analyze the distribution within features without introducing additional modules, activating relevant semantic information to obtain discriminative feature vectors.

SPP-net [8] has demonstrated that filters of deep convolutions can be selectively activated by specific semantic content, and PWA [28] utilizes filters of convolutional layers as part detectors to highlight discriminative regions of samples while suppressing background noise. These approaches show that the layers of feature maps contain valuable cues for understanding objects. As shown in Fig. 1, the variances between two samples from the same class are significantly lower than that from different classes in the same channels, illustrating that the variance of channels can capture the discrepancy in semantic information and exhibit varying degrees of response to class-related semantic information. Motivated by the above experiment, we utilize the channels of feature maps selected based on their variances as Semantic Augmented Activators (SAA). These activators allow us to increase the proportion of distinguishable regions, thereby augmenting the efficacious semantic information and the representational capacity of the features. It is noted that this is an unsupervised process.

SAA focuses on capturing intra-set features. However, the traditional loss function only unidirectionally enforces the model to learn inter-set features and limits the utilization of the given samples. To address this limitation, we propose Self-Cross loss (SC loss) as an auxiliary loss. If the query features can obtain sufficient semantic information from support prototypes, it can also provide classification prediction for support features in the reverse direction. The support prototypes not only interact with query features but also with support features. Similarly, the query prototypes also interact with both query features and support features. This bi-directional loss strengthens the connections between samples and promotes few-shot classification. Furthermore, taking inspiration from transductive methods such as Poodle [10] and EASE [31], which leverage the entire query set, we introduce the transductive module called Map-To-Cluster (MTC) based on cluster theory. MTC reduces the dimension of query features, aiming to filter outlying contents and focus on the essential semantic information. MTC complements existing works in few-shot classification, further improving the performance.

In summary, this paper makes the following contributions:

1. We propose SAA, generated by an unsupervised mechanism, incorporating useful semantic information into the discriminative vectorial representations, without any learnable parameters.
2. We introduce SC loss to extract features not only of the inter-set but also of the intra-set and constraint model to hoist generalization with bi-directional formulas.
3. We utilize the transductive module MTC, which closes the distance between congenetic samples while increasing the separation between samples from different classes.

2 Related Work

2.1 Few-Shot Classification

Few-shot classification aims to predict novel class labels using a limited number of annotated samples. Generally, they are divided into two main branches: optimization-based methods and metric-based methods. Optimization-based approaches [6,16] learn better initial parameters to facilitate subsequent training. On the other hand, metric-based methods [7,9,21,22,24,27] emphasize measuring the distance between samples. Several excellent works make splashes in the metric-based branch. For example, ProtoNet [21] proposes the concept of prototypes and utilizes more representative prototypes for comparing Euclidean distance with the samples, while Matching network [24] introduces LSTM and uses cosine similarity to make predictions. In addition to traditional distance functions, a learnable model [22] can also be introduced.

Furthermore, CTX [5] achieves more generalized feature representations through SimCLR [1]. CAN [9] and FRN [27] reconstruct query samples so that more similar samples have more parallel features between the support set and query set, with the former focusing on attention maps and the latter employing ridge regression. Classification-weight generator [7] effectively incorporates past knowledge through the attention mechanism while learning novel vector representations. In contrast to the above methods that emphasize exploring the relationship between the features of the support set and the query set, our methods focus on analyzing the inherent properties of the intra-set samples, capturing class-related semantic information without the need for complex computations.

2.2 Transductive Inference

Transductive inference involves treating query samples, which are typically considered unlabeled during training, as a whole and making predictions for them. It can boost the performance of few-shot classification. CAN [9] obtains more representative features by augmenting the support samples with the query samples. Poodle [10] and ConFT [4] wield the out-of-distribution samples to improve few-shot generalization. In contrast, our plug-and-play MTC does not introduce extra samples but instead clusters the existing data. This process focuses on the query samples themselves, without relying on support samples and corresponding labels for assistance.

3   Method

In this section, we describe the mechanism of SAA and explain the details of SC loss. Followed by presenting the transductive module MTC used to filtrate SAA-enhanced features. An overview is provided in Fig. 2.

3.1   Preliminaries

We first define some notations used in the few-shot classification scenario. Designating (x_i, y_i) as an image and its corresponding label. The base training set utilized for the pre-training stage is denoted as $D_b = \{(x_i, y_i)\}_{i=1}^{N_b}$. Following we expect to predict labels of unseen samples in the query set $D_q = \{(x_i^q, y_i^q)\}_{i=1}^{N_q}$ using the support set $D_s = \{(x_i^s, y_i^s)\}_{i=1}^{N_s}$. Here, the variables N_b, N_s, and N_q represent the respective number of sample pairs in each set. In a few-shot classification task episode involving N classes with K samples per class, we refer to this as N-way K-shot. Note that the labels of the query set are only used for verification, and therefore we treat the query samples as unlabeled.

While considering scenarios where both training and validation are in the same domain, we also assess the generalization of our methods in a cross-domain scenario. In this scenario, we train the model in one domain and then test its performance in another domain.

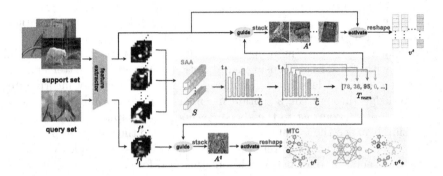

Fig. 2. Flow chart of our methods. First, we calculate the variances of channels and sort them in descending order. This allows us to determine the sequence numbers of channels, which are used to select the top L channels for each sample and stack them to generate the activators. Then, MTC reduces the dimension of SAA-enhanced feature vectors to output the final prediction.

3.2 Semantic Augmented Activators

We propose SAA with the expectation of enhancing feature vectors to stockpile more class-related semantic information for accurate prediction of query labels y^q. Given support samples x^s, we feed them into the embedding module F_θ to obtain feature maps $f^s \in \mathbb{R}^{N_s \times C \times W \times H}$.

First, we perform spatial aggregation of the feature maps for each support sample, resulting in the computation of the summing descriptors $S \in \mathbb{R}^{N_s \times C}$ for each channel. This aggregation is achieved by sum-pooling the feature maps f^s.

$$S = \sum_{w=0}^{W} \sum_{h=0}^{H} f^s(w, h) \tag{1}$$

Variance can measure the degree of differentiation between samples within the same channel, which also indirectly indicates the response of channels to specific categories. A higher variance implies that the corresponding channel contains more discriminative semantic information.

$$T = \frac{1}{N_s} \sum_{i=1}^{N_s} (S_i - \frac{1}{N_s} \sum_{r=1}^{N_s} S_r)^2 \tag{2}$$

According to the calculated variances $T = \{t_1, t_2, ..., t_C\}$, we sort them in descending order to obtain the corresponding channel sequence numbers T_{num}. We then select L channels for each support sample based on T_{num} to stack L-activators, denoted as $A^s \in \mathbb{R}^{N_s \times W \times H}$. These activators highlight the specific class semantic regions while suppressing the background noise, leading to a strong response to the objects.

$$A_r^s = \sum_{l}^{L} f_r^s[T_{num}[l]], r = 1, 2, ..., N_s \tag{3}$$

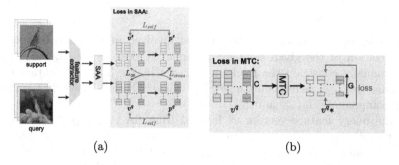

(a)   (b)

Fig. 3. The visual representation of the loss function. (a) describes the SC loss. The SC loss is specifically designed for inductive inference and assists the feature extractor in adequately exploiting the given samples. (b) illustrates the loss function used to optimize the MTC, which leverages transductive inference to boost performance.

It is worth noting that the query activators A^q are selected from the query feature maps based on the same channel sequence numbers T_{num} obtained from the support set. This can enhance the interaction of the same semantic information and the consistency of the same class information between different sets. The support weights $W^s \in \mathbb{R}^{N_s \times W \times H}$ is the normalized coefficient consulted by A^s, while the query weights $W^q \in \mathbb{R}^{N_q \times W \times H}$ incorporates the semantic information learned from the support set, facilitating inter-set knowledge sharing.

$$W^s = \mathbb{M}(\frac{A^s}{\sum_w^W \sum_h^H A^s(w,h)})$$
$$W^q = \mathbb{M}(\alpha \times \mathbb{G}(W^s) + \beta \times \frac{A^q}{\sum_w^W \sum_h^H A^q(w,h)}) \tag{4}$$

where $\mathbb{G}(\cdot)$ is the repeat operator, constraining that the dimensions of the operands are the same, $\mathbb{M}$ denotes a spatial-wise sigmoid, α and β are parameters of power-scaling. Specifically, we set $\alpha = 0.5$ and $\beta = 1.0$ for the *mini*-ImageNet, while $\alpha = 0.1$ and $\beta = 1.0$ for the *tiered*-ImageNet.

Following we weigh the initial representations f^s and f^q with W^s and W^q to highlight the discriminative regions of images and enrich efficacious semantic information for the subsequent feature vectors. We reshape the activated feature maps with global average-pooling to obtain the final augmented feature vectors $v^s \in \mathbb{R}^{N_s \times C}$ and $v^q \in \mathbb{R}^{N_q \times C}$. For getting the corresponding support prototypes $p^s \in \mathbb{R}^{N \times C}$, we calculate the average of v^s belonging to the same class.

$$v^s = \mathbb{P}(W^s f^s)$$
$$p_n^s = \frac{1}{|K|} \sum_{k=1}^K v_{n,k}^s, n = 1, 2, ..., N \tag{5}$$

where $\mathbb{P}(\cdot)$ is the global average-pooling.

3.3 Self-Cross Loss

For considering the features of the intra-set involved in SAA and to maximize the utilization of the given samples, we treat SC loss as an auxiliary loss to bi-directionally constrain the model. As shown in Fig. 3(a), in addition to the traditional loss L_{ce}, we introduce a new loss called SC loss L_{sc}, which is composed of two components: L_{cross} and L_{self}.

If the query feature vectors v^q learned from SAA are representative, the query prototypes p^q generated from them contain sufficient categorical discriminative semantic information, predicting support feature vectors v^s in reverse and deepening the association of samples from the same class.

$$L_{cross} = CE(d(v^s, p^q), y^s) \tag{6}$$

where $CE(\cdot)$ is the cross-entropy loss, $d(\cdot)$ is the Euclidean distance and p^q are the query prototypes. Meanwhile, to encourage alignment between samples from the same class within the same set, we calculate the loss L_{self} by comparing the feature vectors to their corresponding prototypes in each set.

$$L_{self} = CE(d(v^s, p^s), y^s) + CE(d(v^q, p^q), y^q) \tag{7}$$

Traditional loss and SC loss compose the final loss to further improve the performance of the framework.

$$L = L_{ce} + \underbrace{\gamma_1 L_{self} + \gamma_2 L_{cross}}_{L_{sc}} \tag{8}$$

where L_{ce} is the original loss function. $\gamma_1 = 10.0$ and $\gamma_2 = 1.0$ are parameters of power-scaling for SC loss.

3.4 Map-To-Cluster

We introduce the MTC to further bring similar samples closer together and separate dissimilar samples to boost performance. For MTC, we do not rely on the class labels but only leverage the knowledge learned from SAA.

We map the SAA-enhanced query features through MTC to the G-dimensional $(G < C)$ embedding space, denoted as v^q*. MTC is composed of two linear layers with the LeakyReLU function. According to the nearest neighbor [26], the means of query features are taken as the initial cluster centers. The congenetic samples move closer to the centers of the same class in the learned G-dimension embedding space. The final prediction depends on the distances between the samples and cluster centers, and the MTC uses a separate loss function shown in Fig. 3(b) for continuous optimization.

$$P(n|x_i^q) = \frac{e^{\|v_{n,i}^q* - \frac{1}{|N_q|}\sum_{z=1}^{N_s} v_{n,z}^q*\|^2}}{\sum_{n'=1}^{N} e^{\|v_{n,i}^q* - \frac{1}{|N_q|}\sum_{z=1}^{N_s} v_{n',z}^q*\|^2}} \tag{9}$$

Table 1. Performance of selected competitive few-shot classification methods on *mini*-ImageNet and *tiered*-ImageNet. ∘ means transductive inference. ♡ means knowledge distillation.

Model	Backbone	*mini*-ImageNet		*tiered*-ImageNet	
		1-shot	5-shot	1-shot	5-shot
ProtoNet [21,29]	ResNet-12	62.39 ± 0.21	80.53 ± 0.14	68.23 ± 0.23	84.03 ± 0.16
MatchNet [24,30]	ResNet-12	63.08 ± 0.80	75.99 ± 0.60	68.50 ± 0.92	80.60 ± 0.71
MetaOptNet [11]	ResNet-12	62.64 ± 0.61	78.63 ± 0.46	65.99 ± 0.72	81.56 ± 0.53
CAN [9]	ResNet-12	63.85 ± 0.48	79.44 ± 0.34	69.89 ± 0.51	84.23 ± 0.37
MetaBaseline [3]	ResNet-12	63.17 ± 0.23	79.26 ± 0.17	68.62 ± 0.27	83.29 ± 0.18
DSN [20]	ResNet-12	62.64 ± 0.66	78.83 ± 0.45	66.22 ± 0.75	82.79 ± 0.48
Neg-Cosine [13]	ResNet-12	63.85 ± 0.81	81.57 ± 0.56	–	–
Deepemd [30]	ResNet-12	65.91 ± 0.82	82.41 ± 0.56	71.16 ± 0.87	86.03 ± 0.58
FRN [27]	ResNet-12	66.45 ± 0.19	82.83 ± 0.13	71.16 ± 0.22	86.01 ± 0.15
P-Transfer [19]	ResNet-12	64.21 ± 0.77	80.38 ± 0.59	–	–
CAN+T∘ [9]	ResNet-12	67.19 ± 0.55	80.64 ± 0.35	73.21 ± 0.58	84.93 ± 0.38
LaplacianShot∘ [32]	ResNet-18	72.11 ± 0.19	82.31 ± 0.14	78.98 ± 0.21	86.39 ± 0.16
poodle∘ [10]	ResNet-12	74.21	83.71	78.72	86.57
poodle∘ ♡ [10]	ResNet-12	77.56	**85.81**	79.97	86.96
SAA(ours)	ResNet-12	65.39 ± 0.61	81.28 ± 0.43	70.03 ± 0.72	84.53 ± 0.49
+MTC(ours)∘	ResNet-12	76.82 ± 0.41	82.93 ± 0.38	84.41 ± 0.40	86.15 ± 0.36
+MTC(ours)∘ ♡	ResNet-12	**82.79 ± 0.38**	**85.64 ± 0.31**	**86.47 ± 0.37**	**87.28 ± 0.35**

4 Experiments

In this section, we give the implementation details of our methods and conduct extensive experiments to demonstrate the performance of our methods under different scenarios. Finally, we supplement ablation experiments to illustrate the effectiveness of each part of our methods.

4.1 Set up

On the general few-shot classification, we evaluate our methods on *mini*-ImageNet [24] and *tiered*-ImageNet [17]. Instead of training directly from scratch using episodes, we pre-training like [29] on the training set. To verify the generalization of our method, we also test the effectiveness of our method on cross-domain problems in which the model trained on *mini*-ImageNet is evaluated on CUB [25].

   mini-ImageNet chooses 60000 images from ImageNet [18] and 600 images per class, with 100 classes including 64 training, 16 validation, and 20 test classes.

   tiered-ImageNet is the subset of the ImageNet, consisting of 351 training, 97 validation, and 160 test classes with 779165 images.

   CUB is the fine-grained dataset with 11788 images. We randomly split the 200 classes into 100 training, 50 validation, and 50 test.

Implementation Details: We treat ResNet-12 as the feature extractor and ProtoNet as the baseline. During pre-training, we adopt cross-entropy loss to train our model with a batch size of 128. For meta-training, we use SGD optimizer with a fixed learning rate of 3e-4, while in transductive inference, we utilize Adam optimizer with a fixed learning rate of 1e-4. The number of iterations of MTC optimization is consistent with that of the whole model, without repeating iterations. The final performance we evaluate in 5-way 1-shot and 5-shot settings on 600 random sample episodes with 15 query images of size 84 × 84.

Table 2. Performance comparison in the cross-domain setting: *mini*-ImageNet → CUB.

Model	Backbone	5-way 1-shot	5-way 5-shot
ProtoNet [2, 21]	ResNet-18	–	62.02 ± 0.70
MetaOptNet [11, 15]	ResNet-12	44.79 ± 0.75	64.98 ± 0.68
MatchNet+FT [23]	ResNet-10	36.61 ± 0.53	55.23 ± 0.83
RelationNet+FT [23]	ResNet-10	44.07 ± 0.77	59.46 ± 0.71
GNN+FT [23]	ResNet-10	47.47 ± 0.75	66.98 ± 0.68
ConFT [4]	ResNet-10	45.57 ± 0.76	70.53 ± 0.75
SAA(ours)	ResNet-12	48.49 ± 0.60	66.07 ± 0.55
+MTC(ours)	ResNet-12	**63.99 ± 0.47**	**74.20 ± 0.39**

Table 3. The speed comparison per episode (ms) on *mini*-ImageNet.

Methods	meta-training		meta-test	
	5-way 1-shot	5-way 5-shot	5-way 1-shot	5-way 5-shot
ProtoNet	210	250	83	105
Deepemd(qpth)	26,000	>990,000	23,275	>800,000
SAA(ours)	396	451	308	311
+MTC(ours)	424	485	311	325

4.2 Results

General Few-Shot Classification: We show the results of classification with 5-way 1-shot and 5-way 5-shot in Table 1. On *mini*-ImageNet, we find using SAA alone will achieve 3% improvement over amended ProtoNet with 5-way 1-shot, and the performance is even more significant with 14.43% increase when using transductive method MTC to map learned features to G-dimensional embedding space. Although the increase is not very significant on *tiered*-ImageNet, SAA still achieves around 2% improvement, and MTC can even achieve an increase of 16%. 5-way 5-shot shows similar improvement. That means SAA effectively learns discriminative feature vectors with class-related semantic information. And MTC indeed boosts performance by bringing samples of the same classes closer. We also conduct experiments with knowledge distillation, which have outstanding results, matching certain methods using similar additional techniques [10].

Cross-Domain Few-Shot Classification: We evaluate the transferability of our methods on the challenging Cross-domain Few-shot classification. We train the model on *mini*-ImageNet during the pre-train and meta-train stages, and then only test it on the CUB to investigate the discrepancies between the different domains. As shown in Table 2, our methods demonstrate outstanding performance, with MTC being particularly effective in cross-domain variations. We suspect that MTC is trained based on the SAA and not constrained by prior knowledge.

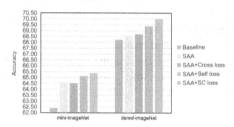

Fig. 4. The accuracy of our method with different SC loss settings on *mini*-ImageNet and *tiered*-ImageNet with 5-way 1-shot.

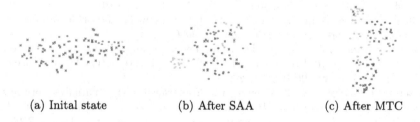

(a) Inital state   (b) After SAA   (c) After MTC

Fig. 5. The visualization of MTC on *mini*-ImageNet with 5-way 1-shot. **The red crosses** representing misclassified samples reduce significantly through MTC.

Ablation Experiments: As shown in Table 3, we do speed experiments to verify that our method is lightweight. Even after incorporating MTC, our method takes only slightly more time per episode than the baseline, but at least 23 s faster than Deepemd.

We conduct ablation experiments on the SC loss to further validate its effectiveness using a 5-way 1-shot setting. Figure 4 demonstrates that both the Self loss and Cross loss components of the SC loss contribute to performance improvement. However, it is worth noting that the performance with the Cross loss in the *mini*-ImageNet seems to betray this result. We suspect that the samples in the *mini*-ImageNet belong to the broader categories, making them more separable. In this case, reusing the inter-set finite samples leads to overfitting and potentially harms the performance.

We adopt t-SNE [14] to visualize the dynamic distribution of correct and incorrect samples. As shown in Fig. 5, the incorrect samples represented by **red crosses** are significantly reduced after the introduction of SAA and MTC.

5   Conclusion

We propose Semantic Augmented Activators, learned by an unsupervised mechanism, to highlight the regions of objects while suppressing noise information. This results in feature vectors that contain more discriminative semantic information. To assist the SAA focused more on intra-set, Self-Cross loss as an auxiliary loss to bi-directionally constraints the model. Additionally, we introduce Map-To-Cluster to extract essential semantic information by reducing the dimension of the feature vectors. Our experiments achieve excellent performance on both general few-shot classification and cross-domain problems.

References

1. Chen, T., Kornblith, S., Norouzi, M., Hinton, G.: A simple framework for contrastive learning of visual representations. In: International Conference on Machine Learning, pp. 1597–1607. PMLR (2020)
2. Chen, W.Y., Liu, Y.C., Kira, Z., Wang, Y.C.F., Huang, J.B.: A closer look at few-shot classification. arXiv preprint arXiv:1904.04232 (2019)
3. Chen, Y., Wang, X., Liu, Z., Xu, H., Darrell, T.: A new meta-baseline for few-shot learning. arXiv preprint arXiv:2003.04390 (2020)
4. Das, R., Wang, Y.X., Moura, J.M.: On the importance of distractors for few-shot classification. In: Proceedings of the IEEE/CVF International Conference on Computer Vision, pp. 9030–9040 (2021)
5. Doersch, C., Gupta, A., Zisserman, A.: CrossTransformers: spatially-aware few-shot transfer. In: Larochelle, H., Ranzato, M., Hadsell, R., Balcan, M., Lin, H. (eds.) Advances in Neural Information Processing Systems, vol. 33, pp. 21981–21993. Curran Associates, Inc. (2020). https://proceedings.neurips.cc/paper/2020/file/fa28c6cdf8dd6f41a657c3d7caa5c709-Paper.pdf
6. Finn, C., Abbeel, P., Levine, S.: Model-agnostic meta-learning for fast adaptation of deep networks. In: International Conference on Machine Learning, pp. 1126–1135. PMLR (2017)
7. Gidaris, S., Komodakis, N.: Dynamic few-shot visual learning without forgetting. In: Proceedings of the IEEE Conference on Computer Vision and Pattern Recognition, pp. 4367–4375 (2018)
8. He, K., Zhang, X., Ren, S., Sun, J.: Spatial pyramid pooling in deep convolutional networks for visual recognition. IEEE Trans. Pattern Anal. Mach. Intell. **37**(9), 1904–1916 (2015)

9. Hou, R., Chang, H., Ma, B., Shan, S., Chen, X.: Cross attention network for few-shot classification. In: Advances in Neural Information Processing Systems 32 (2019)
10. Le, D., Nguyen, K.D., Nguyen, K., Tran, Q.H., Nguyen, R., Hua, B.S.: POODLE: improving few-shot learning via penalizing out-of-distribution samples. Adv. Neural. Inf. Process. Syst. **34**, 23942–23955 (2021)
11. Lee, K., Maji, S., Ravichandran, A., Soatto, S.: Meta-learning with differentiable convex optimization. In: Proceedings of the IEEE/CVF Conference on Computer Vision and Pattern Recognition, pp. 10657–10665 (2019)
12. Lin, T.-Y., et al.: Microsoft COCO: common objects in context. In: Fleet, D., Pajdla, T., Schiele, B., Tuytelaars, T. (eds.) ECCV 2014. LNCS, vol. 8693, pp. 740–755. Springer, Cham (2014). https://doi.org/10.1007/978-3-319-10602-1_48
13. Liu, B., et al.: Negative margin matters: understanding margin in few-shot classification. In: Vedaldi, A., Bischof, H., Brox, T., Frahm, J.-M. (eds.) ECCV 2020. LNCS, vol. 12349, pp. 438–455. Springer, Cham (2020). https://doi.org/10.1007/978-3-030-58548-8_26
14. Van der Maaten, L., Hinton, G.: Visualizing data using t-SNE. J. Mach. Learn. Res. **9**(11), 2579–2605 (2008)
15. Mangla, P., Kumari, N., Sinha, A., Singh, M., Krishnamurthy, B., Balasubramanian, V.N.: Charting the right manifold: manifold mixup for few-shot learning. In: Proceedings of the IEEE/CVF Winter Conference on Applications of Computer Vision, pp. 2218–2227 (2020)
16. Ravi, S., Larochelle, H.: Optimization as a model for few-shot learning. In: International Conference on Learning Representations (2017)
17. Ren, M., et al.: Meta-learning for semi-supervised few-shot classification. arXiv preprint arXiv:1803.00676 (2018)
18. Russakovsky, O., et al.: ImageNet large scale visual recognition challenge. Int. J. Comput. Vision **115**(3), 211–252 (2015)
19. Shen, Z., Liu, Z., Qin, J., Savvides, M., Cheng, K.T.: Partial is better than all: revisiting fine-tuning strategy for few-shot learning. In: Proceedings of the AAAI Conference on Artificial Intelligence, vol. 35, pp. 9594–9602 (2021)
20. Simon, C., Koniusz, P., Nock, R., Harandi, M.: Adaptive subspaces for few-shot learning. In: Proceedings of the IEEE/CVF Conference on Computer Vision and Pattern Recognition, pp. 4136–4145 (2020)
21. Snell, J., Swersky, K., Zemel, R.: Prototypical networks for few-shot learning. In: Advances in Neural Information Processing Systems 30 (2017)
22. Sung, F., Yang, Y., Zhang, L., Xiang, T., Torr, P.H., Hospedales, T.M.: Learning to compare: relation network for few-shot learning. In: Proceedings of the IEEE Conference on Computer Vision and Pattern Recognition, pp. 1199–1208 (2018)
23. Tseng, H.Y., Lee, H.Y., Huang, J.B., Yang, M.H.: Cross-domain few-shot classification via learned feature-wise transformation. arXiv preprint arXiv:2001.08735 (2020)
24. Vinyals, O., Blundell, C., Lillicrap, T., Wierstra, D., et al.: Matching networks for one shot learning. In: Advances in Neural Information Processing Systems 29 (2016)
25. Wah, C., Branson, S., Welinder, P., Perona, P., Belongie, S.: The Caltech-UCSD birds-200-2011 dataset (2011)
26. Weinberger, K.Q., Saul, L.K.: Distance metric learning for large margin nearest neighbor classification. J. Mach. Learn. Res. **10**(2), 207–244 (2009)

27. Wertheimer, D., Tang, L., Hariharan, B.: Few-shot classification with feature map reconstruction networks. In: Proceedings of the IEEE/CVF Conference on Computer Vision and Pattern Recognition, pp. 8012–8021 (2021)

28. Xu, J., Shi, C., Qi, C., Wang, C., Xiao, B.: Unsupervised part-based weighting aggregation of deep convolutional features for image retrieval. In: Proceedings of the AAAI Conference on Artificial Intelligence, vol. 32 (2018)

29. Ye, H.J., Hu, H., Zhan, D.C., Sha, F.: Few-shot learning via embedding adaptation with set-to-set functions. In: Proceedings of the IEEE/CVF Conference on Computer Vision and Pattern Recognition, pp. 8808–8817 (2020)

30. Zhang, C., Cai, Y., Lin, G., Shen, C.: DeepEMD: few-shot image classification with differentiable earth mover's distance and structured classifiers. In: Proceedings of the IEEE/CVF Conference on Computer Vision and Pattern Recognition, pp. 12203–12213 (2020)

31. Zhu, H., Koniusz, P.: EASE: unsupervised discriminant subspace learning for transductive few-shot learning. In: Proceedings of the IEEE/CVF Conference on Computer Vision and Pattern Recognition, pp. 9078–9088 (2022)

32. Ziko, I., Dolz, J., Granger, E., Ayed, I.B.: Laplacian regularized few-shot learning. In: International Conference on Machine Learning, pp. 11660–11670. PMLR (2020)

MixPose: 3D Human Pose Estimation with Mixed Encoder

Jisheng Cheng[1,2,3], Qin Cheng[1,3], Mengjie Yang[4], Zhen Liu[1,3],
Qieshi Zhang[1,3], and Jun Cheng[1,3(✉)]

[1] Guangdong Provincial Key Laboratory of Robotics and Intelligent System,
Shenzhen Institute of Advanced Technology,
Chinese Academy of Sciences, Shenzhen, China
{xjy109188,qin.cheng,zhen.liu1,qs.zhang,jun.cheng}@siat.ac.cn
[2] University of Chinese Academy of Sciences, Beijing, China
[3] The Chinese University of Hong Kong, Hong Kong, China
[4] Shine Technology Co., Ltd., Beijing, China

Abstract. The fusion of spatio-temporal information is crucial for 3D human pose estimation in video. Existing methods usually extract temporal information from the spatially encoded poses, which may lead to limited spatio-temporal information interaction. To address this issue, we propose **MixPose**, a novel network for 3D human pose estimation with mixed encoder in videos. We introduce independent mixed encoders to fuse spatio-temporal information in the sequence, and augment the perception of each point with global information using an attention module. We evaluate MixPose on two public datasets, Human3.6M and HumanEva, experiment results show that MixPose outperforms other state-of-the-art methods in specific scenarios.

Keywords: 3D human pose estimation · Transformer · Mixed encoder

1 Introduction

3D human pose estimation is a significant task in computer vision, aiming to reconstruct the pose information of a human body in 3D space (represented by skeletal structure) from images or videos [1–5]. This task is widely used in action recognition [6–9], action prediction [10], and virtual reality [11]. Mainstream methods for 3D human pose estimation currently employ 2D-to-3D lifting pipeline [4,12–16]. First, a 2D pose detector is employed to obtain the 2D human pose, and then the corresponding 3D human pose is inferred. Due to the excellent performance of 2D detectors [17–20], methods that convert from 2D to 3D frequently achieve favorable results. However, this approach also has some limitations, such as occlusion and non-uniqueness of mapped poses. Some studies [21] have attempted to overcome these challenges by using multiple view input data, but these approaches lack generality as most real-world data typically has a single view. Therefore, the recent research [4,13,16,22] trend is to

Q. Liu et al. (Eds.): PRCV 2023, LNCS 14432, pp. 353–364, 2024.
https://doi.org/10.1007/978-981-99-8543-2_29

directly estimate 3D human poses from monocular videos and leverage temporal features in the videos to improve estimation accuracy and robustness. These approaches take into account temporal information, help to overcome the limitations of 2D-to-3D lifting pipeline, and bring fresh breakthroughs to the task of 3D human pose estimation.

PoseFormer [4] introduces transformer to extract spatio-temporal information in videos, and use the spatio-temporal information of adjacent frames to improve the 3D pose accuracy of the center frame. MixSTE [22] models the motion trajectory of each joint to extract temporal information from the input sequence. While previous studies [4,13,16,22] have partially examined the utilization of temporal information, there is still a limitation in effectively incorporating spatiotemporal information. Typically, temporal information is extracted from spatially encoded sequences, which results in limited interaction between temporal and spatial information. To overcome this shortcoming, we propose a novel transformer-based model, named **MixPose**, which consists of three key modules: temporal encoder, spatial encoder, and mixed encoder. Temporal encoder treats the motion trajectories of each joint as independent features and learns the temporal dependencies between different frames. Spatial encoder places greater emphasis on capturing the relationships between joints in each frame. Moreover, mixed encoder takes these two features as input, fuses and further extracts spatio-temporal features, which enhanced the spatio-temporal information interaction. Our contributions can be summarized in three folds:

- We propose MixPose, a novel approach that fuses spatio-temporal information to offer a fresh perspective on integrating such information in video.
- We present an efficient module for fusing temporal and spatial information, thereby enhancing the interaction between these two dimensions.
- We evaluate our model on the Human3.6M and HumanEva datasets and achieve state-of-the-art performance in most scenarios.

2   Related Work

2.1   2D-to-3D Lifting Pipeline

2D-to-3D pose estimation methods use 2D pose detectors [3,17,20,23] to extract 2D human poses from images or videos, and use them as intermediate representations to generate 3D poses. Therefore, the performance of 2D-to-3D methods is closely related to the 2D detectors used. Benefiting from the excellent performance of 2D detectors, 2D-to-3D methods typically exhibit better performance.

Martinez et al. [12] introduced a straightforward method based on fully-connected networks, which relies on limited spatial information to generate 3D poses. Their method primarily focuses on inferring 3D poses in individual frames and does not consider the temporal dependencies between adjacent frames in video sequences. To fully exploit the temporal dependencies in videos, recent studies [4,16,24] placed significant emphasis on extracting and utilizing temporal features. Pavllo et al. [16] adopted temporal convolution networks to extract temporal features from video frame sequences, and proved that the temporal dependency between adjacent frames can effectively improve the 3D pose accuracy of the center frames. Hossain et al. [24] utilized Long Short-Term Memory networks (LSTM) to extract temporal information from sequences. By leveraging the strengths of LSTM, they were able to effectively capture and model the temporal dependencies present in the data. Zhang et al. [4] designed a spatio-temporal transformer module to extract the global dependencies of poses in sequences, and captured the interactions between poses across different frames, enabling a holistic understanding of the temporal and spatial relationships in the sequence.

While previous studies have made great progress in exploring temporal information, they have not fully considered the spatio-temporal relationship, resulting in limited fusion of temporal and spatial information. In contrast, our approach places a stronger emphasis on the fusion of spatio-temporal information.

2.2   Vision Transformer

Vaswani et al. [25] were the first to propose the transformer architecture based on self-attention mechanism, which achieved remarkable results in natural language processing. Subsequently, Dosovitskiy et al. [26] extended transformer to visual transformer and obtained impressive results in computer vision. Yang et al. [19] introduced transformer into the field of human pose estimation and employed it to investigate the spatial dependency among joints. Li et al. [14] proposed a multi-hypothesis method based on the transformer architecture, to address the challenge of uncertainty in mapping 2D human poses to 3D. Zheng et al. [4] pioneered the application of transformer in extracting temporal information from sequences. To further enhance the expression ability of temporal features, Zhang et al. [22] regarded each joint as a feature and extracted temporal features for each joint. Einfalt et al. [27] proposed a sparse sampling method for sequences, and used limited inputs to infer the 3D poses of the entire sequence. We propose a mixed encoder built upon the transformer architecture to enhance the interaction between spatio-temporal information by capturing the spatio-temporal dependency among joints.

3   Method

As shown in Fig. 1, MixPose employs a hierarchical architecture, comprising temporal, spatial, and mixed encoders in each layer. Our method takes a 2D pose sequence $X_{in} \in \mathbb{R}^{T \times J \times 2}$ as input. J represents the number of body joints in each frame, which is 17 for the Human3.6M [28] dataset and 15 for the HumanEva [29] dataset. T denotes the number of frames in the input sequence.

To enhance the expression ability of spatial information, we use a linear layer to map the 2D spatial information to a high-dimensional space to obtain $X \in \mathbb{R}^{T \times J \times C}$, where C represents the high-dimensional channel after the mapping, named the embedding dimension. Then, we feed X into both the temporal

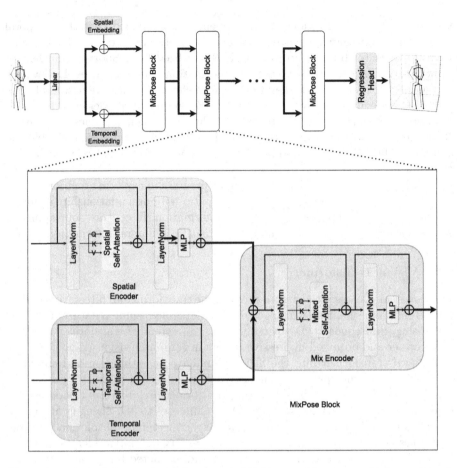

Fig. 1. Overview of the proposed framework. MixPose architecture, which consists of spatial, temporal and mixed encoders.

encoder and spatial encoder simultaneously. The mixed encoder fully integrates the temporal and spatial information and feeds it to the next layer of temporal and spatial encoders.

Finally, we use a simple regression head to map the temporal and spatial sequence to 3D and obtain the 3D pose of the sequence denoted as $X_{out} \in \mathbb{R}^{T \times J \times 3}$.

3.1   Preliminary

The transformer blocks in MixPose follow the design of the scaled dot-product attention, using multi head self attention block as the basic architecture.

Scaled Dot-Product Attention takes the Query matrix Q, Key matrix K, and Value matrix V as its input. The attention function can be described as mapping Q and a set of $K - V$ pairs to the output. It computes the dot product for all pairs of Q and K, and scales the dot product by a factor of $\sqrt{d_k}$, which helps balance the numerical range of the dot product and avoids the issue of the gradient exploding or gradients after applying softmax. The softmax result is used as the weight matrix for V, capturing the dependencies between Q, K, and V. The specific formulation is as follows:

$$Attention(Q, K, V) = Softmax(\frac{QK^T}{\sqrt{d_k}})V \tag{1}$$

Multi Head Self Attention utilizes multiple heads to perform independent attention operations, allowing information at different locations to be modeled in parallel. The results from each head are then linearly transformed back to the original shape using the weight matrix W.

$$MSA = Concat(H_1, H_2, \ldots, H_n)W^O$$
$$where \ H_i = Attention(QW_i^Q, KW_i^K, VW_i^V) \tag{2}$$

3.2   Encoder

Spatial-Temporal Encoder. To enhance the learning of temporal dependencies for each joint, we consider the motion trajectory of each joint as a temporal feature. Specifically, we employ a transformer-based temporal encoder that treats the motion trajectory of each joint as a token, to enhance the interaction of each joint in the temporal dimension. The specific formulation is as follows:

$$X_t = MSA(X') \tag{3}$$

The input token sequence $X' \in \mathbb{R}^{J \times T \times C}$ is obtained by transforming the variable X, which indicates that there are J tokens as input to the temporal encoder. In contrast to methods that extract temporal information from the pose, our approach reduces token size by extracting the motion trajectory of each joint. This reduction in token size helps to decrease computational costs.

Following [4,19,22], we employ a transformer to learn the spatial dependencies among the joints within a single frame of the pose and obtain the spatial features X_s. In addition, we introduce a learnable position embedding [26] in both the temporal encoder and the spatial encoder.

Mixed Encoder. Mixed encoder is a novel module that takes the outputs of temporal and spatial encoders as input, effectively enhancing the interaction between these two dimensions through the utilization of the transformer architecture. This results in a sequence with enriched spatio-temporal information, which is then forwarded to the subsequent layer of the spatio-temporal encoder. This design allows our model to better capture temporal and spatial relationships between joints, leading to improved accuracy in pose estimation.

We perform a simple fusion of the outputs of the temporal and spatial encoders to share their information.

$$X_{mix} = X_t + X_s \tag{4}$$

We model both temporal and spatial dimensions by taking $X_{mix} \in \mathbb{R}^{(T \times J) \times C}$ as input to the Transformer to accurately fuse temporal and spatial features.

$$X = MSA(X_{mix}) \tag{5}$$

X will be used as input to the next layer of the temporal and spatial encoder. Considering computational efficiency, when the number of frames in the input sequence is greater than or equal to 81, we prioritize the representation of temporal and spatial information in specific frames, which significantly reduces the computational cost. This design allows efficient processing of longer sequences while maintaining performance by reducing the computational burden.

3.3 Regression Head

We use a simple linear layer as the regression head to map the joint coordinates from the embedding dimension C to 3D and apply a simple norm layer to the final result to obtain the final result $X_{out} \in \mathbb{R}^{T \times J \times 3}$. After a series of encoding, fusion, and regression operations, we transform the input 2D human pose sequence into the corresponding 3D human pose sequence and finally achieve accurate 3D pose estimation.

3.4 Loss Function

To improve the training of our model, we adopt the loss function from the work of MixSTE [22], which can be expressed as:

$$L = L_w + \lambda_t L_T + \lambda_m L_m \tag{6}$$

L_w represents the Weighted Mean Per Joint Position Error (WMPJPE) [22], L_T represents the Temporal Consistency Loss (TCLoss) [24], and L_m represents the Mean Per Joint Velocity Error (MPJVE) [16]. The coefficients λ_t and λ_m are used to balance the impact of TCLoss and MPJVE to prevent excessive smoothing of the results.

4 Experiment

4.1 Datasets

Human3.6M. [28] is a widely used indoor large-scale dataset in the field of human pose estimation, which contains 3.6 million human poses and corresponding images captured by a high-speed motion capture system and four high-resolution cameras. These samples cover 17 scenarios, such as discussing, smoking, taking photos, performed by 11 professional actors. Due to its extensive data and realistic 3D pose annotations, Human 3.6M has become an essential benchmark for evaluating the performance of human pose estimation models.

Following previous work [4,13,14,16,22], we adopt the same setting. We use five subjects S1, S5, S6, S7 and S8 as training data and test the results on two more challenging subjects S9 and S11. We evaluate results using Mean Per Joint Position Error (MPJPE) [28] as Protocol 1 and Pre Joint Position Error after aligning the root (P-MPJPE) [12] as Protocol 2, both of which compute the Euclidean distance between the predicted and the actual joint positions, but P-MPJPE applies a rotation alignment preprocessing operation to the predictions.

HumanEva. [29] is a tiny public dataset containing human pose data of 4 actors performing 6 common actions. Following previous work [4,16,22,24], we select the "Jog" and "Walk" actions from subjects S1, S2, and S3 for evaluation. Moreover, we use MPJPE and P-MPJPE as our evaluation metrics.

4.2   Implementation Details

We implement our method using the PyTorch framework and train the model on two NVIDIA TESLA P40 GPUs. Our model accepts two types of input data for training and evaluating: 2D joint data extracted by the CPN [3] 2D pose detector and ground truth 2D joint data, to comprehensively evaluate our model.

To validate the effectiveness of our method, we evaluate MixPose on two datasets. For fair comparison, we follow the setting of [2,4,22]. For Human3.6M, we perform experiments using sequence lengths of 27 and 243. For HumanEva, we perform experiments using sequence lengths of 27 and 43. It is worth noting that our method can also accommodates input sequences of different lengths, which is discussed in detail in the ablation study section, where we analyze the effect of different input lengths on the results.

For the hyperparameters, we set batch size to 1024, dropout rate to 0.1, learning rate to 4e-4, and trained the model with Adam optimizer. We also apply simple horizontal flipping to the input data during training for data augmentation. For Human3.6M, $J = 17$, while for HumanEva, $J = 15$. Following [16], we validate the model on Human3.6M using data extracted from the CPN detector and ground truth, and on HumanEva we solely employ ground truth points for validation to ensure an accurate comparison and evaluation with previous work.

4.3   Comparison with State-of-the-Art Methods

Results on Human3.6M. We compare our method with state-of-the-art approaches on the Human3.6M dataset and present the results in Table 1. When using CPN as the 2D detector, our method (T=243) achieves 41.5 mm MPJPE and 32.6 mm P-MPJPE, and obtained the state-of-the-art in most scenarios. To further explore the performance upper bound of MixPose, we use ground truth joint coordinates as input and compare it with previous work under Protocol 1.

Table 1. Detailed results for each scenario of Human3.6M under Protocol 1&2, * represents that using CPN [3] as the 2D detector. T represents the number of frames. Lower is better, the best result is in bold and the second best is underlined.

Protocol 1 (MPJPE)	Dir.	Disc.	Eat	Greet	Phone	Photo	Pose	Pur.	Sit	SitD.	Smoke	Wait	WalkD.	Walk	WalkT.	Avg
Martinez et al. [12] ICCV'17	51.8	56.2	58.1	59.0	69.5	78.4	55.2	58.1	74.0	94.6	62.3	59.1	65.1	49.5	52.4	62.9
Pavlakos et al. [30] CVPR'18	48.5	54.4	54.4	52.0	59.4	65.3	49.9	52.9	65.8	71.1	56.6	52.9	60.9	44.7	47.8	56.2
*Pavllo et al. [16] CVPR'19(T=243)	45.2	46.7	43.3	45.6	48.1	55.1	44.6	44.3	57.3	65.8	47.1	44.0	49.0	32.8	33.9	46.8
*Liu et al. [31] CVPR'20(T=243)	41.8	44.8	41.1	44.9	47.4	54.1	43.4	42.2	56.2	63.6	45.3	43.5	45.3	31.3	32.2	45.1
Zeng et al. [32] ICCV'21(T=243)	43.1	50.4	43.9	45.3	46.1	57.0	46.3	47.6	56.3	61.5	47.7	47.4	53.5	35.4	37.3	47.9
*Chen et al. [2] TCSVT'21(T=243)	41.4	43.5	40.1	42.9	46.6	51.9	41.7	42.3	53.9	60.2	45.4	41.7	46.0	31.5	32.7	44.1
*Zheng et al. [4] ICCV'21(T=81)	41.5	44.8	39.8	42.5	46.5	51.6	42.1	42.0	53.3	60.7	45.5	43.3	46.1	31.8	32.2	44.3
*Zhang et al. [22] CVPR'22(T=81)	39.8	43.0	_38.6_	_40.1_	_43.4_	50.6	_40.6_	41.4	_52.2_	**56.7**	43.8	_40.8_	43.9	_29.4_	30.3	_42.4_
*Li et al. [14] CVPR'22(T=351)	39.2	43.1	40.1	40.9	44.9	51.2	_40.6_	41.3	53.5	60.3	_43.7_	41.1	43.8	29.8	30.6	43.0
*Shan et al. [13] CVPR'22(T=243)	38.9	_42.7_	40.4	41.1	45.6	49.7	40.9	_39.9_	55.5	59.4	44.9	42.2	_42.7_	_29.4_	_29.4_	42.8
*Li et al. [33] TMM'23	40.3	43.3	40.2	42.3	45.6	52.3	41.8	40.5	55.9	66.9	44.2	43.0	44.2	30.0	30.2	43.7
Kang et al. [34] AAAI'23	**36.9**	44.4	41.9	43.3	45.6	**47.8**	43.0	40.7	**50.7**	60.6	44.3	43.6	43.9	33.9	35.0	43.7
*MixPose(Ours, T=27)	42.4	45.3	42.2	44.3	48.0	54.9	43.4	44.0	56.8	64.3	46.4	43.9	47.8	32.1	34.0	46.0
*MixPose(Ours, T=243)	_38.5_	**40.1**	**38.4**	**39.2**	**43.0**	_49.6_	**40.4**	**39.0**	53.6	_59.2_	**42.9**	**40.2**	**41.0**	**28.0**	**28.6**	**41.5**
Protocol 2 (P-MPJPE)	Dir.	Disc.	Eat	Greet	Phone	Photo	Pose	Pur.	Sit	SitD.	Smoke	Wait	WalkD.	Walk	WalkT.	Avg
*Pavllo et al. [16] CVPR'19(T=243)	34.1	36.1	34.4	37.2	36.4	42.2	34.4	33.6	45.0	52.5	37.4	33.8	37.8	25.6	27.3	36.5
*Liu et al. [31] CVPR'20(T=243)	32.3	35.2	33.3	35.8	35.9	41.5	33.2	32.7	44.6	50.9	37.0	32.4	37.0	25.2	27.2	35.6
*Zheng et al. [4] ICCV'21(T=81)	32.5	34.8	32.6	34.6	35.3	39.5	32.1	32.0	_42.8_	48.5	34.8	32.4	35.3	24.5	26.0	34.6
*Zhang et al. [22] CVPR'22(T=81)	32.0	34.2	31.7	33.7	34.4	39.2	32.0	31.8	42.9	46.9	35.5	32.0	34.4	23.6	25.2	_33.9_
*Zhang et al. [22] CVPR'22(T=243)	_30.8_	_33.1_	**30.3**	_31.8_	_33.1_	_39.1_	_31.1_	_30.5_	42.5	**44.5**	34.0	_30.8_	_32.7_	**22.1**	**22.9**	32.6
Li et al. [33] TMM'23	32.7	35.5	32.5	35.4	35.9	41.6	33.0	31.9	45.1	50.1	36.3	33.5	35.1	23.9	25.0	35.2
*MixPose(T=27)	34.0	36.1	34.4	36.5	36.8	43.1	34.2	34.0	46.4	50.7	37.5	33.7	37.5	25.4	28.0	36.6
*MixPose(Ours, T=243)	**30.1**	**32.3**	_30.6_	**31.4**	**33.0**	**38.0**	**30.8**	**29.9**	_42.8_	_47.3_	_34.2_	**30.5**	**32.2**	_22.2_	_23.3_	**32.6**

Table 2. Detailed results for each scenario of Human3.6M under Protocol 1 which uses **ground truth** keypoints as input. T represents the number of frames. Lower is better, the best result is in bold and the second best is underlined.

Protocol 1 (MPJPE)	Dir.	Disc.	Eat	Greet	Phone	Photo	Pose	Pur.	Sit	SitD.	Smoke	Wait	WalkD.	Walk	WalkT.	Avg
Martinez et al. [12] ICCV'17	37.7	44.4	40.3	42.1	48.2	54.9	44.4	42.1	54.6	58.0	45.1	46.4	47.6	36.4	40.4	45.5
Hossain et al. [24] ECCV'2018	35.2	40.8	37.2	37.4	43.2	44.0	38.9	35.6	42.3	44.6	39.7	39.7	40.2	32.8	35.5	39.2
Wang et al. [35] ECCV'2020	23.0	25.7	22.8	22.6	_24.1_	30.6	24.9	24.5	31.1	35.0	25.6	24.3	25.1	19.8	18.4	25.6
Chen er al. [2] TCSVT'21(T=243)	–	–	–	–	–	–	–	–	–	–	–	–	–	–	–	32.3
Zheng et al. [4] CVPR'21(T=81)	30.0	33.6	29.9	31.0	30.2	33.3	34.8	31.4	37.8	38.6	31.7	31.5	29.0	23.3	23.1	31.3
shan et al. [13] ECCV'22(T=243)	28.5	30.1	28.6	27.9	29.8	33.2	31.3	27.8	36.0	37.4	29.7	29.5	28.1	21.0	21.0	29.3
Li et al. [14] CVPR'22	27.7	32.1	29.1	28.9	30.0	33.9	33.0	31.2	37.0	39.3	30.0	31.0	29.4	22.2	23.0	30.5
Zhang et al. [22] CVPR'22(T=81)	25.6	27.8	24.5	25.7	24.9	29.9	28.6	27.4	29.9	29.0	26.1	25.0	25.2	18.7	19.9	25.9
Zhang et al. [22] CVPR'22(T=243)	_21.6_	**22.0**	**20.4**	_21.0_	**20.8**	_24.3_	_24.7_	**21.9**	**26.9**	**24.9**	_21.2_	**21.5**	**20.8**	_14.7_	_15.7_	_21.6_
Li et al. [33] TMM'23	27.1	29.4	26.5	27.1	28.6	33.0	30.7	26.8	38.2	34.7	29.1	29.8	26.8	19.1	19.8	28.5
MixPose(Ours, T=243)	**21.5**	_20.5_	_20.5_	**19.9**	**20.8**	**23.6**	**23.2**	**20.0**	_26.4_	_25.4_	**20.6**	**20.2**	**18.7**	**13.7**	**15.0**	**20.7**

MixPose achieves either the best or second-best results (only being second-best in the eating and sitting down scenario) compared to previous work, the results are presented in Table 2. Our method is mostly distributed in a lower range compared to the other methods, which indicates that our method is more robust in most scenarios. MixPose reduces MPJPE to 20.7 mm and achieves higher performance upper bounds in complex scenarios such as sit, smoke, and photo, where MixPose shows its advantage. This indicates that mixed encoder can effectively combine spatio-temporal information and extract more information. Moreover, some of the more complex scenarios rely more on spatio-temporal information, in which case the advantage of the mixture encoder is more pronounced.

Table 3. Detailed results of HumanEva under Protocol 1 which uses **ground truth** keypoints as input. T represents the number of frames. Lower is better, the best result is in bold and the second best is underlined.

Protocol 1 (MPJPE)	Walk			Jog			Avg
	S1	S2	S3	S1	S2	S3	
Pavllo et al. [16] CVPR'19(T=81)	14.0	12.5	27.1	**20.3**	17.9	17.5	18.2
Zheng et al. [4] ICCV'21(T=43)	14.4	**10.2**	46.6	22.7	<u>13.4</u>	**13.4**	20.1
Zhang et al. [22] CVPR'22(T=43)	**12.7**	<u>10.9</u>	**17.6**	22.6	15.8	17.0	16.1
MixPose(Ours, T=27)	<u>13.7</u>	12.7	19.4	<u>21.6</u>	**13.2**	<u>14.5</u>	**15.8**
MixPose(Ours, T=43)	13.8	11.2	<u>18.8</u>	23.5	13.7	15.4	<u>16.0</u>

Results on HumanEva. Table 3 shows the methods that achieve state-of-the-art performance on HumanEva, all methods are finetuned on large datasets (such as Human3.6M) to achieve the best results. Following [4,16,22], we use ground truth 2D poses as input and without any additional fancy constraints, our average MPJPE reaches 15.7 mm under Protocol 1. Compared to other methods, MixPose exhibits better generalization ability.

4.4 Ablation Study

Effect of Mixed Encoder. To verify the influence of mixed encoder on the experimental results, we evaluate it on Human3.6M with CPN as the 2D pose detector. We keep all parameters consistent with MixPose and only change the network architecture. We replace mixed encoder with a simple matrix addition and normalization layer and compare the results with MixPose under Protocols 1&2 to assess the role of the mixed encoder. The experimental results are shown in Table 4. After applying mixed encoder, MPJPE is reduced by 1.6 mm and P-MPJPE by 1.7 mm, which justifies mixed encoder design.

Table 4. Ablation study for mixed encoder in MixPose, The evaluation is preformed on Human3.6M using CPN as the 2D detector under Protocols 1&2.

Method	MPJPE		P-MPJPE		Avg	
	S9	S11	S9	S11	MPJPE	P-MPJPE
Without Mixed Encoder	44.5	35.5	41.2	32.6	43.1	34.3
MixPose(Ours)	42.6	33.7	40.0	31.0	41.5	32.6

Parameter Analysis. There are three main hyperparameters that effect the results: input sequence length (T), network depth (D), and embedding dimension (C), and we evaluate the experimental results under Protocols 1&2 to explore their impact on the experimental results. We divide the hyperparameters into

Table 5. Ablation study for Parameter Analysis in depth(D_i), input length(T_i) and embedding dimension(C_i), which i represents the optional parameters. The evaluation is performed on HumanEva using ground truth keypoints under Protocol 1.

MPJPE	D_4	D_8	D_{12}	T_9	T_{27}	T_{43}	C_{256}	C_{512}	C_{1024}
Walking	29.0	15.1	19.1	18.2	15.1	14.7	26.0	15.1	12.8
Jog	33.7	16.2	14.9	14.1	16.2	17.3	19.2	16.2	13.7
Avg	31.4	15.7	17.1	16.2	15.7	16.0	22.7	15.7	13.3

three groups for comparison, change the values of only one hyperparameter in each group and keeping the remaining hyperparameters consistent. Our default settings are $C = 512$, $D = 8$, and $T = 27$. The detailed experimental results are presented in Table 5. Although $C = 1024$ can achieve better results, it also brings more overhead, which is the main reason why we set $C = 512$.

5   Conclusion

We propose MixPose, a monocular video-based 3D human pose estimation method that fuses temporal and spatial features based on Transformer. Mixed encoder can effectively fuse the spatial and temporal features in the video, and improve the accuracy of 3D human pose. Comprehensive evaluation shows that our method achieves state-of-the-art performance in most scenarios on Human 3.6M and HumanEva datasets. In the future, we will focus on applying MixPose to other topics, such as temporal action detection and action prediction.

Acknowledgements. The research was supported by National Natural Science Foundation of China (U21A20487), Shenzhen Technology Project (JCYJ2022081810120 6014, JCYJ20220818101211025), Shenzhen Engineering Laboratory for 3D Content Generating Technologies (NO. [2017]476), CAS Key Technology Talent Program.

References

1. Gong, W., et al.: Human pose estimation from monocular images: a comprehensive survey. Sensors **16**(12), 1966 (2016)
2. Chen, T., Fang, C., Shen, X., Zhu, Y., Chen, Z., Luo, J.: Anatomy-aware 3D human pose estimation with bone-based pose decomposition. IEEE Trans. Circuits Syst. Video Technol. **32**(1), 198–209 (2022)
3. Chen, Y., Wang, Z., Peng, Y., Zhang, Z., Yu, G., Sun, J.: Cascaded pyramid network for multi-person pose estimation. In: Proceedings of the IEEE Conference on Computer Vision and Pattern Recognition, pp. 7103–7112 (2018)
4. Zheng, C., Zhu, S., Mendieta, M., Yang, T., Chen, C., Ding, Z.: 3D human pose estimation with spatial and temporal transformers. In: Proceedings of the IEEE/CVF International Conference on Computer Vision, pp. 11656–11665 (2021)
5. Zhan, Y., Li, F., Weng, R., Choi, W.: Ray3D: ray-based 3D human pose estimation for monocular absolute 3D localization. In: Proceedings of the IEEE/CVF Conference on Computer Vision and Pattern Recognition, pp. 13116–13125 (2022)

6. Qin, H., Cheng, J., Song, C., Hao, F., Cheng, Q.: Structure-preserving view-invariant skeleton representation for action detection. In: 2022 26th International Conference on Pattern Recognition (ICPR), pp. 3190–3196. IEEE (2022)

7. Cheng, J., Ren, Z., Zhang, Q., Gao, X., Hao, F.: Cross-modality compensation convolutional neural networks for RGB-D action recognition. IEEE Trans. Circuits Syst. Video Technol. **32**(3), 1498–1509 (2022)

8. Ji, X., Cheng, J., Feng, W., Tao, D.: Skeleton embedded motion body partition for human action recognition using depth sequences. Sig. Process. **143**, 56–68 (2018)

9. Ji, X., Zhao, Q., Cheng, J., Ma, C.: Exploiting spatio-temporal representation for 3D human action recognition from depth map sequences. Knowl.-Based Syst. **227**, 107040 (2021)

10. Diller, C., Funkhouser, T., Dai, A.: Forecasting characteristic 3D poses of human actions. In: Proceedings of the IEEE/CVF Conference on Computer Vision and Pattern Recognition, pp. 15914–15923 (2022)

11. Anvari, T., Park, K., Kim, G.: Upper body pose estimation using deep learning for a virtual reality avatar. Appl. Sci. **13**(4), 2460 (2023)

12. Martinez, J., Hossain, R., Romero, J., Little, J.J.: A simple yet effective baseline for 3D human pose estimation. In: Proceedings of the IEEE International Conference on Computer Vision, pp. 2640–2649 (2017)

13. Shan, W., Liu, Z., Zhang, X., Wang, S., Ma, S., Gao, W.: P-STMO: pre-trained spatial temporal many-to-one model for 3d human pose estimation. In: Avidan, S., Brostow, G., Cissé, M., Farinella, G.M., Hassner, T. (eds.) Computer Vision – ECCV 2022. ECCV 2022. Lecture Notes in Computer Science, vol. 13665, pp. 461–478. Springer, Cham (2022). https://doi.org/10.1007/978-3-031-20065-6_27

14. Li, W., Liu, H., Tang, H., Wang, P., Van Gool, L.: MHFormer: multi-hypothesis transformer for 3D human pose estimation. In: Proceedings of the IEEE/CVF Conference on Computer Vision and Pattern Recognition, pp. 13147–13156 (2022)

15. Zhao, Q., Zheng, C., Liu, M., Wang, P., Chen, C.: PoseFormerV2: exploring frequency domain for efficient and robust 3D human pose estimation. In Proceedings of the IEEE/CVF Conference on Computer Vision and Pattern Recognition, pp. 8877–8886 (2023)

16. Pavllo, D., Feichtenhofer, C., Grangier, D., Auli, M.: 3D human pose estimation in video with temporal convolutions and semi-supervised training. In: Conference on Computer Vision and Pattern Recognition (CVPR) (2019)

17. Cao, Z., Hidalgo Martinez, G., Simon, T., Wei, S., Sheikh, Y.A.: OpenPose: real-time multi-person 2D pose estimation using part affinity fields. IEEE Trans. Pattern Anal. Mach. Intell. **43**, 172–186 (2019)

18. Li, K., et al.: UniFormer: unifying convolution and self-attention for visual recognition. IEEE Trans. Pattern Anal. Mach. Intell. **45**, 12581–12600 (2022)

19. Yang, S., Quan, Z., Nie, M., Yang, W.: Transpose: keypoint localization via transformer. In: Proceedings of the IEEE/CVF International Conference on Computer Vision, pp. 11802–11812 (2021)

20. Xu, Y., Zhang, J., Zhang, Q., Tao, D.: ViTPose: simple vision transformer baselines for human pose estimation. In: Advances in Neural Information Processing Systems (2022)

21. Shuai, H., Wu, L., Liu, Q.: Adaptive multi-view and temporal fusing transformer for 3D human pose estimation. IEEE Trans. Pattern Anal. Mach. Intell. **45**, 4122–4135 (2022)

22. Zhang, J., Tu, Z., Yang, J., Chen, Y., Yuan, J.: MixSTE: seq2seq mixed spatio-temporal encoder for 3D human pose estimation in video. In: Proceedings of the

IEEE/CVF Conference on Computer Vision and Pattern Recognition (CVPR), pp. 13232–13242 (2022)

23. Sun, K., Xiao, B., Liu, D., Wang, J.: Deep high-resolution representation learning for human pose estimation. In: Proceedings of the IEEE/CVF Conference on Computer Vision and Pattern Recognition, pp. 5693–5703 (2019)

24. Hossain, M.R.I., Little, J.J.: Exploiting temporal information for 3D human pose estimation. In: Proceedings of the European conference on computer vision (ECCV), pp. 68–84 (2018)

25. Vaswani, A., et al.: Attention is all you need. In: Guyon, I., et al., (eds.), Advances in Neural Information Processing Systems, vol. 30. Curran Associates Inc. (2017)

26. Dosovitskiy, A., et al.: An image is worth 16×16 words: transformers for image recognition at scale. arXiv preprint arXiv:2010.11929 (2020)

27. Einfalt, M., Ludwig, K., Lienhart, R.: Uplift and upsample: efficient 3D human pose estimation with uplifting transformers. In: Proceedings of the IEEE/CVF Winter Conference on Applications of Computer Vision, pp. 2903–2913 (2023)

28. Ionescu, C., Papava, D., Olaru, V., Sminchisescu, C.: Human3. 6m: large scale datasets and predictive methods for 3D human sensing in natural environments. IEEE Trans. Pattern Anal. Mach. Intell. **36**(7), 1325–1339 (2013)

29. Sigal, L., Balan, A.O., Black, M.J.: HumanEva: synchronized video and motion capture dataset and baseline algorithm for evaluation of articulated human motion. Int. J. Comput. Vis. **87**(1–2), 4 (2010)

30. Pavlakos, G., Zhou, X., Daniilidis, K.: Ordinal depth supervision for 3D human pose estimation. In: Proceedings of the IEEE Conference on Computer Vision and Pattern Recognition, pp. 7307–7316 (2018)

31. Liu, R., Shen, J., Wang, H., Chen, C., Cheung, S.-c., Asari, V.: Attention mechanism exploits temporal contexts: real-time 3D human pose reconstruction. In: Proceedings of the IEEE/CVF Conference on Computer Vision and Pattern Recognition (CVPR) (2020)

32. Zeng, A., Sun, X., Yang, L., Zhao, N., Liu, M., Xu, Q.: Learning skeletal graph neural networks for hard 3D pose estimation. In: Proceedings of the IEEE/CVF International Conference on Computer Vision, pp. 11436–11445 (2021)

33. Li, W., Liu, H., Ding, R., Liu, M., Wang, P., Yang, W.: Exploiting temporal contexts with strided transformer for 3D human pose estimation. IEEE Trans. Multimedia **25**, 1282–1293 (2023)

34. Kang, Y., Liu, Y., Yao, A., Wang, S., Wu, E.: 3D human pose lifting with grid convolution. In: Proceedings of the AAAI Conference on Artificial Intelligence (2023)

35. Wang, J., Yan, S., Xiong, Y., Lin, D.: Motion guided 3D pose estimation from videos. In: Vedaldi, A., Bischof, H., Brox, T., Frahm, J.-M. (eds.) ECCV 2020. LNCS, vol. 12358, pp. 764–780. Springer, Cham (2020). https://doi.org/10.1007/978-3-030-58601-0_45

Image Manipulation Detection Based on Ringed Residual Edge Artifact Enhancement and Multiple Attention Mechanisms

Siyu Chen and Qingfeng Wu[✉]

School of Informatics, Xiamen University, Xiamen, Fujian, China
qfwu@xmu.edu.cn

Abstract. As image editing techniques continue to evolve, concerns over the security risks associated with forged images are growing. Previous studies have suggested that tampering traces hidden at the edges of images are crucial for detecting manipulated regions. To better cope with image tampering post-processing methods, we propose REM-U2-Net, a U-shaped network designed to detect and localize image tampering traces by enhancing edge artifacts. REM-U2-Net extracts the noise distribution of an image by combining it with RGB features as input to capture subtle manipulation traces that may not be visible in the RGB domain. Additionally, ringed residual edge artifact boosting and symmetric attention module designs enable the model to detect both manipulated edges and regions, making it more effective in coping with a wide range of manipulation attacks. We conducted extensive experiments on multiple datasets and demonstrated the effectiveness of our method. Furthermore, REM-U2-Net exhibits excellent robustness to various post-processing methods.

Keywords: Image manipulation detection · Ringed residual edge artifact enhancement · Attention mechanism

1 Introduction

Dispelling the stereotype: seeing is not always believing. With the continuous development of image editing techniques and intelligent processing tools, it has become increasingly easy to generate realistic images. Unfortunately, the misuse of digital image processing methods, such as the use of fake faces [21], for online fraud has caused serious social security risks and significant economic losses. This highlights the importance of detecting manipulated images and locating their tampered regions.

Image manipulation techniques can be broadly categorized into two main types: homologous and heterologous. Copy-move is a common homologous manipulation technique, while removal and splicing are widely used heterologous manipulation techniques. This study aims to detect manipulated regions

Q. Liu et al. (Eds.): PRCV 2023, LNCS 14432, pp. 365–376, 2024.
https://doi.org/10.1007/978-981-99-8543-2_30

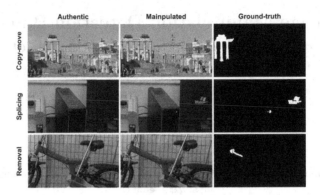

Fig. 1. Examples of images using different manipulation techniques, such as copy-move, splicing, and removal.

in images manipulated by these three techniques, as they have produced a large number of fake images (see Fig. 1).

Many methods have been proposed to detect and localize image manipulation. In recent years, convolutional neural network-based methods have been proposed to detect pixel-level manipulated regions, but most of them detect only one technique [24, 26]. Recently, some general CNN-based methods detect multiple manipulation types [12, 13, 25]. However, two problems still need to be solved for image tampering localization:(1)Image correlation. Standard convolutional layers typically learn input image content features rather than potential forgery traces. Considering image spatial correlation can lead to a more generalized localization solution, but previous studies have mostly ignored this aspect.(2)Edge accuracy. When the manipulated region is subjected to good post-processing, such as image compression and local smoothing, distinguishing between boundary artifacts and natural object edges becomes a major challenge. Therefore, a good edge artifact supervision module can be a powerful solution.

To address the issue mentioned above, we propose REM-U2-Net, a U-shaped network that utilizes ring residual edge artifact enhancement and a multi-attention mechanism. As shown in Fig. 2, REM-U2-Net can detect three common image processing techniques: copy-move, splicing, and removal. We input local noise features and RGB images into the encoder to detect tampering traces generated by homologous and heterologous operations. Additionally, we enhance the "Squeeze-and-Excitation" [9] mechanism in each encoder module and add the Grid-Attention [16] mechanism to the decoder input to exploit image correlation and improve model performance.

To enhance the fine boundary artifacts of fused features, we design the Ring Residual Edge Artifact Enhancement (RRE) module, which enhances edge artifacts through edge detection strategies. By incorporating these designs, REM-U2-Net learns various tampering traces and demonstrates strong performance in publicly available datasets.

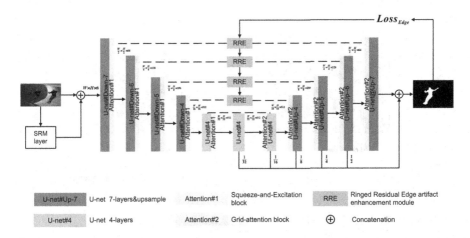

Fig. 2. The framework of our proposed REM-U2-Net is for pixel-level manipulation detection. REM-U2-Net comprises a noise fusion input module, symmetric encoding and decoding sections with an attention mechanism, and a ringed residual edge artifact enhancement module.

We summarize the contributions of this work as follows:(1)A multi-attentive combination is designed to compute the similarity between feature mappings to enhance the representation of regions of interest. (2)A new RRE module is designed to enhance the traces of tampering operations at different scales and still learn edge artifacts effectively even after a series of image post-processing methods.(3)We proposed REM-U2-Net, which extracts noise features and fuses them with RGB maps as feature inputs to detect tampering traces generated by both homologous and heterologous manipulations. By incorporating image correlation and edge artifact enhancement strategies, our approach can accurately detect multiple content alteration manipulation techniques.

2   Related Work

2.1   Noise-Based Methods

Special filters are typically used on original RGB images to generate noise maps that enhance tampering cues and suppress semantic information. RGB-N [28] uses steganalysis-rich model filters [7] to obtain noise maps and employs CNNS to capture local noise anomalies. Meanwhile, Mantra-Net [25] cascades SRM and Bayar filters from noise maps to extract local features.

However, these methods contain redundant parts for capturing noise anomalies. In this paper, we propose a design that uses SRM to acquire the noise map and fuses it with the original RGB image through a lightweight convolutional layer. This design not only addresses the challenges of handling heterogeneous and homogeneous operations but also maximizes network simplification.

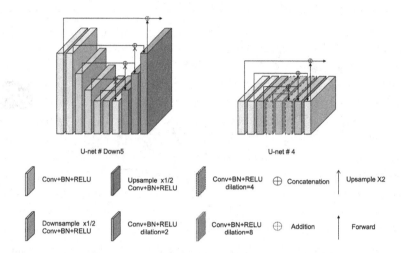

Fig. 3. Two U-Net structures used in REM-U2-Net.

2.2 Edge Artifact-Based Methods

Edge artifacts play a crucial role in detecting image manipulation, as most processed regions have unnatural artifacts at their boundaries. Salloum et al. [20] proposed a multi-task Fully Convolutional Network (FCN) to predict tampered regions and boundaries, with the goal of improving detection results through edge prediction supervision and refining branches. Recently, Zhou et al. [27] designed a three-level architecture that includes edge prediction, refinement, and segmentation branches. However, distinguishing natural edges from boundary artifacts becomes challenging when well-designed post-processing methods, such as local smoothing and adding noise, are used to reduce visual cues. To address this challenge, we propose the Ring Residual Edge artifact enhancement (RRE) module. This module enhances edge artifacts, enabling the learning of more robust features that can distinguish natural edges from boundary artifacts.

2.3 Attention Mechanism

The pioneering work on self-attentive mechanisms was first applied to the field of natural language processing [22] and then gained popularity in the field of vision, where it is now widely used for a variety of visual tasks [2–4]. Recent studies [13] have borrowed the advantages of attention to improve the representation capability of the model. However, due to memory limitations, they can only be applied to high-level features with small spatial sizes. In this work, a symmetric attention mechanism is designed that can further enhance task-specific useful salient features using features of different scales efficiently, thus improving the performance of the model.

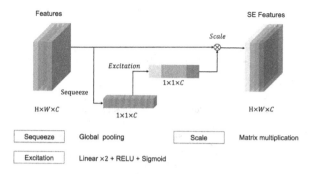

Fig. 4. Self-Attention Block: Image Correlation via Squeeze-and-Excitation Mechanism.

3 Method

This section presents the approach used in our proposed manipulation detection technique and explains the motivation behind our structural design.

3.1 Network Architecture

Encoder. For the encoder, we first pass the RGB image through the SRM filter to obtain local noise distribution features. We then combine the noise features that have passed through the SRM filter with the RGB image as input to the encoder. This approach provides sufficient evidence for both homologous and heterologous manipulation detection. We utilize the U-network architecture from [18] as our encoding module, using two different U-networks (see Fig. 3). This multiscale feature design ensures that each feature contains enough information to predict the manipulation mask at the corresponding scale. To improve our ability to extract forgery-related features, we have added a spatial channel squeezing-excitation mechanism (SE) after the first five U-network structures [9] instead of using the original U-network structure (see Fig. 4).

Decoder. The decoder has a symmetrically similar structure to the encoder. To suppress irrelevant regions in the input image and emphasize salient features useful for a specific task, we enhance the Grid-Attention mechanism [16] at the beginning of the U-shaped decoding module. The Grid-Attention module takes as input a cascade of features that have passed through the symmetric encoder stage of the RRE module and a mapping of upsampled features from the previous stage, and it can automatically learn to focus on target structures of different shapes and sizes.

For the feature mapping $F_x \in \mathbb{R}^{H_x \times W_x \times C_x}$ from the previous stage, the matrix is summed with the feature mapping $F_g \in \mathbb{R}^{H_g \times W_g \times C_g}$ from the encoder via the spatial recalibration module and the weighted matrix $F_s \in \mathbb{R}^{H_s \times W_s \times C_s}$ is generated by the following steps:

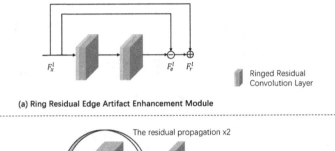

(a) Ring Residual Edge Artifact Enhancement Module

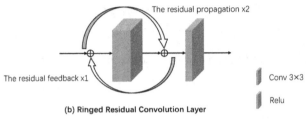

(b) Ringed Residual Convolution Layer

Fig. 5. Framework of (a) RRE module; (b) ringed residual convolution layer in the EAE module.

$$F_s = \tau(\sigma_2(\sigma_1(W_x^T F_x + W_g^T F_g + b_g))) \otimes F_x \tag{1}$$

where W_x and W_g denote the volume machine layer, τ stands for upsampling operation while b_g is the parametric term corresponding to the volume machine layer, and σ_1 and σ_2 denote the Relu activation function and the Sigmoid activation function, respectively.

3.2   RRE

Image post-processing methods can weaken the edge artifacts hidden in an image, making it more difficult to discover subtle edge artifacts. Therefore, we propose CNN-based Ring Residual Edge (RRE) modules, as shown in the Fig. 5, to indirectly enhance the artifacts of fused feature mapping. These modules are located in the middle of the encoder and decoder to improve the accuracy of edge detection.

In each RRE module, the ringed residual convolution layer is trained to eliminate the edge artifacts around the manipulated region. Thus, the residuals between the input of the RRE module and the eliminated feature mapping can represent the edge artifact features. As shown in Fig. 5, our proposed RRE module consists of two annular residual layers. Each ringed residual convolution layer consists of a residual module (RG) and a residual feedback module (RF). The RG module consists of two 3×3 convolution layers with one ReLU activation layer and a residual propagation module. The RF module, on the other hand, consists of a 1×1 convolution layer. Each RRE module has one input F_x^l and two outputs F_e^l and F_r^l, where F_x^l represents the feature input from layer L, F_e^l represents the features with edge artifacts, and F_r^l represents the feature mapping with enhanced edge artifacts as the region input to the layer L decoder.

Fig. 6. Example from our synthetic dataset. The generated images of different operation types and their ground truth masks are displayed.

3.3 Loss Function

To better supervise edges and regions, we use the Dice Loss function in addition to the binary cross-entropy loss function with added edge weights. The loss function equation is:

$$Loss_t = y_e L_{bce}(M_e, G_e) + \delta y_r L_{bce}(M_r, G_r) + y_d(1 - \frac{2\sum_{i=1}^{H \times W} M_i G_i}{\sum_{i=1}^{H \times W} M_i + G_i}) \quad (2)$$

where y_e, y_r, and y_d represent the weights of edge loss, region loss, and Dice loss, respectively. M_i and G_i represent the labeled and predicted values of pixel i.

3.4 Training Data Synthesis

In this paper, we create a comprehensive and high-quality dataset for training and validating REM-U2-Net(see Fig. 6). The dataset consists of four categories: (1) copy-move, (2) splicing, (3) removal, and (4) original, untampered images. To improve the generality of the model, we also add samples from the DEFACTO dataset [33] to the dataset of the removal category.

To generate the copy-shift and stitching types of the dataset, we borrow the method from [28] and use the images from the COCO dataset as the original images. We use segmentation annotations to randomly select objects from COCO [11] and perform corresponding operations to simulate image post-processing operations, such as resizing, adding Gaussian noise, and rotating. To make the generated images more realistic, we also apply the Poisson Image Edit operation to smooth the manipulated edges.

4   Experimental Results

4.1   Experimental Setup

Table 1. Description of the datasets.

Dataset name	Training-testing split		Involved manipulation technique		
	Training	Testing	Splicing	Copy-move	Removal
CASIA	5123	921	✓	✓	✗
Columbia	0	180	✓	✗	✓
Coverage	75	25	✗	✓	✗
NIST	404	160	✓	✓	✓
Ours	147100	36775	✓	✓	✓

Datasets and Metrics. We evaluate our REM-U2-Net on four standard test datasets: CASIA [6], Columbia [15], Coverage [23], and NIST [8], as well as on a dataset that we have created. The details of these datasets are shown in Table 1. To ensure a fair comparison, we refer to the setup ratio of [5] for the training/testing division of these publicly available datasets, which allows for better training and testing of REM-U2-Net. Based on the previous work [5,28], two commonly used pixel classification metrics are used in this paper: F1-Score (F1) and Area Under the receiver operating characteristic Curve (AUC). We will calculate F1 and AUC independently on each image and average them for comparison in subsequent experiments.

Implementation Details. We implemented the proposed method using the PyTorch framework, training the model on an A40 GPU with a batch size of 48. Due to the memory limitations of the computing device, we resize all input images to 224×224. The model was optimized using the ADAM optimizer with an initial learning rate of 1×10^{-4}, and a learning rate scaled down by one-tenth every 20 epochs. Our loss function includes a boundary loss weight of 0.7, a region loss weight of 0.2, and a Dice loss weight of 0.1.

4.2   Comparisons on Localization

Compared Methods. We compared the performance of four SoTA methods for REM-U2-Net, including RRU [1], Mantra [25], MVSS [5], and PSCC [13], in terms of localization. To ensure fairness, we use the results provided in the original papers or public codes of these methods and follow the optimal assignment scheme described in Ref.

Table 2. Quantitative comparison of REM-U2-Net with four SoTA methods on CASIA, Columbia, COVER, NIST and self-made datasets.

Method	CASIA		Columbia		Coverage		NIST		Ours	
	F1	AUC	F1	AUC	F1	AUC	F1	AUC	F1	AUC
RRU-Net [1]	0.318	0.709	0.423	0.764	0.372	0.717	0.583	0.741	0.754	0.819
ManTra-Net [25]	0.373	0.824	0.504	0.832	0.544	0.771	0.739	0.962	0.836	0.873
MVSS-Net [5]	0.492	0.866	0.464	0.918	**0.607**	0.792	0.803	**0.969**	0.903	0.972
PSCC-Net [13]	0.497	0.869	**0.505**	0.848	0.539	**0.805**	0.468	0.707	0.867	0.929
REM-U2-Net(ours)	**0.525**	**0.892**	0.491	**0.929**	0.574	0.789	**0.809**	0.948	**0.939**	**0.984**

Quantitative Results. We evaluate and compare REM-U2-Net with four SoTA methods on five datasets. As shown in Table 2, after fine-tuning, our proposed REM-U2-Net achieves the best performance on most of these datasets, demonstrating its superior localization performance and ability to mitigate domain differences. In addition, when tested on our datasets with different formats, our model outperforms the SoTA methods by a large margin (with an average AUC score improvement of over 9%), which validates the effectiveness of our overall network design.

Qualitative Results. This paper presents the visualization results of all CNN-based manipulation detection methods. As shown in Fig. 7, REM-U2-Net maintains high accuracy in its detection results. Furthermore, the edge detection results of REM-U2-Net are very close to the boundary of the manipulated region, illustrating the effectiveness of our proposed edge enhancement strategy in capturing the tiny edge artifacts hidden in the manipulated image.

4.3 Ablation Study

We performed burn-in implementations for SRM, SE, Grid-attention, and RRE components on a dataset of our own making to investigate the extent to which they contribute to improving the performance of REM-U2-Net. We used the same metrics as in previous experiments, i.e., AUC and F1, we designed 10 variable settings for a comprehensive evaluation, as shown in Table 3.

Our comparison of SRM, SE, Grid-attention, RRE, and REM-U2-Net demonstrates that the inclusion of local noise features can help locate tampered regions, with F1 increasing from 0.793 to 0.847 and AUC improving by 4.1%. Furthermore, the use of spatial and channel attention in the SE and Grid-attention components, respectively, can substantially improve localization based on U2-Net by suppressing irrelevant regions in the input image and emphasizing salient features useful for a specific task. However, noise features and attention mechanisms alone are not sufficient for accurate detection, and effective supervision of edge artifacts is crucial in locating tampered regions as well as boundaries. Our results show that REM-U2-Net, using the RRE module, can extract edge information of the manipulated region more finely, enabling the network to achieve maximum detection performance.

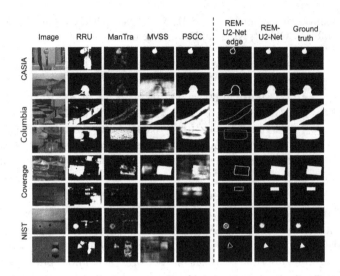

Fig. 7. Qualitative localization evaluation of four standard test datasets and the dataset we produced. From top to bottom, our REM-U2-Net was compared with SOTA on the CASIA, Columbia, Coverage and NIST16 datasets, respectively, with two images for each dataset.

4.4 Robustness Analysis

To verify the robustness of REM-U2-Net we apply image resizing, Gaussian blur, Gaussian noise, and JPEG compression to the Columbia dataset [15]. As shown in Table 4, REM-U2-Net exhibits strong robustness, particularly in the presence of Gaussian noise and compression.

Table 3. Performance comparison of REM-U2-Net setups used in the ablation study.

Ablation Variants	Base	SRM	SE	G	RRE	SRM+SE	SRM+G	SE+G	SRM+SE+G	SRM+SE+RRE	Full
AUC	0.841	0.882	0.893	0.907	0.926	0.936	0.910	0.922	0.941	0.968	**0.984**
F1	0.793	0.847	0.846	0.842	0.871	0.887	0.867	0.873	0.905	0.924	**0.939**

Table 4. Robustness analysis of localization for various distortions is performed. Pixel-level AUCs are reported.

AUC	Resize 0.78x	Resize 0.25x	GSBlur K=3	GSBlur K=15	GSNoise $\sigma=3$	GSNoise $\sigma=15$	JPEGComp q=100	JPEGComp q=50
ManTra [25]	0.732	0.662	0.744	0.631	0.681	0.573	0.792	0.676
PSCC [13]	0.775	0.714	0.741	0.618	0.725	0.631	0.814	0.719
REM-U2NET	0.834	0.783	0.856	0.793	0.879	0.803	0.904	0.873

5 Conclusion

In this paper, we propose a novel model called REM-U2-Net that aims to address manipulation attacks by various image processing techniques. Our approach uses local noise features and RGB image fusion as inputs to enhance the detection of tampered regions. Additionally, we introduce the RRE method to capture edge artifact details and prevent post-processing methods from causing artifact loss. Extensive experimental results demonstrate that REM-U2-Net is advantageous and robust against various post-processing methods and can effectively cope with common manipulation attacks. However, the model has some limitations, including low efficiency and high video memory occupation. Moreover, it may struggle with advanced GANs and real images that require better a priori knowledge to solve. To further resist advanced manipulation techniques, we plan to develop a technique for estimating unpredictable manipulated images.

Acknowledgments. This work was supported by the Industry-University-Research Cooperation Project of Fujian Science and Technology Planning (No:2022H6012); Natural Science Foundation of Fujian Province of China (No.2021J011169,No.2022J 011224,No.2022J011225).

References

1. Bi, X., Wei, Y., Xiao, B., Li, W.: RRU-Net: the ringed residual U-Net for image splicing forgery detection. In: CVPR, pp. 0–0 (2019)
2. Chen, S., Yao, T., Chen, Y., Ding, S., Li, J., Ji, R.: Local relation learning for face forgery detection. In: AAAI, vol. 35, pp. 1081–1088 (2021)
3. Chen, Z., et al.: LCTR: on awakening the local continuity of transformer for weakly supervised object localization. In: AAAI, vol. 36, pp. 410–418 (2022)
4. Dang, H., Liu, F., Stehouwer, J., Liu, X., Jain, A.K.: On the detection of digital face manipulation. In: CVPR, pp. 5781–5790 (2020)
5. Dong, C., Chen, X., Hu, R., Cao, J., Li, X.: MVSS-Net: multi-view multi-scale supervised networks for image manipulation detection. IEEE Trans. Pattern Anal. Mach. Intell. **45**, 3539–3553 (2022)
6. Dong, J., Wang, W., Tan, T.: Casia image tampering detection evaluation database. In: 2013 IEEE China Summit and International Conference on Signal and Information Processing, pp. 422–426. IEEE (2013)
7. Fridrich, J., Kodovsky, J.: Rich models for steganalysis of digital images. IEEE Trans. Inf. Forensics Secur. **7**(3), 868–882 (2012)
8. Guan, H., et al.: Nist nimble 2016 datasets (2016)
9. Hu, J., Shen, L., Sun, G.: Squeeze-and-excitation networks. In: CVPR, pp. 7132–7141 (2018)
10. Kingma, D.P., Ba, J.: Adam: a method for stochastic optimization. arXiv preprint arXiv:1412.6980 (2014)
11. Lin, T.-Y., et al.: Microsoft COCO: common objects in context. In: Fleet, D., Pajdla, T., Schiele, B., Tuytelaars, T. (eds.) ECCV 2014. LNCS, vol. 8693, pp. 740–755. Springer, Cham (2014). https://doi.org/10.1007/978-3-319-10602-1_48
12. Lin, X., et al.: Image manipulation detection by multiple tampering traces and edge artifact enhancement. Pattern Recogn. **133**, 109026 (2023)
13. Liu, X., Liu, Y., Chen, J., Liu, X.: PSCC-Net: progressive spatio-channel correlation network for image manipulation detection and localization. IEEE Trans. Circuits Syst. Video Technol. **32**(11), 7505–7517 (2022)

14. Mahfoudi, G., Tajini, B., Retraint, F., Morain-Nicolier, F., Dugelay, J.L., Marc, P.: DEFACTO: image and face manipulation dataset. In: 2019 27th European Signal Processing Conference (EUSIPCO), pp. 1–5. IEEE (2019)
15. Ng, T.T., Hsu, J., Chang, S.F.: Columbia Image Splicing Detection Evaluation Dataset. Columbia Univ CalPhotos Digit Libr, DVMM lab (2009)
16. Oktay, O., et al.: Attention U-net: learning where to look for the pancreas. arXiv preprint arXiv:1804.03999 (2018)
17. Popescu, A.C., Farid, H.: Exposing digital forgeries in color filter array interpolated images. IEEE Trans. Signal Process. **53**(10), 3948–3959 (2005)
18. Qin, X., Zhang, Z., Huang, C., Dehghan, M., Zaiane, O.R., Jagersand, M.: U2-Net: going deeper with nested U-structure for salient object detection. Pattern Recogn. **106**, 107404 (2020)
19. Ren, S., He, K., Girshick, R., Sun, J.: Faster R-CNN: towards real-time object detection with region proposal networks. In: NIPS 2015: Proceedings of the 28th International Conference on Neural Information Processing Systems - Volume 1, pp. 91–99 (2015)
20. Salloum, R., Ren, Y., Kuo, C.C.J.: Image splicing localization using a multi-task fully convolutional network (MFCN). J. Vis. Commun. Image Represent. **51**, 201–209 (2018)
21. Song, X., Zhao, X., Fang, L., Lin, T.: Discriminative representation combinations for accurate face spoofing detection. Pattern Recogn. **85**, 220–231 (2019)
22. Vaswani, A., et al.: Attention is all you need. In: NIPS, vol. 30 (2017)
23. Wen, B., Zhu, Y., Subramanian, R., Ng, T.T., Shen, X., Winkler, S.: Coverage-a novel database for copy-move forgery detection. In: ICIP, pp. 161–165. IEEE (2016)
24. Wu, Y., Abd-Almageed, W., Natarajan, P.: BusterNet: detecting copy-move image forgery with source/target localization. In: ECCV, pp. 168–184 (2018)
25. Wu, Y., AbdAlmageed, W., Natarajan, P.: ManTra-Net: manipulation tracing network for detection and localization of image forgeries with anomalous features. In: CVPR, pp. 9543–9552 (2019)
26. Zhong, J.L., Gan, Y.F., Vong, C.M., Yang, J.X., Zhao, J.H., Luo, J.H.: Effective and efficient pixel-level detection for diverse video copy-move forgery types. Pattern Recogn. **122**, 108286 (2022)
27. Zhou, P., et al.: Generate, segment, and refine: towards generic manipulation segmentation. In: AAAI, vol. 34, pp. 13058–13065 (2020)
28. Zhou, P., Han, X., Morariu, V.I., Davis, L.S.: Learning rich features for image manipulation detection. In: CVPR, pp. 1053–1061 (2018)

Improving Masked Autoencoders by Learning Where to Mask

Haijian Chen, Wendong Zhang, Yunbo Wang$^{(\boxtimes)}$, and Xiaokang Yang

Shanghai Jiao Tong University, Minhang District, Shanghai, China
{higerchen,diergent}@sjtu.edu.cn

Abstract. Masked image modeling is a promising self-supervised learning method for visual data. It is typically built upon image patches with **random masks**, which largely ignores the variation of information density between them. The question is: Is there a better masking strategy than random sampling and how can we learn it? We empirically study this problem and initially find that introducing object-centric priors in mask sampling can significantly improve the learned representations. Inspired by this observation, we present **AutoMAE**, a fully differentiable framework that uses Gumbel-Softmax to interlink an adversarially-trained mask generator and a mask-guided image modeling process. In this way, our approach can adaptively find patches with higher information density for different images, and further strike a balance between the information gain obtained from image reconstruction and its practical training difficulty. In our experiments, AutoMAE is shown to provide effective pretraining models on standard self-supervised benchmarks and downstream tasks.

Keywords: Self-Supervised Learning · Masked Image Modeling

1 Introduction

Inspired by the masked language modeling framework, *masked autoencoder* (MAE) [12,31], also known as *masked image modeling* (MIM), has become a promising self-supervised learning method based on Vision Transformer (ViT) [9]. The general idea of this framework is to extract semantic representations by learning to maximize the log-likelihood function of unobserved patches for input images.

There are two problems that are coupled and crucial to the quality of the learned representations: *First, how to determine the masks over the images? Second, how to learn from the masked images?* Previous literature has not been able to analyze the first problem sufficiently and satisfactorily in spite of its importance to the learning results. Further, the solutions to *"where to mask"* and *"how to learn from the masked data"* have never been integrated into an end-to-end optimization scheme.

What should be a good masking strategy? In general, the difficulty of pixel reconstruction is related to the information density of corresponding image patches. It is a

---

Supplementary Information The online version contains supplementary material available at https://doi.org/10.1007/978-981-99-8543-2_31.

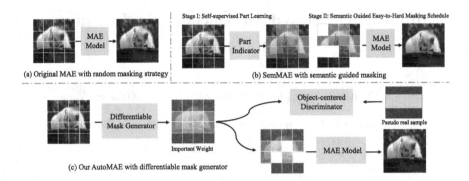

Fig. 1. Comparison of masking strategies. (**a**) The original MAE [12] randomly masks 70% image patches with a uniform probability. (**b**) SemMAE [18] uses a manually designed *easy-to-hard* masking schedule guided by an independently-trained semantic part indicator. (**c**) AutoMAE is a fully differentiable framework that uses an adversarially-trained mask generator.

paradox that if we mask a large number of "hard" patches (such as foreground objects), the model may fail to perceive the high-semantic areas; However, if we mask a large number of "easy" patches (such as the background), the self-supervised training task can be too simple for the model to learn effective representations. In our preliminary experiments, we find that both scenarios result in the degeneration of the learned visual representations, as illustrated in Sect. 2. We refer to such an empirical phenomenon as the *patch selection dilemma* in MAE.

Therefore, we need to weigh between the informative and less-informative patches in the masking strategy, such that during pixel reconstruction, the model can reason about the relations between the missing semantics and the remainders. The random masking approach in the original MAE [12] cannot adaptively perceive the variation of information density and applies a unified masking probability over the entire image. Following this line, SemMAE [18] presents a two-stage MAE training scheme with a manually designed *easy-to-hard* masking scheduler, guided by an independently trained semantic part indicator. However, the existing approaches do not directly optimize the masking strategy, thus leading to sub-optimal solutions to constructing the pretext task.

In this paper, we propose AutoMAE, a pilot study of optimizing the masking strategy in a fully differentiable MAE framework. As shown in Fig. 1, we compare the key differences between AutoMAE and previous work. The key insights are two-folded: First, it exploits an adversarially trained mask generator, in which the output masking probabilities are correlated with sample-specific information density across the patches, and are prone to be higher for "hard-to-reconstruct" regions in the foreground. Second, it jointly optimizes the mask generator and the self-supervised ViT model through Gumbel-Softmax reparameterization. Intuitively, we may prevent the mask generator from overfitting the "hard" regions by back-propagating the reconstruction error of the unobserved patches through the entire model.

Our approach achieves competitive results in the linear probing and finetuning setups on ImageNet-1K [7]. Furthermore, it presents excellent transfer learning ability on CUB-Bird [28], Stanford-Cars [17], iNaturalist 2019 [27], COCO [21], and

ADE20K [34] for fine-grained classification, detection, and segmentation, especially on small datasets with limited finetuning data.

The main contributions of this paper are as follows:

- We demonstrate that the previous random sampling method is a suboptimal masking strategy in MIM, and the pretraining results can be significantly improved by slightly raising the masking rates of the informative foreground image patches.
- Motivated by these findings, we provide an early study of *adversarial mask generation*, which incorporates simple structural priors into the adaptive masking rates.
- Unlike prior work, we propose to *integrate mask generation and image reconstruction in a differentiable framework* using Gumbel-Softmax. It improves the pretraining results by striking a balance between the information gain through image reconstruction and its training difficulty.

2   Preliminaries

2.1   Rethink the Mask Sampling Strategy in MAE

Revisiting MAE [12]. The learning process of MAE can be formalized as the following three steps. First, given an image $I \in \mathbb{R}^{c \times h \times w}$ where h, w and c represent the image height, width, and channel numbers, respectively, MAE first divides the input image into $n = hw/p^2$ non-overlapping patches with spatial size equal to $p \times p$, where p is the patch size. Then these divided patches are embedded by a linear projection layer and added with position embeddings to get the embedded tokens $Z = \{z_i\}_{i=1}^{n}, z_i \in \mathbb{R}^d$ where d represents the token dimension. Next, in the encoding stage, MAE randomly selects only 25% image tokens combined with a learnable [CLS] token $z^c \in \mathbb{R}^d$ to form the input for the transformer-based encoder. These tokens are served as visible hints and processed by the encoder to obtain encoded tokens Z_e. Given these embedded tokens, MAE starts the masked image modeling process which includes two stages: the encoding stage for visible token interactions and the decoding stage for removed content reconstruction. Finally, in the decoding stage, a shared mask token $z^m \in \mathbb{R}^d$ is repeated and used to fill the places of previously dropped tokens. These repeated mask tokens combined with encoded tokens Z_e are further processed by the transformer-based decoder to reconstruct the pixel values for each dropped patch. The model is optimized by minimizing the reconstruction loss $\mathcal{L}_{recon}$ in dropped regions.

Patch Masking Strategies. Among the above three steps, the masking strategy plays an important role in masked image modeling, which determines what kind of information will be learned. Although MAE adopts a high masking ratio, image patches are still dropped with equal probability. The random masking strategy ignores the difference in information density between patches and results in an ineffective learning process.

Two-Stage MAE. To improve the random masking strategy, SemMAE [18] exploits a two-stage framework that first learns the possible semantic parts without supervision and then uses the learned partition to determine which token should be masked. The high-quality part-level partition learned by SemMAE can be seen as local semantic

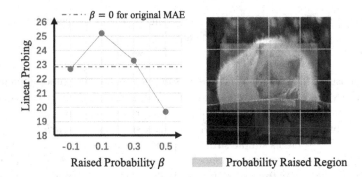

Fig. 2. Effects of raising the masking probability by β on patches within the object bounding boxes. Models are trained on a subset (10%) of the ImageNet dataset. The red dashed line represents the results of using the original random masking strategy. (Color figure online)

guidance to help learn more informative image patches. However, SemMAE still needs a manually designed easy-to-hard masking schedule to regularize the learning process and the semantic number is fixed among different images.

2.2   Key Findings: MAE with Prior Object Hints

Unlike language modeling, the information density of images is more sparse and usually dominated by the objects that appear in the image. Therefore we conduct a preliminary experiment to explore whether introducing the object location priors with high information density can help representation learning. We randomly select a subset (10%) of the ImageNet dataset [7] and use the provided true object bounding boxes to indicate object locations. Instead of random masking, we manually raise the dropped probability 'of patches within the bounding box and use these samples for self-supervised pretraining. We experiment with different values and show the linear probing results in Fig. 2, where β represents the additionally raised probability for patches within the bounding box against those in the background. We have two observations from Fig. 2:

– Compared with random masking, slightly increasing the masking probability for patches within the bounding box significantly improves the linear probing results.
– Excessively raising the masking probability of patches within the bounding box may affect the validity of the pretext task and degrades the performance.

These results indicate that different image patches provide different effects on the learning of visual representations, and learning to reconstruct more foreground objects with higher information density can lead to better pretraining results. Nonetheless, *we need to strike a balance between the information gain through masked image modeling and its training difficulty.* Inspired by this, a natural solution is to integrate mask generation and image reconstruction into a fully differentiable framework. For different images, it can adaptively provide masking weights that reflect prior object hints and are friendly to masked image modeling, that is, *we want to find patches with lower reconstruction difficulty within the obtained areas with higher information density.*

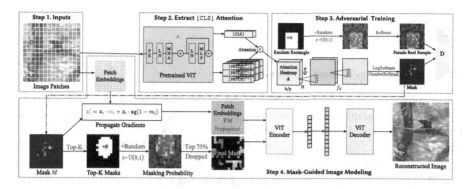

Fig. 3. The end-to-end framework of AutoMAE, which is designed to tackle to *patch selection* dilemma in a fully differentiable manner.

3 Method

3.1 General Design of AutoMAE

Motivated by our observations in Sect. 2.2, we design a novel self-supervised framework that can learn to select more informative foreground patches to mask, in which two key challenges need to be properly handled: First, how to guide the ViT model mining the informative patches without any explicit supervision? Second, how to control the difficulty of the pretext task via an adaptively learned mask strategy?

For the first challenge, we propose a differentiable mask generator G to generate the mask image which contains the important weight for each image patch. The mask generator contains a pretrained ViT encoder (also in a self-supervised manner) and two trainable convolution layers. Given an input image, we use the pretrained ViT encoder to extract the multi-head attention maps from the last transformer block. The obtained attention maps contain indications of different semantic regions of the image and are further processed by the convolution layers for mask generation. Since we cannot obtain explicit supervision to highlight the location of the foreground object, we introduce an object-centered adversarial training strategy to guide the mask generator in producing higher weights on patches within possible foreground objects. Specifically, we introduce a discriminator and generate real mask images by manually raising the masking probabilities of a randomly sampled rectangle area. In this way, we minimize the distribution distance between the generated mask and the sample real mask, which helps the model find the informative foreground patches.

For the second challenge, we directly propagate the gradients from the masked autoencoder back to the mask generator and train these two modules synchronously. In other words, the reconstruction loss applied for the masked autoencoder also constrains the mask generator from generating too hard mask images. By this means, the masked autoencoder encourages the mask generator to produce high weights on patches that can easily infer the masking information, while the discriminator regularizes the mask generator and focuses more on possible foreground objects instead of the background.

These two branches cooperate with each other to facilitate the masked autoencoder to learn more representative patch relations.

3.2   Adversarially-Trained Mask Generator

The framework of AutoMAE is shown in Fig. 3. The differentiable mask generator G aims to generate sample-specific mask images to guide the token selection process in the training of the masked autoencoder, which mainly consists of a pretrained ViT encoder and two trainable convolutional layers. We first exploit an MAE-pretrained [2] ViT encoder[1] to embed the image $I \in \mathbb{R}^{c \times h \times w}$ and extract the multi-head self-attention map $A \in \mathbb{R}^{H \times \frac{h}{p} \times \frac{w}{p}}$ between the query embedding of the [CLS] token and key embeddings of other patch tokens from the last transformer block, where H represents the number of heads. Specifically, the calculation of the i-th attention map $A_i \in \mathbb{R}^{1 \times \frac{h}{p} \times \frac{w}{p}}$ can be formalized as $A_i = \text{softmax}(q_i^c K_i^\top / \sqrt{d'})$, where $q_i^c \in \mathbb{R}^{1 \times d'}$ represents the query embedding of the [CLS] token, $K_i \in \mathbb{R}^{\frac{hw}{p^2} \times d'}$ represents the key embeddings of other patch tokens, and $d' = d/H$ represents the embedding dim. Then we reshape the attention map with spatial size equal to $1 \times \frac{h}{p} \times \frac{w}{p}$. As mentioned in [2], different attention maps in the last transformer block can attend to different semantic regions of the image. Therefore these attention maps can be served as informed initialization for further mask generation. Besides, the parameters of the pretrained ViT encoder are frozen during the training process. Given the multi-head self-attention map A, we further exploit two convolutional layers with ReLU activation f_θ to embed these attention maps $F = f_\theta(A)$, where $F \in \mathbb{R}^{1 \times \frac{h}{p} \times \frac{w}{p}}$. We then sample the final mask image $M \in \mathbb{R}^{1 \times \frac{h}{p} \times \frac{w}{p}}$ given the embedding feature F:

$$f_i' = \log\left(\frac{\exp(f_i)}{\sum_{k=1}^{hw/p} \exp(f_k)}\right) + z, m_i = \frac{\exp(f_i')}{\sum_{k=1}^{hw/p} \exp(f_k')}, \tag{1}$$

where $\log(softmax(\cdot))$ is used to stabilize the training process and z represents a random noise sampled from a Gumbel distribution. The values in M are continuous and can be seen as the important weight of each image patch. We use an extra discriminator and the masked autoencoder itself to supervise the generation process.

Although the attention map A can provide indications for different semantic regions, how to supervise the mask generator to produce a proper pretext task that can focus on foreground objects is still a challenging problem. To tackle this problem, we introduce an extra discriminator and design an object-centered adversarial training strategy to regularize mask generation. The key insight of this approach is to generate pseudo mask images M^p as "real" samples which can imitate the difference between foreground and background patches, and use the adversarial training strategy to minimize the distribution shift between generated mask M and "real" samples M^p.

The way to generate pseudo-mask images is quite simple. Given a zero-initialized image $M^p \in \mathbb{R}^{1 \times \frac{h}{p} \times \frac{w}{p}}$, we first randomly sample a rectangle area within this image

---

[1] Different self-supervised pretrained ViT models can be used for attention map extraction. Their effects are compared in our experiment. In the rest of this paper, we use the MAE-pretrained ViT-Base model without loss of generality.

whose spatial size is between 20%–80% of the whole image area. This sampled rectangle area is assumed to be the bounding box of one pseudo object. Then we flatten the pseudo mask $M^p = \{m_i^p\}_{i=1}^n$ and suppose the index set within the sample rectangle area is X, the values in the pseudo mask are sampled as follows:

$$m_i^p = \epsilon + \begin{cases} \alpha & i \in X \\ 0 & i \notin X \end{cases} \tag{2}$$

where ϵ represents a random noise sampled from a uniform distribution $\epsilon \sim U(0,1)$, and α represents the additional important weight for patches within the rectangle, which is set to 0.5 in our experiments. In this way, we generate the pseudo mask images where foreground patches contain higher weights compared with background patches. We then use the pseudo-mask images as a real sample for adversarial training with the loss function from LSGAN [22]:

$$\begin{aligned} \mathcal{L}_{adv} &= -\mathbb{E}_M\left[(D(M) - c)^2\right], \\ \mathcal{L}_{adv}^D &= \mathbb{E}_{M^p}\left[(D(M^p) - b)^2\right] + \mathbb{E}_M\left[(D(M) - a)^2\right], \end{aligned} \tag{3}$$

where D is the discriminator and we set $a = -1, b = 1, c = 0$ as in previous work.

3.3 Mask-Guided Image Modeling

Given the sampled mask M, we suggest that patches with higher weights are more likely to be foreground objects and more informative than patches with lower weights. Therefore, we raise the masking probabilities of these high-weighted patches and use low-weighted patches as visible hints in masked image modeling. Specifically, we first sort the values in $M = \{m_1, m_2, \cdots, m_n\}$ and then get the top K index set Y. The unnormalized masking probability of each token can be sampled as follows:

$$\gamma_i = \epsilon + \begin{cases} \beta & i \in Y \\ 0 & i \notin Y \end{cases} \tag{4}$$

where ϵ represents a random noise sampled from a uniform distribution $\epsilon \sim U(0,1)$, and β represents the additionally raised probability for tokens with higher weights. In our experiment, we set $K = n/4$ and $\beta = 0.5$ where n represents the number of patch tokens. In this case, patches in set Y are supposed to be more informative than other patches, and models are supposed to pay more attention to these patches.

After that, we use the sampled probability to guide the token selection process where tokens with the top 75% biggest probability are dropped and the remaining tokens with 25% smallest probability are saved as visible hints for masked image modeling. In other words, we still follow the mask ratio used in the original masked autoencoder while using the sampled probabilities to enhance the learning of more important patches.

Particularly, we slightly change the formulation of the input patch tokens to allow the gradients from the masked autoencoder can be propagated back to the mask generator. Given the embedded patch tokens $Z = \{z_i\}_{i=1}^n$ and the generated mask image

$M = \{m_i\}_{i=1}^n$, the new embedded patch tokens $Z' = \{z'_i\}_{i=1}^n$ are obtained via $z'_i = z_i \cdot m_i + z_i \cdot \mathrm{sg}(1 - m_i)$, where $\mathrm{sg}(\cdot)$ represents the stop gradient operation. We conduct token selection on these new embedded tokens Z' and then perform the following encoding and decoding stages which are the same as the original masked autoencoder. In this way, we preserve the values of patch tokens while allowing the masked autoencoder to influence mask generation. Combined with the adversarial loss, the full objective function of the mask generator can be written as $\mathcal{L}_G = \mathcal{L}_{recon} + \lambda\mathcal{L}_{adv}$ where we set $\lambda = 0.2$ by grid search.

4 Experiments

We evaluate AutoMAE in the linear probing and finetuning setups on ImageNet-1K [7], and perform model analyses on its smaller subsets. We also conduct transfer learning experiments on CUB-Bird [28], Stanford-Cars [17], iNaturalist 2019 [27], COCO [21], and ADE20K [34] for fine-grained classification, detection, and segmentation tasks.

We use the same ViT-Base architecture and a 16×16 patch size in all compared models. For AutoMAE, we typically use an 800-epochs-pretrained MAE-Base (called "MAE-800") as the warmup model of the feature extractor in mask generation. We freeze its parameters and train other parts of AutoMAE for 800 epochs. We use Sem-MAE [18] as our primary baseline as its mask generator also uses a warmup model[2]. We use the provided SemMAE pretrained model with a 16×16 patch size.

4.1 Experiments on ImageNet-1K

When we have finished the preliminary model analyses on small subsets, we start to train AutoMAE on the full ImageNet-1K and analyze the experimental results.

Linear Probing. We use 3 alternative models to initialize the feature extractor in the mask generator: (1) the ViT-B network pretrained with MAE [12] (termed as MAE-800), (2) the ViT-B pretrained with iBOT [35] (termed as iBOT-B), and (3) the ViT-S pretrained with DINO [2] (termed as DINO-S). We use the LARS optimizer to train the model for 90 epochs. The base learning rate is set to 0.1, the momentum is 0.9, and the batch size is 16,384. We use random resized cropping and horizontal flipping for data augmentation. In Table 1, AutoMAE consistently achieves better performance than the vanilla MAE model, regardless of the base models used for mask generation.

Finetuning. In the finetuning stage, the base learning rate is set to 5×10^{-4} and the AdamW momentum is configured as $\beta_1, \beta_2 = 0.9, 0.999$. The effective batch size is 1,024. Similar to previous literature, we use the RandAug technique [6] for data augmentation and employ layer-wise learning rate decay (0.75), Mixup [33] (0.8), Cut-Mix [32] (1.0), and DropPath [15] (0.1) in the finetuning stage. All compared models are trained for 100 epochs in the same setting. As shown in Table 2, our approach achieves competitive results to MAE and SemMAE. It is reasonable that SemMAE

---

[2] SemMAE [18] uses an 800-epochs iBOT model in the mask generator.

Table 1. Linear probing results of models pretrained on the full ImageNet for 800 epochs. Regardless of the methods used in the mask generator to warm up its feature extractor, AutoMAE consistently achieves better performance.

Model	Warmup in G	Acc-1(%)
MAE [12]	N/A	63.7
SemMAE [18]	iBOT-800	65.0
AutoMAE	MAE-800	<u>66.7</u>
AutoMAE	iBOT-B [35]	<u>66.7</u>
AutoMAE	DINO-S [2]	**68.8**

Table 2. Finetuning results on ImageNet-1K. "Ratio" represents the percentage of the data we used for finetuning. All methods use ViT-Base as the encoder with a patch size of 16×16.

Model	Warmup in G	Ratio	Acc-1(%)
MAE [12]	N/A	100	83.26
SemMAE [18]	iBOT-800	100	**83.34**
AutoMAE	MAE-800	100	<u>83.32</u>
MAE [12]	N/A	30	78.95
SemMAE [18]	MAE-800	30	79.00
AutoMAE	MAE-800	30	**79.14**

has a slightly better performance on full-set finetuning, as it uses an iBOT-based mask generator which is pretrained by contrastive learning. It is also worth noting that Sem-MAE with a 16×16 patch size obtains similar finetuning results, indicating that the non-linear features for big tokens in MAE are strong enough for image classification. Furthermore, we perform an additional experiment to finetune the models on the 30% subset. As shown in Table 2, AutoMAE outperforms MAE and SemMAE by large margins, suggesting that it is more effective in scenarios with limited data.

4.2 Downstream Tasks

Fine-Grained Image Classification. We first finetune the models on fine-grained image classification datasets, including CUB-Bird [28], Stanford-Cars [17], and iNaturalist 2019 [27]. As shown in Table 3, AutoMAE achieves the best performance on all datasets. Similarly, we experiment with random-sampled subsets (50% and 10%), where AutoMAE shows more significant improvements compared with other models. These results demonstrate the transfer learning ability of our method to small datasets.

Object Detection and Instance Segmentation. Table 4 gives the detection and instance segmentation results on COCO. We follow the previous work [19] to finetune the Mask-RCNN [14] with a ViT-based FPN backbone. Due to the memory limit, we set the total batch size to 32, half of the original value, and set the learning rate to 4×10^{-5}. We

Table 3. Fine-grained classification results. All methods exploit ViT-Base as the encoder with a patch size equal to 16 × 16. "Ratio" is the percentage of data used in fine-tuning. Ratio 10% for CUB and Cars is omitted because the two datasets are not large enough.

Method	Ratio(%)	CUB	Cars	iNat-19
MAE [12]	100	83.3	92.7	79.50
SemMAE [18]	100	82.1	92.4	79.60
AutoMAE	100	**83.7**	**93.1**	**79.93**
MAE [12]	50	70.6	84.2	73.76
SemMAE [18]	50	70.6	82.1	73.47
AutoMAE	50	**73.4**	**84.6**	**74.03**
MAE [12]	10	-	-	49.14
SemMAE [18]	10	-	-	48.65
AutoMAE	10	-	-	**50.13**

Table 4. Results of object detection and instance segmentation on the COCO dataset. * indicates results from the original literature.

Method	Pretraining Epochs	AP^{box}	AP^{mask}
BEiT* [1]	800	49.8	44.4
MAE* [12]	1600	50.3	44.9
AutoMAE	800	**50.5**	**45.0**

Table 5. Results of semantic segmentation on ADE20K.

Model	Ratio(%)	mIoU
Supervised Pretraining	100	45.3
BEiT* [1]	100	45.8
MAE [12]	100	46.1
SemMAE [18]	100	46.3
AutoMAE	100	**46.4**
MAE [12]	50	41.7
SemMAE [18]	50	41.9
AutoMAE	50	**42.4**

finetune AutoMAE for 100 epochs which is the same as the compared models. It shows that even though AutoMAE is pretrained for half epochs (800), it still outperforms the original MAE by 0.2 in AP^{box} and 0.1 in AP^{mask}.

Semantic Segmentation. We use the ADE20K dataset for semantic segmentation, which contains 150 semantic categories and 25K images. Similar to SemMAE [34], we use UperNet [30] as the network backbone to finetune the pretrained ViT-B network for 160K iterations on ADE20K. The learning rate is set to 5×10^{-5} and the batch size is 8. We also finetune our pretrained model for 100 epochs which is the same as other compared methods. As shown in Table 5, AutoMAE outperforms MAE by 0.3 mIoU and outperforms SemMAE by 0.1 mIoU. We then use fewer data (50%) in the finetuning stage. We observe that AutoMAE achieves a more significant advantage over MAE and SemMAE in this case, increasing mIoU results by 0.7 and 0.5, respectively.

4.3 Mask Visualization

To verify whether the mask generator can perceive meaningful areas, we conduct experiments on ImageNet-9 [29], a dataset with foreground objects only. As shown in Fig. 4, the most high-weighted masks in AutoMAE mainly focus on meaningful patches, such as the back and eyes of the jaguar, and gradually involve less significant patches as the masking ratio increases. More results are provided in supplementary materials.

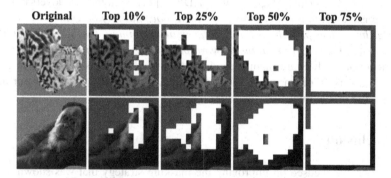

Fig. 4. High-weighted masks on **ImageNet-9** produced by the mask generator. The highlighted areas are obtained from the mask before adding random noise by K-largest values.

5 Related Work

The majority of recent studies of self-supervised learning for visual representations of deep networks can be roughly grouped into two categories, *i.e.*, contrastive learning and masked image modeling.

Contrastive Learning. Contrastive learning methods typically distinguish positive samples from other samples. Early work like MoCo [13] and SimCLR [4] focuses on distinguishing the negative pairs, while BYOL [11], SimSiam [5] and DINO [2] eliminates the need for negative pairs. DINO is trained by self-distillation, and it observes that ViT features contain explicit information about the semantic segmentation of an image.

Masked Image Modeling. Recent studies in masked image modeling (MIM) [1,3,9,10, 26] are largely inspired by the success of BERT [8]. According to different regression targets, MIM methods can be divided into two groups: pixel-level reconstruction and feature-level prediction. Feature-level prediction usually requires an external model called "Tokenizer" to generate reconstruction targets. BEiT [1] and PeCo [23] are pretrained to predict visual tokens generated by discrete variational autoencoder (dVAE) [23]. iBOT [35] presents an online tokenizer as the teacher network to produce the target and perform self-distillation. On the other hand, pixel-level reconstruction methods directly use raw pixel values as the regression targets. MAE [12] and SimMIM [31] show that masking a high ratio of patches and directly predicting RGB values can achieve BEiT-level performance. The group of work most relevant to AutoMAE includes methods that also improve masking strategies in MIM. SemMAE [18] adopts a two-stage framework that first learns a part indicator and then fixes it in the MIM training stage. AttMask [16] presents an independently-trained self-attention module to mask high-attended patches, while MST [20] proposes to mask low-attended patches. *In the above discussions, we empirically show the superiority of the proposed differentiable framework over two-stage training.* ADIOS [24] combines an adversarial MIM with contrastive learning. *In contrast, AutoMAE does not rely on contrastive learning but rather introduces prior hints through adversarial training. Furthermore, AutoMAE does not require expensive multiple forward passes in a single training iteration.* The experiments of ADIOS are mainly conducted on small datasets such as STL10 and a downsized version of ImageNet-100 [25], which supports the validity of our model analysis setup on the ImageNet subsets.

6   Conclusion

In this paper, we focused on improving the masking strategy that was shown to play an important role in the masked image modeling framework. We illustrated that introducing object-centric priors to guide the masking strategy can significantly improve the learned visual representations. Starting from this point, we proposed AutoMAE, which integrates a differentiable mask generator to provide the sample-specific categorical distribution of masking probabilities across all image patches. The mask generator is jointly trained with the ViT model with an adversarial learning constraint. We validated the effectiveness of AutoMAE on different benchmarks and downstream tasks.

References

1. Bao, H., Dong, L., et al.: Beit: BERT pre-training of image transformers. In: ICLR (2022)
2. Caron, M., Touvron, H., et al.: Emerging properties in self-supervised vision transformers. In: ICCV (2021)
3. Chen, M., Radford, A., et al.: Generative pretraining from pixels. In: ICML (2020)
4. Chen, T., Kornblith, S., Norouzi, M., et al.: A simple framework for contrastive learning of visual representations. In: ICML (2020)
5. Chen, X., He, K.: Exploring simple siamese representation learning. In: CVPR (2021)
6. Cubuk, E.D., Zoph, B., et al.: Randaugment: practical automated data augmentation with a reduced search space. In: CVPR Workshops (2020)
7. Deng, J., et al.: Imagenet: a large-scale hierarchical image database. In: CVPR (2009)
8. Devlin, J., Chang, M., et al.: BERT: pre-training of deep bidirectional transformers for language understanding. In: NAACL-HLT (2019)
9. Dosovitskiy, A., Beyer, L., et al.: An image is worth 16x16 words: transformers for image recognition at scale. In: ICLR (2021)
10. Feichtenhofer, C., Fan, H., et al.: Masked autoencoders as spatiotemporal learners. In: NeurIPS (2022)
11. Grill, J.B., Strub, F., et al.: Bootstrap your own latent-a new approach to self-supervised learning. In: NeurIPS (2020)
12. He, K., Chen, X., et al.: Masked autoencoders are scalable vision learners. In: CVPR (2022)
13. He, K., Fan, H., et al.: Momentum contrast for unsupervised visual representation learning. In: CVPR (2020)
14. He, K., Gkioxari, G., et al.: Mask R-CNN. In: ICCV (2017)
15. Huang, G., Sun, Yu., Liu, Z., Sedra, D., Weinberger, K.Q.: Deep networks with stochastic depth. In: Leibe, B., Matas, J., Sebe, N., Welling, M. (eds.) ECCV 2016. LNCS, vol. 9908, pp. 646–661. Springer, Cham (2016). https://doi.org/10.1007/978-3-319-46493-0_39
16. Kakogeorgiou, I., Gidaris, S., et al.: What to hide from your students: Attention-guided masked image modeling. In: Avidan, S., Brostow, G., Cissé, M., Farinella, G.M., Hassner, T. (eds.) ECCV 2022. LNCS, vol. 13690, pp. 300–318. Springer, Cham (2022). https://doi.org/10.1007/978-3-031-20056-4_18
17. Krause, J., Stark, M., et al.: 3D object representations for fine-grained categorization. In: ICCV Workshops (2013)
18. Li, G., Zheng, H., et al.: Semmae: semantic-guided masking for learning masked autoencoders. In: NeurIPS (2022)
19. Li, Y., Xie, S., et al.: Benchmarking detection transfer learning with vision transformers. CoRR (2021)
20. Li, Z., Chen, Z., et al.: MST: masked self-supervised transformer for visual representation. In: NeurIPS (2021)
21. Lin, T.-Y., et al.: Microsoft COCO: common objects in context. In: Fleet, D., Pajdla, T., Schiele, B., Tuytelaars, T. (eds.) ECCV 2014. LNCS, vol. 8693, pp. 740–755. Springer, Cham (2014). https://doi.org/10.1007/978-3-319-10602-1_48
22. Mao, X., Li, Q., et al.: Least squares generative adversarial networks. In: ICCV (2017)
23. Ramesh, A., et al.: Zero-shot text-to-image generation. In: ICML (2021)
24. Shi, Y., et al.: Adversarial masking for self-supervised learning. In: ICML (2022)
25. Tian, Y., Krishnan, D., Isola, P.: Contrastive multiview coding. In: Vedaldi, A., Bischof, H., Brox, T., Frahm, J.-M. (eds.) ECCV 2020. LNCS, vol. 12356, pp. 776–794. Springer, Cham (2020). https://doi.org/10.1007/978-3-030-58621-8_45
26. Tong, Z., Song, Y., et al.: Videomae: masked autoencoders are data-efficient learners for self-supervised video pre-training. In: NeurIPS (2022)

27. Van Horn, G., Mac Aodha, O., et al.: The inaturalist species classification and detection dataset. In: CVPR (2018)
28. Wah, C., Branson, S., et al.: The Caltech-UCSD birds-200-2011 dataset (2011)
29. Xiao, K., Engstrom, L., Ilyas, A., Madry, A.: Noise or signal: the role of image backgrounds in object recognition. In: ICLR (2021)
30. Xiao, T., Liu, Y., Zhou, B., Jiang, Y., Sun, J.: Unified perceptual parsing for scene understanding. In: Ferrari, V., Hebert, M., Sminchisescu, C., Weiss, Y. (eds.) ECCV 2018. LNCS, vol. 11209, pp. 432–448. Springer, Cham (2018). https://doi.org/10.1007/978-3-030-01228-1_26
31. Xie, Z., Zhang, Z., et al.: Simmim: a simple framework for masked image modeling. In: CVPR (2022)
32. Yun, S., Han, D., et al.: Cutmix: regularization strategy to train strong classifiers with localizable features. In: ICCV (2019)
33. Zhang, H., Cissé, M., et al.: mixup: beyond empirical risk minimization. In: ICLR (2018)
34. Zhou, B., Zhao, H., et al.: Semantic understanding of scenes through the ADE20K dataset. In: ICCV (2019)
35. Zhou, J., Wei, C., et al.: iBOT: image BERT pre-training with online tokenizer. In: ICLR (2022)

An Audio Correlation-Based Graph Neural Network for Depression Recognition

Chenjian Sun[1,2] and Yihong Dong[1,2(✉)]

[1] Faculty of Electrical Engineering and Computer Science, Ningbo University,
Ningbo 315211, China
dongyihong@nbu.edu.cn
[2] Zhejiang Key Laboratory of Mobile Network Application Technology,
Ningbo University, Ningbo 315211, China

Abstract. Depression is a prevalent mental health disorder. The diagnosis of depression hinges largely on the medical practitioner's subjective assessment of the patient's diagnostic process. The involvement of multiple subjective factors during this process can further complicate the diagnosis. In this paper, we propose a novel approach for depression recognition using the graph neural network that incorporates potential connections within and between audio signals. Specifically, we first extract time series information between frame-level audio signal features through GRU. We then construct two graph neural network modules to explore the potential connections of inter-audio and inter-audio. In the first graph module, we construct a graph using the frame-level features of each audio as nodes and embed the output graph into a feature vector representation. In the second graph module, we represent the graph embedding feature vector as a node and encode the potential relationships between audio signals through node neighbourhood information propagation. Additionally, we extract emotional features related to depression using a pre-trained emotion recognition network and enhance the connection between coded audio signals through a self-attention mechanism to further improve the model's performance. We conducted extensive experiments on three depression datasets, and our proposed model outperformed all benchmark models, demonstrating its effectiveness.

Keywords: automatic depression detection · graph neural network · audio recognition

1 Introduction

Major Depression Disorder (MDD) is a prevalent mental disorder that may result from various psychological, social, or physical factors. In our fast-paced society,

This work has been supported by the Natural Science Foundation of Ningbo (No. 2023J114).

the number of individuals with depression has increased significantly, leading to suicidal tendencies [12]. Currently, the diagnosis of depression mainly relies on clinical interviews and questionnaires to assess the patient's depression. However, the diagnostic process is subjective and may result in misdiagnosis. Thus, developing an automatic depression recognition model can help reduce misdiagnosis by clinicians.

Studies have shown that there is a strong relationship between MDD and language behavior [3]. Using audio recordings to assess depression not only maintains patients' privacy but also encourages their willingness to voluntarily seek diagnosis and treatment. Therefore, depression detection based on audio signals has garnered increasing attention from scholars. Convolutional neural networks (CNN) and recurrent neural networks (RNN) can capture the spatial and temporal characteristics of features in spectral, Mel-frequency cepstral coefficients (MFCCs), or other acoustic features of audio signals. By exploring the spatial and temporal information of audio features, CNN, RNN, and their variants have achieved promising results in classification tasks based on audio signals. However, in essence, CNN and RNN can only learn the local information of the feature matrix and fail to consider global information.

To explore the global correlation of features, Shirian et al. [19] constructed a graph convolutional neural network by using the acoustic features of each time frame as nodes and performed emotion recognition tasks. Niu et al. [15] proposed a graph attention model based on the form of question-answer pairs in clinical interview data for depression detection. The graph classification model of a graph neural network for a single audio sample ignores the similarity and differences between different audio classes. Therefore, Chen et al. [4] constructed a graph neural network model by using the characteristics of the entire audio sample as node features. However, existing research only flattens or averages the different time frame features of the entire audio to obtain the node feature representation, ignoring the potential correlation between different time frames within the audio. Consequently, the majority of methodologies are inadequate thoroughly extract depression-related information from audio. To address the issue, we propose a graph neural network method based on audio signal depression detection. Our method constructs a graph network between the frame-level features inside the audio and between all audios, which combines the characteristics of two graph convolutional networks to better capture depression-related cues. The main contributions of this paper are as follows:

1) In this study, we introduce a novel graph neural network model for detecting depression from audio signals. Our model simultaneously considers potential associations among frame-level features within audio signals, while also accounting for inter-class similarities and differences between audios. To the best of our knowledge, few studies have investigated MDD detection tasks by considering both intra- and inter-audio associations.

2) To better explore depression-related cues in audio signals, emotional features are extracted through a trained emotion recognition network. These emotional features are then fused using a self-attention mechanism to further improve the performance of the model.

3) Our model has achieved quite good results on three depression datasets DAIC-WOZ, MODMA, and D-Vlog, which further proves the effectiveness and rationality of our model.

2   Related Work

Convolutional neural networks and recurrent neural networks can capture spatial and temporal features. Niu et al. [16] proposed a CNN model based on the attention mechanism, which combines CNN and RNN to capture temporal and spatial features, with a focus on the time stamp, frequency band, and channel related to depression detection. Huang et al. [11] proposed the FVTC-CNN model, which uses an extended CNN to extract channel coordination features for predicting depression. Long Short-Term Memory (LSTM) [10] has a long-term memory function, which mitigates the problem of gradient disappearance or explosion during training. Du et al. [8] proposed a new LSTM module to adapt to the irregular occurrence of bipolar disorder in different periods. Seneviratne et al. [18] designed a two-layer neural network architecture to extend CNN-LSTM, which addresses the problems of repeated sampling and discontinuous boundary of adjacent sub-matrices in traditional methods.

To better handle non-European data, Thomas et al. [13] proposed a graph convolutional neural network (GCN) that can directly process graph-structured data. Subsequently, significant results have been achieved in processing graph data using the GNN method, such as social networks, transportation networks, and citation networks. Researchers have also started exploring the application of graph neural networks in the audio field. Ghadiri et al. [7] proposed to enrich audio analysis for depression detection by combining graph transformation of audio signals with representation learning of natural language processing. Niu et al. [15] proposed the hierarchical context-aware model based a graph attention network (HCAG) for depression detection. HCAG transformed a clinical interview into several question-and-answer pairs, reflecting the structure of depression assessment. However, for each interviewee, the questions to be answered are the same, making it worthwhile to consider whether the same questions can distinguish the effectiveness of the interviewee's depression. Most of the existing methods only consider the correlation within the audio, ignoring the heterogeneity between audio samples. Chen et al. [4] proposed a graph neural network model (MS2-GNN) and used the node classification method to explore the potential relationship between subjects. Most of the above methods unilaterally explore the potential relationship between patients and diseases from audio signals, do not consider the correlation between audio and audio at the same time, and do not fully explore the MDD-related clues in the subject's audio signals.

3   Methods

We propose a graph neural network method for detecting depression in audio signals, as shown in Fig. 1. Firstly, we preprocess the audio data and extract low-

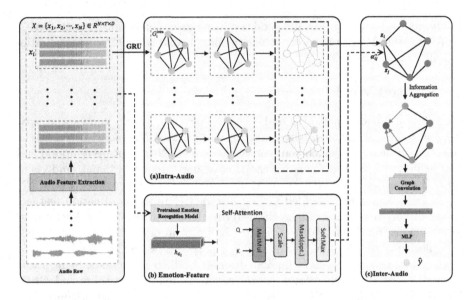

Fig. 1. The overall frame model structure of the proposed method.

level audio features. The Intra-Audio module consists of gated recurrent units (GRU) and GCNs, which are used to extract time series information within the audio features and the potential associations of frame-level features and output an embedded feature vector representation for each audio. The Emotion-Feature module is a pre-trained network for extracting depression-related emotional features from low-level audio features. Then, we construct a graph neural network using the output from the Intra-Audio and Emotion-Feature modules to explore the intra-class similarity and inter-class differences among audios. Finally, the extracted high-level features are input into a multilayer perceptron (MLP) layer for the final depression prediction task.

3.1   Data Preprocessing

The low-level audio feature extraction process involved sampling the audio signal using Librosa, an audio signal processing library in Python, with a fixed frame window and a frame interval at the original audio sampling rate. To ensure the duration of all discourses was consistent, we extracted time-fixed audio features for each audio. We extracted frame-level audio features for each audio, including MFCCs, logarithmic fundamental frequency (log F0), and constant-Q transform (CQT) features.

3.2   Intra-audio Correlation

In this paper, the initial audio features are defined as $X = \{x_1, x_2, \cdots, x_N\} \in R^{N \times T \times D}$, where N denotes the number of samples, T denotes the timestamp of each audio sequence, and D denotes the audio feature dimension of each frame. First, we use the RNN to extract the time series information of each audio feature. To avoid the gradient disappearance and explosion problems of traditional RNNs in the backpropagation process, this paper uses the GRU to extract the time series information of audio signals. The input of the GRU-based time series extraction network is as follows:

$$x_i' = GRU\left([x_{i,1}, x_{i,2}, \cdots, x_{i,T}]\right), \quad i \in \{1, 2, \cdots, N\}. \tag{1}$$

where i denotes the i-th sample. $x_i' \in R^{T \times D'}$ is the feature matrix extracted by the GRU network. T is the number of frames in the feature sequence. D' is the dimension of the characteristics of the audio.

To explore the correlation between different time frame-level features within the audio. In this paper, the graph convolutional neural network is used to establish the connection between different time frames, to mine their potential connection. Since there is no predefined adjacency matrix in the audio data, this paper establishes the connection of different time frames according to the time series of the frames and determines the connection strength through the attention mechanism. A graph $G_i^{intra} = \left(V_i^{intra}, E_i^{intra}\right)$ is constructed for the i-th audio, where the nodes are composed of frame-level features. For each discourse vertex v_i, it is connected with the first p time frame nodes and the last s time frame nodes. To fully explore the impact of neighborhood frame nodes at different times, we use a graph attention network (GAT) to propagate neighborhood node information. Specifically, GAT takes all nodes as input and updates the target node features through the adjacent nodes $\mathcal{N}_i$ in the graph. The process of updating the node h_i^{l+1} is as follows:

$$h_i^{l+1} = \sigma\left(\sum_{j \in \mathcal{N}_i} w_{ij} W_1^l h_j^l\right), \quad i \in \{1, 2, \cdots, T\}. \tag{2}$$

where W_1^l is the trainable weight matrix, l represents the l-th graph convolution layer, h_i^{l+1} is the node representation of the i-th time-frame node of the current audio after the update of the l-th graph convolution layer, σ is the nonlinear activation function Relu, and w_{ij} is the attention coefficient between node i and node j. The calculation process is as follows:

$$w_{ij} = \frac{\exp\left(\text{LeakyReLU}\left(a\left[W_2 h_i \| W_2 h_j\right]\right)\right)}{\sum_{k \in \mathcal{N}_i} \exp\left(\text{LeakyReLU}\left(a\left[W_2 h_i \| W_2 h_k\right]\right)\right)}. \tag{3}$$

where $a \in R^{2D' \times 1}$ and $W_2 \in R^{D' \times D'}$ are the learnable training parameters, which are calculated and normalized to obtain the weight coefficients between node i and node j.

After the graph convolution layer, each frame node obtains enough information from its neighboring time-frame nodes, and to obtain the feature representation of the whole graph, we use graph averaging pooling to obtain the graph embedding representation. The embedding representation z_i for each audio sample is as follows:

$$z_i = \text{mean}\left(\left\{ h^L \mid h^L \subseteq G_i^{\text{intra}} \right\}\right), \quad i \in \{1, 2, \cdots, N\}. \tag{4}$$

3.3 Emotional Features

Psychological studies suggest that depression can directly affect an individual's emotional expression and perception. Cognitive biases and defects caused by depression can affect emotional regulation, such as habitually up-regulating negative emotions in emotional expression. Considering the impact of depression on emotions, we pre-trained a compact graph-structured emotion recognition network CompactSER [19] to extract emotional features from audio signals. After the embedding layer is a fully connected layer, as the classification layer of the model. In this paper, we remove the classification layer from the pre-trained model and use it as the emotional feature extractor of our model.

3.4 Inter-audio Correlation

After going through the Intra-Audio module, the embedding representation of each sample was obtained. To explore the potential relationships between different samples, we reconstruct a graph $G^{inter} = (V^{inter}, E^{inter})$ with the embedding features of N samples. The adjacency matrix of G^{inter} is represented as $A' \in R^{N \times N}$, where α'_{ij} is the element corresponding to the adjacency matrix A', which denotes the weights of the edges between node i and node j. To obtain the topology of the feature space and to be able to better capture the dependencies between different nodes in the graph structure and avoid unrelated neighborhood interference. In this paper, we use cosine similarity to construct the graph structure, and the process of generating a graph based on cosine similarity is as follows:

$$\begin{aligned} \alpha'_{ij} &= \frac{z_i^T z_j}{\|z_i\| \times \|z_j\|}, \\ \alpha'_{ij} &= \begin{cases} \alpha'_{ij}, & \text{if } \alpha'_{ij} \geq \varepsilon \\ 0, & \text{otherwise} \end{cases}. \end{aligned} \tag{5}$$

where z_i denotes the feature vector representation of the i-th audio, ε is the threshold hyperparameter. And α'_{ij} is the cosine similarity of node i to node j, both the weights of node i and node j edges.

To better integrate emotional features, we use the self-attention mechanism to calculate the potential relationship between emotional features in different samples. Compared with simple feature splicing, the weight fusion method we adopted can not only fully mine depression cues and emotional cues between different samples but also flexibly adjust the weight ratio of the two, to more fully integrate depression features and emotional features. The process of the self-attention mechanism is as follows:

$$A'' = \text{soft max}\left(\frac{QK^T}{\sqrt{d_k}}\right),$$
$$Q = [h_{G_1}, h_{G_2}, \cdots, h_{G_N}]^T W_Q,$$
$$K = [h_{G_1}, h_{G_2}, \cdots, h_{G_N}]^T W_K.$$

(6)

Where W_Q and W_K are learnable parameters. h_{G_i} is the emotional feature embedding representation of the i-th audio. Q, K and d_k are the Query and Key vectors and their feature dimensions. A'' is the weight matrix of the feature vectors, α''_{ij} denotes is the value of A'' corresponding to row i and column j, both the weight between node i and node j. '

The next key step is features aggregation, where a node aggregates information from its neighboring nodes to the target node. The graph convolution layer receives the node feature matrix H^l and the association matrix A, and then update the node features as follows:

$$H^{l+1} = \sigma\left(AH^lW^l\right),$$
$$A = \beta A' + (1 - \beta)A''.$$

(7)

where W^l is a learnable parameter, A' and A'' are the weight matrices of depression features and emotion features, respectively, and β is an importance coefficient ranging from 0 to 1, a hyperparameter. After the multilayer graph convolution layers, the learned features are fed into our classifier for depression detection. In this paper, the MLP layer is finally used as our final classifier:

$$\hat{y} = \text{softmax}\left(MLP\left(H^L\right)\right).$$

(8)

The depression detection in this task is a binary classification task, so we use cross entropy as the loss function to train the model parameters:

$$\text{loss} = \sum_{c \in [0,1]} P(c \mid y) \log P(c \mid \hat{y}).$$

(9)

where 0 and 1 denote the sample labels of health and depression, respectively, $P(c \mid y)$ is the true label distribution, and $P(c \mid \hat{y})$ is the estimated probability distribution of label c.

4 Experiments

4.1 Datasets

DAIC-WOZ [9]: DAIC-WOZ is a clinical interview data set of depression, including 189 subjects, 133 depression (PHQ-8 $\geq$ 10), and 56 health (PHQ-8 < 10). In this experiment, we extracted the subjects' discourse fragments from each interview record, and a total of 4903 individual subjects' audio fragments were obtained. To eliminate the adverse effects caused by the imbalance of depression and health in the data, all audio fragments were randomly down-sampled, and the audio discourse sample data of 2914 subjects were finally obtained.

MODMA [2]: MODMA is an open data set for the analysis of mental disorders released by Lanzhou University. Among 23 MDD subjects and 29 healthy subjects were included. Each subject completed 29 recording tasks in different emotional states through interviews, reading, and picture descriptions, with time ranging from seconds to tens of seconds. A total of 1502 samples (661 depression and 841 health) were obtained by eliminating 6 abnormal sample files.

D-Vlog [21]: D-Vlog consists of 961 Vlogs published on YouTube between 2020 and 2021. And ensure that the sample video is in Vlog format (a person speaks directly to the camera), and check the transcript automatically generated by the video voice content to determine whether the speaker has depression, including 555 depression samples and 406 healthy samples.

IEMOCAP [1]: IEMOCAP is a sentiment database widely used in automatic emotion recognition, which is collected by the SAIL laboratory of the University of Southern California. Each of these binary dialogues is further divided into utterances, and each of these utterances is labeled with an emotional tag. In this paper, we use this dataset to train our emotion feature extractor.

4.2   Implementation Details

The details and hyperparameters of our experiment are as follows. For the number of edge connections of nodes in different time frames inside the audio, we set the window p to 10 and s to 12 respectively. For the dataset, DAIC-WOZ, the weight matrix fusion hyperparameter β of depression features and emotional features is 0.5. The convolution layers of the two graphs are 3 and 2, respectively. For the dataset, MODMA, the weight matrix fusion hyperparameter β is 0.6. The convolution layers of the two graphs are 2 and 2, respectively. For the dataset, D-Vlog, the weight matrix fusion hyperparameter β of depression features and emotional features is 0.6. The convolution layers of the two graphs are 2 and 3, respectively. All experiments used the Adam optimizer training iteration 300 times, and the initial learning rate was 0.005.

4.3   Comparison with Existing Methods

We evaluated the performance of our model on depression data from three datasets: DAIC-WOZ, MODMA, and D-Vlog, and compared it to existing methods. To ensure the stability and reliability of our model, we used ten-fold cross-validation to allocate the training, validation, and test data in an 8:1:1 ratio. Among all the comparison methods, SVM [20] and Chen et al. [5] used traditional machine learning methods. Most of the neural network methods used in Depaudionnet [14], Rejaibi et al. [17], Yoon et al. [21], MSCDR [6], DEPA [22],

TAMFN [23], CAIINET [24] are based on CNN or RNN and their combinations or variants. The methods belonging to graph neural networks are Ghadiri et al. [7], HCAG [15], and MS2-GNN [4].

Table 1. Depression detection results of the method proposed in this paper and other comparative methods on the data set DAIC-WOZ, MODMA, and D-Vlog, where the best results are shown in bold.

Dataset	Method	Acc(%)	Pre(%)	Rec(%)	F1(%)
DAIC-WOZ	SVM [20]	-	31.6	85.7	46.2
	Depaudionnet [14]	-	**100.0**	35.0	52.0
	Rejaibi et al. [17]	76.0	69.0	35.0	46.0
	Yoon et al. [21]	-	62.57	52.63	55.45
	MSCDR [6]	77.1	-	-	66.0
	DEPA [22]	-	91.0	89.0	90.0
	Ghadiri et al. [7]	61.0	61.1	66.7	63.4
	HCAG [15]	-	77.0	83.0	80.0
	MS2-GNN [4]	89.13	80.0	85.71	82.76
	Ours	**92.21**	92.36	**92.18**	**92.23**
MODMA	Chen et al. [5]	83.4	83.5	76.8	80.0
	MSCDR [6]	85.7	-	-	84.0
	MS2-GNN [4]	86.49	82.35	87.5	84.85
	Ours	**90.35**	**88.25**	**90.33**	**89.15**
D-Vlog	Yoon et al. [21]	-	65.4	65.57	63.5
	TAMFN [23]	-	66.02	66.5	65.82
	CAIINET [24]	-	66.57	66.98	66.56
	Ours	**93.91**	**91.9**	**98.48**	**95.05**

It presents the performance of our proposed method and existing comparison methods on different datasets in Table 1. Our method outperforms existing methods on all three datasets. (i) Comparison of non-graph neural network methods: For the DAIC-WOZ dataset, the accuracy and recall rates in SVM [20], Depaudionnet [14], and Rejaibi et al. [17] differ significantly. Although the accuracy rate in Depaudionnet [14] reaches 100%, the recall rate and F1-score are only 35% and 52%, respectively. Compared to the best non-graph neural network method, our proposed method improves the F1-score by 2.23% and 5.15% on the DAIC-WOZ and MODMA datasets, respectively. The performance improvement on the D-Vlog dataset is as high as 31.92%. It is worth noting that our work is the first to apply the graph neural network method on the D-Vlog dataset, illustrating the effectiveness of our approach. (ii) Compared to graph neural network methods such as Ghadiri et al. [7], HCAG [15], and MS2-GNN [4], our proposed method also demonstrates improved performance on depression classification tasks. Our model performs 3.08% and 3.86% better than the best comparison method MS2-GNN [4] on the DAIC-WOZ and MODMA datasets, respectively.

4.4 Ablation Experiments

Table 2. Ablation experiments on each module, the best results are shown in bold.

Method	Acc(%)	Pre(%)	Rec(%)	F1(%)
w/o Emotion	89.6	87.57	89.43	88.35
w/o audio_inter	84.75	86.28	78.8	81.81
w/o audio_intra	88.4	86.56	88.52	87.53
ours	**90.35**	**88.25**	**90.33**	**89.15**

We have studied the Intra-Audio module, Emotion-Feature module, and Inter-Audio module of the model proposed in this paper in detail, and carried out ablation experiments between different modules in the data set MODMA to verify the role of different modules in our proposed model. Table 2 presents the comparison results between the full model and different versions of the model in terms of various performance metrics. It is evident that the full model significantly outperforms the other versions in all metrics. Specifically, compared to the "w/o Emotion", "w/o audio_inter", and "audio_intra" versions, the full model improves by 0.75%, 5.6%, and 1.95% in accuracy, and 0.8%, 7.34%, and 1.62% in F1-score, respectively. Our proposed model not only considers potential connections between frame-level features within and between audio signals but also takes into account mood features that are highly correlated with depression, thus enhancing the ability of the model to capture depression-related cues in audio signals.

4.5 Visualization

We performed feature visualization analysis on three datasets and visualized the original features of the three datasets and the embedding representations after model learning in this paper by the t-SNE algorithm, respectively. These visualizations are presented in Fig. 2. The discrete distribution of the raw features of the three datasets is evident from (a), (b), and (c), with different types of node features overlapping and interspersed without clear delineation. This reflects the difficulty in distinguishing between the features of the depressed group and the features of the healthy control group. However, the visualization of the embedding representation of the original features after learning by our model is shown in (d), (e), and (f), and it is clear that the distribution of features in the depression group and the healthy control group has significantly improved. This improvement increases the intra-class similarity and inter-class separability to a large extent, enabling clearer division of nodes from different classes. These results demonstrate the effectiveness of our proposed model.

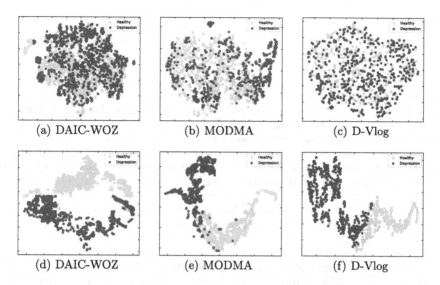

Fig. 2. The original features (a, b, c) of the three datasets DIAC-WOZ, MODMA, and D-Vlog and the features after our model learning (d, e, f) visualized by t-Distributed Stochastic Neighbor Embedding (t-SNE) projection onto a two-dimensional space.

5   Conclusion

We proposed a novel graph neural network method for recognizing MDD. Our proposed intra-audio and inter-audio modules not only considered the internal correlation of audio signals but also fully exploited the potential links between audio signals to effectively explore the depression-related cues in audio signals. Additionally, we used an emotional feature extractor to extract the emotional information of the patient's audio signal that was highly related to depression. This strengthened the connection between patients with depression, thereby improving the performance of the model. We conducted extensive experiments on multiple datasets to verify the effectiveness of our proposed model. However, our models were trained, verified, and tested on a single dataset. In practical applications, patients may come from different regions and speak different languages, presenting challenges for cross-database, cross-cultural, or cross-language depression recognition. Therefore, our next research direction will focus on addressing these challenges and achieving cross-cultural and cross-lingual depression recognition.

References

1. Busso, C., et al.: IEMOCAP: interactive emotional dyadic motion capture database. Lang. Resour. Eval. **42**, 335–359 (2008)
2. Cai, H., et al.: MODMA dataset: a multi-modal open dataset for mental-disorder analysis. arXiv preprint arXiv:2002.09283 (2020)

3. Chen, H., Jiang, D., Sahli, H.: Transformer encoder with multi-modal multi-head attention for continuous affect recognition. IEEE Trans. Multimedia **23**, 4171–4183 (2020)
4. Chen, T., Hong, R., Guo, Y., Hao, S., Hu, B.: MS^2-GNN: exploring GNN-based multimodal fusion network for depression detection. IEEE Trans. Cybern. (2022)
5. Chen, X., Pan, Z.: A convenient and low-cost model of depression screening and early warning based on voice data using for public mental health. Int. J. Environ. Res. Public Health **18**(12), 6441 (2021)
6. Du, M., et al.: Depression recognition using a proposed speech chain model fusing speech production and perception features. J. Affect. Disord. **323**, 299–308 (2023)
7. Ghadiri, N., Samani, R., Shahrokh, F.: Integration of text and graph-based features for detecting mental health disorders from voice. arXiv preprint arXiv:2205.07006 (2022)
8. Gong, Y., Poellabauer, C.: Topic modeling based multi-modal depression detection. In: Proceedings of the 7th Annual Workshop on Audio/Visual Emotion Challenge, pp. 69–76 (2017)
9. Gratch, J., et al.: The distress analysis interview corpus of human and computer interviews. Technical report, University of Southern California Los Angeles (2014)
10. Hochreiter, S., Schmidhuber, J.: Long short-term memory. Neural Comput. **9**(8), 1735–1780 (1997)
11. Huang, Z., Epps, J., Joachim, D.: Exploiting vocal tract coordination using dilated CNNs for depression detection in naturalistic environments. In: ICASSP 2020-2020 IEEE International Conference on Acoustics, Speech and Signal Processing (ICASSP), pp. 6549–6553. IEEE (2020)
12. Kessler, R.C., et al.: The epidemiology of major depressive disorder: results from the national comorbidity survey replication (NCS-R). JAMA **289**(23), 3095–3105 (2003)
13. Kipf, T.N., Welling, M.: Semi-supervised classification with graph convolutional networks. arXiv preprint arXiv:1609.02907 (2016)
14. Ma, X., Yang, H., Chen, Q., Huang, D., Wang, Y.: Depaudionet: an efficient deep model for audio based depression classification. In: Proceedings of the 6th International Workshop on Audio/Visual Emotion Challenge, pp. 35–42 (2016)
15. Niu, M., Chen, K., Chen, Q., Yang, L.: HCAG: a hierarchical context-aware graph attention model for depression detection. In: ICASSP 2021-2021 IEEE International Conference on Acoustics, Speech and Signal Processing (ICASSP), pp. 4235–4239. IEEE (2021)
16. Niu, M., Liu, B., Tao, J., Li, Q.: A time-frequency channel attention and vectorization network for automatic depression level prediction. Neurocomputing **450**, 208–218 (2021)
17. Rejaibi, E., Komaty, A., Meriaudeau, F., Agrebi, S., Othmani, A.: MFCC-based recurrent neural network for automatic clinical depression recognition and assessment from speech. Biomed. Signal Process. Control **71**, 103107 (2022)
18. Seneviratne, N., Espy-Wilson, C.: Speech based depression severity level classification using a multi-stage dilated CNN-LSTM model. arXiv preprint arXiv:2104.04195 (2021)
19. Shirian, A., Guha, T.: Compact graph architecture for speech emotion recognition. In: ICASSP 2021-2021 IEEE International Conference on Acoustics, Speech and Signal Processing (ICASSP), pp. 6284–6288. IEEE (2021)
20. Valstar, M., et al.: AVEC 2016: depression, mood, and emotion recognition workshop and challenge. In: Proceedings of the 6th International Workshop on Audio/Visual Emotion Challenge, pp. 3–10 (2016)

21. Yoon, J., Kang, C., Kim, S., Han, J.: D-vlog: multimodal vlog dataset for depression detection. In: Proceedings of the AAAI Conference on Artificial Intelligence, vol. 36, pp. 12226–12234 (2022)
22. Zhang, P., Wu, M., Dinkel, H., Yu, K.: DEPA: self-supervised audio embedding for depression detection. In: Proceedings of the 29th ACM International Conference on Multimedia, pp. 135–143 (2021)
23. Zhou, L., Liu, Z., Shangguan, Z., Yuan, X., Li, Y., Hu, B.: TAMFN: time-aware attention multimodal fusion network for depression detection. IEEE Trans. Neural Syst. Rehabil. Eng. **31**, 669–679 (2022)
24. Zhou, L., Liu, Z., Yuan, X., Shangguan, Z., Li, Y., Hu, B.: CAIINET: neural network based on contextual attention and information interaction mechanism for depression detection. Digital Signal Process. **137**, 103986 (2023)

Dynamic Gesture Recognition Based on 3D Central Difference Separable Residual LSTM Coordinate Attention Networks

Jie Chen[1], Yun Tie[1(✉)], Lin Qi[1], and Chengwu Liang[2]

[1] School of Electrical and Information Engineering, Zhengzhou University,
Zhengzhou 450001, China
ieytie@zzu.edu.cn

[2] School of ELectrical and Control Engineering, Henan University of Urban
Construction, Pingdingshan 467036, China

Abstract. The recognition of dynamic gestures has garnered significant attention in the field of human-computer interaction. However, several factors unrelated to the gestures, such as background, and spatial scale, pose significant challenges in improving the accuracy of recognition, despite the flexibility of the gestures themselves. In this paper, we propose an end-to-end recognition network, named 3D central difference separable residual long and short-term memory (LSTM) coordinate attention (3D CRLCA), which addresses these issues. Our network utilizes 3D central difference separable convolution (3D CDSC) to extract fine-grained spatiotemporal feature information, facilitating the recognition and classification of dynamic gestures. We have also incorporated a residual module in the network to enhance the discriminative ability between gesture categories. To further extract semantic and action information of the gestures, we have combined the LSTM-CA attention mechanism, which enables the network to focus on the gesture area and the temporal and spatial characteristics of gestures, to facilitate gesture recognition. Our experiments on the ChaLearn Large-scale Gesture Recognition Dataset (IsoGD) and IPN dataset demonstrate that our approach outperforms other methods.

Keywords: 3D CDSC · residual module · LSTM-CA · Dynamic gesture recognition

1 Introduction

Gesture recognition has long been a significant research field within computer vision, dynamic gesture recognition also plays a very important role in the field

This work was funded by the National Natural Science Foundation of China under Grant 62176086.

of 3D vision. Improving the recognition accuracy of dynamic gestures requires the support of computer graphics and other related technologies. In the realm of human-computer interaction (HCI), gesture recognition based on vision is crucial in various applications, such as automatic driving [14], virtual reality [11], augmented reality [8], and medical assistance [13]. Gesture recognition can be classified into two categories: static and dynamic. While significant advancements have been made in the methods for static gesture recognition, dynamic gesture recognition still faces several challenges that need to be addressed.

The limited spatial scale of gestures, or small regions of gestures in video frames, makes elements unrelated to gestures, such as the background and lighting in the input video sequence, capable of significantly affecting the recognition accuracy during dynamic gesture recognition. The high flexibility of gestures means that their spatial location can affect the conveyed message, and even subtle differences in gestures can convey vastly different meanings. Additionally, as the number of dynamic gesture categories increases, there will be more overlap between categories, leading to less differentiation between them. Currently, popular networks such as ResNet [5], SENet [7], I3D [2], C3D [17], and SlowFast [4] are commonly used as the backbone of dynamic gesture recognition networks. While these networks have achieved some success in gesture recognition, they still struggle to effectively focus on gesture regions and intra-class differences in region classification. Therefore, it is essential to develop methods to enable the network to focus on gesture regions in both space and time.

Based on the insights we have discussed earlier, we present our proposed solution for dynamic gesture recognition - the 3D CRLCA network. Our approach utilizes the 3D CDSC technique to effectively extract the spatiotemporal features of gestures in videos while minimizing resource usage. To better classify gestures, we have incorporated a residual network that enables the identification of intra-class and inter-class variations in gestures. In addition, we introduce the LSTM-CA network that employs the CA attention mechanism to guide the network's focus towards the gesture region. By processing the input video sequence through the LSTM module, we allow the network to concentrate on features with higher weight to recognize gestures accurately, enhance the overall recognition effect, and mitigate the impact of background noise. Since the input video sequence is a long time series, we ensure that the network pays attention to its long-term spatiotemporal features.

In general, the primary contributions in this work can be summarized as follows: (1) We propose a novel spatiotemporal feature extraction method called 3D Central Difference Separable Convolution (3D CDSC), which leverages the concept of central difference to extract spatial and temporal differences within frames and across frames. The 3D CDSC is designed to be separable, which reduces the number of parameters and improves resource utilization. (2) To better differentiate intra-class and inter-class variations within gestures and achieve more accurate classification, we introduce a residual module in our network. It learns residual representations and employs skip connections to alleviate the vanishing gradient problem. (3) To direct the network's attention to the gesture region, we design an LSTM-Coordinate Attention (LSTM-CA) module in our model. The module processes the input video sequence using an LSTM and employs CA attention to emphasize high-weight features for gesture recognition.

2   Related Work

2.1   ConvNets-Based Models

Simonyan, Karen and Zisserman, Andrew [15] first used a 2D convolutional neural network (CNN) for the gesture and action recognition tasks, [10,16] used 3D CNN for gesture recognition, and [22] proposed 3D separable convolution network for dynamic gesture recognition. [3] proposed 3DDSN method can be aggregated in parallel to capture embedded information that complements RGB modes. Although some of these methods have produced positive results in gesture recognition, most of them require additional structure and trainable parameters to modify the initial spatiotemporal features. In this paper, we propose a 3D-CDSC to model the rich temporal context that is essential for characterizing accurate hand and arm movements. By exploiting the concept of central difference, the 3D CDSC can efficiently extract fine-grained spatiotemporal difference information, as well as intra-frame spatial, inter-frame temporal, and spatiotemporal difference information. Moreover, the parameters can be reduced and the resource utilization can be improved by being separable.

2.2   Attention-Based Models

To improve the recognition network's attention to the gesture, researchers have proposed the use of attention mechanisms. Both [12] and [23] proposed methods for dynamic gesture recognition that incorporate attention mechanisms. [12] used a CBAM-C3D model that employs CBAM (Convolutional Block Attention Module) to focus on the gesture regions in the input video frames. On the other hand, [23] proposed the Dynamic Static Attention (DSA) module. Both [20] and [24] use AttentionLSTM method to increase the weight of gesture area to improve the recognition accuracy. These attention mechanisms help the recognition network focus on the most relevant information for accurate recognition of dynamic gestures. However, these methods require additional data and add complexity to the network. In contrast, the LSTM-CA module in our approach only requires RGB and Depth data from the original dataset and can be learned end-to-end within the recognition network. This makes it simpler to use the network in multiple contexts, and the benefits of temporal and spatial attention can be improved as LSTM-CA emphasizes two body parts in each frame and moves through subsequent frames.

3   Methods

The 3D CRLCA network is designed for dynamic gesture recognition and consists of three main components: (1) 3D CDSC, (2) Residual module, (3) LSTM-CA module. This improves the overall recognition effect and lessens the impact of background noise. Figure 1 shows the architecture of the 3D CRLCA network, which combines these three components for dynamic gesture recognition.

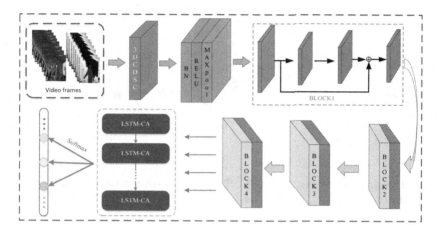

Fig. 1. The structure of the 3D CRLCA. This block diagram is mainly composed of three parts, namely 3D CDSC, residual module and LSTM-CA.

3.1 3D Central Difference Separable Convolution

CDC, or central difference convolution [19], introduces a central difference step between the sampling and aggregation processes of standard convolution, thereby enhancing its representation and generalization ability. While the sampling procedure is similar to vanilla convolution, the aggregation procedure differs significantly. Central difference convolution has a tendency to aggregate the central gradient of the sampled values. The characteristic calculation method of CDC can be expressed as follows:

$$Out_{x,y,z} = \sum w_{i,j,k} \cdot (x_{i+x,j+y,k+z} - x_{x,y,z}) \tag{1}$$

When $(i, j, k) = (0, 0, 0)$, the gradient value with respect to the central position (x, y, z) is always zero. To effectively recognize dynamic gestures, it is crucial to consider both the gradient level differential message, which emphasizes the local appearance and motion details of gestures, and the spatiotemporal intensity level semantic information, which facilitates global gesture modeling. Therefore, combining CDC with 3D convolution may offer a practical solution to enhance the robustness and distinctness of the modeling capabilities. The 3D CDC can be defined as follows:

$$Out_{x,y,z} = \sum w_{i,j,k} \cdot (K_{i+x,j+y,k+z} - K_{x,y,z}) - \theta \left(K_{x,y,z} \sum w_{i,j,k} \right) \tag{2}$$

The contribution of the information at the intensity level and gradient level is weighed by the hyperparameter $\theta \in [0,1]$. It is important to note that the weight parameter $w_{i,j,k}$ is shared by both the CDC and 3D convolution entries, and thus no additional parameters are introduced.

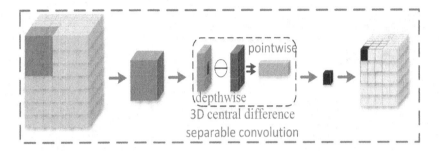

Fig. 2. The structure of the 3D CDSC. The input is treated with central difference and separated convolution and then output.

Unlike conventional 3D convolution, 3D CDSC involves primarily two steps, as depicted in Fig. 2. The 3D CDSC is a type of separable convolution that incorporates the central difference operation to improve spatiotemporal feature extraction. The central difference operation is introduced between the sampling and aggregation processes of the separable convolution. The central difference step is used to compute the difference between two adjacent frames, which captures the dynamic changes of the gesture. The central difference step is then aggregated through the separable convolution to generate the final feature map.

The characteristic calculation method of the 3D CDSC is shown as follows: (1) For the input tensor X, apply a central difference operation along the time dimension to obtain a tensor Y. (2) Apply a separable convolution to Y to obtain the output tensor Z. The separable convolution consists of a depthwise convolution followed by a pointwise convolution. The central difference operation is defined as:

$$\boldsymbol{Y}_{i,j,k,l} = \boldsymbol{X}_{i,j,k+1,l} - \boldsymbol{X}_{i,j,k-1,l} \tag{3}$$

where i, j, k, l are the batch index, channel index, time index, and spatial index, respectively. The depthwise convolution in the separable convolution is defined as:

$$\boldsymbol{Z}_{i,j,k,l} = \sum_{p=1}^{\mathcal{C}_{in}} \boldsymbol{Y}_{i,p,k,l} \cdot \boldsymbol{W}_{j,p} \tag{4}$$

where C_{in} is the number of input channels, $W_{j,p}$ is the depthwise convolution kernel for the j-th output channel and the p-th input channel. The pointwise convolution in the separable convolution is defined as:

$$\boldsymbol{O}_{i,j,k,l} = \sum_{q=1}^{C_{out}} \boldsymbol{Z}_{i,q,k,l} \cdot \boldsymbol{V}_{j,q} \tag{5}$$

where C_{out} is the number of output channels, $V_{j,q}$ is the pointwise convolution kernel for the j-th output channel and the q-th input channel. The output tensor O is then passed through a residual module and the LSTM-CA module to extract spatiotemporal features for dynamic gesture recognition.

3.2   Residual Module

This article utilizes residual networks to directly transmit information from the time and spatial dimensions of the upper layer to the next layer. The direct mapping component, represented by X, and the residual component, $F(x)$, which typically consists of two or three convolutional operations, form the residual block. The above residual blocks can be expressed as follows: $X_{l+1} = X_l + F(x)$. In order to train deep networks, the residual module transfers the gradient obtained from the back layer network to the front layer network, thereby avoiding the problem of gradient vanishing caused by excessive network depth. In addition to enabling deeper network training, this algorithm successfully captures the temporal characteristics of the relevant frames. The shallow features extracted by the residual module can fully represent the target information of the original data, facilitating differentiation between classification categories. Furthermore, deepening the network can significantly improve its ability to adapt to relevant networks, without considering model parameters.

3.3   Long Short-Term Memory Coordinate Attention

Traditional LSTM can only generate fixed-length vectors as outputs, without distinguishing the importance of information. Attention has been proposed to solve these problems. However, LSTM Attention cannot capture positional information, meaning it cannot learn the sequential order of a sequence. This results in the loss of spatial information during dynamic gesture recognition, leading to a decrease in recognition accuracy. The LSTM-CA module that we designed can effectively address these issues.

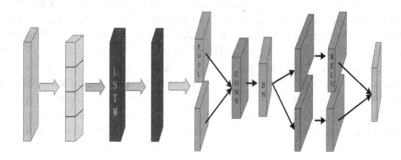

Fig. 3. The architecture of the proposed LSTM-CA model.

The LSTM-CA module is designed to capture both high-speed and low-speed gesture information during dynamic gesture detection by using a coordinate attention mechanism in addition to the long short-term memory network. While the gate mechanism in LSTM cannot accurately reflect the feature area, the CA method records channel relationships and location data to improve feature representation. This attention network was originally presented by [6] for two-dimensional picture detection and recognition, but has been enhanced for our 3D video processing and integrated with LSTM for long-time series feature extraction and concentration on gesture regions. The LSTM-CA module, shown in Fig. 3, uses the input data and node states of certain regions in the previous sequence to identify the specific shape of future associated regions. The attention mechanism is then applied to enhance performance and address the issue of spatial feature loss. The related formulas for the module are as follows:

$$Q_t = \omega \cdot (\tanh(W_{xi} * X_t + W_{hi} * H_{t-1} + b_i)) \tag{6}$$

$$A_t^{i,j} = e^{Q_t^{i,j}} / \sum \sum e^{Q_t^{i,j}} \tag{7}$$

$$Y_t = A_t \circ X_t \tag{8}$$

$$i_t = \sigma \cdot (W_{xi} * Y_t + W_{hi} * H_{t-1} + b_i) \tag{9}$$

$$t_t = \sigma \cdot (W_{xt} * Y_t + W_{ht} * H_{t-1} + b_t) \tag{10}$$

$$o_t = \sigma \cdot (W_{xo} * Y_t + W_{ho} * H_{t-1} + b_o) \tag{11}$$

$$C_t = t_t \circ C_{t-1} + i_t \circ \tanh(W_{xc}{}^* Y_t + W_{tc}{}^* H_{t-1} + b_c) \tag{12}$$

$$H_t = o_t \circ \tanh(C_t) \tag{13}$$

Where $X_1, \ldots, X_{t-1}, X_t$ is the input data, $C_1, \ldots, C_{t-1}, C_t$ is each cell's output state, $h_1, \ldots, h_{t-1}, h_t$ is the hidden state, and input gate i_t, transmission gate t_t, and output gate o_t are the gates of the convolution LSTM. σ is the sigmoid function to prevent the disappearance of gradient, $W_{x\sim}$ and $W_{h\sim}$ is the two-dimensional convolution kernel. w is the input encoding weight. "$*$" stands for convolution and "$\circ$" for the Hadamard product.

Coordinate Attention is a method that can incorporate the positional information of each element in a sequence into the attention mechanism, which can help the model better focus on important information at different positions in the sequence. In this paper, Coordinate Attention is used to weight the hidden state sequence H of LSTM, further improving the performance of the model. After applying the Coordinate Attention method to the LSTM hidden state sequence, the weighted feature sequence $Z = z_1, z_2, ..., z_T$ is obtained, where z_t represents the weighted feature representation of the first t elements in the input sequence. Specifically, the calculation formula of Coordinate Attention is as follows:

$$e_{t,i,j} = w_e^T [f_t]_{i,j} + b_e \tag{14}$$

$$\alpha_{t,i,j} = \frac{\exp(e_{t,i,j})}{\sum_{k=1}^{H} \sum_{l=1}^{W} \exp(e_{t,k,l})} \tag{15}$$

$$[z_t]_{i,j} = \sum_{k=1}^{H} \sum_{l=1}^{W} \alpha_{t,k,l} [f_t]_{k,l} \cdot [f_t]_{i,j} \tag{16}$$

Here, $[f_t]_{i,j}$ represents the feature in the i-th row and j-th column of the t-th frame in the feature sequence, and w_e and b_e are the learnable parameters in the Coordinate Attention. In the given equation, $e_{t,i,j}$ represents the attention score of the t-th element at position (i, j). Here, t denotes the time step, and i and j represent the position coordinates. Additionally, $\alpha_{t,i,j}$ signifies the attention weight of the t-th element at position (i, j), while $z_{t,i,j}$ denotes the weighted feature representation of the t-th element at position (i, j).

The weighted feature sequence is then input to a fully connected layer, yielding the final output y: $y = W_{out} z_T + b_{out}$. Where W_{out} and b_{out} represent the parameters of the fully connected layer. By combining the Coordinate Attention and LSTM modules, the model can utilize both the position information and hidden state information of each element in the sequence, thereby improving the performance of the model.

4   Experiment

To evaluate the effectiveness of our approach, we compared it to the state-of-the-art methods on the IsoGD dataset [18] using accuracy as evaluation metrics. We also performed ablation studies to analyze the contribution of each component of our network. The experimental results demonstrated that our approach outperformed the state-of-the-art methods on the IsoGD dataset, achieving higher accuracy. Moreover, the ablation studies showed that each component of our network contributed to the improvement of the recognition performance, indicating the effectiveness of our proposed approach.

4.1   Datasets

IsoGD [18] is a significant gesture dataset that has been derived from the ChaLearn Gesture Dataset (CGD), and is comprised of 47,933 RGB and Depth gesture movies. These recordings showcase 249 unique gestures, each of which has been performed by 21 distinct individuals, encompassing a range of motions, including signs, Hindi mudras, and other gestures. It is impressive that IsoGD contains a large number of RGB and Depth gesture movies, covering a wide range of gestures and participants. The variety of motions and people included in the dataset can help to evaluate the robustness and generalizability of gesture recognition methods.

The IPN gesture dataset [1] consists of RGB videos captured in 28 different scenarios, with a pixel resolution of 640×480 and a frame rate of 30 frames per second. To simulate real-world usage environments and enhance the robustness of the trained model, the background captured during data collection was intentionally designed to have lighting changes and cluttered characteristics. Moreover, the dataset includes 14 commonly used touch gesture types, with different numbers of instances allocated to each gesture type based on their frequency of occurrence in real-world usage.

4.2   Implementation

The PyTorch platform is utilized in this article to perform experimental validation. We conducted experiments on NVIDIA GTX-1080TI GPU. As there is no pre-existing model that aligns with the proposed structure, the model must be trained from scratch. Due to the copious amounts of data, the convergence rate of the model is slowed, thereby necessitating the use of small batch processing and batch processing norms to expedite training convergence. The input of the network is $12 \times 3 \times 32 \times 112 \times 112$, and the training process is calibrated using the relevant modal data derived from the IsoGD dataset. The learning rate follows a polynomial decay from 0.001 to 0.000001 within a total of 200 epochs when training on IsoGD and IPN dataset.

4.3   Analysis of Results

Table 1. Results on IsoGD dataset.

Accuracy(%)		
Methods	Modality	
	RGB	Depth
3DDSN [3]	46.08	54.95
3DCNN + ConvLSTM + 2DCNN [21]	51.31	49.81
3D Separable ConvLSTM Network [22]	52.21	51.73
AttentionLSTM [20]	55.98	53.28
Redundancy+AttentionLSTM [24]	57.42	54.18
NAS [19]	58.88	55.68
Improved YOLOv4 [9]	55.77	–
3D CRLCA (ours)	**58.52**	**56.15**

Table 1 provides a comparison of our approach with recent methods, utilizing various modal validation sets obtained from the IsoGD dataset. The results show that the proposed architecture works better than other architectures, due to its incorporation of 1) temporal and spatial characteristics of gestures on recognition, 2) the intricacies involved in acquiring the IsoGD gesture dataset, 3) long sequence gesture features, and other factors. As a result, the proposed architecture outperforms other methods in dynamic gesture recognition challenges. As shown in Table 1, the proposed method exhibits the best performance when using single RGB/Depth data. Specifically, for RGB data, our approach demonstrates approximately 0.3% lower accuracy than the NAS approach, but our approach demonstrates approximately 1% greater accuracy than the Redundancy AttentionLSTM approach. While the performance of depth data is slightly lower than that of RGB data, this may be attributed to the absence of finger detail texture in depth data. Nonetheless, the recognition results of depth data are similar to those of RGB data, and our approach outperforms the second-best approach, NAS, by approximately 0.5%.

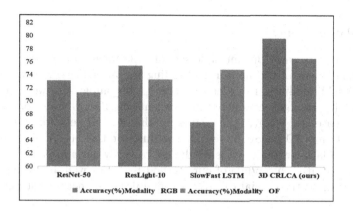

Fig. 4. Comparison of accuracy of IPN dataset.

To validate the effectiveness of the dynamic gesture recognition structure, experiments were conducted and compared on the IPN gesture dataset in this section. Since the IPN gesture dataset was recorded using only a computer's camera, only the RGB modality information is available for direct use. However, considering the importance of temporal information for dynamic gesture data and the fact that RGB data is not sufficient to provide motion information, the creators of the dataset provided optical flow information as well to supplement the motion information of the gestures. The comparison of recognition accuracy and baseline accuracy under different modalities is shown in Fig. 4.

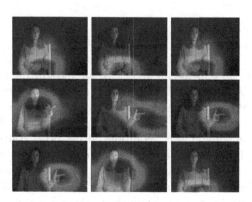

Fig. 5. Feature map in network extraction process.

In addition, the impact of the network architecture is illustrated in Fig. 5, highlighting the characteristics of the network. The visualization results indicate that the 3D CRLCA network structure considerably diminishes the response of the background and other regions unrelated to the gesture in the feature map.

4.4 Ablation Study

In this section, we carry out multiple sets of experiments to validate the effects of each component of the network, including the 3D CDSC, residual modules and LSTM-CA modules. In order to make a fair comparison, we removed the other modules and only conducted experiments on 3D CNN network and took this as the baseline. We give the results for the original network and when the network only has 3D CDSC or residual modules or LSTM-CA. As depicted in Table 2, an ablation experiment is conducted to verify our approach. The results demonstrate that each module of our method performs well in the experiment, but also highlight the crucial role of each module in the network structure.

Table 2. Performance of different modules on IsoGD dataset.

Accuracy(%)		
Methods	Modality	
	RGB	Depth
3D CNN	37.30	40.50
Res3D	45.07	48.44
3D CDSC	54.32	52.19
LSTM-CA	52.61	49.26
3D CRLCA	58.52	56.15

5 Conclusion

This paper presents an end-to-end structure for recognizing dynamic gestures, based on 3D CRLCA networks. The ResNet50 network is employed as the backbone, and 3D CDSC is used instead of 3D convolution, resulting in the network being better suited to extract precise spatiotemporal features from the video sequence. The separable convolution outperforms conventional 3D convolution due to its lower redundancy. Furthermore, an attention module called LSTM-CA was developed to help the network focus on the arm region and the motion trajectory between video sequences, while preserving early features of the long series. Experimental results demonstrate the effectiveness of the proposed method, which is compared to recent methods in the IsoGD dataset and IPN dataset.

References

1. Benitez-Garcia, G., Olivares-Mercado, J., Sanchez-Perez, G., et al.: IPN hand: a video dataset and benchmark for real-time continuous hand gesture recognition. In: 2020 25th International Conference on Pattern Recognition (ICPR), pp. 4340–4347. IEEE (2021)
2. Carreira, J., Zisserman, A.: Quo vadis, action recognition? A new model and the kinetics dataset. In: CVPR, pp. 6299–6308 (2017)
3. Duan, J., Wan, J., Zhou, S., Guo, X., Li, S.Z.: A unified framework for multi-modal isolated gesture recognition. ACM TOMM **14**(1s), 1–16 (2018)
4. Feichtenhofer, C., Fan, H., Malik, J., He, K.: Slowfast networks for video recognition. In: CVPR, pp. 6202–6211 (2019)
5. He, K., Zhang, X., Ren, S., Sun, J.: Deep residual learning for image recognition. In: CVPR, pp. 770–778 (2016)
6. Hou, Q., Zhou, D., Feng, J.: Coordinate attention for efficient mobile network design. In: CVPR, pp. 13708–13717 (2021). https://doi.org/10.1109/CVPR46437.2021.01350
7. Hu, J., Shen, L., Sun, G.: Squeeze-and-excitation networks. In: CVPR, pp. 7132–7141 (2018)
8. Jinyu, L., Bangbang, Y., Danpeng, C., et al.: Survey and evaluation of monocular visual-inertial slam algorithms for augmented reality. VRIH **1**(4), 386–410 (2019)
9. Li, D., Zhang, Z., Zhao, H.: Dynamic gesture recognition based on YOLOv4 and deep-sort methodological research. J. Intell. Fuzzy Syst. **43**, 1–11 (2022)
10. Li, Y., Miao, Q., Tian, K., et al.: Large-scale gesture recognition with a fusion of RGB-D data based on optical flow and the C3D model. Pattern Recogn. Lett. **119**, 187–194 (2019)
11. Liu, Y., Peng, M., Swash, M.R., et al.: Holoscopic 3D microgesture recognition by deep neural network model based on viewpoint images and decision fusion. THMS **51**(2), 162–171 (2021)
12. Liu, Y., Jiang, D., Duan, H., et al.: Dynamic gesture recognition algorithm based on 3D convolutional neural network. Comput. Intell. Neurosci. **2021** (2021)
13. Luo, B., Sun, Y., Li, G., Chen, D., Ju, Z.: Decomposition algorithm for depth image of human health posture based on brain health. Neural Comput. Applic. **32**(10), 6327–6342 (2020)
14. Miao, Y., Shi, E., Lei, M., et al.: Vehicle control system based on dynamic traffic gesture recognition. In: 2022 5th ICCSS, pp. 196–201. IEEE (2022)
15. Simonyan, K., Zisserman, A.: Two-stream convolutional networks for action recognition in videos. In: NIPS, vol. 27 (2014)
16. Tang, X., Yan, Z., Peng, J., et al.: Selective spatiotemporal features learning for dynamic gesture recognition. Expert Syst. Appl. **169**, 114499 (2021)
17. Tran, D., Bourdev, L., Fergus, R., et al.: Learning spatiotemporal features with 3D convolutional networks. In: CVPR, pp. 4489–4497 (2015)
18. Wan, J., Zhao, Y., Zhou, S., Guyon, I., Escalera, S., Li, S.Z.: Chalearn looking at people RGB-D isolated and continuous datasets for gesture recognition. In: CVPR Workshops, pp. 56–64 (2016)
19. Yu, Z., Zhou, B., Wan, J., et al.: Searching multi-rate and multi-modal temporal enhanced networks for gesture recognition. TIP **30**, 5626–5640 (2021)
20. Zhang, L., Zhu, G., Mei, L., et al.: Attention in convolutional LSTM for gesture recognition. In: NIPS, vol. 31 (2018)

21. Zhang, L., Zhu, G., Shen, P., et al.: Learning spatiotemporal features using 3DCNN and convolutional LSTM for gesture recognition. In: CVPR Workshops, pp. 3120–3128 (2017)
22. Zhang, X., Tie, Y., Qi, L.: Dynamic gesture recognition based on 3D separable convolutional LSTM networks. In: 11th ICSESS, pp. 180–183. IEEE (2020)
23. Zhou, B., Li, Y., Wan, J.: Regional attention with architecture-rebuilt 3D network for RGB-D gesture recognition. In: AAAI, vol. 35, pp. 3563–3571 (2021)
24. Zhu, G., Zhang, L., Yang, L., et al.: Redundancy and attention in convolutional LSTM for gesture recognition. IEEE Trans. Neural Netw. Learn. Syst. 31(4), 1323–1335 (2019)

QESAR: Query Effective Decision-Based Attack on Skeletal Action Recognition

Zi Kang⑩, Yumei Zhang⑩, Rui Zhang⑩, Yanan Jiang⑩, and Hui Xia$^{(\boxtimes)}$⑩

Computer Science and Technology, Ocean University of China, Qingdao 266100,
China
{kangzi,zym8004,zhangrui0504}@stu.ouc.edu.cn, xiahui@ouc.edu.cn,
jyniw0923@163.com

Abstract. Generating high-quality adversarial examples in a black-box setting often requires larger query volumes and adds more noticeable perturbations in skeleton action recognition tasks. We propose a query effective decision-based attack on skeletal action recognition (QESAR) to address this. We use the current gradient direction and successful historical information to reduce query volumes to update the sampling distribution, enabling a quick estimation of the next gradient direction. To ensure the imperceptibility of adversarial examples, we propose a hierarchical joint perturbation method to estimate the direction of perturbation accurately. Additionally, we design an objective function that satisfies joint angle and bone length constraints to minimize the magnitude of perturbation, ensuring that the generated adversarial examples do not exhibit noticeable distortions. Finally, in experiments, we find that QESAR can generate adversarial examples that satisfy skeletal action constraints with lower query volume. On the HDM05 and NTU datasets targeted at the ST-GCN and SGN models, QESAR achieves a 100% attack success rate while reducing the query volume by hundreds to thousands. Specifically, in the untargeted attack scenario, the QESAR scheme outperforms in terms of four metrics on the HDM05 dataset against the SGN model: average joint position deviation, average joint position acceleration deviation, average joint angle deviation, reducing them by 0.0078, 0.0679, and 6.9311, respectively.

Keywords: Deep Learning · Adversarial Attack · Adversarial Example · Decision-based Attack · Skeleton-based Action Recognition

1 Introduction

The development of deep neural networks has led to significant advancements in the research of skeleton-based action recognition models, showcasing their immense potential and value. These models analyze the movement patterns of

Supplementary Information The online version contains supplementary material available at https://doi.org/10.1007/978-981-99-8543-2_34.

human skeletal joints to accurately recognize actions, offering practical applications in various fields such as human-computer interaction, motion analysis, and health monitoring. However, skeleton-based action recognition models have also become targets for adversarial attacks. Adversarial attacks are malicious operations aimed at deceiving the model and causing it to generate incorrect predictions. In skeleton action recognition tasks, deep neural networks are often susceptible to adversarial attacks, leading to misclassifications or erroneous recognition results. This presents challenges to the reliability and security of skeleton-based action recognition models. To address these challenges, it is crucial to actively explore the potential vulnerabilities of skeleton-based action recognition models. This exploration includes designing new attack methods, which can provide valuable insights and guidance for building more reliable and secure skeleton-based action recognition systems.

Adversarial attacks can be classified into white-box attacks and black-box attacks based on the attacker's knowledge of the target model. In white-box attacks [3,7,11,12], the attacker has full knowledge of the target model's structure, parameters, and training data, and can directly generate adversarial examples against the model's internal information. However, this approach is difficult to apply to real scenarios. In black-box attacks, the attacker can only access the target model's input and output and cannot directly obtain detailed information about the model. Black-box attacks only require access to the inputs and outputs of the target model and cannot generate adversarial examples with direct knowledge of the model's details, which are further divided into transfer-based attacks and query-based attacks. Transfer-based attacks [7,12,17] use training data to train a substitute model to generate adversarial examples, but cannot guarantee the attack success rate. Query-based attacks are further divided into score-based attacks and decision-based attacks. Score-based attacks [5] attack by accessing the target model's confidence scores. Decision-based attacks [2,13,15] only rely on the final decision to attack, which is more challenging and practical. Therefore, designing a strong adversarial attack scheme with low query volume and imperceptibility is still an unsolved problem.

We propose a novel black-box attack scheme for skeleton-based action recognition, called QESAR, to reduce the query volume and enhance the imperceptibility of the attack. Experimental results show that QESAR outperforms the state-of-the-art black-box attacks on skeleton-based action recognition regarding the query volume and imperceptibility of the generated adversarial examples. The main contributions are as follows:

- QESAR is the first black-box attack scheme that generates skeletal adversarial examples based on bias sampling. Compared to other methods, it designs a bias sampling distribution that does not rely on accurate gradients, but samples the direction of the biased gradient, thereby reducing the query volume required for the attack.
- We propose a hierarchical joint perturbation method that uses different perturbation directions for different skeletal components to more accurately esti-

mate the gradient direction, thereby further improving the imperceptibility of skeletal motion.
- We design an objective function with constraints on joint angles and skeletal lengths, ensuring the imperceptibility of the attack.
- We compare the BASAR and SMART attack schemes targeting the ST-GCN and SGN target models on the HDM05 and NTU datasets. This comparison validated the QESAR low query requirement and strong imperceptibility. Specifically, for the SGN target model on the HDM05 dataset, QESAR reduces 2102 queries compared to BASAR. Regarding average joint position deviation, average joint position acceleration deviation, and average joint angle acceleration deviation, QESAR achieves reductions of 0.0078, 0.0679, and 6.9311, respectively. Furthermore, the attack success rate of QESAR reaches 100%.

2   Related Work

This section mainly introduces skeleton-based action recognition and adversarial attacks on skeleton-based action recognition from two aspects.

2.1   Skeleton-Based Action Recognition

With the advancement of deep learning technologies, the field of skeleton-based action recognition has made significant progress, finding crucial applications in computer vision, human-computer interaction, and motion analysis. Consequently, researchers increasingly want to delve deeper into skeleton-based action recognition methods. Ke et al. [4] segmented skeleton sequences into segments and incorporated long-term temporal information and spatial relationships using deep neural networks. Liu et al. [6] transformed joint sequences into image-like forms using convolutional neural networks to capture spatial and spatiotemporal information to address long-term dependencies. They also leveraged joint velocities and positions' micro and macro temporal relationships. Shiraki et al. [10] combined spatial and temporal feature networks to predict static relationships and actions' dynamic importance. Cheng et al. [1] introduced shift-map operations and lightweight point-wise convolutions to achieve a flexible receptive field for spatial and temporal graphs.

In summary, skeleton-based action recognition holds significant promise in computer vision. However, it is susceptible to adversarial attacks, necessitating further in-depth research for mitigation and resolution.

2.2   Adversarial Attacks on Skeleton-Based Action Recognition

In the context of action recognition tasks, researchers have explored various adversarial attack methods, with most focusing on the white-box setting, where the attacker has complete knowledge of the target model's information. Liu et al. [7] proposed a CIASA scheme to perturb joint positions of skeleton actions with multiple physical constraints and regularize adversarial skeletons using a generative network to ensure the naturalness of the skeletons. Zheng *et al.* [17] used ADMM to optimize the unconstrained problem of generating adversarial skeleton actions. Wang *et al.* [12] designed a perceptual loss based on the dynamics of motion and skeleton structure to ensure the imperceptibility of the attack. Tanaka *et al.* [11] perturbed the length of the skeleton in a low-dimensional environment. While white-box attacks often demonstrate better performance, the challenge for attackers to obtain all model information in reality makes black-box attacks, which do not require detailed model information, more attractive to researchers. Wei *et al.* [13] designed a heuristic black-box attack scheme that selects important frames and salient regions. Diao *et al.* [2] considered the complexity of decision boundaries and data manifolds, proposing the first black-box attack for skeleton action recognition. However, the large number of queries required by existing decision-based attacks for skeleton action recognition reduces their applicability.

We propose QESAR to generate high imperceptibility adversarial examples with low query volume. This scheme reduces the query volume for generating skeleton adversarial examples. It enhances adversarial examples imperceptibility through biased gradient direction estimation and a target function with joint angle and bone length constraints.

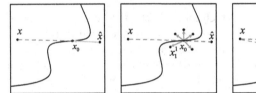

Fig. 1. A figure of the pipeline of QESAR. Where x denotes the original example and $\hat{x}$ denotes a randomly selected example different from the original example class. Firstly, perform a binary search to obtain the point x_0 on the boundary. Secondly, perform bias sampling and bias gradient direction estimation, and update the example $x_0 \rightarrow x_1^1$. Thirdly, perform a binary search again to obtain example x_1^2. Finally, apply joint angle and bone length constraints to example x_1^2 to obtain example x_1^3.

3   Scheme

This section introduces a QESAR scheme from attack setting, bias sampling, and bias gradient direction estimation. Figure 1 shows a figure of the pipeline of QESAR. The detailed algorithm is shown in Algorithm 1.

3.1   Attack Setting

The given skeletal action sequence $x = \{v_{t,j} \mid t = 1, \ldots, T, j = 1, \ldots, J\}$, where T represents the number of frames of skeletal actions, and J represents the number of skeletal joints. Let $f : \mathbb{R}^d \to \mathbb{R}^m$ be the action recognition model that predicts the label of input skeletal actions, where $x \in \mathbb{R}^d$ represents input skeletal actions, $y \in \mathbb{R}^m$ represents the true label, $f_m(x)$ represents the class probability that x belongs to the m-th class, and $c(x) = \arg\max_{i \in m} f_m(x)$ is the predicted label. Given a true skeletal action sequence x, the attacker's goal is to find an adversarial example x_{adv} that is as close as possible to the original example while causing the model to misclassify, i.e., under the untargeted setting, $c(x_{adv}) \neq y$, or under the targeted setting, $c(x_{adv}) = y_t (y_t \neq y)$.

We propose an objective function that can deceive the action recognizer while minimizing the difference between the original examples and the adversarial examples. This objective function incorporates the first-order derivatives to ensure the imperceptibility of the skeletal motion. In the untargeted setting, we can express the objective function in the joint position space as follows:

$$
\begin{aligned}
&\min_{x'_{adv}} G(x_{adv}, x) + G(x'_{adv}, x'), \\
&\text{s. t. } \rho^j_{adv} \in [\rho^j_{\min}, \rho^j_{\max}] \text{ and } b^j_{adv} = b^j \\
&c(x_{adv}) \neq y.
\end{aligned}
\tag{1}
$$

where $G(\cdot, \cdot)$ is the distance metric of joint positions, x and x_{adv} denote the original example and adversarial example, x' and x'_{adv} are the first-order derivatives of the original and adversarial examples, ρ^j_{adv} is the j-th joint angle of the adversarial example, $\rho^j_{\min}$ and $\rho^j_{\max}$ are the minimum and maximum values of the j-th joint angle, b^j and b^j_{adv} are the j-th bone length of the original and adversarial examples, respectively. Under target setting, the label $c(x_{adv}) \neq y$ is replaced with $c(x_{adv}) = y_t (y_t \neq y)$.

3.2 Bias Sampling

Most sampling-based attacks sample perturbations from a normal distribution, which means that the distribution is unbiased and independent of previous information. Although this approach is very flexible, unbiased sampling has a very low efficiency. To address this issue, we use historical gradient information to guide current sampling by adding the historical gradient information as a bias term to the sampling distribution, resulting in a new biased sampling distribution $\delta \sim N(\alpha, I)$. Here, I represents δ as a unit perturbation, α is the bias parameter, and the following formula is used to obtain:

$$
\alpha = (1 - \beta)\alpha + \beta \hat{g},
\tag{2}
$$

where $\beta \in (0, 1)$ is a key parameter that controls the degree of movement towards historical gradient information.

3.3 Bias Gradient Direction Estimation

In decision-based attacks, we use the characteristic that the classes inside and outside the decision boundary are different to estimate the gradient. By querying

---

Algorithm 1. QESAR

Input: action classifier c, original example x, original example class label y, number of iterations I, number of examples H;

Output: adversarial example x_{adv};

1: Initialize $\hat{x} \in R^d$, with $c(\hat{x}) \neq y$ (Untargeted attack) or $c(\hat{x}) = y_t$ (Targeted attack);

2: $x_0 \leftarrow$ BinarySearch$(x, \hat{x}, \varepsilon)$;

3: **for** $t = 1$ to N **do**

4:     Sampling H units of perturbation $\delta \sim N(\alpha, I)$;

5:     According to Equation (8), the estimated gradient $\hat{g}$;

6:     Set $\lambda = \frac{\|x_{i-1} - x\|_2}{\sqrt{i}}$;

7:     **while** $\varphi(x_{i-1} + \lambda\hat{g}) \neq 1$ **do**

8:         $\lambda \leftarrow \frac{\lambda}{2}$;

9:     **end while**

10:     $x_i \leftarrow$ BinarySearch$(x, x_{i-1} + \lambda\hat{g}, \varepsilon)$;

11:     **if** $\varphi(x_i) = 1$ **then**

12:         $\alpha \leftarrow (1 - \beta)\alpha + \beta g$;

13:     **end if**

14:     Update example x_i according to joint angle constraint and bone length constraint;

15: **end for**

---

the gradient direction, we can avoid estimating the exact gradient and thus reduce the query volume. We estimate the gradient direction of the example based on a Boolean function.

First, we define the function as follows:

$$Z(\tilde{x}) = \begin{cases} \max_{c \neq c^*} f_c(\tilde{x}) - f_{c^*}(\tilde{x}) & \text{(Untargeted)}, \\ f_{c_t}(\tilde{x}) - \max_{c \neq c_t} f_c(\tilde{x}) & \text{(Targeted)}. \end{cases} \quad (3)$$

where c is the predicted label, c^* is the original example label, and c_t is the target example label.

After that, we introduce a Boolean function to determine whether the attack was successful:

$$\varphi(\tilde{x}) = sign(Z(\tilde{x})) = \begin{cases} +1 & \text{if } Z(\tilde{x}) > 0, \\ -1 & \text{Otherwise}. \end{cases} \quad (4)$$

if $\varphi(\tilde{x}) = +1$, it indicates a successful attack, else $\varphi(\tilde{x}) = -1$ indicates an unsuccessful attack.

The gradient direction can be estimated based on the biased sampling distribution by sampling H times and using the following formula:

$$g = \frac{1}{H-1} \sum_{h=1}^{H} \left(\varphi(x + \delta_h) - \left(\frac{1}{H} \sum_{h=1}^{H} \varphi(x + \delta_h) \right) \right) \delta_h,$$

$$\hat{g} = \frac{g}{\|g\|_2}. \quad (5)$$

It only requires H queries. Then update the step size λ in the direction of the estimated gradient. If the attack is unsuccessful $\varphi\left(x_{i-1}+\lambda\hat{g}\right)\neq 1$, then λ is halved until the attack succeeds or stops.

Hierarchical Joint Perturbation (HJP): During the bias gradient direction estimation process, specific perturbation directions are applied to different skeletal parts to estimate the gradient direction accurately. We determine appropriate perturbation directions based on the functional roles of each skeletal segment. We categorize the skeletal structure into five parts: trunk, left upper limb, right upper limb, left lower limb, and right lower limb.

4   Experiments

This section mainly validates the effectiveness of QESAR from five aspects: experimental setting, impact of Hierarchical Joint Perturbation, untargeted attack, targeted attack, and comparison of attack success rates.

4.1   Experimental Setting

Datasets and Target Models. We compare the performance of the attack schemes on two skeleton action datasets: the HDM05 dataset [8] and the NTU60 dataset [9]. We selected two action recognition models, ST-GCN [14] and SGN [16], as our target models for evaluation on the HDM05 and NTU60 datasets.

Compare Methods. Our approach is compared with BASAR [2] and SMART [12]. BASAR is the first decision-based attack targeting skeleton motions. SMART is effective in both white-box and black-box settings but requires a substitute model in the black-box setting. We choose HCN as the substitute model.

Evaluation Metrics. We evaluate the performance of different approaches by measuring the attack success rate and average queries. Moreover, to further validate the effectiveness of our proposed method, we use four metrics, including average joint position deviation Δl, average joint position acceleration deviation $\Delta l''$, and average joint angular acceleration deviation $\Delta\rho''$, to measure the imperceptibility of adversarial examples.

$$\Delta l = \frac{1}{NTJD}\sum_{i=1}^{N}\left\|x^{(i)}-x_{adv}^{(i)}\right\|_2$$
$$\Delta l'' = \frac{1}{NTJ}\sum_{i=1}^{N}\left\|x''^{(i)}-x_{adv}''^{(i)}\right\|_2 \tag{6}$$
$$\Delta\rho^m = \frac{1}{NTJD}\sum_{i=1}^{N}\left\|\rho''^{(i)}-\rho_{adv}''^{(i)}\right\|_2$$

where N represents the number of adversarial examples, T represents the total number of frames in skeletal motion, J represents the total number of joints in the skeleton, and D represents the number of features in the skeleton.

Others. The experiments randomly select 20 actions to attack the ST-GCN and SGN action recognition models. Different iteration numbers are used to attack different action recognition models, which can better explore the performance of the attack scheme on different models and datasets. The QESAR scheme is applied with 50 iterations on the HDM05 and NTU datasets. In the experiment, during the first iteration, the hierarchical joint perturbation method is not used to obtain an initial gradient direction. However, in subsequent iterations, the hierarchical joint perturbation method is employed to more accurately estimate the gradient direction based on the initial gradient direction.

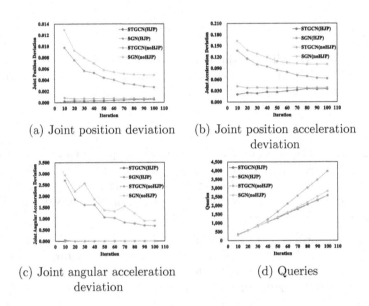

(a) Joint position deviation

(b) Joint position acceleration deviation

(c) Joint angular acceleration deviation

(d) Queries

Fig. 2. Under the untargeted attack settings, we compare the joint position deviation, joint position acceleration deviation, joint angle acceleration deviation, and query of adversarial examples generated by QESAR with and without hierarchical joint perturbation for the ST-GCN and SGN action recognition models on the HDM05 dataset. In the comparison, 'HJP' refers to QESAR with hierarchical joint perturbation, and 'noHJP' refers to QESAR without hierarchical joint perturbation.

4.2   Impact of Hierarchical Joint Perturbation

In Fig. 2, we present a performance comparison of the QESAR scheme with and without hierarchical joint perturbation, targeting the ST-GCN and SGN action recognition models under untargeted attack settings. From the graph, it can be observed that the curves of QESAR with hierarchical joint perturbation are lower than those without hierarchical joint perturbation. This indicates that using hierarchical joint perturbation improves the imperceptibility of adversarial examples at lower queries. The results in Fig. 2 demonstrate the effectiveness of the hierarchical joint perturbation method in enhancing the imperceptibility of adversarial examples for action recognition models. This method achieves more deceptive adversarial examples through accurate estimation of the perturbation gradient direction and fine adjustments in the optimization process.

Table 1. The comparison of BASAR and QESAR in terms of average joint position deviation, average joint position acceleration deviation, average joint angle acceleration deviation, and average queries in the untargeted attack.

Datasets	Models	Schemes	$\triangle l$	$\triangle l''$	$\triangle \rho''$	Q
HDM05	ST-GCN	BASAR	0.0015	0.0528	0.0000	3577
		QESAR	**0.0011**	**0.0466**	0.0000	**1270**
	SGN	BASAR	0.0097	0.1208	7.2933	3500
		QESAR	**0.0019**	**0.0529**	**0.3622**	**1398**
NTU	ST-GCN	BASAR	0.0045	0.0886	0.0616	3642
		QESAR	**0.0040**	**0.0669**	**0.0000**	**1268**
	SGN	BASAR	0.0082	0.1140	0.0000	3383
		QESAR	**0.0030**	**0.0715**	0.0222	**1623**

4.3   Untargeted Attack

Table 1 compares the performance of the QESAR and BASAR schemes in generating adversarial examples for the ST-GCN and SGN models under an untargeted attack scenario. From the table, it can be observed that QESAR outperforms BASAR in terms of average joint position deviation, and average joint angular acceleration deviation, indicating that QESAR's adversarial examples are more difficult to detect by human eyes. Specifically, on the HDM05 dataset, QESAR reduces the average joint position deviation by 0.0078, the average joint angular acceleration deviation by 0.0679 for the SGN model, and the average joint position acceleration deviation by 6.9311. Moreover, QESAR is more efficient in terms of queries, reducing the average queries by 2102 for the ST-GCN model on the HDM05 dataset, compared to BASAR. These improvements are due to the deviation gradient direction estimation method adopted by the QESAR attack scheme, which does not require accurate gradient estimation and reduces the number of queries. At the same time, QESAR uses a manifold function with joint angle constraints and bone length constraints to ensure the imperceptibility of the adversarial examples.

4.4   Targeted Attack

Table 2 shows the performance comparison of QESAR and BASAR attack schemes in terms of four evaluation metrics, including average queries, average joint position deviation, and average joint angular acceleration deviation, on the adversarial examples generated by these schemes for ST-GCN and SGN action recognition models in the HDM05 and NTU datasets under the targeted attack setting. The table indicates that QESAR outperforms BASAR in four imperceptible indicators, including average joint position deviation, and average joint angular acceleration deviation, indicating that the adversarial examples generated by QESAR are more difficult to be detected by human eyes. Particularly, on the NTU dataset, QESAR's average joint position deviation for the SGN model is 0.1447 less than that of BASAR, the average joint position acceleration deviation is reduced by 0.6854, and the average joint angle acceleration deviation is reduced by 5.8969. Moreover, QESAR is more efficient in terms of query volume, reducing the average queries by 799 times for BASAR on the HDM05 dataset against the ST-GCN model. Comparing Table 1 and Table 2, it can be seen that in untargeted attacks, QESAR outperforms BASAR in terms of average joint position deviation, average joint position acceleration deviation, and average joint angular acceleration deviation, while also requiring fewer average queries. It is worth noting that targeted attacks require generating adversarial examples specific to certain target classes, whereas untargeted attacks do not have such a requirement.

Table 2. The comparison of BASAR and QESAR in terms of average joint position deviation, average joint position acceleration deviation, average joint angle acceleration deviation, and average queries in the targeted attack.

Datasets	Models	Schemes	$\triangle l$	$\triangle l''$	$\triangle \rho''$	Q
HDM05	ST-GCN	BASAR	0.2412	0.6116	1.4378	4043
		QESAR	**0.1928**	**0.5174**	11.579	**2165**
	SGN	BASAR	0.2748	1.1064	6.9994	3478
		QESAR	**0.1301**	**0.4210**	**1.1025**	**2679**
NTU	ST-GCN	BASAR	0.0500	0.1971	0.0911	3728
		QESAR	**0.0364**	**0.1532**	**0.0512**	**2286**
	SGN	BASAR	0.0390	0.2508	0.0000	3440
		QESAR	**0.0200**	**0.1920**	0.3334	**2431**

4.5 Comparison of Attack Success Rates

In Table 3, we present the comparison of attack success rates for three attack schemes, BASAR, SMART, and QESAR, targeting the ST-GCN and SGN action recognition models in both the HDM05 and NTU datasets under untargeted and targeted attack settings. From the table, it can be observed that QESAR achieves a success rate of 100%, indicating that the generated adversarial examples by QESAR are more difficult to perceive by human eyes. For instance, on the HDM05 dataset, both QESAR and BASAR achieve a 100% attack success rate, while SMART has lower success rates of 25% and 31% respectively. This demonstrates that QESAR exhibits a significant advantage in terms of attack success rates against the ST-GCN and SGN models on the HDM05 and NTU datasets.

Table 3. Comparison of attack success rates of three attack schemes, BASAR, SMART, and QESAR, on HDM05 and NTU datasets for ST-GCN and SGN action recognition models under untargeted and targeted attack settings.

T/U	Datasets	Models	Schemes		
			SMART	BASAR	QESAR
Untargeted	HDM05	ST-GCN	25%	100%	**100%**
		SGN	31%	100%	**100%**
	NTU	ST-GCN	24%	100%	**100%**
		SGN	88%	100%	**100%**
Targeted	HDM05	ST-GCN	1%	100%	**100%**
		SGN	2%	100%	**100%**
	NTU	ST-GCN	1%	100%	**100%**
		SGN	2%	100%	**100%**

5   Conclusion

We propose a novel black-box attack scheme called QESAR, which can reduce the query volume of the attack and increase the imperceptibility for skeleton action recognition tasks. We design a bias sampling method that can effectively limit the range of gradient directions, thus reducing the query volume required for attacking skeleton action examples. Additionally, we propose a hierarchical joint perturbation method that employs different perturbation directions for different skeletal parts, optimizing the estimated gradient direction to enhance the imperceptibility of skeletal motion further. At the same time, we propose an objective function with joint angle and bone length constraints, which can coordinate skeleton actions to ensure the imperceptibility of the attacked skeleton actions. Experimental results show that the QESAR performs better than the current state-of-the-art black-box attack schemes for skeleton action recognition regarding query volume and the imperceptibility of generated adversarial examples. We discuss the vulnerability of skeleton action recognition models, hoping to provide insights for future security research.

Acknowledgements. This research is supported by the National Natural Science Foundation of China (NSFC) [grant numbers 62172377, 61872205], the Shandong Provincial Natural Science Foundation [grant number ZR2019MF018], and the Startup Research Foundation for Distinguished Scholars No. 202112016.

References

1. Cheng, K., Zhang, Y., He, X., Chen, W., Cheng, J., Lu, H.: Skeleton-based action recognition with shift graph convolutional network. In: Proceedings of the IEEE/CVF Conference on Computer Vision and Pattern Recognition, pp. 183–192 (2020)
2. Diao, Y., Shao, T., Yang, Y.L., Zhou, K., Wang, H.: BASAR: black-box attack on skeletal action recognition. In: Proceedings of the IEEE/CVF Conference on Computer Vision and Pattern Recognition, pp. 7597–7607 (2021)
3. Hwang, J., Kim, J.H., Choi, J.H., Lee, J.S.: Just one moment: structural vulnerability of deep action recognition against one frame attack. In: Proceedings of the IEEE/CVF International Conference on Computer Vision, pp. 7668–7676 (2021)
4. Ke, Q., Bennamoun, M., An, S., Sohel, F., Boussaid, F.: A new representation of skeleton sequences for 3d action recognition. In: Proceedings of the IEEE Conference on Computer Vision and Pattern Recognition, pp. 3288–3297 (2017)
5. Li, J., et al.: Projection & probability-driven black-box attack. In: Proceedings of the IEEE/CVF Conference on Computer Vision and Pattern Recognition, pp. 362–371 (2020)
6. Liu, J., Akhtar, N., Mian, A.: Skepxels: spatio-temporal image representation of human skeleton joints for action recognition. In: CVPR Workshops, pp. 10–19 (2019)
7. Liu, J., Akhtar, N., Mian, A.: Adversarial attack on skeleton-based human action recognition. IEEE Trans. Neural Netw. Learn. Syst. **33**(4), 1609–1622 (2020)
8. Müller, M., Röder, T., Clausen, M., Eberhardt, B., Krüger, B., Weber, A.: Documentation mocap database HDM05. Technical report CG-2007-2 (2007)
9. Shahroudy, A., Liu, J., Ng, T.T., Wang, G.: NTU RGB+ D: a large scale dataset for 3d human activity analysis. In: Proceedings of the IEEE Conference on Computer Vision and Pattern Recognition, pp. 1010–1019 (2016)
10. Shiraki, K., Hirakawa, T., Yamashita, T., Fujiyoshi, H.: Spatial temporal attention graph convolutional networks with mechanics-stream for skeleton-based action recognition. In: Proceedings of the Asian Conference on Computer Vision (2020)
11. Tanaka, N., Kera, H., Kawamoto, K.: Adversarial bone length attack on action recognition. In: Proceedings of the AAAI Conference on Artificial Intelligence, vol. 36, pp. 2335–2343 (2022)
12. Wang, H., et al.: Understanding the robustness of skeleton-based action recognition under adversarial attack. In: Proceedings of the IEEE/CVF Conference on Computer Vision and Pattern Recognition, pp. 14656–14665 (2021)
13. Wei, Z., et al.: Heuristic black-box adversarial attacks on video recognition models. In: Proceedings of the AAAI Conference on Artificial Intelligence, vol. 34, pp. 12338–12345 (2020)
14. Yan, S., Xiong, Y., Lin, D.: Spatial temporal graph convolutional networks for skeleton-based action recognition. In: Thirty-Second AAAI Conference on Artificial Intelligence (2018)
15. Ye, M., Miao, C., Wang, T., Ma, F.: TextHoaxer: budgeted hard-label adversarial attacks on text. In: Proceedings of the AAAI Conference on Artificial Intelligence, vol. 36, pp. 3877–3884 (2022)

16. Zhang, P., Lan, C., Zeng, W., Xing, J., Xue, J., Zheng, N.: Semantics-guided neural networks for efficient skeleton-based human action recognition. In: Proceedings of the IEEE/CVF Conference on Computer Vision and Pattern Recognition, pp. 1112–1121 (2020)
17. Zheng, T., Liu, S., Chen, C., Yuan, J., Li, B., Ren, K.: Towards understanding the adversarial vulnerability of skeleton-based action recognition. arXiv preprint arXiv:2005.07151 (2020)

A Closer Look at Few-Shot Object Detection

Yuhao Liu, Le Dong[✉], and Tengyang He

Department of Computer Science and Technology, University of Electronic Science and Technology of China, Chengdu, China
202221080910@std.uestc.edu.cn, ledong@uestc.edu.cn

Abstract. Few-shot object detection, which aims to detect unseen classes in data-scarce scenarios, remains a challenging task. Most existing works adopt Faster RCNN as the basic framework and employ fine-tuning paradigm to tackle this problem. However, the intrinsic concept drift in the Region Proposal Network and the rejection of false positive region proposals hinder model performance. In this paper, we introduce a simple and effective task adapter in RPN, which decouples it from the backbone network to obtain category-agnostic knowledge. In the last two layers of the task adapter, we use large-kernel spatially separable convolution to adaptively detect objects at different scales. In addition, We design an offline structural reparameterization approach to better initialize box classifiers by constructing an augmented dataset to learn initial novel prototypes and explicitly incorporating priors from base training in extremely low-shot scenarios. Extensive experiments on various benchmarks have demonstrated that our proposed method is significantly superior to other methods and is comparative with state-of-the-art performance.

Keywords: Few-shot learning · Object detection · Few-shot object detection · Transfer learning

1 Introduction

Deep learning has witnessed remarkable achievements over the past decade [8,14,36]. However, the performance of deep learning models heavily relies on the availability of annotated training data. In contrast, human intelligence has the remarkable capability to learn new concepts with little supervision. This motivates the research in Few-shot Learning, which aims to develop models capable of learning from a small amount of data [6]. Few-shot object detection (FSOD) aims at detecting unseen classes with few annotated examples. Current FSOD methods employ two primary methodologies: meta-learning and transfer learning. Meta-learning approaches seek to enhance model adaptation to unseen categories by constructing diverse tasks. However, meta-learning is not computationally friendly, requiring simultaneous processing of both support and query

T. He—Supported by National Key R&D Program of China (2022ZD0114900).

Q. Liu et al. (Eds.): PRCV 2023, LNCS 14432, pp. 430–447, 2024.
https://doi.org/10.1007/978-981-99-8543-2_35

images. On the other hand, transfer learning provides a more feasible approach by utilizing fine-tuning paradigms to address data distributions between seen and unseen categories. Due to data scarcity, transfer learning is prone to overfitting unseen categories, leading to suboptimal performance.

FSOD methods usually adopt Faster RCNN [26] as the basic framework [1,10,29,35]. Region Proposal Networks (RPN) are utilized for foreground object detection, while main networks perform fine-grained classification in traditional object detection systems. Typically, RPN is assumed to be category-agnostic, as their training and testing share the same class space. However, there is a significant shift in the definition of foreground concepts, which we define as "**concept drift**". Unseen categories need to be rectified as foreground during fine-tuning despite being considered as background during pre-training. The limited availability of samples poses challenges for RPN to adapt effectively to this concept drift through fine-tuning.

Another challenge in object detection lies in preventing RPN's detection of false positives. Once a category-agnostic RPN is obtained, the main network needs to leverage prior knowledge to facilitate learning new classes, effectively minimizing false positives within current categories.

Based on the analysis, we introduce a simple yet effective task adapter to decouple the main network and RPN tasks to address the "concept drift" issue. The task adapter incorporates a series of residual modules and two large-kernel separable convolutions to enhance the detection of foreground objects at different scales. We integrate the task adapter with the region proposal network to form a unified framework called **D**ecoupled large kernel **R**egion **P**roposal **N**etwork (**DearRPN**). To leverage the knowledge acquired by the detector on seen categories, we propose an **O**ffline **S**tructural **R**eparameterization (**OSR**) method. This method constructs an augmented dataset to learn prototypes for novel categories and explicitly reparameterize knowledge from the base class to initialize novel class classifiers.

- To alleviate the inherent "concept drift" problem in few-shot object detection, we propose a simple yet effective large-kernel task adapter, which improves RPN discrimination ability for unseen foreground objects.
- We design an offline structural reparameterization method to utilize few images from unseen categories and learned prior knowledge, which enhances detector's generalization capability in the extremely low-shot scenario.
- Extensive experiments on different benchmark datasets demonstrate the effectiveness of our approach, especially in extremely low-shot settings.

2   Related Work

2.1   Conventional Object Detection

Object detection is a crucial and formidable task in computer vision. Its primary objective is to accurately identify the locations and categories of all objects of interest within an image. Currently, object detection methods are commonly classified into two major categories. Two-stage detectors, such as Faster R-CNN [26] and Mask R-CNN [13], typically employ Region Proposal Networks (RPN) to generate initial bounding box proposals for foreground regions. These proposals are then subjected to regions-of-interest (RoI) pooling, enabling the extraction of box features for subsequent classification and bounding box regression tasks on each proposed region. Another widely adopted concept is the one-stage detector. Another widely embraced approach is the one-stage detector. In contrast to two-stage detectors, one-stage detectors, such as YOLO (You Only Look Once) [25] and RetinaNet [20], offer the advantages of simplicity in implementation and faster detection speed. These methods directly scan the image in a dense manner, extracting the object's category probability and location coordinates. However, they may trade off some accuracy in comparison to two-stage detectors.

2.2   Few-Shot Learning

Few-shot learning aims to develop models capable of generalizing quickly to new tasks with only a few samples. Over the past decade, significant efforts have been devoted to this field [2,6,7,12,23,28]. There are two primary approaches: data-based methods and model-based methods. Data-based methods are based on the simple idea of performing data augmentations during training [7,12,23]. Model-based methods are one of the most widely studied few-shot learning methods. Among them, meta-learning methods use prototype learning [28] and metric learning [30] to achieve few-shot classification by constructing a framework that can train in an end-to-end fashion. Few-shot learning is a promising research field, which is meaningful for some highly specialized fields.

2.3   Few-Shot Object Detection

To ease dependence on data, few-shot object detection aims to learn from few object instances of new categories in the domain of unseen categories. Recent literature has proposed several approaches [3,10,16,21,24,31,35,39] in this field. Meta-learning-based few-shot object detection mainly adopts a siamese architecture to encode the support and query images and then uses a query-based pipeline to perform a densely visual search on support features with query features, such as Meta Faster R-CNN [10]. Transfer learning-based few-shot object detection approaches [3,24,31,35] have also shown their potential. They are usually built on Faster RCNN and employ pre-training and fine-tuning paradigms to solve few-shot object detection. LSTD [3] devises a background suppression and knowledge transfer regularization term to penalize noise in model training so that the model pays more attention to regions of interest. MPSR [35] concentrates on the scale variation problem in the same class of objects. It incorporates Feature Pyramid Network (FPN) into the Faster R-CNN framework and proposes Multi-Scale Positive Sample Refinement. This method employs an auxiliary branch to extract and resize samples to different sizes, refining the detections.

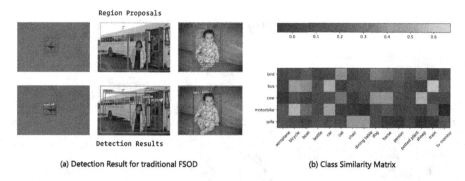

Fig. 1. Visualizations of detection results and category similarity. Figure (a) visualizes the detection results of the traditional fine-tuning-based Faster RCNN. Green boxes represent GTs, red boxes are the region proposals of the RPN, and blue boxes denote detection results. Figure (b) represents category similarity between base and novel classes of Pascal VOC Split 1, where the y-axis is the novel classes, and the x-axis denotes the base classes. (Color figure online)

3   Approach

Our proposed Revised RCNN (ReRCNN) is based on Faster R-CNN, which consists of a main network and a region proposal network (RPN). RPN extracts foreground objects, while the main network comprises a feature extractor and a detection head. The feature maps extracted by the feature extractor are shared by both RPN and the detection head. The detection head is used for fine-grained classification of the features obtained from the region proposal network. The overall framework of ReRCNN is shown in Fig. 2. We first pre-train the entire ReRCNN network on base classes with abundant data and perform offline structural reparameterization. We then finetune ReRCNN on few novel samples.

3.1   Problem Definition

Few-shot object detection aims to detect novel classes unseen during training with little supervision [32]. Formally, for a given dataset $\mathcal{D}$ contains $\mathcal{C}$ classes, we divide it into two subsets $\mathcal{D}_{base}$ ($y_i \in \mathcal{C}_{base} \subset \mathcal{C}$) and $\mathcal{D}_{novel}$ ($y_i \in \mathcal{C}_{novel} \subset \mathcal{C}$) whose class spaces are disjoint, namely $\mathcal{C}_{base} \cap \mathcal{C}_{novel} = \emptyset$. The base set $\mathcal{D}_{base}$ contains abundant data commonly used for model pre-training, while novel set $\mathcal{D}_{novel}$ contains only few annotated samples. It is worth mentioning that since an image may contain multiple objects, the number of objects is not necessary to match the number of images exactly. For novel objects in the base set or base objects in the novel set, common practice usually view them as background.

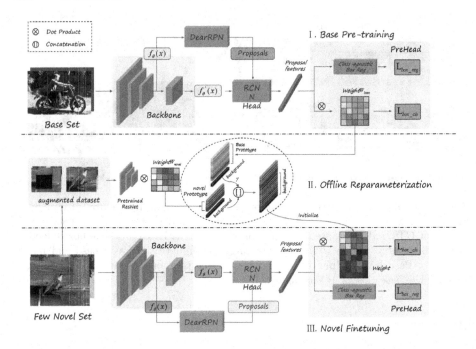

Fig. 2. Flowchart of proposed **Revised RCNN (ReRCNN)**. Only the first few layers of the backbone (the yellow box) are frozen during pre-training, and the box classifier (the green box) is unfrozen during fine-tuning. (Color figure online)

3.2   DEcoupled lArge KeRnel RPN

As discussed above, concept drift causes RPN to falsely detect base objects as an object of interest (false positive) while regarding novel objects as background in the fine-tuning stage, particularly in data-scarcity scenarios. As shown in Fig. 1(a), due to the presence of airplanes and people in the base class, the detector stubbornly detects airplanes and people as foreground objects. During base training phase, shared feature maps misled RPN to learn base objects as the foreground. During fine-tuning, it is difficult to correct this inductive bias due to data limitations. In common FSOD practice, feature extractors are usually pre-trained on classification tasks, resulting in extracted features with strong category-aware priors, which poses challenges for RPN to learn category-agnostic knowledge.

We devised a Decoupled large kernel RPN, denoted as DearRPN, to detect natural foreground objects instead of just objects of the base classes. DearRPN introduces a task adapter to extract category-agnostic semantics, thus achieving decoupling with the main network. Instead of employing cumbersome feature pyramid networks, we turn to two large kernel convolutions for objects with extreme scales as [4, 22] pointed out that large convolution kernels do not increase FLOPs much but can significantly improve the performance of the model. Large

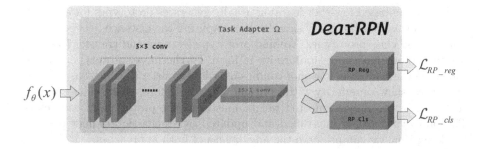

Fig. 3. Illustration of proposed DearRPN.

convolution kernels have a larger receptive field and can better deal with objects of different aspect ratios. We illustrate the structure of DearRPN in Fig. 3.

More formally, for a given image $x_i \in \mathbb{R}^{H \times W \times C}$, we first extract feature maps with feature extractor f_θ parameterized by θ. The feature maps are then fed into task adapter Ω, a twenty-two-layer residual convolutional network, to obtain category-agnostic semantic features. In the last two layers of the adapter, spatial-separable large kernel convolution of kernel size 15×1 and 1×15 are utilized to replace the 15×15 convolution kernel, significantly reducing learnable parameters. We experiment with different kernel sizes in Appendix D. Then we use region proposal regressor and region proposal classifier to obtain the coordinate shift $o_i = (x_i, y_i, w_i, h_i)$ and confidence of proposals p_i as in Faster RCNN [26].

$$
\begin{aligned}
p_i &= cls(\Omega(f_\theta(x_i))) \\
o_i &= reg(\Omega(f_\theta(x_i)))
\end{aligned}
\tag{1}
$$

We alternate Smooth L1 loss for the regression branch and binary cross-entropy for classification branch. Finally, the total loss of DearRPN is defined as follows:

$$
\mathcal{L}_{DearRPN} = \lambda \mathcal{L}_{RP\_reg} + \mathcal{L}_{RP\_cls}
\tag{2}
$$

λ is an adjustment coefficient that adjusts the regression contribution.

3.3 Offline Structural Reparameterization

Due to disjoint class space, the parameters of box classifier are usually initialized randomly during fine-tuning. Due to data scarcity, training these parameters from scratch is prone to overfitting. Meta-learning claims that it enables models to learn by better initialization with prior knowledge so that detectors can generalize to new tasks with fewer gradient descent steps. On the other hand, we noticed from Fig. 1(b) that the base class and novel classes are always related, so we devise an Offline Structural Reparameterization (OSL) to better initial the fine-tuning parameters with prior knowledge from base set.

Prototype-based few-shot learning usually generates prototypes for each category and classifies the query samples by computing the distance between query samples and class prototypes. We find prototype learning and transfer learning-based methods have some uniformity in form. Furthermore, prototype learning usually performs well in low-shot conditions, while transfer learning-based methods always show superiority with sufficient data. Therefore, we consider each row vector w_i of the fully connected weight matrix as the prototype of each class. The prototype is then fine-tuned and updated by stochastic gradient descent. By doing this, we can preserve the advantages of both approaches.

Our proposed offline structural reparameterization treats fine-tuning as a prototype learning process. Through pre-training on base classes, we obtain prototypes for base classes. Also, we can get prototypes of novel classes by pre-training on few novel samples. Concretely, we construct an augmented dataset with few novel samples. In this dataset, objects from novel classes cropped from images are labeled as their corresponding classes, while objects from base classes are classified into background classes. We also mask all objects from the original image and regard the masked image as background. We then use a pre-trained ResNet, the same structure as the backbone, and a randomly initialized classifier to perform classification on the augmented dataset. Formally, we first pre-train classifier ϕ_ω on the augmented dataset $\mathcal{D}_{aug} = \{(x'_1, y'_1), (x'_2, y'_2), \ldots, (x'_i, y'_i)\}$, $y = 0, 1, \ldots, K$, and the K-th category is background. Then we concatenate the pre-trained classifier weights W_{novel} with the base-trained box classifier weights W_{base}. When fine-tuning, we initialize the box classifier of ReRCNN with reparameterized classifier W_{rep} and fine-tune ReRCNN with reparameterized box classifier. According to the few-shot setting Sect. 4.1, we consider all base prototypes as background. This can be summarized as:

$$argmin \, \mathcal{L}_{CE}(y_i, W_{novel} \cdot f_{\theta'}(x'_i))$$
$$W_{rep} = Concat(W_{base}, \gamma \cdot W_{novel}) \tag{3}$$

where $f_{\theta'}$ is a pre-trained ResNet parameterized by θ'. γ is a temperature parameter used to balance the contribution of the base classes. Under the traditional few-shot learning setting, we explicitly classify all base classes into the background.

4 Experiments

4.1 Experimental Setup

Datasets. We follow previous work [29,32,35] and conduct experiments on Pascal VOC and MS COCO. We randomly divide the 20 classes three times to form Split 1, 2, and 3, and k is set to 1, 2, 3, 5, and 10 on Pascal VOC. As for MSCOCO, we divide it into the base set with 60 classes and the novel set with 20 classes and also conduct experiments on sampled 20 way-K shot tasks, where $K = 1, 2, 3, 5, 10, 30$. To follow the setting of most previous work, we report AP50 on Pascal VOC and mAP on MSCOCO, respectively.

Table 1. Experimental results on the PASCAL VOC dataset. We evaluate the AP50 of our proposed ReRCNN on different splits of Pascal VOC and different shots of each split. The best results are shown in **bold** and the second is underlined.

Method	Backbone	Novel set Split 1					Novel set Split 2					Novel set Split 3				
		1	2	3	5	10	1	2	3	5	10	1	2	3	5	10
LSTD [3]	VGG-16	8.2	1.0	12.4	29.1	38.5	11.4	3.8	5.0	15.7	31.0	12.6	8.5	15.0	27.3	36.3
RepMet [17]	InceptionV3	26.1	32.9	34.4	38.6	41.3	17.2	22.1	23.4	28.3	35.8	27.5	31.1	31.5	34.4	37.2
YOLO-ft [33]	YOLO V2	6.6	10.7	12.5	24.8	38.6	12.5	4.2	11.6	16.1	33.9	13.0	15.9	15.0	32.2	38.4
Meta R-CNN [37]	FRCN-R101	19.9	25.5	35.0	45.7	51.5	10.4	19.4	29.6	34.8	45.4	14.3	18.2	27.5	41.2	48.1
TFA w/ cos [32]	FRCN-R101	39.8	36.1	44.7	55.7	56.0	23.5	26.9	34.1	35.1	39.1	30.8	34.8	42.8	49.5	49.8
FSOD-KT [18]	FRCN-R101	27.8	41.4	46.2	55.2	56.8	19.8	27.9	38.7	38.9	41.5	29.5	30.6	38.6	43.8	45.7
MPSR [35]	FRCN-R101	41.7	–	51.4	55.2	61.8	24.4	–	39.2	39.9	47.8	35.6	–	42.3	48.0	49.7
FSOD-UP [34]	FRCN-R101	43.8	47.8	50.3	55.4	61.7	31.2	30.5	41.2	42.2	48.3	35.5	39.7	43.9	50.6	53.5
DCNet [15]	FRCN-R101	33.9	37.4	43.7	51.1	59.6	23.2	24.8	30.6	36.7	46.6	32.3	34.9	39.7	42.6	50.7
SRR-FSD [43]	FRCN-R101	47.8	50.5	51.3	55.2	56.8	32.5	35.3	39.1	40.8	43.8	40.1	41.5	44.3	46.9	46.4
TFA+DR+DC [42]	FRCN-R101	46.7	53.1	53.8	61.0	62.1	30.1	34.2	41.6	41.9	44.8	41.0	46.0	47.2	55.4	55.6
KFSOD [40]	FRCN-R101	44.6	–	54.4	60.9	65.8	**37.8**	–	43.1	48.1	50.4	34.8	–	44.1	52.7	53.9
FCT [11]	PVTv2-B2-Li	49.9	57.1	57.9	63.2	**67.1**	27.6	34.5	43.7	**49.2**	51.2	39.5	**54.7**	52.3	57.0	**58.7**
HTRPN [27]	FRCN-R101	47.0	48.8	53.4	62.9	65.2	29.8	32.6	**46.3**	47.7	**53.0**	40.1	45.9	49.6	**57.0**	**59.7**
PTF+KI [38]	FRCN-R101	41.5	48.7	49.2	57.6	60.9	26.0	34.4	39.6	40.1	46.4	37.1	44.4	44.6	53.2	54.9
ReRCNN	FRCN-R101	**56.9**	**59.9**	**62.8**	**64.6**	65.4	31.4	**36.5**	39.3	40.6	43.0	**49.1**	51.8	**53.7**	54.9	55.9

Implementation Details. Our method adopts Faster RCNN as the basic framework, which adopts ResNet-101 pre-trained on ImageNet as the backbone. We use stochastic gradient descent with a weight decay of $5e-5$ and a base learning rate of 0.02 to optimize our parameters. The batch size of the mini-Batch is set to 16 in our experiments. For γ, we use 0.7 for Pascal VOC 1 and 0.47 for MSCOCO. The setting of gamma we conduct experiments in Appendix E.

4.2   Experimental Results

Result on Pascal VOC. Our experimental results on Pascal VOC are shown in Table 1. Under the few-shot object detection setting, the results demonstrate that our approach achieves excellent performance, particularly in extremely low data scenarios, such as 1-shot. Notably, on 1 shot of Split 1 and 3, our method (ReRCNN) achieves an improvement of 7% and 8.1%, respectively, significantly surpassing the second-place method. However, we also observe that AP50 does not increase linearly as the number of shots increases. Take novel set Split 1 as an example, AP50 increased by 3% from 1 shot to 2 shot, but from 5 shot to 10 shot, AP50 only increase 0.9%. We attribute this to two factors: Firstly, the performance of fine-tuning primarily depends on the quality of available support samples. Secondly, as the number of shots increases, the box classifier can learn meaningful semantics without overfitting the novel set. Different class splits greatly influence detector performance. For example, the performance of Split 1 is always better than that of Split 2. This may be due to the different class similarities between different splits. When there is a large disparity between base classes and novel classes, the detector may struggle to generalize effectively.

Table 2. Experimental results on the MS COCO dataset. We evaluate the nAP of our proposed ReRCNN on different shots of MSCOCO. The superscript * indicates that the results refer from [24]

Method	Backbone	Novel set					
		1	2	3	5	10	30
LSTD [3]	VGG-16	–	–	–	–	3.2	6.7
FSRW [16]	YOLO V2	–	–	–	–	5.6	9.1
FRCN-ft [33]	FRCN-R101	1.0*	1.8*	2.8*	4.0*	6.5	11.1
Meta R-CNN [37]	FRCN-R101	–	–	–	–	8.7	12.4
Meta-RetinaNet [19]	RetinaNet R-18	–	–	–	–	9.7	13.1
TFA w/cos [32]	FRCN-R101	4.4*	5.4*	6.0*	7.7*	10.0	13.7
FSCE [29]	FRCN-R101	–	–	–	–	11.9	16.4
MPSR [35]	FRCN-R101	5.1*	6.7*	7.4*	8.7*	9.8	14.1
GenDet [21]	FRCN-R101	–	–	–		9.9	14.0
QA-FewDet [9]	FRCN-R101	4.9	7.6	8.4	9.7	11.6	16.5
SRR-FSD [43]	FRCN-R101	–	–	–	–	11.3	14.7
CoRPNs [41]	FRCN-R101	4.4	5.6	7.2	–	–	–
Meta FRCN [10]	FRCN-R101	5.1	7.6	9.8	10.8	12.7	16.6
TFA+DR+DC [42]	FRCN-R101	5.7	7.1	8.6	–	12.5	_17.1_
FCT [11]	PVTv2-B2-Li	_5.6_	_7.9_	_11.1_	**14.0**	**17.1**	**21.4**
PTF+KI [38]	FRCN-R101	–	–	–	–	11.8	16.0
ReRCNN	FRCN-R101	**7.9**	**9.1**	**11.4**	_11.5_	_13.8_	16.7

Result on MSCOCO. MSCOCO poses greater difficulty for few-shot detection due to its larger number of classes. Our experimental results are shown in Table 2. Experimental results demonstrate the effectiveness of our approach. In the case of an extremely low-shot setting, our method has a 6.9% improvement over the fine-tuned Faster RCNN. Even compared to the second-placed method, our method still shows a 2.8% improvement. The FCT method performs better as it employs a larger backbone and incorporates additional priors. As the number of accessible samples increases, so does the AP50 of the detector. From the parameter point of view, we reckon that our proposed method, ReRCNN, has fewer parameters to be fine-tuned from scratch in the fine-tuning stage than other methods, which prevents the model parameters from overfitting on limited samples. Moreover, our method achieves state-of-the-art cross-domain performance, which is discussed in detail in Appendix C.

We further provide a performance comparison for each novel class on 1 shot of Pascal VOC Split 1 in Appendix A and experiments under G-FSOD setting detailed in Appendix B.

Table 3. Effectiveness of different modules in ReRCNN. Our results are reported on 1 shot of the Pascal VOC dataset Split 1.

FRCN	DR	CAR	OSR	Base		Novel 1shot	
				AP50	AP	AP50	AP
✓				79.75	53.16	14.82	6.44
✓			✓			20.77	12.59
✓		✓		79.51	53.78	36.46	16.39
✓		✓	✓			40.54	23.96
✓	✓			80.78	54.28	32.89	19.71
✓	✓		✓			51.87	29.45
✓	✓	✓		80.36	54.43	49.05	29.05
✓	✓	✓	✓			55.13	32.81

Contribution of Large Kernel Convolution in DearRPN. In our proposed method, we claim that large kernel convolution can enhance the discrimination ability of the region proposal network by increasing the receptive field of the detectors, thereby improving the performance of the model. Our ablation experiments, as shown in Table 3, support this claim, demonstrating that DearRPN indeed provides an additional boost to the final performance. However, the specific impact of large kernel convolutions on performance remains unknown. In other words, it is unclear whether larger convolution kernels always lead to better results. To address this question, we conduct in-depth experiments to study the impact of the size of the convolution kernels in the region proposal network on the final performance and detailed in Appendix D. After extensive evaluation, we choose the convolution kernel size of $1 \times 15 + 15 \times 1$, which not only has a strong ability to detect large objects but also minimizes the impact on the detection of small objects by avoiding the introduction of excessive background.

4.3 Ablation Studies

As shown in Table 3, we build on Faster RCNN and gradually add DearRPN, a category-agnostic box regressor, and a structure-reparameterized box classifier, where DR stands for Decoupled large kernel Region proposal network, CAR, and OSR stand for category-agnostic box Regressor and structurally Offline Structural Reparameterization respectively.

For fine-tuning Faster RCNN, we use the training strategy in [5]. We have found that when the parameters that fix the parameters of the RoI head can effectively improve the detector's generalization capability on novel classes. For example, the experimental results that adopt the CAR or OSR module are significantly better than the corresponding experiments that unfreeze the RoI head. Decoupled large kernel region proposal network (DearRPN) dramatically improves the generalization capacity of new categories 12%–14% in the AP50 on the base class, also increased by 1%. The category-agnostic Box Regressor also achieves 3%–5% performance improvement. From the table, we can conclude that even though the detection performance improvement on the base class is unnecessary to improve the detection performance on the novel class. Additionally, we conducted experiments to test the impact of hyperparameter γ and the number of epochs in the algorithm, and the results are presented in Appendix E.

5   Conclusion

In this paper, we revisit the two-stage fine-tuning paradigm for few-shot object detection and analyze the inherent "concept shift" problem of this widely embraced fashion. To alleviate these problems, we propose a region proposal network with a large kernel task adapter to decouple the region proposal network from the main network and utilize large kernel spatially separable convolutions to adaptively detect objects of different scales. To improve the generalization performance of the detector under extremely low data volume, we design an offline structural reparameterization approach to initialize the boxes classifier by utilizing prior knowledge learned from the base set and prototypes of novel classes learned on the augmented dataset. Our experimental evaluations demonstrate the effectiveness and superiority of our approach compared to existing methods.

Appendix

A   Performance of Each Novel Class

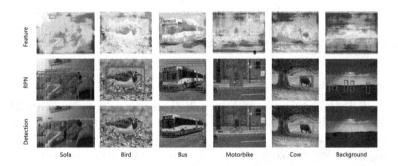

Fig. 4. We visualize the detection results of our model: regions of interest that the region proposal network and the RCNN head focus on (RED means larger value) when inference is performed and detection results. For more clarity, we further visualize the feature maps output by the RPN to illustrate the RPN's regions of interest.

We can see our method for each novel class on 1 shot of Pascal VOC Split 1, such as Fig. 5. Compared to the fine-tuned Faster RCNN, our method enables

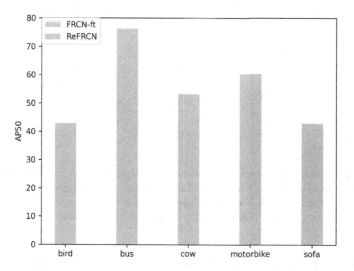

Fig. 5. Comparison of performance improvement for each category on 1 shot of Pascal VOC Split 1. FRCN-ft denotes fine-tuned Faster RCNN.

detectors to exploit known knowledge to facilitate the learning of new categories. For example, ReRCNN significantly improves the detection ability of cars and motorbikes because there are very similar categories in the base class (refer to Fig. 1), such as trains and bicycles, respectively. However, for the bird, the improvement is not as obvious as the former because its semantics are not so related to objects in the base class (Fig. 4).

B   G-FSOD Experiments

Traditional few-shot object detection settings only focus on the generalization performance of the detector on novel classes. However, the learning of new classes has the potential to harm the learned knowledge, resulting in catastrophic forgetting. Therefore, we introduce the Generalized few-shot object detection setting (G-FSOD) according to the setting in [5]. In this setting, the detector learns the knowledge of the novel class and retains the detection ability of the base class as it generalizes to the novel class. That is to say, detectors need to detect both the novel objects and the base objects in the image.

For the G-FSOD setting, ReRCNN no longer regards the base prototypes in the reparameterized weights W_{rep} as backgrounds but as their original categories. With this minor modification, ReRCNN quickly adapts to the G-FSOD setting. We conducted experiments on different shots and different splits on Pascal VOC, and the experimental results are shown in Fig. 6. The experimental results show an interesting phenomenon across three different splits. On Split 1 and Split 2, our method does not exhibit catastrophic forgetting, and even on Split 1, our method performs well in the G-FSOD scenario while retaining the ability to

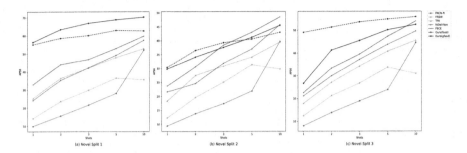

Fig. 6. Experimental results under generalized few-shot object detection. We use AP50 as the performance measure. The dotted lines represent the experimental results of our method under conventional few-shot object detection. Our reported results are the average of multiple runs on different splits and shots of the Pascal VOC dataset.

detect base classes. However, on Split 3, catastrophic forgetting hits our method, especially in scenarios with insufficient data. In general, catastrophic forgetting eases as the volume of data increases.

C   Cross Domain Adaption

Table 4. Experimental performance comparison of cross-domain few-shot object detection. Our results are reported on the test set of Pascal VOC 2007.

Method	FRCN-ft	FSRW	MetaDet	MetaRCNN	MPSR	DeFRCN	**Ours**
mAP	31.2	32.3	33.9	37.4	42.3	55.9	**56.8**

Different classes have different distributions. Since the few-shot learning task has disjoint class spaces at training and testing time, we can view few-shot learning as a domain adaptation problem with small distribution shifts. We conduct cross-domain few-shot object detection experiments on the Pascal VOC 2007 test set. Following previous work [16,35], we take the 60 classes on the MS COCO dataset as base classes and then fine-tune the novel classes from Pascal VOC. We report the performance of ReRCNN on 10 shots of Pascal VOC Split 1. Our experimental results are shown in Table 4. As can be seen from the table, our method achieves state-of-the-art cross-domain performance. Compared to the original fine-tuned Faster RCNN, our method achieves a 25.6% improvement and a 14.5% improvement over MPSR.

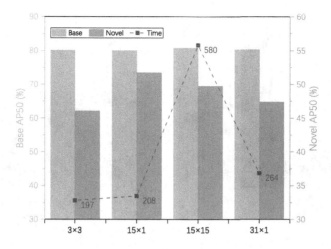

Fig. 7. Comparison of different convolution kernel sizes. We report the base-class training time, base-class performance, and novel performance on Pascal VOC for convolutional kernels of different sizes.

D   Ablation on Large Kernel Sizes

We carry out experiments on two convolution kernels of 3×3, a set of $1 \times 15 + 15 \times 1$, a set of $1 \times 31 + 31 \times 1$, and one 15×15, respectively. The experimental results are shown in Fig. 7. We report the AP50 of different kernels size on Pascal VOC and the time cost of pre-training on the base class. Our training time is reported on four NVIDIA RTX 3090 s. As we detailed above, with the increase in kernel size, the learnable parameters increase non-linearly. So we alternate two smaller convolution kernels of different sizes instead of large ones to reduce the parameter burden. As can be seen from Fig. 7, compared to the 15×15 convolution kernels, a set of $1 \times 15 + 15 \times 1$ convolution kernels significantly reduces pre-training time from 9 h to about 3 h, thanks to an 86.7% reduction in learnable parameters in the category-agnostic Decoupled Module. We also notice that the increase of the convolution kernel size does not necessarily equal the improvement to the pre-training performance on the base classes, which is in line with the fact that the leverage of large-kernel convolution on traditional object detection [22] does not bring significant performance improvement. However, when generalizing to novel classes, a set of large convolution kernels of $1 \times 15 + 15 \times 1$ significantly improve compared to regular 3×3 convolutions. Moreover, large-kernel convolutions do not always improve the performance of the detectors. We observe that when the convolution kernel changes from $1 \times 15 + 15 \times 1$ to $1 \times 31 + 31 \times 1$, the detectors degrade, which may be because the volume of data can no longer satisfy so many parameters to search for the optimal generalization solution in the hypothesis space.

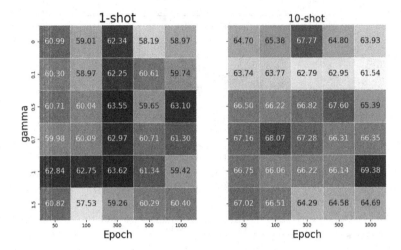

Fig. 8. The Effectiveness of different epochs for classifier pre-training and the contribution of base weights controlled by γ. We evaluate on the Pascal VOC dataset, and AP50 are reported on 1 shot and 10 shot on Split 1. The gamma of the y-axis represents the hyperparameter γ.

E   Effectiveness of Hyperparameters

In experiments, we found that novel classifier pre-training is a less stable process, leading to the box classifier's unstable performance after prototypical offline reparameterization. For different sample sizes and pre-training parameter settings, different γ and different epochs are adopted in our experiments. Inappropriate γ may degrade the model even more. In order to further understand the effect of γ on the final result, we then conducted experiments on Pascal VOC to explore different settings of γ. The experimental results are shown in Fig. 8. We found that in the 1-shot scenario when training for 300 epochs with γ set to 1, offline reparameterization of the weights leads to the best final performance. Nevertheless, in the 10-shot scenario, there is no clear trend. We claim this may be because the 10-shot has a large amount of data. The influence of the initial parameters gradually becomes weaker as the gradient descent steps increase.

References

1. Cao, Y., et al.: Few-shot object detection via association and discrimination. Adv. Neural Inf. Process. Syst. **34** (2021)
2. Chen, H., et al.: Harnessing multi-semantic hypergraph for few-shot learning. In: Pattern Recognition and Computer Vision, pp. 232–244 (2022)
3. Chen, H., Wang, Y., Wang, G., Qiao, Y.: LSTD: a low-shot transfer detector for object detection. In: Proceedings of the AAAI Conference on Artificial Intelligence, vol. 32 (2018)

4. Ding, X., Zhang, X., Han, J., Ding, G.: Scaling up your kernels to 31x31: revisiting large kernel design in CNNs. In: Proceedings of the IEEE/CVF Conference on Computer Vision and Pattern Recognition (CVPR), pp. 11963–11975, June 2022
5. Fan, Z., Ma, Y., Li, Z., Sun, J.: Generalized few-shot object detection without forgetting. In: Proceedings of the IEEE/CVF Conference on Computer Vision and Pattern Recognition, pp. 4527–4536 (2021)
6. Finn, C., Abbeel, P., Levine, S.: Model-agnostic meta-learning for fast adaptation of deep networks. In: International Conference on Machine Learning, pp. 1126–1135. PMLR (2017)
7. Gao, H., Shou, Z., Zareian, A., Zhang, H., Chang, S.F.: Low-shot learning via covariance-preserving adversarial augmentation networks. Adv. Neural Inf. Process. Syst. **31** (2018)
8. Gu, T.: Multilabel convolutional network with feature denoising and details supplement. IEEE Trans. Neural Netw. Learning Syst. **34**, 8349–8361 (2022)
9. Han, G., He, Y., Huang, S., Ma, J., Chang, S.F.: Query adaptive few-shot object detection with heterogeneous graph convolutional networks. In: Proceedings of the IEEE/CVF International Conference on Computer Vision, pp. 3263–3272 (2021)
10. Han, G., Huang, S., Ma, J., He, Y., Chang, S.F.: Meta faster R-CNN: towards accurate few-shot object detection with attentive feature alignment. In: Proceedings of the AAAI Conference on Artificial Intelligence, vol. 36, pp. 780–789 (2022)
11. Han, G., Ma, J., Huang, S., Chen, L., Chang, S.F.: Few-shot object detection with fully cross-transformer. In: Proceedings of the IEEE/CVF Conference on Computer Vision and Pattern Recognition (CVPR), pp. 5321–5330, June 2022
12. Hariharan, B., Girshick, R.: Low-shot visual recognition by shrinking and hallucinating features. In: Proceedings of the IEEE International Conference on Computer Vision, pp. 3018–3027 (2017)
13. He, K., Gkioxari, G., Dollár, P., Girshick, R.: Mask R-CNN. In: Proceedings of the IEEE International Conference on Computer Vision, pp. 2961–2969 (2017)
14. He, K., Zhang, X., Ren, S., Sun, J.: Deep residual learning for image recognition. In: Proceedings of the IEEE Conference on Computer Vision and Pattern Recognition, pp. 770–778 (2016)
15. Hu, H., Bai, S., Li, A., Cui, J., Wang, L.: Dense relation distillation with context-aware aggregation for few-shot object detection. In: Proceedings of the IEEE/CVF Conference on Computer Vision and Pattern Recognition, pp. 10185–10194 (2021)
16. Kang, B., Liu, Z., Wang, X., Yu, F., Feng, J., Darrell, T.: Few-shot object detection via feature reweighting. In: Proceedings of the IEEE/CVF International Conference on Computer Vision, pp. 8420–8429 (2019)
17. Karlinsky, L., et al.: RepMet: representative-based metric learning for classification and few-shot object detection. In: Proceedings of the IEEE/CVF Conference on Computer Vision and Pattern Recognition, pp. 5197–5206 (2019)
18. Kim, G., Jung, H.G., Lee, S.W.: Few-shot object detection via knowledge transfer. In: 2020 IEEE International Conference on Systems, Man, and Cybernetics (SMC), pp. 3564–3569. IEEE (2020)
19. Li, S., Song, W., Li, S., Hao, A., Qin, H.: Meta-retinaNet for few-shot object detection. In: BMVC (2020)
20. Lin, T.Y., Goyal, P., Girshick, R., He, K., Dollár, P.: Focal loss for dense object detection. In: Proceedings of the IEEE International Conference on Computer Vision, pp. 2980–2988 (2017)
21. Liu, L., et al.: GenDet: meta learning to generate detectors from few shots. IEEE Trans. Neural Netw. Learning Syst. **33**, 3448–3460 (2021)

22. Lu, H.-F., Du, X., Chang, P.-L.: Toward scale-invariance and position-sensitive region proposal networks. In: Ferrari, V., Hebert, M., Sminchisescu, C., Weiss, Y. (eds.) ECCV 2018. LNCS, vol. 11212, pp. 175–190. Springer, Cham (2018). https://doi.org/10.1007/978-3-030-01237-3_11

23. Qi, H., Brown, M., Lowe, D.G.: Low-shot learning with imprinted weights. In: Proceedings of the IEEE Conference on Computer Vision and Pattern Recognition, pp. 5822–5830 (2018)

24. Qiao, L., Zhao, Y., Li, Z., Qiu, X., Wu, J., Zhang, C.: DeFRCN: decoupled faster R-CNN for few-shot object detection. In: Proceedings of the IEEE/CVF International Conference on Computer Vision, pp. 8681–8690 (2021)

25. Redmon, J., Divvala, S., Girshick, R., Farhadi, A.: You only look once: unified, real-time object detection. In: Proceedings of the IEEE Conference on Computer Vision and Pattern Recognition, pp. 779–788 (2016)

26. Ren, S., He, K., Girshick, R., Sun, J.: Faster R-CNN: towards real-time object detection with region proposal networks. Adv. Neural Inf. Process. Syst. **28** (2015)

27. Shangguan, Z., Rostami, M.: Identification of novel classes for improving few-shot object detection. arXiv preprint arXiv:2303.10422 (2023)

28. Snell, J., Swersky, K., Zemel, R.: Prototypical networks for few-shot learning. Adv. Neural Inf. Process. Syst. **30** (2017)

29. Sun, B., Li, B., Cai, S., Yuan, Y., Zhang, C.: FSCE: few-shot object detection via contrastive proposal encoding. In: Proceedings of the IEEE/CVF Conference on Computer Vision and Pattern Recognition, pp. 7352–7362 (2021)

30. Sung, F., Yang, Y., Zhang, L., Xiang, T., Torr, P.H., Hospedales, T.M.: Learning to compare: Relation network for few-shot learning. In: Proceedings of the IEEE Conference on Computer Vision and Pattern Recognition, pp. 1199–1208 (2018)

31. Tong, J., Chen, T., Wang, Q., Yao, Y.: Few-shot object detection via understanding convolution and attention. In: Pattern Recognition and Computer Vision, pp. 674–687 (2022)

32. Wang, X., Huang, T., Gonzalez, J., Darrell, T., Yu, F.: Frustratingly simple few-shot object detection. In: III, H.D., Singh, A. (eds.) Proceedings of the 37th International Conference on Machine Learning. Proceedings of Machine Learning Research, vol. 119, pp. 9919–9928. PMLR, 13–18 July 2020

33. Wang, Y.X., Ramanan, D., Hebert, M.: Meta-learning to detect rare objects. In: Proceedings of the IEEE/CVF International Conference on Computer Vision, pp. 9925–9934 (2019)

34. Wu, A., Han, Y., Zhu, L., Yang, Y.: Universal-prototype enhancing for few-shot object detection. In: Proceedings of the IEEE/CVF International Conference on Computer Vision, pp. 9567–9576 (2021)

35. Wu, J., Liu, S., Huang, D., Wang, Y.: Multi-scale positive sample refinement for few-shot object detection. In: Vedaldi, A., Bischof, H., Brox, T., Frahm, J.-M. (eds.) ECCV 2020. LNCS, vol. 12361, pp. 456–472. Springer, Cham (2020). https://doi.org/10.1007/978-3-030-58517-4_27

36. Xu, Y., et al.: Gliding vertex on the horizontal bounding box for multi-oriented object detection. IEEE Trans. Pattern Anal. Mach. Intell. **43**(4), 1452–1459 (2020)

37. Yan, X., Chen, Z., Xu, A., Wang, X., Liang, X., Lin, L.: Meta R-CNN: towards general solver for instance-level low-shot learning. In: Proceedings of the IEEE/CVF International Conference on Computer Vision, pp. 9577–9586 (2019)

38. Yang, Z., Zhang, C., Li, R., Lin, G.: Efficient few-shot object detection via knowledge inheritance. IEEE Trans. Image Process. **32**, 321–334 (2023)

39. Zhang, G., Luo, Z., Cui, K., Lu, S.: Meta-DETR: few-shot object detection via unified image-level meta-learning. arXiv preprint arXiv:2103.11731 **2**(6) (2021)

40. Zhang, S., Wang, L., Murray, N., Koniusz, P.: Kernelized few-shot object detection with efficient integral aggregation. In: 2022 IEEE/CVF Conference on Computer Vision and Pattern Recognition (CVPR), pp. 19185–19194 (2022)
41. Zhang, W., Wang, Y.X.: Hallucination improves few-shot object detection. In: Proceedings of the IEEE/CVF Conference on Computer Vision and Pattern Recognition, pp. 13008–13017 (2021)
42. Zhao, Z., Liu, Q., Wang, Y.: Exploring effective knowledge transfer for few-shot object detection. In: Proceedings of the 30th ACM International Conference on Multimedia (2022)
43. Zhu, C., Chen, F., Ahmed, U., Shen, Z., Savvides, M.: Semantic relation reasoning for shot-stable few-shot object detection. In: Proceedings of the IEEE/CVF Conference on Computer Vision and Pattern Recognition, pp. 8782–8791 (2021)

Learning-Without-Forgetting via Memory Index in Incremental Object Detection

Haixin Zhou[1], Biaohua Ye[1], and JianHuang Lai[1,2,3,4(✉)]

[1] School of Computer Science and Engineering, Sun Yat-sen University, Guangzhou,
China
{zhouhx8,yebh3}@mail2.sysu.edu.cn
[2] Pazhou Lab (HuangPu), Guangdong 510000, China
[3] Guangdong Province Key Laboratory of Information Security Technology,
Guangzhou, China
stsljh@mail.sysu.edu.cn
[4] Key Laboratory of Machine Intelligence and Advanced Computing,
Ministry of Education, Guangzhou, China

Abstract. Object detection has made significant progress in recent years. However, when the training data is continuous and dynamic, notorious catastrophic forgetting will occur. In addition, training together with old class data requires more storage space and training time, and even the old data is not available. To address the above challenges in incremental object detection (IOD), this paper proposes an one-stage and anchor-based IOD paradigm via parameter isolation with memory index (LwF-MI) that does not require any old data. The core of LwF-MI is to design the anchor-based classification masks for memory indexing to guide the incremental network output. Besides, we propose a multi-scale fusion method for the preprocessing of the above masks, which addresses the poor performance of large-scale masks. Extensive experiments on MS COCO demonstrate that our approach achieves state-of-the-art results, which even exceed the full training results.

Keywords: incremental learning · object detection · parameter isolation

1 Introduction

In nature, the biological visual system can continuously learn new knowledge. For them, learning is incremental. In contrast, most of the existing object detection models do not have the ability of incremental learning [14]. Common learning paradigms implicitly assume that the training data is fixed or static, but in the real world, the data flow is continuous and dynamic. When the model acquires training data in batches from a continuous data stream for training, the new data will interfere with the training effect of the old data, which is the notorious catastrophic forgetting that means that in the absence of old class data, an old model fine-tuned only with new class data will perform poorly on the old class. Therefore, appropriate measures must be taken to avoid catastrophic forgetting. Based on whether task identities are provided or inferred, incremental learning is classified into three types: task/domain/class increment. In this paper, we will tackle the most intractable incremental type of object detection: class incremental object detection.

In IOD, effective methods to overcome catastrophic forgetting include knowledge distillation, replay, and parameter isolation. Among them, knowledge distillation and replay have been fully explored. Shmelkov et al. [23] perform the first work on incremental object detection, which applies LwF [15] to Fast-RCNN. Feng et al. [5] proposed a response-based incremental distillation method to obtain elastically learning responses. Liu et al. [17] propose a general EWC-based [11] incremental object detection framework under both Fast R-CNN and Faster R-CNN. Acharya et al. [1] use a novel memory replay mechanism that efficiently replays entire scenes. Shieh et al. [22] proposed the use of experience replay with different buffer sizes and the YOLO-V3 architecture for the problem of adding multiple classes at once to an object detector. However, little research has done on parameter isolation [19] and none of the current methods can fully overcome the catastrophic forgetting problem.

In order to solve the above problems, we choose one-stage and anchor-base object detector [2] as the base model to achieve high efficiency and light weight of the model. Secondly, we design a new and old classifier mask based on anchor (anchor mask), which build the memory index to realize the new and old classification of each corresponding anchor, and then guide the corresponding network for output. The above mask is constructed through the anchor-point-based classifier module, which consists of a 1*1 convolutional layer followed with a normalization layer. In addition, we propose a multi-scale fusion module to improve the performance of the above mask, which only uses one feature map to obtain the all preprocessing feature maps of the mask.

Our contributions can be summarized as follows,

1) Based on one-stage and anchor-base object detector, this paper proposes an incremental learning paradigm via parameter isolation method with memory index for IOD.
2) An anchor-based classifier mask (anchor mask) is proposed to implement memory indexing function and a multi-scale fusion module is used in the preprocessing of the above masks to improve their performance.
3) Extensive experiments demonstrate LwF-MI achieves SOTA performance, and the detection speed on A100 can reach 140 FPS, which can realize real-time object detection.

2   Related Work

Catastrophic forgetting is the main problem of incremental learning. In the field of incremental image classification task, there are three main methods used to overcome catastrophic forgetting: parameter constraints [11], knowledge distillation [15] and network expansion [24]. Compared with incremental classification, the current research on IOD is less, and the complexity of class IOD also increases the difficulty of research. The current research mainly has two directions, based on knowledge distillation and replay. The rest is based on parametric isolation, which is still not fully studied.

2.1   Knowledge Distillation Methods

Knowledge distillation methods have made great progress. Shmelkov et al. [23] is the first work on incremental object detection, which applies LwF to Fast-RCNN. Chawla et al. [3] propose a data-free distillation technique for object detection by synthesizing images conditioned on classes by inverse mapping. Based on Faster-RCNN, Hao et al. [8] propose feature-changing loss to generate accurate bbox foreground RPN branches, and improved the classification head. Thinking that both foreground and background play an important role in object detection, Guo et al. [6] propose a knowledge distillation method that decouples foreground and background. Based on one-stage object detection, Yang et al. [25] propose a novel multi-view correlation distillation method, which explores the intra-feature correlations in the feature space of the object detector. Chen et al. [4] innovatively use confidence loss to extract the confidence information of the initial model to avoid forgetting learned knowledge from old datasets. But these research cannot solve catastrophic forgetting, and it is difficult to strike a balance between new knowledge and old knowledge.

2.2   Replay Methods

Replay methods keep the representative samples of the original task and continuously uses these samples to stimulate the neurons while training new tasks to prevent the catastrophic forgetting of neurons. Hao et al. [7] employed the use of a small buffer of samples along with logits distillation to perform better than its competitors in the incremental learning of common objects from vending machines. Liu et al. [18] presented the use of an adaptive exemplar sampling for selecting replay instances and proposed different ways of applying the attention mechanism within the feature distillation procedure as a strategy to hinder forgetting. Yang et al. [26] proposed to use a pretrained language model to constrain the topology of the feature space within the model and capture the nuances of the semantic relations associated with each class name. Kj et al. [10] produced a hybrid strategy that relied on knowledge distillation, replay, and meta-learning to avoid forgetting. Joseph et al. [9] propose a open world object detector based on contrastive clustering and energy based unknown identification, in which identifying and characterising unknown instances helps to reduce confusion in an incremental object detection setting. However, these methods have a strong dependence on old samples, and the selection of old samples is also a problem.

2.3   Parameter Isolation Methods

Parameter isolation methods are less studied and not fully explored at present. Li et al. [13] mines memory neurons in original model, and then the rest neurons are updated and employed to detect new class. Zhang et al. [27] proposed a compositional architecture based on the mixture of compact expert detectors based on YOLOv3. Shieh et al. [21]utilize incremental branches on a one-stage object detection framework to avoid catastrophic forgetting, and utilize episodic memory to save finite samples of data and replay them during incremental learning. However, the above research requires a certain amount of old data.

3   Method

3.1   Motivation

The core purpose of IOD is to avoid catastrophic forgetting. Both knowledge distillation and replay have their own drawbacks. In the former, memorizing old knowledge and learning new knowledge are not compatible. In the latter, saving old samples somewhat violates the premise of incremental learning. Therefore, if we completely freeze old neurons to memorize old knowledge and add a new model to learn new knowledge, the key lies in how we output on the basis of these two models. We design a anchor-base classifier module acting as memory index to identify new and old classes on every anchor of feature maps and guide the selection of network output. Besides, the model always perform better using multiscale fusion feature. We explore a variety of multi-scale fusion strategies to find the best way.

3.2   Overall Structure

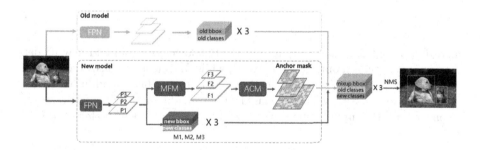

Fig. 1. The overall framework of LwF-MI.

The overall framework of the proposed method is shown in Fig. 1. First, our model consists of two major modules, which we refer to as the old model and the new model. The parameters of the old model come from the model trained on the old class data. Second, the framework mainly adopts the classic one-stage and anchor-based object detector structure: an input image is sent to the FPN to obtain three multi-dimensional feature maps of different sizes (P1, P2, P3). At each point (called an anchor) of the output (M1, M2, M3) of the two model, there are two locators and logits of the corresponding class. Under the guide of anchor-based classifier module, which implements the memory index function, the two locators and class logits concat in order. Finally, the results are outputted through NMS (non-max-suppression). Each module is introduced separately below.

3.3 Anchor-Based Classifier Module (ACM)

In our object detection framework, the inference results come from each anchor point of the three feature maps on the M layer. The classification results on each anchor point, together with the bias of the localizer, realize the identification and localization of the objects. Therefore, we can distinguish between old and new classes by constructing a mask corresponding to the size of the feature map to build memory index to identify new and old classes at each anchor point, and then correctly guide the final output of the network.

The above mask is constructed through the anchor-point-based classifier module, which consists of a 1*1 convolutional layer plus a normalization layer. The 1*1 convolution is used to reduce the dimension of the feature map, and also realizes the combination of cross-channel information, and then adds a normalization layer, which can greatly increase the nonlinear characteristics. The process of ACM guiding the network to output can be expressed by the following formula as,

$$Lct_{out} = (Lct_{old} \odot \overline{M_{ACM}}) \,\&\, (Lct_{new} \odot M_{ACM}) \tag{1}$$

$$Cls_{out} = Cls_{old} \oplus Cls_{new} \tag{2}$$

$$Y_{out} = Lct_{out} \oplus Cls_{out} \tag{3}$$

where Lct_{old} and Lct_{new} are separately the two locators of old and new model. M_{ACM} is anchor mask outputted by ACM, while $\overline{M_{ACM}}$ is the bitwise inverse of M_{ACM}. $\odot$ refers to bitwise multiplication, and $\&$ means bitwise addition. Cls_{old} and Cls_{new} are separately the two class logits from old and new model. $\oplus$ means to concatenate the input on both sides. Y_{out} is the final output before NMS.

3.4 Multi-scale Fusion Module (MFM)

In the process of implementing the ACM module, at the beginning, the corresponding anchor point mask was simply obtained directly from each feature map (P1, P2, P3), but it was found that the high-resolution mask (from F1, F2) did not accurately distinguish between old and new classes. Therefore, we tried various mask extraction and fusion algorithms, as shown in the Fig. 2.

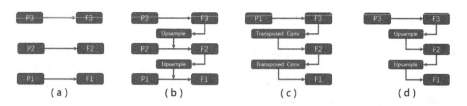

Fig. 2. Four implement methods of MFM.

(a) The graph is simply extracted directly without any processing. (b) On the basis of (a), a FPN-like [16] network is used for context fusion. (c) is to directly use the feature map (P3) with the largest receptive field in the FPN network to obtain F3, and then obtain F1 and F2 through the deconvolution method. (d) is based on (c), using the nearest upsampling method to obtain F1 and F2.

A series of experiments demonstrate (Table 4) (d) method achieve optimal results. The underlying reason may be that for the network, identifying whether an object is new class requires higher-dimensional feature information, and too much noise in low-dimensional feature interfere with the classification results.

3.5   Loss

Given an old object detection model trained on m classes and a training dataset of n new classes, our goal is to incrementally train an object detection model which preforms object detection on the complete set of $m + n$ classes. Common Loss constraints in object detection include classification loss, localization regression loss (bbox loss), and confidence loss. But in our work, There is an auxiliary loss in our framework, the training loss of the anchor mask. The auxiliary loss is defined as,

$$L_{aux} = \sum_{n=1}^{N} \ell(F_{O_n}, F_{T_n}) = mean\{l_1, \ldots, l_N\} \tag{4}$$

$$l_n = \frac{-1}{W \cdot H} \sum_{w=0}^{W} \sum_{h=0}^{H} [y_n(w,h) \cdot \log x_n(w,h) + (1 - y_n(w,h)) \cdot \log(1 - x_n(w,h))] \tag{5}$$

In the Eq. (4) F_{O_n} is anchor mask outputted by ACM, and F_{T_n} is anchor mask constructed from the provided ground-truth labels. While in the Eq. (5) W and H are the width and height of the anchor mask respectively, and $y_n(w,h)$ is the old or new label corresponding to F_{T_n} at the position of (w,h) and so is $x_n(w,h)$.

Therefore, the total training loss can be defined as,

$$Loss_{total} = L_{cls} + L_{bbox} + L_{conf} + L_{aux} \tag{6}$$

In our work, the $Loss_{total}$ is only used to adjust the parameters of the new model, and the parameters of the old model are frozen in order to completely avoid catastrophic forgetting.

4   Experiments and Discussions

In this section, we perform experiments on Pascal VOC 2007 and MS COCO 2017 using the baseline detector YOLOv5 to validate LwF-MI. Then we perform ablation studies to prove the effectiveness of each component. Finally, we discuss the incremental scenario of LwF-MI.

4.1   Datasets and Experiment Setting

MS COCO 2017. MS COCO 2017 is a challenging object detection data set, including 80 classes, a total of more than 100,000 images. We use the train set for training and the minival set for testing. We performed experiments for the following 4 scenarios: 20 old classes + 60 new classes, 40 old classes + 40 new classes, 60 old classes + 20 new classes, 70 old classes + 10 new classes. In our experiments, old class samples or auxiliary datas are not needed at all.

Pascal VOC 2007. The PASCAL Visual Object Classes (VOC) dataset is a benchmark in visual object category recognition and detection. The dataset contains 20 classes and contains 10,011 pictures. We performed experiments for the following 3 scenarios: 5 old classes + 15 new classes, 10 old classes + 10 new classes, 15 old classes + 5 new classes.

Evaluation Criterion. We use the most representative mAP_{50} as one of the evaluation criterion. In addition, most of the work is still using the difference between the upper and lower limits of the baseline as the evaluation indicator, but this standard is one-sided. Because it cannot reflect the performance of the model on the old and new classes. So a related indicator is proposed: G_{up}, AOLR (Average Old Class Learning Rate) and ANLR (Average New Class Learning Rate). G_{up} as the comparison standard between different method, is the ratio of the gap towards the upper bound to the upper bound.

$$AOLR = \frac{1}{I_{old}} \sum_{i=0}^{I_{old}} AP50_i \qquad (7)$$

$$ANLR = \frac{1}{I_{new}} \sum_{i=0}^{I_{new}} AP50_i \qquad (8)$$

where I_{old} is the number of the old classes, while I_{new} is the new classes and $AP50_i$ is the $AP50$ of the i class. AOLR represents the degree of forgetting of the network for the old class, while ANLR shows the learning ability of the network for new classes.

4.2 Overall Performance

Experiments on the COCO Dataset

We report the experimental results on the COCO dataset (see Table 1). First of all, LwF-MI is SOTA. We observed that in the 40 classes + 40 classes scenario, if the old detector and new data are directly used for fine-tuning, compared with 56.1% of full data training, mAP_{50} drops to 25.9%, which shows that during the fine-tuning process, the learning process of new class data interferes with the old memory. Remarkably, LwF-MI avoids the above situation very well, and can reach 59.5% on mAP_{50}, which is even higher than the upper limit (56.1%). AOLR is −0.1%, indicating that the old classes memory is well preserved, and the ANLR

Table 1. Incremental results (%) on COCO benchmark under different scenarios.

Scenarios	Method	mAP_{50}	G_{up}	AOLR	ANLR
Full data	Upper Bound	56.1	0.0	-	-
20 + 60	**Our**	**58.9**	**+5.0**	**+0.08**	**+1.37**
40 + 40	Catastrophic Forgetting	25.9	−53.8	-	-
	LwF [3] (2021)	25.4	−54.7	-	-
	RILOD [12] (2019)	45.0	−19.8	-	-
	SID [20] (2021)	51.4	−8.4	-	-
	ERD [5] (2022)	54.5	−2.9	-	-
	Our	**59.5**	**+6.1**	**−0.1**	**+2.5**
50 + 30	Catastrophic Forgetting	20.6	−63.3	-	-
	LwF [3] (2021)	9.5	−83.1	-	-
	RILOD [12] (2019)	43.2	−23.0	-	-
	SID [20] (2021)	51.0	−9.1	-	-
	ERD [5] (2022)	54.0	−3.7	-	-
	Our	**59.2**	**+5.5**	**+2.3**	**+3.5**
60 + 20	Catastrophic Forgetting	14.0	−75.0	-	-
	LwF [3] (2021)	10.8	−80.7	-	-
	RILOD [12] (2019)	38.8	−30.8	-	-
	SID [20] (2021)	49.8	−11.2	-	-
	ERD [5] (2022)	52.9	−5.7	-	-
	Our	**59.1**	**+5.4**	**+1.8**	**+2.1**
70 + 10	Catastrophic Forgetting	6.5	−88.4	-	-
	LwF [3] (2021)	12.4	−77.9	-	-
	RILOD [12] (2019)	37.9	−32.4	-	-
	SID [20] (2021)	49.0	−12.7	-	-
	ERD [5] (2022)	51.9	−7.5	-	-
	Our	**58.3**	**+3.9**	**+1.3**	**−2.2**

is +2.5%, indicating that the learning quality of the new class has exceeded the upper bound. While in the three scenarios, the results are universal, indicating that LwF-MI is robust and suitable for a variety of incremental scenarios. In addition, our model inference speed can reach 140FPS (Origin model is 160FPS).

Table 2. Per-class accuracy (%) on VOC benchmark under different scenarios.

Scenarios	mAP_{50}	ANLR	AOLR
Full data	57.2		
5	48.6		
5+15	51.8	−10.5	−2.1
10	56.4		
10+10	54.1	−3.6	−3.3
15	58.8		
15+5	50.3	−3.0	−20.0

Experiments on the VOC Dataset

We show the incremental results for each category on the VOC dataset in Table 2. We find that in all three scenarios, the upper limit cannot be exceeded as in the COCO dataset. At the same time, we found that AOLR and ANLR in three different scenarios are quite different. In the 5 classes + 15 classes Scenario, the AOLR is obviously larger than the average value, reaching -10.5%, while in 15+5, the situation is reversed, where the ANLR value is abnormal, which is −20.0%. Besides, the maximum value of mAP_{50} (54.1%) is in the 10 + 10 scenario. These results show that in different incremental scenarios, the network forms different learning results for different learning processes of old and new classes. This part will be discussed in discuss section in detail.

4.3 Ablation Study

We validate the effectiveness of each component of the proposed method on COCO, as shown in Table 3. In the 70 + 10 scenario, in the table, only 52.6% of mAP_{50} can be achieved without using the ACM and MFM modules. According to the large negative values of AOLR, it can be inferred that the model does

Table 3. Ablation study (%) about ACM and MFM using the COCO benchmark under first 70 classes + last 10 classes.

Method	ACM	MFM	mAP_{50}	AOLR	ANLR
Upper Bound	-	-	56.1	-	-
Catastrophic Forgetting	-	-	6.5	-	-
Ours	-	-	52.6 (−3.5)	+2.3	−8.2
	✓	-	57.5 (+1.4)	+2.7	−4.1
	✓	✓	58.3 (+2.2)	+1.3	−2.2

Table 4. Ablation study (%) about four implement methods of MFM using the COCO benchmark under first 70 classes + last 10 classes.

Method	mAP_{50}	AOLR	ANLR
(a) Direct	52.6	+2.3	−8.2
(b) FPN-like	57.2	+2.5	−4.6
(c) Only-P3-deconvolution	56.1	+2.7	−4.2
(d) Only-P3-nearest-upsampling	58.3	+1.3	−2.2

not perform well on new classes. When using ACM, the mAP_{50} came to 57.5%. Then using the proposed MFM, mAP_{50} came to 58.3%, AOLR is +1.3% while ANLR is only −2.2%, indicating that the model using the little forgetting of old knowledge to get more learning of new knowledge. Besides, we evaluate four implement methods of MFM in the same scenario, as shown in Table 4.

4.4 Visualization

See Fig. 3.

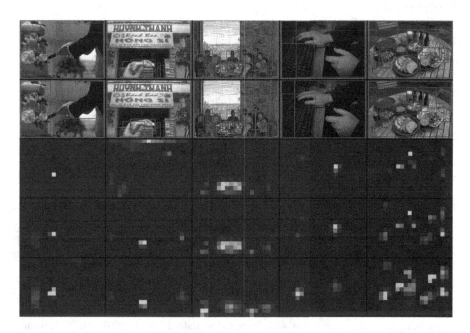

Fig. 3. The visualization of LwF-MI. Five examples are displayed, with the following sequence from top to bottom: the original dataset label, the model prediction result, and three different scale masks outputted by ACM from F1, F2, F3. Old class objects are represented by red boxes, while new class objects are by green boxes. (Color figure online)

5 Conclusion

In this paper, we propose a incremental paradigm via parameter isolation with memory index for one-stage and anchor-base object detector, which significantly solves catastrophic forgetting. We design an anchor-based classifier module (ACM), which implements memory indexing functions to guide the network output. Besides, we introduce multi-scale fusion module (MFM) which integrates features at various scales to enable ACM's optimal performance. Extensive experiments demonstrate the effectiveness of LwF-MI. Finally, a detailed analysis of the interpretability of LwF-MI, as well as the essential differences between incremental image classification tasks and incremental object detection, provides insights for further exploration in IOD.

Acknowledgement. This work was supported in part by the Key Areas Research and Development Program of Guangzhou under Grant 2023B01J0029; and in part by the Key Area Research and Development Program of Guangdong Province, China, under Grant 2018B010109007.

References

1. Acharya, M., Hayes, T.L., Kanan, C.: Rodeo: replay for online object detection. arXiv preprint arXiv:2008.06439 (2020)
2. Bochkovskiy, A., Wang, C.Y., Liao, H.Y.M.: Yolov4: optimal speed and accuracy of object detection. arXiv preprint arXiv:2004.10934 (2020)
3. Chawla, A., Yin, H., Molchanov, P., Alvarez, J.: Data-free knowledge distillation for object detection. In: Proceedings of the IEEE/CVF Winter Conference on Applications of Computer Vision, pp. 3289–3298 (2021)
4. Chen, L., Yu, C., Chen, L.: A new knowledge distillation for incremental object detection. In: 2019 International Joint Conference on Neural Networks (IJCNN), pp. 1–7. IEEE (2019)
5. Feng, T., Wang, M., Yuan, H.: Overcoming catastrophic forgetting in incremental object detection via elastic response distillation. In: Proceedings of the IEEE/CVF Conference on Computer Vision and Pattern Recognition, pp. 9427–9436 (2022)
6. Guo, J., Han, K., Wang, Y., Wu, H., Chen, X., Xu, C., Xu, C.: Distilling object detectors via decoupled features. In: Proceedings of the IEEE/CVF Conference on Computer Vision and Pattern Recognition, pp. 2154–2164 (2021)
7. Hao, Y., Fu, Y., Jiang, Y.G.: Take goods from shelves: a dataset for class-incremental object detection. In: Proceedings of the 2019 on International Conference on Multimedia Retrieval, pp. 271–278 (2019)
8. Hao, Y., Fu, Y., Jiang, Y.G., Tian, Q.: An end-to-end architecture for class-incremental object detection with knowledge distillation. In: 2019 IEEE International Conference on Multimedia and Expo (ICME), pp. 1–6. IEEE (2019)
9. Joseph, K., Khan, S., Khan, F.S., Balasubramanian, V.N.: Towards open world object detection. In: Proceedings of the IEEE/CVF Conference on Computer Vision and Pattern Recognition, pp. 5830–5840 (2021)
10. Joseph, K., Rajasegaran, J., Khan, S., Khan, F.S., Balasubramanian, V.N.: Incremental object detection via meta-learning. IEEE Trans. Pattern Anal. Mach. Intell. **44**(12), 9209–9216 (2021)

11. Kirkpatrick, J., et al.: Overcoming catastrophic forgetting in neural networks. Proc. Natl. Acad. Sci. **114**(13), 3521–3526 (2017)

12. Li, D., Tasci, S., Ghosh, S., Zhu, J., Zhang, J., Heck, L.: Rilod: near real-time incremental learning for object detection at the edge. In: Proceedings of the 4th ACM/IEEE Symposium on Edge Computing, pp. 113–126 (2019)

13. Li, W., Wu, Q., Xu, L., Shang, C.: Incremental learning of single-stage detectors with mining memory neurons. In: 2018 IEEE 4th International Conference on Computer and Communications (ICCC), pp. 1981–1985. IEEE (2018)

14. Li, X., Wang, W., Hu, X., Li, J., Tang, J., Yang, J.: Generalized focal loss V2: learning reliable localization quality estimation for dense object detection. In: Proceedings of the IEEE/CVF Conference on Computer Vision and Pattern Recognition, pp. 11632–11641 (2021)

15. Li, Z., Hoiem, D.: Learning without forgetting. IEEE Trans. Pattern Anal. Mach. Intell. **40**(12), 2935–2947 (2017)

16. Lin, T.Y., Dollár, P., Girshick, R., He, K., Hariharan, B., Belongie, S.: Feature pyramid networks for object detection. In: Proceedings of the IEEE Conference on Computer Vision and Pattern Recognition, pp. 2117–2125 (2017)

17. Liu, L., Kuang, Z., Chen, Y., Xue, J.H., Yang, W., Zhang, W.: IncDet: in defense of elastic weight consolidation for incremental object detection. IEEE Trans. Neural Netw. Learn. Syst. **32**(6), 2306–2319 (2020)

18. Liu, X., Yang, H., Ravichandran, A., Bhotika, R., Soatto, S.: Multi-task incremental learning for object detection. arXiv preprint arXiv:2002.05347 (2020)

19. Menezes, A.G., de Moura, G., Alves, C., de Carvalho, A.C.: Continual object detection: a review of definitions, strategies, and challenges. Neural Netw. (2023)

20. Peng, C., Zhao, K., Maksoud, S., Li, M., Lovell, B.C.: SID: incremental learning for anchor-free object detection via selective and inter-related distillation. Comput. Vis. Image Underst. **210**, 103229 (2021)

21. Shieh, J.L., Haq, M.A., Haq, Q.M.U., Ruan, S.J., Chondro, P.: Utilizing incremental branches on a one-stage object detection framework to avoid catastrophic forgetting. Mach. Vis. Appl. **33**(2), 28 (2022)

22. Shieh, J.L., et al.: Continual learning strategy in one-stage object detection framework based on experience replay for autonomous driving vehicle. Sensors **20**(23), 6777 (2020)

23. Shmelkov, K., Schmid, C., Alahari, K.: Incremental learning of object detectors without catastrophic forgetting. In: Proceedings of the IEEE International Conference on Computer Vision, pp. 3400–3409 (2017)

24. Yan, S., Xie, J., He, X.: DER: dynamically expandable representation for class incremental learning. In: Proceedings of the IEEE/CVF Conference on Computer Vision and Pattern Recognition, pp. 3014–3023 (2021)

25. Yang, D., et al.: Multi-view correlation distillation for incremental object detection. Pattern Recogn. **131**, 108863 (2022)

26. Yang, S., et al.: Objects in semantic topology. arXiv preprint arXiv:2110.02687 (2021)

27. Zhang, N., Sun, Z., Zhang, K., Xiao, L.: Incremental learning of object detection with output merging of compact expert detectors. In: 2021 4th International Conference on Intelligent Autonomous Systems (ICoIAS), pp. 1–7. IEEE (2021)

SAMDConv: Spatially Adaptive Multi-scale Dilated Convolution

Haigen Hu, Chenghan Yu, Qianwei Zhou, Qiu Guan, and Qi Chen[✉]

Zhejiang University of Technology, Hangzhou, China
chenqi@zjut.edu.cn

Abstract. Dilated convolutions have received a widespread attention in popular segmentation networks owing to the ability to enlarge the receptive field without introducing additional parameters. However, it is unsatisfactory to process multi-scale objects from different spatial positions in an image only by using multiple fixed dilation rates based on the structure of multiple parallel branches. In this work, a novel spatially-adaptive multi-scale dilated convolution (SAMDConv) is proposed to adaptively adjust the size of the receptive field for different scale targets. Specifically, a Spatial-Separable Attention (SSA) module is firstly proposed to personally select a reasonable combination of sampling scales for each spatial location. Then a recombination module is proposed to combine the output features of the four dilated convolution branches according to the attention maps generated by SSA. Finally, a series of experiments are conducted to verify the effectiveness of the proposed method based on various segmentation networks on various datasets, such as Cityscapes, ADE20K and Pascal VOC. The results show that the proposed SAMDConv can obtain competitive performance compared with normal dilated convolutions and depformable convolutions, and can effectively improve the ability to extract multi-scale information by adaptively regulating the dilation rate.

Keywords: SAMDConv · Image Segmentation · Receptive Field

1 Introduction

In recent years, deep learning has made significant breakthroughs in various computer vision tasks due to the powerful representation capabilities of neural networks. Especially, the convolution operation in Convolutional Neural Networks (CNNs) is particularly important owing to its strong feature extraction capabilities. The potential of convolution kernels has been widely studied and explored to fully exploit their power. As a classic variant of convolution, dilated convolution [26] was firstly applied in the feature extraction stage of image segmentation tasks, aiming to solve the problem of information loss caused by pooling or downsampling when reducing image resolution. The idea of dilated convolution is to add holes (padding '0') in the standard convolution kernel to achieve a larger receptive field without adding extra parameters. Currently, it has been

Q. Liu et al. (Eds.): PRCV 2023, LNCS 14432, pp. 460–472, 2024.
https://doi.org/10.1007/978-981-99-8543-2_37

common sense that dilated convolution has been widely applied in the backbone networks of various segmentation networks, becoming an indispensable part of segmentation networks.

Fig. 1. the sampling area of the dilated convolution on different targets.

However, objects in an image usually have different scales due to various factors such as spatial distance and object itself size. Traditional dilated convolutions can only use a single-scale convolution kernel to extract features, and such single-scale can naturally lead to a decrease in the ability to extract small targets. If dilated convolutions with the same dilation rate are used to recognize the objects with different sizes in an image as shown in Fig. 1, the convolution kernel can still correctly capture the main features of large objects like trees. While for small targets like cows, the extracted features are easily mixed with background information, thereby becoming ambiguous. This can cause significant interference to the understanding of semantic information, especially for a large number of objects with different scales in segmentation task, which can greatly lead to the degradation of model performance.

To solve the above problem, one common way is to use multiple convolution kernels with different scales to extract features, such as parallel branch structures with dilated convolutions [8,16], or different dilation rates at different levels of the network [2,3,21]. These methods can enable the network to extract multi-scale information. For example, Wang et al. [21] adopted a set of consecutive dilated convolutions with different dilation rates for semantic segmentation under the constraint that their greatest common divisor cannot be 1. However, these methods have a common issue that the multi-scale extraction capability is obtained through structural adjustments, rather than the convolutional kernels themselves. As shown in Fig. 1, the receptive field required for cows and trees is completely different. However, conventional methods only apply multi-scale feature extraction to these targets, and do not meet the different requirements of different target regions in a personalized way. Thus, these conventional approaches have no such ability to dynamically adjust the receptive field according to different targets.

To address this issue, a novel spatially adaptive multi-scale dilated convolution, called SAMDConv, is proposed to dynamically adjust the size of the receptive field for different scale targets by adaptively regulating the dilation rate of dilated convolutions in this paper. Specially, the proposed SAMDConv firstly uses multiple parallel dilated convolutions with different scales to extract multi-scale features at each spatial position of the feature map. Then, a Spatial-Separable Attention (SSA) module is proposed to personally select a reasonable combination of sampling scales for each spatial location by filtering out ambiguous sampling regions and enhancing the correct sampling information. As shown in Fig. 2, the proposed SAMDConv firstly use multi-scale convolution kernels to sample each spatial position, and then adaptively selects a reasonable sampling region size based on the size and shape of the object at the current location. When a smaller object like a cow is sampled, higher weights are selected for the convolutional kernels with smaller dilation rates and vice versa. In this way, the optimal dynamic combination can make the feature semantics extracted by convolution more clear and eliminate unnecessary interference information through the attention mechanism.

Fig. 2. the sampling area of SAMDConv on different targets.

Our contributions can be summarized into twofold as follows:

- A SSA module is proposed to personally select a reasonable combination of sampling scales for each spatial location.
- A recombination module is proposed to combine the output features of the four dilated convolution branches according to the attention maps generated by SSA module.
- A SAMDConv module is proposed to adaptively adjust the size of the receptive field for different scale targets by dynamically regulating the dilation rate of dilated convolutions.

2 Related Work

2.1 Dilated Convolution

In recent years, dilated convolution has been widely applied in computer vision tasks, such as object detection [12], image segmentation [19], and crowd density estimation [18]. Dilated convolution effectively increases the receptive field of the model, but it also introduces certain issues such as the Gridding Effect [16] and the possibility of long-range information being irrelevant. When using dilated convolution repeatedly, since a large number of holes are introduced into the convolution kernel, the kernel is not continuous, thereby causing some pixels to be left out of the calculation. In other words, some information is lost during the convolution process, which can be fatal for pixel-level dense prediction tasks. Additionally, although dilated convolutions were initially designed to obtain long-range information, it may only be effective for large objects to use a high dilation rate in the segmentation task, while being detrimental for smaller objects. To address these issues, researchers have proposed various solutions [16,21,26], including the High-dimensional Convolution (HDC) structure. This structure can ensure that the dilation rates used in the stacked convolutions do not have a common factor greater than one, and the dilation rates of consecutive layers are set in a zigzag pattern like [1,2,5,1,2,5]. This approach can to some extent meet the segmentation requirements for both large and small objects, while also preserving the continuity of information.

2.2 Attention Mechanism

Since 2017, the transformer, proposed by Vaswani et al. [20], has become one of the most advanced methods in many NLP tasks. With the impressive performance of the transformer in the NLP field, Alexey Dosovitskiy et al. [5] proposed the Vision Transformer (ViT), which extends the transformer architecture to the computer vision field and achieves excellent performance. The attention module used in the transformer is a key factor for its unique advantages in various fields. Compared with convolutional operations, attention mechanisms can better obtain global information. Therefore, some researchers have applied attention mechanisms in convolutional neural networks by combining the global information extraction capability of attention mechanisms with the local information extraction capability of convolutions [17]. Compared with transformers, the receptive field of convolutional neural networks is local, and it is necessary to accumulate many layers to associate different parts of the image. Therefore, Hu et al. proposed SENet [9] to statistically capture global information from the feature channel level of the image. Furthermore, CBAM [22] was proposed for a lightweight feature optimization general module by combining channel attention with spatial attention, and it can greatly improve the performance of various visual tasks.

3 Method

3.1 Preliminary

Traditional image segmentation methods input the image to a CNN (typical examples include FCN [15]). There are two key factors in image segmentation FCN: pooling to reduce image size and increase receptive field, and upsampling to expand image size. In the process of reducing and then increasing size, some information is inevitably lost. Can a new operation be designed to see more information without using pooling? The answer is dilated convolution. The main contribution of dilated convolution is to remove the pooling and downsampling operation without reducing the network's receptive field.

The dilated convolution has gained widespread application in the backbone of image segmentation. A 3×3 dilated convolution with rate $r = 2$ can achieve the same receptive field as a 5×5 standard convolution, while not introducing extra parameters with a larger receptive field. The receptive field size of a dilated convolution can be expressed using the following Eq. 1:

$$s = (k - 1) \times (r - 1) + k \tag{1}$$

where s represents the receptive field size of the dilated convolution, k represents the kernel size, and r represents the dilation rate.

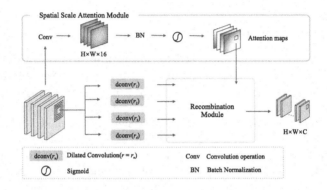

Fig. 3. Overall architecture of the proposed SAMDConv.

3.2   Overall Architecture

The proposed SAMDconv consists of SSA module and Recombination module, and it is an aggregation of four convolutional kernels that share the same kernel size and input/output dimensions but have different dilation rates. They are weightedly aggregated by the Spatial-Separable Attention (SSA) module, and the overall architecture is illustrated in Fig. 3. The proposed SAMDConv can replace the conventional dilated convolutions in the backbone of image segmentation, but it has a more powerful expressive ability compared to the ordinary dilated convolutions.

3.3   Spatial Scale Attention (SSA) Module

Spatial attention mechanisms [11] have been widely used in various computer vision networks and have become one of the hot research directions. However, the existing spatial attention mechanisms have fatal flaws in the task of image segmentation: while focusing on important objects, other objects are inevitably ignored. For pixel-level prediction tasks, image segmentation networks need to retain more effective information from the original image, which is important and cannot be ignored for segmentation results. Traditional attention weighting methods cannot simultaneously pay attention to the important and secondary objects.

To address the shortcomings of spatial attention mechanisms, a Spatial Scale Attention (SSA) mechanism is proposed, and the corresponding structure is illustrated in Fig. 3. The input feature map goes through convolutional layers, BN layers, and sigmoid activation function to obtain attention maps with the same spatial dimensions as the output feature maps of the parallel dilated convolution branches. The output attention maps contain 16 channels, which are evenly divided into 4 groups. The attention feature maps on the 4 channels within each group correspond to 4 different scales of dilated convolution branches, and the size of each attention map is the same as the output feature map of the branch. Each pixel on the attention map represents the importance of the corresponding scale branch at the current spatial position. The 4 groups of 4-channel attention maps represent 4 combinations of dilation rates, and it can reduce the information loss caused by the misjudgment of attention maps and increase the diversity of the output features. Significantly, the SSA module uses the sigmoid activation function to obtain a weight distributed between 0 and 1. The reason why sigmoid is used instead of softmax is that SAMDConv needs to obtain combinations, and the larger inputs in softmax will suppress the smaller inputs. Therefore, an independent sigmoid activation function is used for each pixel, and the mathematical expression of SSA is illustrated in Eq. 2.

$$Attention = Sigmoid(BN(Conv(x))) \qquad (2)$$

where, Conv, BN, and Sigmoid represent convolution operation, batch normalization layer, and Sigmoid activation function, respectively. Attention represents the output attention feature map.

Different from traditional spatial attention mechanisms, SSA does not simply weight the targets in the image, but rather selects a suitable set of weights for a specific point on the feature map in a combination form. It can play an important role in selecting appropriate scales and filtering out interference scale information. The proposed SAMDConv applies the appropriate scale combination to this point based on the selected weights. It is worth noting that different combinations are applied to different points on the feature map. By this way, SAMDConv can achieve equal treatment of different points on the feature map, without any absolute important points. Each point is applied with the most suitable scale combination.

3.4   Recombination Module

Recombination module is used to combine the output features of the four dilated convolution branches according to the attention maps generated by SSA,

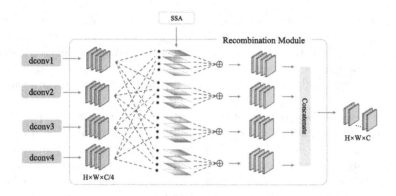

Fig. 4. An illustration of Recombination Module.

and the corresponding architecture is illustrated in Fig. 4. The 16-channel attention maps generated by SSA are evenly divided into four groups. The attention
maps within each group are multiplied with the corresponding feature maps
of the branches in sequence. The results of the multiplications are then added
together to obtain one set of output features after scale selection. There are a
total of four sets of output features generated, and the process of feature fusion
is shown in Eq. 3.

$$Output_j = \sum_{i=1}^{4} F_i \times Attention(4j + i - 4) \tag{3}$$

where $Output_j$ and F_i respectively represent the jth blue output feature map
and the ith purple output feature map obtained by the ith dilated convolution.
$Attention(4j + i - 4)$ represents the ith Attention map from the top down in the
j^{th} group of Attention maps. Finally, the four groups of output feature maps are
concatenated together to obtain the final output Out.

$$Out = Concat(Output_1, Output_2, Output_3, Output_4) \tag{4}$$

Table 1. Comparison results between normal dilated convolution (Baseline) and the proposed SAMDConv for image segmentation on CityScapes dataset

Methods	Methods	Backbone	Mean IoU(%)	Mean Acc(%)	Pixel Acc(%)
Dilated Convolution (Baseline)	GCNet [1]	ResNet-50	77.44	84.75	95.95
	APCNet [7]	ResNet-50	77.43	85.21	95.98
	UperNet [24]	ResNet-50	77.28	86.24	96.04
	ISANet [25]	ResNet-50	77.80	85.69	96.05
	PSPNet [27]	ResNet-50	78.47	87.38	96.03
	CCNet [10]	ResNet-50	78.20	85.75	96.21
	PointRend [14]	ResNet-50	75.95	82.97	95.96
	Semantic FPN [13]	ResNet-101	77.94	85.63	96.02
	PointRend [14]	ResNet-101	77.72	84.94	96.25
SAMDConv	GCNet [1]	ResNet-50	**78.05($\uparrow$0.61)**	**85.35($\uparrow$0.60)**	**96.07($\uparrow$0.12)**
	APCNet [7]	ResNet-50	**78.44($\uparrow$1.01)**	**86.30($\uparrow$1.09)**	**96.07($\uparrow$0.09)**
	UperNet [24]	ResNet-50	**78.58($\uparrow$1.30)**	**87.20($\uparrow$0.96)**	**96.22($\uparrow$0.18)**
	ISANet [25]	ResNet-50	**78.74($\uparrow$0.94)**	**86.21($\uparrow$0.52)**	**96.21($\uparrow$0.16)**
	PSPNet [27]	ResNet-50	**78.98($\uparrow$0.51)**	**87.68($\uparrow$0.30)**	**96.13($\uparrow$0.10)**
	CCNet [10]	ResNet-50	**79.16($\uparrow$0.94)**	**86.32($\uparrow$0.57)**	**96.26($\uparrow$0.05)**
	PointRend [14]	ResNet-50	**79.17($\uparrow$3.22)**	**86.49($\uparrow$3.52)**	**96.30($\uparrow$0.34)**
	Semantic FPN [13]	ResNet-101	**78.57($\uparrow$0.61)**	**86.25($\uparrow$0.62)**	**96.13($\uparrow$0.11)**
	PointRend [14]	ResNet-101	**79.98($\uparrow$2.26)**	**86.98($\uparrow$2.04)**	**96.40($\uparrow$0.15)**

4   Experiments and Results

4.1   Datasets and Experimental Settings

To validate the effectiveness of the proposed SAMDConv, a series of comparative experiments were conducted on several commonly used models, including GCNet [1], APCNet [7], PointRend [14], UperNet [24], Semantic FPN [13]. The proposed SAMDConv was used to replace the last three layers of the backbone part, and the weights of the last three layers were randomly initialized. To ensure fairness, all methods were tested under the same parameter settings, operating environment, and training strategies. To demonstrate the generalization ability of the convolution module, various tasks were tested on various public datasets, such as CityScapes [4], VOC [6], and ADE20k [28]. All segmentation tasks were performed on a desktop computer equipped with an i9-10900X 3.70GHz 10-core processor and four GeForce GTX 2080 graphics cards, using the SGD optimizer and the Pytorch framework for all experiments.

4.2   Performance on Image Segmentation

Table 2. Comparison results between normal dilated convolution (Baseline), Deformable convolution and the proposed SAMDConv for image segmentation on ADE20K dataset

Methods	Methods	Backbone	Mean IoU(%)	Mean Acc(%)	Pixel Acc(%)
Dilated Convolution (Baseline)	Semantic FPN [13]	ResNet-50	34.73	43.44	77.08
	PSPNet+JUP [23]	ResNet-50	39.00	50.52	79.28
	PointRend [14]	ResNet-50	38.54	49.48	78.22
	Semantic FPN [13]	ResNet-101	36.28	44.49	78.19
	PointRend [14]	ResNet-101	40.02	50.56	79.09
Depformable convolution	Semantic FPN [13]	ResNet-50	35.64	44.68	78.12
	PSPNet+JUP [23]	ResNet-50	38.79	50.31	79.12
	PointRend [14]	ResNet-50	39.09	49.59	78.05
	Semantic FPN [13]	ResNet-101	36.98	44.88	78.37
	PointRend [14]	ResNet-101	39.74	50.43	79.00
SAMDConv	Semantic FPN [13]	ResNet-50	**39.11(↑4.38)**	**50.16(↑6.72)**	**79.00(↑1.92)**
	PSPNet+JUP [23]	ResNet-50	**39.59(↑0.59)**	**51.06(↑0.54)**	**79.70(↑0.42)**
	PointRend [14]	ResNet-50	**40.91(↑2.37)**	**52.47(↑2.99)**	**79.56(↑1.34)**
	Semantic FPN [13]	ResNet-101	**40.81(↑4.53)**	**51.52(↑7.03)**	**79.54(↑1.35)**
	PointRend [14]	ResNet-101	**42.04(↑2.02)**	**53.67(↑3.11)**	**80.28(↑1.19)**

Performance on Cityscapes. Table 1 shows the comparative experimental results between standard dilated convolution and the proposed SAMDConv based on eight mainstream segmentation networks on the Cityscapes dataset. It can be clearly seen that the proposed SAMDConv can comprehensively outperforms dilated convolution, with significant improvements in mIoU, mAcc, and Pixel Acc. These results demonstrate the proposed SAMDConv has significant advantages over the baseline (using Dilated Convolution) in all eight methods, with noticeable improvements in each method, even in PSPNet, where the improvement in mIoU was only 0.51%. In PointRend, SAMDConv showed strong adaptability, with improvements of 3.22% mIoU, 3.52% mAcc, and 0.34% PixelAcc when using ResNet50 as the backbone network, while improvements of 2.26% mIoU, 2.04% mAcc, and 0.15% PixelAcc when using ResNet101. This indicates that the proposed SAMDConv has provided extremely helpful support for improving segmentation accuracy and pixel prediction accuracy. From the whole table, the proposed SAMDConv can provide an average growth of 1.27% mIoU, 1.14% mAcc, and 0.12% PixelAcc, which demonstrates its strong generalization performance. The proposed SAMDConv has the potential to be widely applied as a basic module in various segmentation methods.

Performance on ADE20K. Besides accuracy, network parameter number and computational speed are also important evaluation metrics in segmentation tasks, and some lightweight networks are designed for this purpose. However, to reduce network parameters and speed up computation, a decrease in accuracy is inevitable. Most existing lightweight segmentation networks have extremely fast processing speeds, but their segmentation accuracy and semantic understanding accuracy are often lower than conventional segmentation methods. Therefore, improving the segmentation accuracy of lightweight networks is a highly meaningful task in practical applications.

To verify the effectiveness of lightweight segmentation networks, a series of comparative experiments are conducted based on lightweight segmentation networks on the ADE20K dataset. Table 2 illustrates the comparison results by respectively using normal dilated convolution (Baseline), depformable convolution and the proposed SAMDConv. It is very clear from the table that the proposed SAMDConv performs remarkably well in these lightweight networks, substantially improving their accuracy, and its degree of adaptation to these networks even exceeds that of the conventional segmentation methods in the previous experiments.

Compared to normal dilated convolution (Baseline), the proposed SAMD-Conv can respectively increase by 0.59%–4.53% in mIoU, by 0.54%–7.03% in mAcc, and by 0.42%–1.92% in Pixel Acc. On the other hand, deformable convolution has no significant improvement in segmentation performance owing to the fact that the spatial attention mechanism is implicitly introduced into depformable convolution by learning the position of the sampling points. Among these methods, Semantic FPN has the most obvious improvement effect when using the proposed SAMDConv module. For example, mIoU, mAcc and Pixel Acc can respectively increase from 34.73% to 39.11%, from 43.44% to 50.16%, and from 77.08% to 79.00% when ResNet-50 is selected as the backbone in Semantic FPN. When using ResNet-101 as the backbone, Semantic FPN can respectively increase by 4.53% in mIoU, by 7.03% in mAcc, and by 1.35% in Pixel Acc. Such a significant improvement fully demonstrates that the proposed SAMDConv is far superior to dilated convolution (Baseline) in terms of accuracy in feature extraction and semantic understanding, making it extremely suitable for image segmentation tasks.

Performance on Pascal VOC. To further verify the strong generalization performance of the proposed SAMDConv on various datasets, a series of semantic segmentation experiments are conducted on the Pascal VOC dataset. Some segmentation networks like GCNet [1], CCNet [10], ISANet [25] and ANNNet [29] are used for comparison between the default setting (using dilated convolution) and the proposed SAMDConv module. ResNet-50 is selected as the backbone for all networks, and the experimental settings and running environments were the same.

Table 3 illustrates the comparison results between the proposed SAMDConv and dilated convolutions, and the results show that the proposed SAMDConv can obtain competitive performance on Pascal VOC. For instance, the improvements of mIoU and mAcc respectively range from 0.15% to 1.20%, and from 0.33% to 1.33%. Overall, the proposed SAMDConv can also outperform the normal dilated convolution in various semantic segmentation networks on Pascal VOC.

Table 3. Comparison results of various methods between normal dilated convolution (Baseline) and the proposed SAMDConv for image segmentation on Pascal VOC.

Methods	SAMDConv	Mean IoU(%)	Mean Acc(%)
GCNet [1]	×	76.22	85.88
	✓	**76.37(↑0.15)**	**86.21(↑0.33)**
CCNet [10]	×	76.04	84.43
	✓	**76.39(↑0.35)**	**85.62(↑1.19)**
ISANet [25]	×	74.91	84.40
	✓	**75.56(↑0.65)**	**85.08(↑0.68)**
ANNNet [29]	×	74.58	83.34
	✓	**75.68(↑1.20)**	**84.77(↑1.33)**

× represents the default setting (using dilated convolution).
✓ represents replacing dilated convolution with the proposed SAMDConv module.

5   Conclusions

A novel spatially-adaptive multi-scale dilated convolution (SAMDConv) is proposed in this work. The proposed SAMDConv can dynamically adjust the size of the receptive field for different scale targets by adaptively regulating the dilation rate of dilated convolutions. Meanwhile, a SSA module is proposed to personally select a reasonable combination of sampling scales for each spatial location. Finally, a series of experiments are conducted to verify the effectiveness of the proposed SAMDConv module based on various segmentation networks on various datasets, such as Cityscapes, ADE20K and Pascal VOC. The results show that the proposed methods can obtain competitive performance compared with normal dilated convolutions and depformable convolutions, and can effectively improve the ability to extract multi-scale information by adaptively regulating the dilation rate.

Acknowledgments. This work was supported in part by National Natural Science Foundation of China (Grant No. 62373324, 62271448 and U20A20171), in part by Zhejiang Provincial Natural Science Foundation of China (Grant No. LGF22F030016 and LY21F020027), and in part Key Programs for Science and Technology Development of Zhejiang Province (2022C03113).

References

1. Cao, Y., Xu, J., Lin, S., Wei, F., Hu, H.: GCNet: non-local networks meet squeeze-excitation networks and beyond. In: Proceedings of the IEEE/CVF International Conference on Computer Vision Workshops (2019)
2. Chen, L.C., Zhu, Y., Papandreou, G., Schroff, F., Adam, H.: Encoder-decoder with atrous separable convolution for semantic image segmentation. In: Proceedings of the European Conference on Computer Vision (ECCV), pp. 801–818 (2018)
3. Chen, Q., Wang, Y., Yang, T., Zhang, X., Cheng, J., Sun, J.: You only look one-level feature. In: Proceedings of the IEEE/CVF Conference on Computer Vision and Pattern Recognition, pp. 13039–13048 (2021)
4. Cordts, M., et al.: The cityscapes dataset for semantic urban scene understanding. In: Proceedings of the IEEE Conference on Computer Vision and Pattern Recognition, pp. 3213–3223 (2016)
5. Dosovitskiy, A., et al.: An image is worth 16x16 words: transformers for image recognition at scale. arXiv preprint arXiv:2010.11929 (2020)
6. Everingham, M., Van Gool, L., Williams, C.K., Winn, J., Zisserman, A.: The pascal visual object classes (VOC) challenge. Int. J. Comput. Vision **88**, 303–338 (2010)
7. He, J., Deng, Z., Zhou, L., Wang, Y., Qiao, Y.: Adaptive pyramid context network for semantic segmentation. In: Proceedings of the IEEE/CVF Conference on Computer Vision and Pattern Recognition, pp. 7519–7528 (2019)
8. He, K., Zhang, X., Ren, S., Sun, J.: Spatial pyramid pooling in deep convolutional networks for visual recognition. IEEE Trans. Pattern Anal. Mach. Intell. **37**(9), 1904–1916 (2015)
9. Hu, J., Shen, L., Sun, G.: Squeeze-and-excitation networks. In: Proceedings of the IEEE Conference on Computer Vision and Pattern Recognition, pp. 7132–7141 (2018)
10. Huang, Z., Wang, X., Huang, L., Huang, C., Wei, Y., Liu, W.: CCNet: criss-cross attention for semantic segmentation. In: Proceedings of the IEEE/CVF International Conference on Computer Vision, pp. 603–612 (2019)
11. Jaderberg, M., Simonyan, K., Zisserman, A., et al.: Spatial transformer networks. In: Advances in Neural Information Processing Systems, vol. 28 (2015)
12. Jiang, W., Liu, M., Peng, Y., Wu, L., Wang, Y.: HDCB-Net: a neural network with the hybrid dilated convolution for pixel-level crack detection on concrete bridges. IEEE Trans. Industr. Inf. **17**(8), 5485–5494 (2020)
13. Kirillov, A., Girshick, R., He, K., Dollár, P.: Panoptic feature pyramid networks. In: Proceedings of the IEEE/CVF Conference on Computer Vision and Pattern Recognition, pp. 6399–6408 (2019)
14. Kirillov, A., Wu, Y., He, K., Girshick, R.: Pointrend: image segmentation as rendering. In: Proceedings of the IEEE/CVF Conference on Computer Vision and Pattern Recognition, pp. 9799–9808 (2020)
15. Long, J., Shelhamer, E., Darrell, T.: Fully convolutional networks for semantic segmentation. In: Proceedings of the IEEE Conference on Computer Vision and Pattern Recognition, pp. 3431–3440 (2015)
16. Mehta, S., Rastegari, M., Caspi, A., Shapiro, L., Hajishirzi, H.: Espnet: efficient spatial pyramid of dilated convolutions for semantic segmentation. In: Proceedings of the European Conference on Computer Vision (ECCV), pp. 552–568 (2018)
17. Pan, X., et al.: On the integration of self-attention and convolution. In: Proceedings of the IEEE/CVF Conference on Computer Vision and Pattern Recognition, pp. 815–825 (2022)

18. Son, H., Lee, J., Cho, S., Lee, S.: Single image defocus deblurring using kernel-sharing parallel atrous convolutions. In: Proceedings of the IEEE/CVF International Conference on Computer Vision, pp. 2642–2650 (2021)

19. Takahashi, N., Mitsufuji, Y.: Densely connected multi-dilated convolutional networks for dense prediction tasks. In: Proceedings of the IEEE/CVF Conference on Computer Vision and Pattern Recognition, pp. 993–1002 (2021)

20. Vaswani, A., et al.: Attention is all you need. In: Advances in Neural Information Processing Systems, vol. 30 (2017)

21. Wang, P., et al.: Understanding convolution for semantic segmentation. In: 2018 IEEE Winter Conference on Applications of Computer Vision (WACV), pp. 1451–1460. IEEE (2018)

22. Woo, S., Park, J., Lee, J.-Y., Kweon, I.S.: CBAM: convolutional block attention module. In: Ferrari, V., Hebert, M., Sminchisescu, C., Weiss, Y. (eds.) ECCV 2018. LNCS, vol. 11211, pp. 3–19. Springer, Cham (2018). https://doi.org/10.1007/978-3-030-01234-2_1

23. Wu, H., Zhang, J., Huang, K., Liang, K., Yu, Y.: FastFCN: rethinking dilated convolution in the backbone for semantic segmentation. arXiv preprint arXiv:1903.11816 (2019)

24. Xiao, T., Liu, Y., Zhou, B., Jiang, Y., Sun, J.: Unified perceptual parsing for scene understanding. In: Proceedings of the European Conference on Computer Vision (ECCV), pp. 418–434 (2018)

25. Xu, Z., Ren, H., Zhou, W., Liu, Z.: ISANET: non-small cell lung cancer classification and detection based on CNN and attention mechanism. Biomed. Signal Process. Control 77, 103773 (2022)

26. Yu, F., Koltun, V., Funkhouser, T.: Dilated residual networks. In: Proceedings of the IEEE Conference on Computer Vision and Pattern Recognition, pp. 472–480 (2017)

27. Zhao, H., Shi, J., Qi, X., Wang, X., Jia, J.: Pyramid scene parsing network. In: Proceedings of the IEEE Conference on Computer Vision and Pattern Recognition, pp. 2881–2890 (2017)

28. Zhou, B., Zhao, H., Puig, X., Fidler, S., Barriuso, A., Torralba, A.: Scene parsing through ADE20K dataset. In: Proceedings of the IEEE Conference on Computer Vision and Pattern Recognition, pp. 633–641 (2017)

29. Zhu, Z., Xu, M., Bai, S., Huang, T., Bai, X.: Asymmetric non-local neural networks for semantic segmentation. In: Proceedings of the IEEE/CVF International Conference on Computer Vision, pp. 593–602 (2019)

SADD: Generative Adversarial Networks via Self-attention and Dual Discriminator in Unsupervised Domain Adaptation

Zaiyan Dai[1,2], Jun Yang[1,2(✉)], Anfei Fan[1,2], Jinyin Jia[1,2], and Junfan Chen[1,2]

[1] College of Computer Science, Sichuan Normal University, Chengdu, China
daizaiyan@163.com
[2] Visual Computing and Virtual Reality Key Laboratory of Sichuan Province, Chengdu, China

Abstract. Image classification is an active research in the field of computer vision. There are still significant challenges in the classification accuracy of cross-domain images due to the privacy, human, and material cost issues involved in labeled data collection, and the distribution differences among the collected images of the same category. To address the above problems, unsupervised domain adaptation (UDA) methods emerge, which transfer prior knowledge from the labeled source domain to the unlabeled target domain. In this work, we propose a new UDA architecture, SADD, which performs feature-level and pixel-level discrimination in a self-attention generative adversarial network. Specifically, we use the self-attention mechanism in extracting features to obtain globally dependent embeddings. In addition, we apply pixel-level distribution consistency loss on the embedding-generated images to mitigate the pixel-level distribution shifts due to unstable image style shifts. Further, we use discriminators for embedding reconstruction to assist the feature extractor in aligning features and enhancing the classification ability of the classifier. We evaluate our approach on the DIGITS classification dataset and the OFFICE-31 recognition dataset, and the results demonstrate the robustness and superiority of our approach.

Keywords: Unsupervised Domain Adaptation · Self-Attention · Dual Discriminator · Generative Adversarial Networks

1 Introduction

Convolutional neural networks (CNNs), i.e., AlexNet [1], VGGNet [2], InceptionNet [3], and ResNet [4] and their variants, demonstrate superior advantages

Supported by Natural Science Foundation of Sichuan Province under Grant 2022NSFSC0552, and National Natural Science Foundation of China (62006165).
Z. Dai and J. Yang—Authors contribute equally to this work.

in classification tasks by learning discriminative feature representations of data from large amounts of labeled data. However, in practical scenarios, the quantity of data labels is insufficient, the labeling process is complicated and the cross-domain data distribution lacks consistency, so the classification results of convolutional neural networks are not satisfactory when testing unseen datasets. To tackle these issues, unsupervised domain adaptation has become popular due to powerful generalization capabilities. In this work, we propose a novel unsupervised domain adaptive architecture based on generative adversarial networks via self-attention and double discriminator (SADD), which uses supervised data sampled from the source domain and unsupervised data sampled from the target domain for joint training to obtain shared feature embeddings. The proposed generative adversarial network with dual discriminators is employed to ensure the maximum similarity between the synthetic image generated from the source embedding and the target embedding and the ground image in the source domain, further minimizing the distribution distance between the domains.

SADD can produce globally dependent feature embeddings while retaining fine foreground detail. Different from USADA [5] and GTA [6], neither does SADD choose to train the classifier with unstable images after style migration as USADA did, nor does SADD use features extracted by the feature extractor as GTA did. SADD trains the classifier interactively using a global dependency embedding and a reconstruction embedding that relies on generating adversarial networks on pixel-level and feature-level dual discriminators for implementation. Furthermore, to enhance the robustness of the model in the face of style changes, pixel-level distribution consistency loss is applied as an additional constraint in the generated images.

The main contributions of our work are as follows:

- A new unsupervised domain adaptive network named SADD extracts effective embedding while preserving foreground information.
- A novel distribution consistency loss to enhance the pixel-level details, which improves the stylistic consistency among the adapted images.
- Extensive experiments demonstrate that SADD achieves outstanding effects and is capable of producing good results while preserving fine content details.

2   Related Work

Existing unsupervised domain adaptive methods are divided into two research directions, one of which is to define a suitable metric to measure domain differences [7,8], and the other is adversarial training. The Former focuses on confusing the source and target domains by minimizing the loss of the domain difference metric between the two domains. Tzeng et al. proposed the DDC [9] algorithm to measure the distance between domains using MMD for the first time, which improved to utilize MK-MMD [10] the following year. Liu et al. used CORAL loss to compute the covariance matrix to mix source and target domain features. Recently Wasserstein [12] distance and KL [13] scatter measurement have also emerged, and joint computation on multilayer neural networks to obtain domain invariant features have been attempted, such as JAN [7]. However, in some cases, the limitation of the size of the dataset and the number of layers of the network leads to inaccurate classification, so defining a suitable metric to measure domain discrepancy is frequently combined with other approaches to achieve an effective result.

The adversarial training approach adopts the idea of a "zero-sum game" in game theory to achieve Nash equilibrium in the adversarial training, and to accomplish domain adaptive classification, which mainly includes two methods: feature-level adaptive and pixel-level adaptive. The former employs adversarial discrimination to discriminate which domain the features come from by using various network structures. Except for the typical CNN-based methods such as DANN [14] and ADDA [15], the Transformer-based methods DAT [16] and SUDA [17] also appeared lately. Other methods like DRCN [18] and MCD [19] use the idea of encoder-decoder and decision boundary, respectively, for this purpose. Pixel-level adaptation is the mapping of source and target images to each other with adversarial generation methods. COGAN [11] is similar to [6], which is a classical approach to generating resembling image pairs. In recent years adaptive methods based on style transfers [20–22] have gradually emerged. Due to the instability of GAN training, combining feature-level adaptive and pixel-level adaptive methods can alleviate this problem to some extent, as in [23,24]. It is worth pointing out that Zhu and Chen's work is more like ours in that they rely heavily on the quality of the generated images, and the generation quality directly affects the feature alignment effect later. In contrast, the pixel-level alignment of the generative adversarial network in SADD is assisted for feature-level alignment. In other words, when SADD is used to deal with complex datasets with huge domain differences, the model can still classify the target domain accurately in a relatively short period even if the generation effect is slightly worse.

3   Approach

As shown in Fig. 1, the SADD proposed in this paper is the architecture of self-attention and dual discriminator generative adversarial network, which is improved based on SAGAN [25] and ACGAN [26]. Firstly, the source domain embedding f_s and the target domain embedding f_t are extracted from the source domain image x_s and the target domain image x_t by the feature extractor F, and then, f_s, f_t will be sent to the D_1 discriminator to discriminate whether the embedding is from the source or the target domain. z is the random sampling noise from $N(0, 1)$, l is the one-hot encoding of class labels. The embedding, z and l are concatenated as the input of the generator, The source domain fake image $\widehat{x}_s$ and the target domain fake image $\widehat{x}_t$ generated from the generator will be fed to the D_2 discriminator for true-false discrimination and embedding reconstruction $\widehat{f}_s, \widehat{f}_t$. Ultimately, the embeddings obtained by F and reconstructed by the discriminator D_2 are used to train the classifier. We will describe the specific network structure in detail later.

Feature Extractor(F). F mainly extracts the image features in the source and target domains, and it consists of a basic architecture and a self-attention module. The basic structure of F is varied for images of different sizes and complexity. $32 \times 32 \times 1$ digital recognition image F uses the basic CNN architecture, which contains three convolutional layers, three Relu activation functions, and two max-pooling layers. F captures the features of 128; F adopts the swin-transformer structure [27], which is pre-trained on ImageNet to handle $256 \times 256 \times 3$ images and outputs 2048 features, which has a more enhanced feature extraction capability than most experiments using pre-trained resnet50.

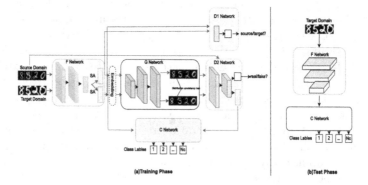

Fig. 1. Explanation of the proposed method. (a) has a total of five parts, a feature extractor, a generator, two discriminators, and a classifier, where the red flow indicates the propagation reverse of the labeled data; the blue flow indicates the propagation direction of the unlabeled data; and the black flow is the common propagation direction. (b) has only a feature extractor and a classifier for feature extraction and embedding classification separately.

The calculation process of self-attention is illustrated in Fig. 2, where it is on the last layer of F. The features converge this module to get 128 or 2048 embeddings which are more prominent in foreground information. Experiments demonstrate that the model with SA is significantly better than the model without SA (See Fig. 3).

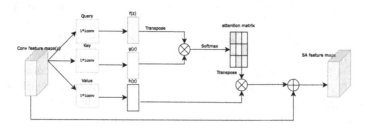

Fig. 2. Overview of self-attention mechanism.

Generator(G). The inputs of G are $E_s = [f_s, z, l]$, $E_t = [f_t, z, l]$, which is the combination of embedding z, l. $l \in \{0, 1\}^{(N_c+1)}$ is the one-hot encoding of the class label, where N_c is the number of real categories while $\{N_c + 1\}$ being the fake class, which assigned to unlabeled target domain data. The network structure in G contains four Blocks, where each Block has a transposed convolutional layer, a batch norm layer, and a Relu activation function, and the final output of G is similar to the source domain image with $\widehat{x_s}, \widehat{x_t}$.

Discriminator(D). The inputs of $D1$ are f_s, f_t, trying to discriminate the embedding from the source domain or the target domain at the feature level, and finally achieves to confuse the embedding distribution and reduce the domain difference in the adversarial training. $D1$ network contains three full connection layers and two Relu activation functions. Finally, the sigmoid outputs the probability from the source domain and models it as a binary classifier. The inputs of $D2$ are $\widehat{x_s}, \widehat{x_t}$. It attempts to distinguish between real and fake images at the pixel level and finally generates images similar to the source domain in the confrontation training further reduce domain differences. $D2$ network contains four blocks, each of which has a convolution layer, a batch norm layer, a LeakyRelu activation function and a max-pooling layer. $D2$ has two outputs: (1) D_{2rf} is real-fake output which modeled as a binary classifier; (2) D_{2re}: rebuilding $\widehat{f_s}$ and $\widehat{f_t}$. As shown in Table 4, the feasibility of the dual discriminator method was demonstrated in the ablation experiment.

Classifier(C). The inputs of C are $f_s, f_t, \widehat{f_s}$ and $\widehat{f_t}$, while the output is the probability of each class, modeled as a multi-valued classifier. C consists of two fully connected layers and a Relu activation function. Compared with [6], we additionally train the classifier with the reconstructed embedding $\widehat{f_s}, \widehat{f_t}$ with a view to C accurately classifying different positional embeddings of the same data and enhancing the generalization ability of C. Since the target domain data are unlabeled, we use $f_s, \widehat{f_s}$ with labeled data for training C.

We now describe our optimization process where we alternate between optimizing $D2, G, F, C$ and $D1$ networks.

Given a source domain image as input, $D1, F$ is optimized by minimizing a binary cross-entropy against loss L_{1adv_src} for better feature alignment.

$$L_{1adv\_src} = \mathbf{E}_{x_s \sim X_s} \min_F \max_D \log \left(D_1 \left(F \left(x_s \right) \right) \right) \tag{1}$$

In the case of the source image input, the $D2$ discriminator has two outputs: the true/false output D_{2rf} and the reconstructed output D_{2re}. The labeled reconstructed embedding is fed into C for classification. Thus D_{2rf} is optimized by a minimizing binary cross-entropy against loss L_{2adv_src}. D_{2re} uses the gradient from C and minimizes the cross-entropy classification loss L_{2cls_src} for optimization.

$$L_{2adv\_src} = \mathbf{E}_{x_s \sim X_s, e_s \sim E_s} \max_{D2} \log \left(D_{2rf} \left(x_s \right) \right) + \log \left(1 - D_{2rf} \left(G \left(e_s \right) \right) \right) \tag{2}$$

$$L_{2cls\_src} = \mathbf{E}_{x_s \sim X_s, e_s \sim E_s} \max_{D2} \log C \left(D_{2re} \left(G \left(e_s \right) \right) \right) \tag{3}$$

In the case of source image input, the gradient from $D2$ is used to update G using a combination of source-domain adversarial loss and source-domain classification loss L_{G_fromD2}, to produce realistic class-consistent source images.

$$L_{G\_fromD2} = \mathbf{E}_{x_s \sim X_s, e_s \sim E_s} \min_{G} -log\left(C\left(D_{2re}\left(G\left(e_s\right)\right)\right)\right) + log\left(1 - D_{2rf}\left(G\left(e_s\right)\right)\right)$$
(4)

F and C are updated in a supervised manner at the source image, the image generated by the source embedding, and the source label, using a classification loss that minimizes cross-entropy to achieve this.

$$L_{C\_cls} = \mathbf{E}_{x_s \sim X_s, e_s \sim E_s} \min_{C} -log\left(C\left(D_{2re}\left(G\left(e_s\right)\right)\right)\right) - log\left(C\left(F\left(x_s\right)\right)\right)$$
(5)

$$L_{F\_cls} = \mathbf{E}_{x_s \sim X_s, e_s \sim E_s} \min_{F} -log\left(C\left(D_{2re}\left(G\left(e_s\right)\right)\right)\right) - log\left(C\left(F\left(x_s\right)\right)\right)$$
(6)

Given the target domain image as input. $D1, F$ are optimized by minimizing a binary cross-entropy against loss L_{1adv_tgt}.

$$L_{1adv\_tgt} = \mathbf{E}_{x_t \sim X_t} \min_{F} \max_{D1} log\left(1 - D_1\left(F\left(x_t\right)\right)\right)$$
(7)

Given the target domain image as input, F is also updated using the adversarial gradient from $D2$ so that the feature learning and image generation processes proceed smoothly and simultaneously.

$$L_{Fadv\_tgt} = \mathbf{E}_{e_t \sim E_t} \min_{F} log\left(1 - D_{2rf}\left(G\left(e_t\right)\right)\right)$$
(8)

Given the target domain image as input, the real/fake output D_{2rf} and the reconstructed output D_{2re} of the D2 discriminator. Since the target domain data is unlabeled and D_{2re} is not considered, D_{2rf} is optimized against the loss L_{2adv_tgt} by a minimizing binary cross-entropy.

$$L_{2adv\_tgt} = \mathbf{E}_{e_t \sim E_t} \max_{D2} log\left(1 - D_{2rf}\left(G\left(e_t\right)\right)\right)$$
(9)

Distribution Consistency Loss:The advantage of our approach is the use of $L2$ loss to define the pixel-level distribution differences. Given a batch of x_s and x_t as input, we update G using the following distribution consistency loss.

$$L_{G\_mse} = \mathbf{E}_{e_s \sim E_s, e_t \sim E_t} \min_{G} \|G\left(e_s\right) - G\left(e_t\right)\|_2$$
(10)

Algorithm 1. Iterative training procedure of our approach

---

for epoch in 1:N **do**

    Sample k images from source domain $X_s = \left\{x_s^i, y_s^i\right\}_{i=1}^k$

    let $f_s = F(x_s)$ embeddings computed for the target images

    Sample noise $z = \left\{z_i\right\}_i^k \sim N(0,1)$

    let e_s, e_t be the input of the generator

    Update $D2$ using $L_{D_2} = L_{2adv\_src} + L_{2cls\_src} + L_{2adv\_tgt}$

    Update G using $L_G = L_{G\_formD2} + L_{G\_MSE}$

    Update F using $L_F = L_{F\_cls} + L_{Fadv\_tgt} + \alpha\left(L_{1adv\_src} + L_{1adv\_tgt}\right)$

    Update C $using L_C = L_{c\_cls}$

    Update D1 using $L_{D1} = L_{1adv\_src} + L_{1adv\_tgt}$

α is the weight added to the loss function. During the experiment, we found that the alignment features obtained by the feature extractor at the beginning of the iteration do not align well with the source domain features, therefore the loss value of the feature extractor is relatively high, so we add a "penalty" to the F part of the loss, which can make the initial loss smaller so that the model can gradually converge during the iteration.

4 Experiments

4.1 Implementation Details

Datasets. This section reports the experimental validation of our method. In this paper, we set up two adaptation experiments (1) a small-scale simple data distribution: the DIGITS dataset with input images randomly cropped to 32×32 (2) a large-size complex data distribution: the OFFICE-31 dataset with input images resized to 256×256, where the experimental dataset contains both pixel variability and sample number variability. Our method performs well in both cases and thus proving the effectiveness of our approach.

Training Information. Pytorch framework is used to implement SADD. The network was trained on a 3090 GPU using the Adam optimizer [28] with an initial learning rate of 5e-4 and a learning rate decay parameter of 1e-4, where the exponential decay parameter β_1 was 0.8 and β_2 was 0.999 batches were set to 100 and 10 on the DIGITS dataset and the OFFICE dataset, respectively.

4.2 Digit Experiments

We conducted our experiments under the following settings: MNIST⟶USPS and USPS⟶MNIST using the entire training set of MNIST and USPS, and SVHN⟶MNIST using the labeled 73257 SVHN images and the unlabeled 60,000 MNIST images to train our model. Table 1 shows the results of our proposed method compared with other methods. We can see that our approach is practically valid. For SVHN⟶MNIST experiments with significant domain differences, our self-attention mechanism is effective, as shown in Fig. 3, where self-attention can better highlight the foreground information and ignore the

Table 1. Precision (mean ± std%) values for five independent runs of the cross-domain recognition task on the DIGITS dataset. The best numbers are marked in bold, - indicates unreported results. MN: mnist, US: usps, SV: svhn.

Method	MN→US	US→MN	SV→MN	Average
Source only	76.1 ± 1.2	60.2 ± 1.1	62.2 ± 1.2	66.2
DANN [14]	89.4 ± 0.2	90.1 ± 0.8	76.0 ± 1.8	85.2
DDC [9]	79.1 ± 0.5	66.5 ± 3.3	68.1 ± 0.3	71.2
DRCN [18]	91.8 ± 0.09	73.7 ± 0.04	82.0 ± 0.16	82.5
ADDA [15]	89.4 ± 0.2	90.1 ± 0.8	76.0 ± 1.8	85.2
CycADA [22]	94.8 ± 0.2	95.7 ± 0.2	88.3 ± 0.2	92.9
GTA [6]	92.8 ± 0.9	90.8 ± 0.9	92.4 ± 0.9	92.0
Pixel DA [20]	95.9	-	-	-
UDAG [29]	97.5 ± 0.1	**98.3 ± 0.1**	94.3 ± 1.6	96.7
Ours	**98.1 ± 0.1**	97.7 ± 0.1	**95.2 ± 0.6**	**97.0**

non-transferable features between domains in the face of invalid domain information. As visualized in the t-SNE plot in Fig. 4, the two domains adapt very well.

Fig. 3. The top and bottom rows are the generation images of the same epoch in the SVHN→MNIST experiment, in which the top row does not add the attention mechanism while the bottom row adds it. The discriminative features of the image's texture features are more prominent with the addition of self-attention.

4.3   OFFICE Experiments

The OFFICE-31 dataset is a small sample dataset containing 31 classes of images from three domains, Amazon, Webcam, and DSLR, each containing 2817, 795, and 498 images, respectively. In the experiments, retraining the network with fewer samples of large images would lead to overfitting and low test accuracy. Therefore, we chose to fine-tune the Swin-Transformer model [27] pre-trained on ImageNet, and G generated 64*64 images. Experiments show that this network is more advantageous than the previously pre-trained resnet-50 [6]. Table 2 reports the performance of our method compared with other methods.

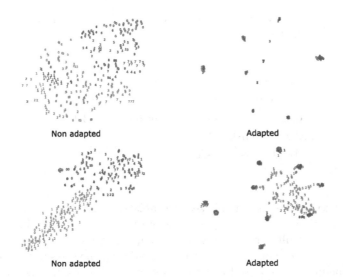

Fig. 4. The top (bottom) row shows the t-SNE before and after the adaptation of MN⟶US (SV⟶MN), respectively, and the number of data in the graph we choose is 300.

4.4 Comparison and Ablation Experiments

Comparison Experiments: To verify the effectiveness and superiority of the pixel-level distribution alignment loss (L2 loss) proposed in this paper, the loss in [9,11,12] is used as a comparison for the experiments, which the final accuracy is shown in Table 3. Mainly because pixel-level distribution alignment is to perform, and L2 loss is not only smooth and microscopic but also based on the pixel-by-pixel difference comparison for the trade-off, the peak signal-to-noise ratio (PSNR) [30] is commonly used in the experiments as the image evaluation index, which is the counterpart of L2 loss.

Table 2. Precision (mean ± std%) values on the OFFICE dataset. Results are reported for an average of more than 5 independent runs. The best numbers are marked in bold. A: amazon, W: webcam, D: dslr.

Method	A→W	D→W	W→D	A→D	D→A	W→A	Average
ResNet-Source only	68.4 ± 0.2	96.7 ± 0.1	99.3 ± 0.1	68.9 ± 0.2	62.5 ± 0.3	60.7 ± 0.3	76.1
DDC [9]	75.6 ± 0.2	76.0 ± 0.2	98.2 ± 0.1	76.5 ± 0.3	62.2 ± 0.4	61.5 ± 0.5	78.3
DAN [10]	80.5 ± 0.4	97.1 ± 0.2	99.6 ± 01	78.6 ± 0.2	63.3 ± 0.3	62.8 ± 0.2	80.4
JAN [7]	85.4 ± 0.3	97.4 ± 0.2	99.8 ± 0.2	84.7 ± 0.3	68.8 ± 0.3	70.0 ± 0.4	84.3
GTA [6]	89.5 ± 0.5	97.9 ± 0.3	99.8 ± 0.4	87.7 ± 0.5	72.8 ± 0.3	71.4 ± 0.4	86.5
Ours	**95.1 ± 0.2**	**99.0 ± 0.2**	**100.0 ± 0.1**	**92.0 ± 0.5**	**81.1 ± 0.3**	**80.5 ± 0.3**	**91.3**

Table 3. Comparison experiments, accuracy (mean ± std%) values obtained using different losses on the DIGITS dataset.

Method	MN⟶US	US⟶MN	SV⟶MN
SADD_No loss	95.0	91.9	91.9
SADD_MMD	95.8`	92.7	93.8
SADD_CORAL	96.0	95.6	93.4
SADD_WDGRL	97.1	96.4	94.0
SADD_L2 loss	**98.1**	**97.7**	**95.2**

Ablation Experiments: In this section, ablation experiments are conducted on the DIGITS dataset to verify the necessity of each module in the SADD model, and the results are in Table 4. (1) The experimental results show that when the model does not reconstruct the embedding, the classifier can only rely on the embedding extracted by the feature extractor for training without the interaction information with the GAN, which eventually leads to poor generalization and low classification accuracy of the classifier. (2) When the model uses only the $D2$ discriminator, it is not enough to rely on the adversarial generation approach to help F extract the embedding, and the $D1$ discriminator can perform adversarial discrimination from the feature level. F combines the results of these two discriminations to effectively reduce the discrimination error and eventually achieve efficient domain adaptation. (3) The method enhances the robustness of the model to style changes by incorporating the distribution consistency loss between the generated images so that the distribution of the generated dummy images and the source domain images are as close as possible. (4) The method enhances the robustness of the model to style changes by incorporating the distribution consistency loss between the generated images so that the distribution of the generated dummy images and the source domain images are as close as possible. When the model does not incorporate the self-attention mechanism, the embedding obtained by F retains poor foreground information, and the embedding affects the classification accuracy to some extent.

Table 4. Ablation experiments accuracy (mean ± std%) values obtained using different losses on the DIGITS dataset.

Method	MN⟶US	US⟶MN
Rebuild Embedding	95.7 ± 0.2	94.9 ± 0.2
Add D1 Discriminator	96.4 ± 0.1	95.3 ± 0.2
Add L2 Loss	97.5 ± 0.3	96.7 ± 0.3
Add Self Attention	**98.1 ± 0.1**	**97.7 ± 0.1**

5   Conclusion

In this work, we propose SADD, a UDA method. SADD introduces a self-attention mechanism in the feature extractor to capture the remote information between feature embeddings. And adversarial discrimination at the feature level while adversarial generation at the pixel level. The model adds pixel-level distributional consistency loss and employs feature embedding reconstruction on the discriminator to aid in feature alignment and enhance the classifier's generalization capabilities. Experiments on two datasets demonstrate the superiority of our approach, and we prepare to use more powerful network architectures to solve more challenging domain adaptive problems in the future.

References

1. Krizhevsky, A., Sutskever, I., Hinton, G.E.: ImageNet classification with deep convolutional neural networks. Commun. ACM **60**, 84–90 (2012)
2. Simonyan, K., Zisserman, A.: Very Deep Convolutional Networks for Large-Scale Image Recognition. CoRR, abs/1409.1556 (2014)
3. Szegedy, C., Ioffe, S., Vanhoucke, V., Alemi, A.A.: Inception-V4, Inception-ResNet and the Impact of Residual Connections on Learning. arXiv, abs/1602.07261 (2016)
4. He, K., Zhang, X., Ren, S., Sun, J.: Deep residual learning for image recognition. In: 2016 IEEE Conference on Computer Vision and Pattern Recognition (CVPR), pp. 770–778 (2015)
5. Li, Y., Lin, C., Li, H., et al.: Unsupervised domain adaptation with self-attention for post-disaster building damage detection. Neurocomputing **415**, 27–39 (2020)
6. Sankaranarayanan, S., Balaji, Y., Castillo, C.D., et al.: Generate to adapt: aligning domains using generative adversarial networks. In: Proceedings of the IEEE Conference on Computer Vision and Pattern Recognition, pp. 8503–8512 (2018)
7. Long, M., Zhu, H., Wang, J., et al.: Deep transfer learning with joint adaptation networks. In: International Conference on Machine Learning, pp. 2208–2217. PMLR (2017)
8. Shen, J., Qu, Y., Zhang, W., et al.: Wasserstein distance guided representation learning for domain adaptation. In: Proceedings of the AAAI Conference on Artificial Intelligence, vol. 32, no. 1 (2018)
9. Tzeng, E., Hoffman, J., Zhang, N., et al.: Deep domain confusion: maximizing for domain invariance. arXiv preprint arXiv:1412.3474 (2014)
10. Long, M., Cao, Y., Wang, J., et al.: Learning transferable features with deep adaptation networks. In: International Conference on Machine Learning, pp. 97–105. PMLR (2015)
11. Sun, B., Saenko, K.: Deep CORAL: correlation alignment for deep domain adaptation. In: Hua, G., Jégou, H. (eds.) ECCV 2016. LNCS, vol. 9915, pp. 443–450. Springer, Cham (2016). https://doi.org/10.1007/978-3-319-49409-8_35
12. Chen, P., Zhao, R., He, T., Wei, K., Qidong, Y.: Unsupervised domain adaptation of bearing fault diagnosis based on join sliced Wasserstein distance. ISA Trans. **129**, 504–519 (2022)
13. Nguyen, A., Tran, T., Gal, Y., Torr, P.H., Baydin, A.G.: KL Guided Domain Adaptation. arXiv, abs/2106.07780 (2021)
14. Ganin, Y., Ustinova, E., Ajakan, H., et al.: Domain-adversarial training of neural networks. J. Mach. Learn. Res. **17**(1), 2096-2030 (2016)

15. Tzeng, E., Hoffman, J., Saenko, K., et al.: Adversarial discriminative domain adaptation. In: Proceedings of the IEEE Conference on Computer Vision and Pattern Recognition, pp. 7167–7176 2017

16. Lee, J., Hwang, K., Kwak, M., et al.: Domain adaptation training of a transformer. In: 2022 IEEE International Conference on Consumer Electronics-Asia (ICCE-Asia), pp. 1–5. IEEE (2022)

17. Zhang, J., Huang, J., Tian, Z., et al.: Spectral unsupervised domain adaptation for visual recognition. In: Proceedings of the IEEE/CVF Conference on Computer Vision and Pattern Recognition, pp. 9829–9840 (2022)

18. Ghifary, M., Kleijn, W.B., Zhang, M., Balduzzi, D., Li, W.: Deep reconstruction-classification networks for unsupervised domain adaptation. In: Leibe, B., Matas, J., Sebe, N., Welling, M. (eds.) ECCV 2016. LNCS, vol. 9908, pp. 597–613. Springer, Cham (2016). https://doi.org/10.1007/978-3-319-46493-0_36

19. Saito, K., Watanabe, K., Ushiku, Y., et al.: Maximum classifier discrepancy for unsupervised domain adaptation. In: Proceedings of the IEEE Conference on Computer Vision and Pattern Recognition, pp. 3723–3732 (2018)

20. Bousmalis, K., Silberman, N., Dohan, D., et al.: Unsupervised pixel-level domain adaptation with generative adversarial networks. In: Proceedings of the IEEE Conference on Computer Vision and Pattern Recognition, pp. 3722–3731 (2017)

21. Tran, L., Sohn, K., Yu, X., Liu, X., Chandraker, M.: Gotta adapt 'em all: joint pixel and feature-level domain adaptation for recognition in the wild. In: 2019 IEEE/CVF Conference on Computer Vision and Pattern Recognition (CVPR), pp. 2667–2676 (2018)

22. Hoffman, J., et al.: CyCADA: cycle-consistent adversarial domain adaptation. In: International Conference on Machine Learning (2017)

23. Zhu, H., Yin, H., Xia, D., Wang, D., Liu, X., Zhu, S.: Joint pixel-level and feature-level unsupervised domain adaptation for surveillance face recognition. In: Chinese Conference on Pattern Recognition and Computer Vision (2022)

24. Chen, Z., Zhao, L., He, Q., Kuang, G.: Pixel-level and feature-level domain adaptation for heterogeneous SAR target recognition. IEEE Geosci. Remote Sens. Lett. **19**, 1–5 (2022)

25. Poojary, A., Phapale, A., Salpekar, R., Balpande, S.: Self-Attention Generative Adversarial Network: The Latest Advancement in GAN (2020)

26. Odena, A., Olah, C., Shlens, J.: Conditional Image Synthesis with Auxiliary Classifier GANs. Presented at the (2016)

27. Liu, Z., et al.: Swin transformer: hierarchical vision transformer using shifted windows. In: 2021 IEEE/CVF International Conference on Computer Vision (ICCV), pp. 9992–10002 (2021)

28. Kingma, D.P., Ba, J.: Adam: a method for stochastic optimization. arXiv preprint arXiv:1412.6980 (2014)

29. Wang, G.G., Guo, T., Yu, Y., Su, H.: Unsupervised domain adaptation classification model based on generative adversarial network. Acta Electonica Sinica **48**(6), 1190 (2020)

30. Poobathy, D., Chezian, R.M.: Edge detection operators: peak signal to noise ratio based comparison. Int. J. Image Graph. Sig. Process. **6**, 55–61 (2014)

ELFLN: An Efficient Lightweight Facial Landmark Network Based on Hybrid Knowledge Distillation

Shidong Chen[1,2,3], Yalun Wang[1,2,3], Huicong Bian[1,2,3], and Qin Lu[1,2,3]([✉])

[1] Key Laboratory of Computing Power Network and Information Security,
Ministry of Education, Shandong Computer Science Center,
Qilu University of Technology (Shandong Academy of Sciences), Jinan, China
luqin@qlu.edu.cn
[2] Shandong Engineering Research Center of Big Data Applied Technology,
Faculty of Computer Science and Technology,
Qilu University of Technology (Shandong Academy of Sciences), Jinan, China
[3] Shandong Provincial Key Laboratory of Computer Networks,
Shandong Fundamental Research Center for Computer Science, Jinan, China

Abstract. A facial landmark detector based on coordinate regression has minimal parameters and memory consumption, making it suitable for deployment on mobile devices. Typically, it employs a lightweight network as the backbone, but this network is not capable of effectively extracting features. Knowledge distillation is a promising methodology for developing a precise and lightweight network. The existing lightweight student networks do not have corresponding teacher networks, thereby hindering their ability to leverage the knowledge distillation technique. To tackle the inefficiency of the lightweight network in extracting features, we present an efficient lightweight network in this study, named ELFLN. In addition, we propose a novel hybrid knowledge distillation (HKD) framework to address the problem that the inadequacy of the lightweight network in carrying out features knowledge distillation. Finally, we augment our ELFLN by integrating facial landmark detection with head pose estimation, thereby enhancing the network generalization capability. To verify the efficacy of our proposed approach, we perform comprehensive experiments on 300W and WFLW datasets, achieving NME reach of 3.20% on 300W and 4.12% on WFLW with ELFLN+HKD. The number of parameters of ELFLN is observed to reduce by 62%, and FLOPs diminish by 87% compared to state-of-the-art SLPT.

Keywords: Facial landmark detection · Coordinate-based regression · Lightweight network · Knowledge distillation

Supported by Shandong Province Key R&D Program (Major Science and Technology Innovation Project) Project under Grants 2020CXGC010102.

Q. Liu et al. (Eds.): PRCV 2023, LNCS 14432, pp. 485–497, 2024.
https://doi.org/10.1007/978-981-99-8543-2_39

1   Introduction

Facial landmark detection is a crucial step in many facial image analyses and applications. It aims to automatically localize a group of pre-defined fiducial points, e.g., eye, mouth, and nose. In the past 20 years, great progress has been made toward improving facial landmark detection algorithms accuracy. Although current models have relative high precision, these models are hard to run on resource-constrained devices. Therefore, how to improve the facial landmark detection efficiency and quality is still a meaningful but challenging work for the facial landmark detection.

Deep Convolutional Neural Networks have considerably advanced face landmark detection. These methods mainly fall into two categories, i.e., the coordinate-based regression model and the heatmap-based regression model. The coordinate-based regression model directly predicts the facial landmark coordinates from the input image. The heatmap-based regression model generates the likelihood heatmaps for each facial landmark, and then the network is trained to generate these heatmaps for each input image. Wu et al. [1] suggested that facial boundary lines contain valuable information that can help detect facial landmarks. Xia et al. [2] proposed a method that involves learning the intrinsic relationships between facial landmarks, using Transformers to generate heatmaps. Huang et al. [3] proposed ADNet, which comprises the anisotropic direction loss (ADL) and anisotropic attention module (AAM) that perform coordinate and heatmap regression, respectively.

The concept of knowledge distillation was first proposed by Hinton [4]. Knowledge distillation defines a learning manner where a bigger teacher network is employed to guide the training of a smaller student network for the image classification task. Knowledge distillation has two types: logits knowledge distillation and features knowledge distillation. In logits knowledge distillation, the focus is on transferring the output of the teacher network to the student network. Current feature knowledge distillation has proven more effective than logits knowledge distillation as it teaches the student model to learn the teacher model feature map, rather than just the final output. Heo et al. [5] suggested using the activation boundary of hidden layer neurons to transfer knowledge. Kim et al. [6] proposed the factor transfer approach. The comparison between logits knowledge distillation and features knowledge distillation is shown in Fig. 1.

Despite the strong spatial generalization ability and high precision of heatmap-based regression models, they are challenging to run on resource-constrained devices and are not end-to-end solutions. Moreover, all heatmap-based regression networks suffer from quantization errors. The coordinate-based regression models adopt end-to-end solutions. Typically, this approach provides fast inference speed, low parameters, and low memory consumption. However, the following two challenges hinder the adoption of coordinate-based regression methods. The main issue is that coordinate-based regression models commonly use lightweight backbones, such as MobileNetV2 and ShuffleNetV2, which have insufficient feature extraction capabilities. The other issue is that these lightweight models lack corresponding high parameter models for features knowl-

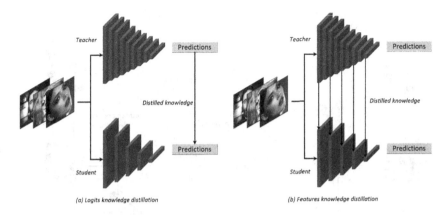

Fig. 1. The comparison between logits knowledge distillation and features knowledge distillation.

edge distillation. Although Zhang et al. [7] utilized logits knowledge distillation to improve detection accuracy, their model only transfers output results, and its performance is still suboptimal.

To address the above issues, we developed an efficient lightweight facial landmark network (ELFLN) and proposed a hybrid knowledge distillation (HKD) framework to attain accuracy comparable to state-of-the-art methods. ELFLN comprises a backbone network, an upsample module, and a feature learning module. The feature learning module includes feature learning block (FLB) and coordinate learning block (CLB). ELFLN has a relatively small number of parameters but comparable accuracy to larger models. To utilize features knowledge distillation in lightweight networks, we propose the HKD framework. Our approach involves training a three-branch lightweight network as the teacher network and a single-branch lightweight network as the student network. In both models, FLB is used to transfer feature information, while CLB transfers the keypoint coordinate data. The teacher network has the same architecture as the student network.

Extensive experiments on two academic datasets are demonstrated the superior performance of our method to other solutions. The contributions of this study are listed as follows:

- In this paper, we propose a novel efficient lightweight facial landmark network named ELFLN.
- To enable feature knowledge distillation for lightweight models, we propose hybrid knowledge distillation framework.
- In order to enhance the network generalization capability, we apply multi-task learning to combine facial landmark detection with head pose estimation.
- Extensive experiments validate the effectiveness of our proposed method on two popular benchmark datasets 300W and WFLW.

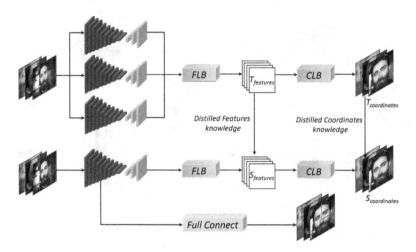

Fig. 2. The specific framework of the hybrid knowledge distillation. The three-branch network is the teacher network, and the single-branch network is the student network. FLB is features learning block, CLB is coordinates learning block.

2   Method

2.1   Overview

First, we describe ELFLN, followed by an explanation of our hybrid knowledge distillation framework for facial landmark detection. Finally, we introduce how we incorporate multi-task learning in our ELFLN. The network framework can be seen in Fig. 2.

2.2   ELFLN

ELFLN consists of a backbone network, an upsample module, and a feature learning module. ShuffleNetV2 is a well-known lightweight deep neural network that has been popular for its ability to operate on embedded and mobile devices. ShuffleNetV2 is shown to cope with complex tasks such as image classification, object detection, and semantic segmentation. Therefore, ELFLN adopts ShuffleNetV2 as the backbone network. However, compared to larger networks, ShuffleNetV2 has limited capability in extracting features. To enhance the feature extraction capability of ShuffleNetV2, three upsample layers are added as an upsample module after ShuffleNetV2, which boosts the spatial and semantic richness of the features, based on inspiration from Feature Pyramid Networks (FPN) [8]. Table 4 demonstrates the improved performance of ShuffleNetV2 on the WFLW and 300W datasets for facial landmark detection after the upsample module is integrated. In the upsample module, PixelShuffle [9] is utilized as the upsample method, commonly used in super-resolution, and applied to facial landmark detection for the first time to address some limitations problems in traditional interpolation upsample and deconvolution upsample methods. Furthermore, this method does not require many parameters compared to deconvolution convolution. After adding upsample module, to further improve the detection accuracy, we build the feature learning block (FLB) and the coordinate learning block (CLB) as feature learning module. The FLB and CLB blocks serve two distinct purposes for the student model: FLB employs convolutions to produce a set of learnable features maps, while CLB uses convolutions and softmax to generate the coordinates of facial landmarks. These blocks allow the student model to learn from the feature maps and facial landmarks coordinates generated by the teacher model. ELFLN has fewer parameters and FLOPs compared to other Convolutional Neural Network (CNN) models, such as HRNet. Additionally, it provides accurate detection. The specific ELFLN network is presented in Fig. 2.

2.3 Hybrid Knowledge Distillation

As the lightweight model doesn't have a corresponding high-parameter model, feature knowledge distillation cannot be applied to it. We proposed HKD to solve the above issue. The teacher network is a three-branch architecture that incorporates three homogeneous student networks, each having the same configuration. In teacher network, each branch applies the ShuffleNetV2 2x version. Following this network augmentation, the teacher network shares a similar architecture with the student network, but it has a more powerful feature learning ability than the student network. Figure 2 displays the framework of hybrid knowledge distillation. The teacher network generates $T_{features}$ and $T_{coordinates}$, while the student network generates $S_{features}$ and $S_{coordinates}$. In the process of distillation training, the loss function is defined as Eq. (1):

$$L_{face} = MSE_S + MSE_{KD} + JS_S + JS_{KD}. \tag{1}$$

To minimize the mean squared error between $S_{coordinates}$ generated by the student network and the ground-truth coordinates $GT_{coordinates}$, we use MSE_S as the loss function. The specific loss functions are defined as follows:

$$MSE_S = \frac{1}{n} \sum_{i=1}^{n} \left(S^i_{\text{coordinates}}, GT^i_{\text{coordinates}} \right)^2, \tag{2}$$

where n represents the total number of facial landmarks. Similarly, the MSE_{KD} calculates the loss between $S_{coordinates}$ and $T_{coordinates}$, given by the following equation:

$$MSE_{KD} = \frac{1}{n} \sum_{i=1}^{n} \left(S^i_{\text{coordinates}}, T^i_{\text{coordinates}} \right)^2. \tag{3}$$

The loss functions JS_S and JS_{KD} are employed to calculate the Jensen-Shannon divergences of $S_{features}$ and $GT_{features}$ and $S_{features}$ and $T_{features}$, respectively. Equations (4) and (5) show the respective equations.

$$JS_S = \frac{1}{2}KL_S \left(S_{\text{features}} \mid \frac{S_{\text{features}} + GT_{\text{features}}}{2} \right), \qquad (4)$$

$$JS_{KD} = \frac{1}{2}KL_{KD} \left(S_{\text{features}} \mid \frac{S_{\text{features}} + T_{\text{features}}}{2} \right), \qquad (5)$$

where KL_S represents the KL divergence. A smaller value of KL_S indicates $S_{features}$ and $GT_{features}$ are similar and a smaller value of KL_{KD} indicates $S_{features}$ and $T_{features}$ are similar..

2.4   Multi-task Learning

Previous works [10,11] have demonstrated that employing multi-task learning can improve the stability and robustness of landmark localization. The primary task of this study is the facial landmark detection, while the secondary task is the head pose estimation. The secondary task utilizes some of the same shallow features as the primary task, resulting in complementary domain-related information. This complementing process facilitates enhanced generalization through shared learning between the tasks. As the 300W and WFLW datasets lack head pose angle labels, we utilized 6DRepNet [12] that is a state-of-the-art algorithm for head pose estimation to generate head pose angle information. We employed its generated results as soft labels for the head pose angles in the datasets. ELFLN directly integrates the soft labels into its training process. Head pose estimation is obtained from the feature maps of the ShuffleNetV2 backbone via a full-connection layer. More details are presented in Fig. 2. The smooth L1 loss function is utilized to calculate the difference between the predicted angles and soft ground-truth angles, as shown in the following equation:

$$L_{head} = \begin{cases} 0.5(x - x_i)^2 & \text{if } |x - x_i| < 1 \\ |x - x_i| - 0.5 & \text{otherwise}, \end{cases} \qquad (6)$$

where x is the predicted angles and x_i is the soft ground-truth angles. The final loss function is presented below:

$$L_{total} = L_{face} + \alpha \times L_{head}, \qquad (7)$$

where α is set to 0.2. For the setting of α, we refer to work [13]. The task of facial landmark detection is regarded as the primary task, and the weight should be greater than that of the task of head pose estimation. After doing extensive experiments, we identify an optimal value and set α to 0.2.

3   Experimental Results

3.1   Datasets

300W. The 300W [14] dataset is comprised of 3148 training images and 689 test images, all of which contain 68 manually annotated landmarks. For testing, 300W has three subsets: a common subset consisting of 554 images, a challenging subset containing 135 images, and a full subset that includes both the common and challenging subsets.

WFLW. The WFLW [1] dataset is also commonly utilized and includes 7500 training images and 2500 testing images, each containing 98 manually annotated landmarks. To evaluate performance under varying conditions, the WFLW dataset comprises six subsets with 314 images having expressions, 326 images with extreme poses, 206 images with makeup, 736 images with occlusions, 698 images with varied illumination, and 773 images with blur.

3.2   Evaluation Metrics

Normalized mean error (NME) metric is utilized to assess the accuracy of our proposed model, as it has been done in previous studies. The "inter-ocular" distance, which represents the distance between the outer eye corners, is defined as the normalizing factor. Equation (8) demonstrates how the NME can be mathematically expressed:

$$nme = \frac{\sum_{i=1}^{N} \|x_{\text{pre}} - x_{gt}\|}{N \times d_{IOD}} \times 100\%, \tag{8}$$

where x_{pre} is the predicted coordinates and x_{gt} is the true coordinates, d_{IOD} is the distance between the outer eye corners. The calculation of the Area Under the Curve (AUC) is based on the cumulative error distribution (CED) curve. A higher AUC score indicates a significant portion of the test dataset has been accurately predicted. Parameters (Params) refer to the total number of trainable parameters in a deep learning model. The number of parameters in a model is a measure of its capacity or complexity, and larger models typically have more parameters. Floating Point Operations (FLOPs) is a measure of the number of floating-point operations performed by a deep learning model during inference. It is a measure of the computational complexity of the model and is related to the model accuracy and efficiency. The model becomes more computationally efficient as the FLOPs decrease.

3.3   Implementation Details

The facial images are resized to a 256×256. We utilize various data augmentation techniques to augment training set, including brightness, contrast, color modification, rotation, horizontal flipping. In the training phase, the teacher network is independently trained for 200 epochs using a batch size of 16, followed by the student network, which is trained using the proposed hybrid knowledge distillation, with batch size 16. The optimizer is AdamW with a learning rate of 10^{-4}. All experiments are conducted on an NVIDIA A100 GPU. ShuffleNetV2 $n \times$ ($n = 0.5, 1.0, 1.5, 2.0$) represents four backbone models of different levels of complexities. ELFLN $n \times$ ($n = 0.5, 1.0, 1.5, 2.0$) refers to levels of complexities of the ShuffleNetV2 backbone models. The ELFLN model uses ShuffleNetV2 2.0× unless otherwise state.

(a)   (b)

Fig. 3. Facial Landmark detection using ELFLN+HKD on 300W and WFLW test dataset. The left image (a) shows the comparison results on the 300W test dataset, while the right image (b) shows the comparison results on the WFLW test dataset.

Evaluation on 300W. The 300W dataset is a challenging benchmark for facial landmark detection. Table 1 compares our proposed method to other methods.

Table 1. NME (in %) of 68-point landmarks detection on 300W test dataset. The best and our results are marked in colors of red and blue, respectively.

Method	Year	Params	FLOPs	Normalized Mean Error		
				Full	Com.	Cha.
LAB [1]	2018	12.26M	18.96G	3.49	2.98	5.19
AVS [15]	2019	15.30M	24.27G	3.86	3.21	6.49
Awing [16]	2019	6.33M	8.78G	3.07	2.72	4.52
HRNet [17]	2020	9.66M	4.79G	3.32	2.87	5.15
SDFL [18]	2021	24.22M	4.94G	3.28	2.88	4.93
ADNet [3]	2021	13.37M	17.05G	2.93	2.53	4.58
SLPT [2]	2022	19.45M	8.14G	3.17	2.75	4.90
Teacher	2023	22.18M	3.03G	3.19	2.78	4.90
ELFLN	2023	7.40M	1.03G	3.34	2.94	5.26
ELFLN+HKD	2023	**7.40M**	1.03G	**3.20**	**2.84**	**5.18**

Although the performance of ELFLN+HKD does not outperform the state-of-the-art method SLPT, considering the number of parameters and FLOPs, the accuracy of our model is acceptable. Figure 3 (a) shows some examples of facial landmark detection using ELFLN+HKD on the challenging subset of the 300W dataset. The results indicate that the proposed ELFLN+HKD performs well even in challenging facial images, and experiments highlight the effectiveness of this method on the 300W dataset.

Evaluation on WFLW. Table 2 compares the performance of the state-of-the-art method to our proposed method on the WFLW dataset and its six subsets. The performance of ELFLN+HKD is comparable to the performance of the state-of-the-art method SLPT. However, ELFLN+HKD is a lightweight model with fewer parameters and FLOPs. Compared with SLPT, the number of parameters of ELFLN decrease by 62% and FLOPs decrease by 87%. Figure 3(b) shows some examples of facial landmark detection using ELFLN+HKD on the WFLW dataset. The results demonstrate that ELFLN+HKD performs well, with acceptable performance under various circumstances, such as occlusion, large pose, expression, illumination, blur, and make-up.

Table 2. NME (in %) and AUC of 98-point landmarks detection on WFLW dataset. The best and our results are marked in colors of red and blue, respectively.

Metric	Method	Year	Params	FLOPs	Full	Pose	Exp.	Illu.	Mu.	Occ.	Blur
NME	LAB [1]	2018	12.26M	18.96G	5.27	10.24	5.51	5.23	5.15	6.79	6.32
	AVS [15]	2019	15.30M	24.27G	4.39	8.42	4.68	4.24	4.37	5.6	4.86
	AWing [16]	2019	6.33M	8.78G	4.36	7.38	4.58	4.32	4.27	5.19	4.96
	HRNet [17]	2020	9.66M	4.79G	4.60	7.94	4.85	4.55	4.29	5.44	5.42
	SDFL [18]	2021	24.22M	4.94G	4.35	7.42	4.63	4.29	4.22	5.19	5.08
	ADNet [3]	2021	13.37M	17.05G	4.14	6.96	4.38	4.09	4.05	5.06	4.79
	SLPT [2]	2022	19.45M	8.14G	4.12	6.99	4.37	4.02	4.03	5.01	4.79
	Teacher	2023	22.18M	3.03G	4.17	7.45	4.23	4.30	4.06	5.08	4.80
	ELFLN	2023	7.40M	1.03G	4.25	7.78	4.36	4.56	4.26	5.21	4.86
	ELFLN+HKD	2023	**7.40M**	1.03G	4.12	**7.29**	**4.26**	**4.33**	3.96	5.00	4.77
AUC	LAB [1]	2018	12.26M	18.96G	0.532	0.235	0.495	0.543	0.539	0.449	0.463
	AVS [15]	2019	15.30M	24.27G	0.591	0.311	0.549	0.609	0.581	0.516	0.551
	AWing [16]	2019	6.33M	8.78G	0.572	0.312	0.515	0.578	0.572	0.502	0.512
	HRNet [17]	2020	9.66M	4.79G	0.524	0.251	0.510	0.533	0.545	0.459	0.452
	SDFL [18]	2021	24.22M	4.94G	0.576	0.315	0.550	0.585	0.583	0.504	0.515
	ADNet [3]	2021	13.37M	17.05G	0.602	0.344	0.523	0.580	0.601	0.530	0.548
	SLPT [2]	2022	19.45M	8.14G	0.595	0.348	0.573	0.603	0.608	0.520	0.537
	Teacher	2023	22.18M	3.03G	0.605	0.334	0.589	0.614	0.604	0.530	0.555
	ELFLN	2023	7.40M	1.03G	0.600	0.332	0.582	0.611	0.593	0.527	0.552
	ELFLN+HKD	2023	**7.40M**	1.03G	0.610	**0.346**	0.591	0.620	0.609	0.536	0.563

3.4   Ablation Study

Evaluation on Different Upsample Methods. Table 3 compares the effectiveness of various upsample methods in the upsample module, and demonstrates that PixelShuffle, which is commonly utilized in super-resolution, achieves the most favorable results in our task. Therefore, we have opted to use it in the upsample module.

Table 3. Comparisons about different upsample methods on the 300W and WFLW test datasets.

Upsample Method	NME (300W)	NME (WFLW)
ShuffleNetV2 2.0x + Interpolation	3.64	4.53
ShuffleNetV2 2.0x + Deconvolution	3.69	4.59
ShuffleNetV2 2.0x + PixelShuffle	3.46	4.43

Evaluation on Upsample Module and Feature Learning Module. The ablation studies presented in Table 4 demonstrate the efficacy of the upsample

Table 4. Comparisons about upsample module and feature learning module on the 300W and WFLW test datasets.

Method	NME (300W)	NME (WFLW)
ShuffleNetV2 0.5x	3.87	5.07
ShuffleNetV2 1.0x	3.62	4.68
ShuffleNetV2 1.5x	3.59	4.55
ShuffleNetV2 2.0x	3.58	4.52
ShuffleNetV2 0.5x+upsample	3.78	4.90
ShuffleNetV2 1.0x+upsample	3.55	4.60
ShuffleNetV2 1.5x+upsample	3.50	4.54
ShuffleNetV2 2.0x+upsample	3.46	4.43
ELFLN 0.5x	3.77	4.64
ELFLN 1.0x	3.53	4.37
ELFLN 1.5x	3.40	4.28
ELFLN 2.0x	3.34	4.25
SLPT	3.17	4.12

module and feature learning module in all versions of ShuffleNetV2. The results indicate a significant reduction in the NME after implementing the upsample module, and the generated feature maps contain both spatial and semantic information. Furthermore, the addition of the feature learning module further reduces the NME, validating its merit and preparing for the subsequent hybrid knowledge distillation. ELFLN 2.0x achieves comparable NME to the state-of-the-art SLPT, as depicted in Table 4.

Evaluation on KD and HKD. To assess the effectiveness of HKD, we compared it to conventional KD. We found that ELFLN became less effective after independently using KD because KD was initially used only for image classification task, where the final output of the network is usually a probability distribution after the softmax function, while the facial landmark detection task is a regression task, where the result is usually a determined coordinate value. It is easier for the ELFLN to learn a probability distribution generated by the teacher network than to learn a numerical value. Therefore, for the ELFLN, learning the coordinate values predicted by the teacher network will degrade the performance. Conversely, HKD enables ELFLN to learn the feature maps and coordinates of the teacher network simultaneously, leading to superior performance. As demonstrated in Fig. 4, HKD significantly outperforms traditional KD.

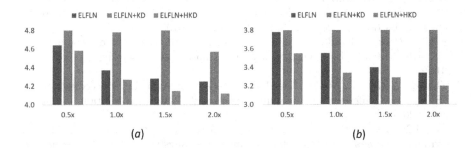

Fig. 4. Comparisons about HKD and KD on the 300W and WFLW test datasets. The left image (a) shows the comparison results on the 300W test dataset, while the right image (b) shows the comparison results on the WFLW test dataset.

4   Conclusion

This paper introduces a novel lightweight network, ELFLN, and the HKD framework, which were proposed to solve the limitations of lightweight models. ELFLN addresses the problem of lightweight models that cannot extract features effectively, while HKD addresses the problem of lightweight models' inability to use features in knowledge distillation. The combination of these two methods improves the accuracy of facial landmark detection. Not only does ELFLN+HKD perform comparably to the state-of-the-art method SLPT, but also it has fewer parameters and FLOPs, reducing them by 62% and 87%, respectively. Moreover, ELFLN and HKD can be applied to other computer vision tasks, such as body-joint tracking. Therefore, future research will explore the possibility of using this method in similar applications.

References

1. Wu, W., Qian, C., Yang, S., Wang, Q., Cai, Y., Zhou, Q.: Look at boundary: a boundary-aware face alignment algorithm. In: Proceedings of the IEEE Conference on Computer Vision and Pattern Recognition, pp. 2129–2138 (2018)
2. Xia, J., Qu, W., Huang, W., Zhang, J., Wang, X., Xu, M.: Sparse local patch transformer for robust face alignment and landmarks inherent relation learning. In: Proceedings of the IEEE/CVF Conference on Computer Vision and Pattern Recognition, pp. 4052–4061 (2022)
3. Huang, Y., Yang, H., Li, C., Kim, J., Wei, F.: Adnet: leveraging error-bias towards normal direction in face alignment. In: Proceedings of the IEEE/CVF International Conference on Computer Vision, pp. 3080–3090 (2021)
4. Hinton, G., Vinyals, O., Dean, J., et al.: Distilling the knowledge in a neural network. arXiv preprint arXiv:1503.02531 2(7) (2015)
5. Heo, B., Lee, M., Yun, S., Choi, J.Y.: Knowledge transfer via distillation of activation boundaries formed by hidden neurons. In: Proceedings of the AAAI Conference on Artificial Intelligence, vol. 33, pp. 3779–3787 (2019)
6. Kim, J., Park, S., Kwak, N.: Paraphrasing complex network: Network compression via factor transfer. In: Advances in Neural Information Processing Systems, vol. 31 (2018)
7. Zhang, F., Zhu, X., Ye, M.: Fast human pose estimation. In: Proceedings of the IEEE/CVF Conference on Computer Vision and Pattern Recognition, pp. 3517–3526 (2019)
8. Lin, T.Y., Dollár, P., Girshick, R., He, K., Hariharan, B., Belongie, S.: Feature pyramid networks for object detection. In: Proceedings of the IEEE Conference on Computer Vision and Pattern Recognition, pp. 2117–2125 (2017)
9. Shi, W., et al.: Real-time single image and video super-resolution using an efficient sub-pixel convolutional neural network. In: Proceedings of the IEEE Conference on Computer Vision and Pattern Recognition, pp. 1874–1883 (2016)
10. Teichmann, M., Weber, M., Zoellner, M., Cipolla, R., Urtasun, R.: Multinet: real-time joint semantic reasoning for autonomous driving. In: 2018 IEEE Intelligent Vehicles Symposium (IV), pp. 1013–1020. IEEE (2018)
11. Standley, T., Zamir, A., Chen, D., Guibas, L., Malik, J., Savarese, S.: Which tasks should be learned together in multi-task learning? In: International Conference on Machine Learning, pp. 9120–9132. PMLR (2020)
12. Hempel, T., Abdelrahman, A.A., Al-Hamadi, A.: 6D rotation representation for unconstrained head pose estimation. arXiv preprint arXiv:2202.12555 (2022)
13. Vandenhende, S., Georgoulis, S., Van Gansbeke, W., Proesmans, M., Dai, D., Van Gool, L.: Multi-task learning for dense prediction tasks: a survey. IEEE Trans. Pattern Anal. Mach. Intell. **44**(7), 3614–3633 (2021)

14. Sagonas, C., Tzimiropoulos, G., Zafeiriou, S., Pantic, M.: 300 faces in-the-wild challenge: the first facial landmark localization challenge. In: Proceedings of the IEEE International Conference on Computer Vision Workshops, pp. 397–403 (2013)
15. Qian, S., Sun, K., Wu, W., Qian, C., Jia, J.: Aggregation via separation: boosting facial landmark detector with semi-supervised style translation. In: Proceedings of the IEEE/CVF International Conference on Computer Vision, pp. 10153–10163 (2019)
16. Wang, X., Bo, L., Fuxin, L.: Adaptive wing loss for robust face alignment via heatmap regression. In: Proceedings of the IEEE/CVF International Conference on Computer Vision, pp. 6971–6981 (2019)
17. Wang, J., et al.: Deep high-resolution representation learning for visual recognition. IEEE Trans. Pattern Anal. Mach. Intell. **43**(10), 3349–3364 (2020)
18. Lin, C., et al.: Structure-coherent deep feature learning for robust face alignment. IEEE Trans. Image Process. **30**, 5313–5326 (2021)

Enhancing Continual Noisy Label Learning with Uncertainty-Based Sample Selection and Feature Enhancement

Guangrui Guo[1,2], Zhonghang Wei[3], and Jinyong Cheng[1,2(✉)]

[1] Key Laboratory of Computing Power Network and Information Security, Ministry of Education, Shandong Computer Science Center (National Supercomputer Center in Jinan, Qilu University of Technology (Shandong Academy of Sciences), Jinan, China
cjy@qlu.edu.cn
[2] Shandong Provincial Key Laboratory of Computer Networks, Shandong Fundamental Research Center for Computer Science, Jinan, China
[3] College of Computing and Informatics, Universiti Tenaga Nasional, Putrajaya Campus, Jalan Kajang - Puchong, Kajang, Malaysia

Abstract. The task of continual learning is to design algorithms that can address the problem of catastrophic forgetting. However, in the real world, there are noisy labels due to inaccurate human annotations and other factors, which seem to exacerbate catastrophic forgetting. To tackle both catastrophic forgetting and noise issues, we propose an innovative framework. Our framework leverages sample uncertainty to purify the data stream and selects representative samples for replay, effectively alleviating catastrophic forgetting. Additionally, we adopt a semi-supervised approach for fine-tuning to ensure the involvement of all available samples. Simultaneously, we incorporate contrastive learning and entropy minimization to mitigate noise memorization in the model. We validate the effectiveness of our proposed method through extensive experiments on two benchmark datasets, CIFAR-10 and CIFAR-100. For CIFAR-10, we achieve a performance gain of 2% under 20% noise conditions.

Keywords: Catastrophic forgetting · Noisy data · Replay

1 Introduction

Deep learning models have shown impressive performance in various tasks of computer vision, machine intelligence, and natural language processing. However, deep neural networks (DNNs) still face challenges in continual learning of new tasks. While some research has addressed the issue of continual learning, exploration of continual learning and noisy label classification in a single framework is relatively limited continual learning aims to enable models to maintain

This work was supported by the Natural Science Foundation of Shandong Province, China, under Grant Nos. ZR2020MF041 and ZR2022MF237, and the National Natural Science Foundation of China under Grant No. 11901325.

their learning capabilities in a continually changing environment, addressing the problem of catastrophic forgetting [15]. On the other hand, noisy label learning is used to address issues such as noise and missing labels in labeled data [29]. Since continual learning and noisy label tasks are inevitable in real-world scenarios, it is likely that they can occur simultaneously. Therefore, this study investigates a feasible deep learning approach that can overcome the challenges of catastrophic forgetting and noisy label data.

We first employ a replay-based approach to mitigate catastrophic forgetting in neural networks. Directly replaying noisy data can significantly degrade model performance, as deep learning models tend to remember previously learned knowledge. SPR [10] proposes that maintaining a clean replay buffer can greatly improve performance. It uses self-supervised learning to create a clean buffer, but this learning process requires long training times and high computational resources. CNLL [7] separates clean and noisy buffers by introducing a mask-based purification technique, which significantly improves classification accuracy, but there is still ample room for further development.

To this end, we propose an innovative framework that combines continual learning with noisy label learning [7,10]. Specifically, we use a sample selection method based on sample uncertainty to separate the data stream [1]. Analyzing the uncertainty of samples allows us to identify crucial samples for the model's learning process, improving its performance and generalization capability. Additionally, we integrate saliency features [21,23] into the data stream to enhance the model's representation ability and robustness. To mitigate the memorization of noisy data, we employ contrastive learning techniques [4,24] that compare the model's output with the expected output, adjusting parameters to reduce overfitting to noisy data and enhance stability and accuracy. Furthermore, we introduce entropy minimization [2] to encourage the model to rely more on input information during output generation, increasing interpretability and reliability in predictions.We verify our claims through extensive experimental evaluation.

In summary, the contributions of this research are as follows:

1. Introducing saliency features in the data stream to highlight the most important regions in images for model training.
2. Introducing a way to incorporate sample uncertainty into the sample separation stage for separating noisy data.
3. Mitigating the model's noise memorization through contrastive learning and entropy minimization, thereby enhancing the model's representation capability and robustness. Our experimental results demonstrate the effectiveness and superiority of our approach, opening up new possibilities for deep learning research and practical applications.

2   Related Work

continual learning addresses the catastrophic forgetting problem in neural networks, aiming to enable models to continually learn new tasks in evolving environments while retaining previously acquired knowledge [5]. Methods such as

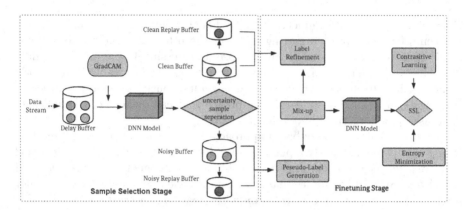

Fig. 1. Overall structure diagram. As the data stream progresses, we start with model pretraining. At this stage, we perform significant feature extraction. Then, utilizing sample uncertainty, we separate the noisy data and divide the data stream into two buffers: a clean buffer C for storing low-uncertainty samples and a noise buffer N for storing noisy samples. Simultaneously, we select the most representative data from both buffers for replay. Next, we perform semi-supervised learning-based fine-tuning, where pseudo-labels are generated for the noisy data, and contrastive learning and entropy minimization techniques are incorporated to further enhance the model's performance.

replay [1,16,18], where historical samples are stored and replayed during training, and regularization [6,11], which adds additional terms to control parameter updates and reduce forgetting, have been commonly used. Recent advances in continual learning include attention-based methods, generative models, and meta-learning, which offer more flexible and robust solutions.

Recent work in the field of continual learning has also focused on addressing other challenges, such as noisy labels, class imbalance, and concept drift. The issue of noisy labels refers to the presence of mislabeled data in the training data, which can degrade the performance of the model. Therefore, developing effective continual learning methods that can handle noisy labels is of great practical significance.

Noisy label learning focuses on addressing noise and missing label problems in labeled data [13,22]. The current approaches to handling noise can be divided into two subsets. One subset is label cleaning, which involves identifying samples with incorrect labels and removing them from the training data. Some methods use multi-objective interpolation training [17], sample separation based on noise distribution [27], or a unified selection mechanism to ensure class balance [8]. The differences among these methods lie in the criteria used to select clean samples. The other subset is loss regularization, where noise-robust loss functions and deep noise-cleaning models are employed to improve model robustness [14,30]. These methods reduce sensitivity to noisy data by considering noise characteristics and distributions. However, existing approaches for handling noisy labels may not be directly applicable to continual noisy learning.

3 Methods

In this work, we aim to address the challenges of online noise data stream learning. We propose a framework that effectively handles noisy data while retaining important information for future learning tasks. The specific workflow is illustrated in Fig. 1. Next, we will provide a detailed description of the components in our framework.

3.1 Sample Separation

During the continual learning process, to help the model adapt to new data distributions, we perform a warm-up on the model $F()$. Specifically, we conduct a short-term, fully supervised training on the data stream. During this period, we gradually reduce the standard cross-entropy (CE) loss with a very low learning rate to conduct the warm-up. We divide the data stream into two buffers during the warm-up phase: a clean buffer C and a noisy buffer N.

To select clean samples, we employ a method based on sample uncertainty. We believe that if the predictions of a sample remain stable after multiple perturbations, it indicates high confidence in that sample. This method helps us in choosing clean samples, reducing the impact of noise and mislabeled samples on the model, thereby improving its robustness and generalization ability.

First, we define the predictive probability as $p(y = c \mid x)$, where y represents the class label and x represents the sample. Then, we apply multiple perturbations to the sample x, resulting in a perturbed sample set $\tilde{x}_r = \{\tilde{x}_1, \tilde{x}_2, \ldots, \tilde{x}_r\}$, where R is the number of perturbations. For each perturbed sample $\tilde{x}_r$, we make predictions using the model F and obtain the predictive probability distribution $\tilde{p}(y = c \mid \tilde{x}_r)$.

$$\tilde{x}_r = F_r(x \mid \theta_r), r = 1, \ldots, R \tag{1}$$

$$\tilde{p}(y = c \mid \tilde{x}_r) = F(\tilde{x}_r \mid \theta) \tag{2}$$

Next, we calculate the uncertainty of the samples using a normalization method. Specifically, we compare whether the predicted category of each perturbed sample is equal to the target category and calculate the number of times they are equal, denoted as S_c. Finally, we calculate the uncertainty of a sample by subtracting the normalized maximum stability from 1. The specific calculation method is as follows:

$$S_c = \sum_{r=1}^{R} \mathbb{I}_c \, \text{argmax}_c \left(\tilde{p}(y = c \mid \tilde{x}_r) \right) \tag{3}$$

where r represents the number of perturbations, $\mathbb{I}_c$ is a discriminator used to determine whether the category of the perturbed sample is equal to the category of the original sample.

$$u(x) = 1 - \frac{1}{R} \max_c S_c \tag{4}$$

where $u(x)$ represents the uncertainty of a sample. A smaller uncertainty value indicates higher confidence in the sample, while a larger uncertainty value indicates lower confidence in the sample.

After sample selection, we divide the data stream into two buffers. Similar to the approach mentioned in CNLL [7], we replay samples with the highest confidence from the clean buffer and samples with the lowest confidence from the noise buffer.

3.2   Feature Extraction for Salient Features

To enhance the accuracy of our model, we adopted the GradCAM technique to extract salient features from input images. GradCAM computes gradient information with respect to the target class, identifying the most crucial regions for the model's classification decision.

(a) Original Image   (b) Guided Backprop 'Cat'   (c) Grad-CAM 'Cat'

Fig. 2. (a) Original image with a cat and a dog. (b) Guided Backpropagation: highlights all contributing features. (c) GradCAM: localizes class-discriminative regions.

Specifically, we extract feature heatmaps for each data during the warm-up phase. Then, we create a mask based on the heat map and fuse it with the original image to highlight areas with high activation values in the original data. The visualization of the results is shown in Fig. 2. By extracting salient features, we can capture important visual features from the data, enabling the model to better understand and classify images.

3.3   Fine-Tuning Phase

After the pre-training phase, the data stream is divided into two buffers: a Clean Buffer C and a Noise Buffer N. The Clean Buffer contains samples with low uncertainty, while the Noise Buffer contains the remaining noisy data stream. In the subsequent fine-tuning phase, the focus is on effectively handling the noisy data stream. To achieve this, we employ unsupervised feature learning techniques. In this case, the data in C is considered to have high confidence, and their labels are preserved. On the other hand, the labels of the data in N are discarded, and pseudo-labels are generated for them. In the initial phase, the standard cross-entropy loss function Lx is employed to address the clean labels. By minimizing the disparity between the model's predicted probabilities and the true labels, the model is enabled to acquire a more precise understanding of the correct class categorizations. The minimization of Lx enhances the model's ability to accurately classify the training data.

$$Lx = -\frac{1}{N}\Sigma\left(\log\left(p_i\left(y_i = c \mid x_i\right)\right)\cdot y_i\right) \tag{5}$$

where N represents the number of samples, p represents the predicted probability of the sample, and y represents the label of the sample. This loss function is used to measure the difference between the model's predicted probabilities and the true labels.

For unlabeled data, we regenerate pseudo-labels and utilize the Mean Squared Error (MSE) loss function, which is a commonly used loss function that measures the squared difference between predicted values and true values.

$$Lu = \frac{1}{N}\Sigma\left(\left(p_i\left(\hat{y}_i = c \mid x_i\right) - \hat{y}_i\right)^2\right) \tag{6}$$

where $\hat{y}_i$ is the pseudo-label.

3.4 Contrastive Learning

In addition, we have incorporated contrastive learning into our semi-supervised learning training pipeline. This allows us to learn more discriminative features by pulling representations of similar samples closer together and pushing representations of different samples apart. Contrastive learning is an unsupervised learning method that learns feature representations by comparing the similarity between samples [4,24], without relying on perfect separation of data streams or accurate pseudo-labels. In our approach, we use the SupConLoss as the contrastive learning loss function. By pulling representations of similar samples closer and pushing representations of different samples apart, the model can learn more discriminative and generalizable feature representations.

Contrastive learning can improve the quality of feature representations, making them more discriminative and thus more effective in downstream tasks. It can also reduce the risk of memorizing noisy labels during the semi-supervised learning training process. In this way, our approach achieves better performance and effectiveness in continual learning tasks.

$$L_{cl} = -\log\frac{\exp\left(\varphi\left(x_i\right),\varphi\left(x_j\right)\cdot\tau\right)}{\sum_{x\in C}\exp\left(\varphi\left(x_i\right),\varphi\left(x_j\right)\right)\cdot\tau\right)} \tag{7}$$

where $\varphi()$ represents the feature extractor and τ represents the temperature coefficient.

3.5   Minimization of Entropy

In the final stage of our framework, we incorporate the objective of entropy minimization. It is a common information theory method widely used in machine learning and pattern recognition tasks. The main idea is to optimize the learning process of the model by minimizing the entropy or conditional entropy of the samples. It encourages the model to rely more on the input information during the output generation, and helps reduce overfitting and improve the model's generalization on unseen samples. The specific formula is described as follows:

$$L_{mi} = \mathcal{H}\left(E_{x \in C} F(\varphi(x))\right) - E_{x \in C} \mathcal{H}(F(\varphi(x))) \tag{8}$$

where $\mathcal{H}$ denotes the entropy function, E denotes the expectation operation.

The final overall loss function equation is:

$$L_{\text{total}} = L_x + \lambda_u L_u + \lambda_{cl} L_{cl} + \lambda_{mi} L_{mi} \tag{9}$$

where λ_u, λ_{cl}, λ_{mi} are hyperparameters.

4   Experiments

In this section, we compare our method with other state-of-the-art (SOTA) models and evaluate its performance on the CIFAR-10 [12] and CIFAR-100 [12] datasets.

4.1   Experimental Design

In our experiment, our aim was to evaluate the performance of our proposed method compared to other state-of-the-art models on the CIFAR-10 and CIFAR-100 datasets.

For the CIFAR-10 dataset, we randomly selected two classes for each of the five tasks. Similarly, for the CIFAR-100 dataset, we selected five classes for each of the 20 tasks in two different ways: based on superclasses and random selection. We introduced two noise models to create a noise dataset. The first one is symmetric label noise, which evenly distributes some samples of a specific class to other classes. Then, we randomly group these classes to form five tasks. The second is asymmetric label noise [12], which inserts similar class labels into specific classes. In order to evaluate our proposed method, we pre trained the model for 30 periods with a low learning rate of 0.0005 and a batch size of 128 to fully warm up the model, and performed fine-tuning for 60 periods. We set the size of the delay buffer to 500. For the playback buffer, we have the same settings as in CNLL, with a clean replay buffer size of 25 and a noise replay buffer size of 50. We used two A100 NVIDIA graphics cards for model training in the experiment. For the optimizer, we use SGD.

We compared our method with the most advanced models currently available, which combine noise label learning and continual learning.

4.2 Baseline Methods

This study explores the scenario of continual learning with noisy labeled data. Therefore, we designed baselines by combining state-of-the-art methods for noise-label learning and continual learning. For continual learning, we chose CRS [20], PRS [9], and Gdumb [18]. We selected six methods for noise-label learning from the literature, namely SL [25], JoCoR [26], L2R [19], Pencil [28]. In addition, we also compared the methods proposed in SPR [10] and CNLL [7], and to our knowledge, SPR [10] was the first study to simultaneously handle continual learning and noise-label learning.

- Multitask [3]: We perform i.i.d offline training for 50 epochs with uniformly sampled mini-batches.
- Finetune: We run online training through the sequence of tasks.
- Gdumb [18]: As an advantage to GDumb, we allow CutMix with $p = 0.5$ and $\alpha = 1.0$. We use the SGDR schedule with $T_0 = 1$ and $T_{\mathrm{mult}} = 2$. Since access to a validation data in task-free continual learning is not natural, the number of epochs is set to 100 for CIFAR-10.
- L2R [19]: We use meta update with $\alpha = 1$, and set the number of clean data per class as 100 and the clean update batch size as 100.
- Pencil [28]: We use $\alpha = 0.4$, $\beta = 0.1$, $stage1 = 70$, $stage2 = 200$, $\lambda = 600$.
- SL [25]: We use $\alpha = 1.0$, $\beta = 1.0$.
- JoCoR [26]: We set $\Lambda = 0.1$.
- SPR [10]: Self-supervised batch size is 500 for CIFAR-10 and CIFAR-100. Among other parameters, we set $\beta_1 = 0.9$, $\beta_2 = 0.999$, $\epsilon = 0.0002$, $E_{max} = 5$.
- CNLL [7]: The delay buffer size is 500. The Clean buffer have a size of 500. N has a size of 1000. N_1 has a size of 25. N_2 has a size of 50.

4.3 Results

For CIFAR-10, we have compared the performance of our method against the baselines listed in Table 1. Our method outperforms other approaches under low-noise conditions for both symmetric and asymmetric noise types. For 20% symmetric noise, our performance shows a 2.0% improvement compared to CNLL. We achieve similar improvements in the more realistic scenario of asymmetric noise. Even under high-noise conditions, our results outperform most methods. These performance gains can be attributed to our carefully designed framework. For CIFAR-100, we also achieve similar improvements, as shown in Table 2.

506   G. Guo et al.

Table 1. The overall accuracy of the CIFAR-10 dataset. The results are generated after continual learning of noise labeling for all tasks. Here, the size of the delay buffer is 500. The results are the average of five experiments.

CIFAR-10					
Noise rate (%)	Symmetric			Asymmetric	
	20	40	60	20	40
Multitask 0% noise	**84.7**				
Finetune	18.5	18.1	17.0	15.3	12.4
EWC	18.4	17.9	15.7	13.9	11.0
CRS	19.6	18.5	16.8	28.9	25.2
CRS+L2R	29.3	22.7	16.5	39.2	35.2
CRS+Pencil	23.0	19.3	17.5	36.2	29.7
CRS+SL	20.0	18.8	17.5	32.4	26.4
CRS+JoCoR	19.4	18.6	21.1	30.2	25.1
PRS	19.1	18.5	16.7	25.6	21.6
PRS+L2R	30.1	21.9	16.2	35.9	32.6
PRS+Pencil	19.8	18.3	17.6	29.0	26.7
PRS+SL	20.1	18.8	17.0	29.6	24.0
PRS+JoCoR	19.9	18.6	16.9	28.4	21.9
MIR	19.6	18.6	16.4	26.4	22.1
MIR+L2R	28.2	20.0	15.6	35.1	34.2
MIR+Pencil	22.9	20.4	17.7	35.0	30.8
MIR+SL	20.7	19.0	16.8	28.1	22.9
MIR+JoCoR	19.6	18.4	17.0	27.6	23.5
GDumb	29.2	22.0	16.2	33.0	32.5
GDumb+L2R	28.2	25.5	18.8	30.5	30.4
GDumb+Pencil	26.9	22.3	16.5	32.5	29.7
GDumb+SL	28.1	21.4	16.3	32.7	31.8
GDumb+JoCoR	26.3	20.9	15.0	33.1	32.2
SPR	43.9	43.0	40.0	44.5	43.9
CNLL	68.7	**65.1**	**52.8**	67.2	**59.3**
OURS	**70.7**	54.7	33.9	**68.4**	50.9

Table 2. The reported results here are obtained on CIFAR-100. These results are the average of five experiments.

CIFAR-100						
Noise rate (%)	Superclass			Random		
	20	40	60	20	40	60
GDumb + L2R	15.7	11.3	9.1	16.3	12.1	10.9
GDumb + Pencil	17.5	11.6	6.8	16.7	12.5	4.1
GDumb + SL	18.6	13.9	9.4	19.3	13.8	8.8
GDumb + JoCoR	15.0	9.5	5.9	16.1	8.9	6.1
SPR	20.5	19.8	16.5	21.5	21.1	18.1
CNLL	39.0	**32.6**	**27.5**	38.7	**32.1**	**26.2**
OURS	**39.8**	28.2	18.1	**39.4**	27.8	17.7

4.4 Ablation Studys

In this section, we evaluated the impact of sample uncertainty, saliency features, Contrastive learning, and entropy minimization on our approach, as shown in Table 3. The experiments were conducted in both symmetric and asymmetric noise scenarios. Introducing sample uncertainty, saliency features, Contrastive learning, and entropy minimization consistently improved performance compared to the baseline model. The combination of all components $(+sap+cl+mi)$ achieved the highest accuracy. Particularly, the inclusion of Contrastive learning and entropy minimization objective proved to be beneficial as they further enhanced the model's ability to handle noisy data. The combination of these techniques yielded superior results compared to the individual components or the baseline approach. These findings highlight the importance of incorporating Contrastive learning and entropy minimization objective in our approach to effectively address the challenges of noise data stream learning.

Table 3. The impact of Contrastive learning and entropy minimization objective formulations on our approach. "+" indicates the inclusion of a specific component in the experiment.

Noise rate (%)	Symmetric			Asymmetric	
	20	40	60	20	40
+un	68.9	53.4	32.5	67.6	49.1
+sap	69.7	53.7	33.0	67.9	49.8
+sap+cl	70.6	54.4	33.8	68.4	50.5
+sap+cl+mi	**70.7**	**54.7**	**33.9**	**68.4**	**50.9**

5   Conclusion

In this study, we propose a simple and powerful solution that addresses both the noise labeling problem and the catastrophic forgetting problem. Our method includes two key stages: sample selection and fine-tuning. After extracting salient features, we divide the data stream into two parts based on the uncertainty of the samples, and select samples with high noise and high confidence for playback simultaneously. Subsequently, we conducted semi supervised fine-tuning of the image classification task, supplemented by comparative learning, to reduce the model's vulnerability to memory decline caused by noise. Finally, we introduced the goal of minimizing entropy to encourage the model to generate more input dependent outputs. Numerous experiments have demonstrated the effectiveness of our method, especially under low noise conditions.

References

1. Bang, J., Kim, H., Yoo, Y., Ha, J.W., Choi, J.: Rainbow memory: continual learning with a memory of diverse samples. In: Proceedings of the IEEE/CVF Conference on Computer Vision and Pattern Recognition, pp. 8218–8227 (2021)
2. Berthelot, D., et al.: Remixmatch: semi-supervised learning with distribution alignment and augmentation anchoring. arXiv preprint arXiv:1911.09785 (2019)
3. Caruana, R.: Multitask learning. Mach. Learn. **28**, 41–75 (1997)
4. Chen, T., Kornblith, S., Norouzi, M., Hinton, G.: A simple framework for contrastive learning of visual representations. In: International Conference on Machine Learning, pp. 1597–1607. PMLR (2020)
5. Ferdinand, Q., Clement, B., Oliveau, Q., Le Chenadec, G., Papadakis, P.: Attenuating catastrophic forgetting by joint contrastive and incremental learning. In: Proceedings of the IEEE/CVF Conference on Computer Vision and Pattern Recognition, pp. 3782–3789 (2022)
6. Guo, Y., Hu, W., Zhao, D., Liu, B.: Adaptive orthogonal projection for batch and online continual learning. In: Proceedings of the AAAI Conference on Artificial Intelligence, vol. 36, pp. 6783–6791 (2022)
7. Karim, N., Khalid, U., Esmaeili, A., Rahnavard, N.: CNLL: a semi-supervised approach for continual noisy label learning. In: Proceedings of the IEEE/CVF Conference on Computer Vision and Pattern Recognition, pp. 3878–3888 (2022)
8. Karim, N., Rizve, M.N., Rahnavard, N., Mian, A., Shah, M.: UNICON: combating label noise through uniform selection and contrastive learning. In: Proceedings of the IEEE/CVF Conference on Computer Vision and Pattern Recognition, pp. 9676–9686 (2022)
9. Kim, C.D., Jeong, J., Kim, G.: Imbalanced continual learning with partitioning reservoir sampling. In: Vedaldi, A., Bischof, H., Brox, T., Frahm, J.-M. (eds.) ECCV 2020. LNCS, vol. 12358, pp. 411–428. Springer, Cham (2020). https://doi.org/10.1007/978-3-030-58601-0_25
10. Kim, C.D., Jeong, J., Moon, S., Kim, G.: Continual learning on noisy data streams via self-purified replay. In: Proceedings of the IEEE/CVF International Conference on Computer Vision, pp. 537–547 (2021)

11. Kirkpatrick, J., et al.: Overcoming catastrophic forgetting in neural networks. Proc. Natl. Acad. Sci. **114**(13), 3521–3526 (2017)
12. Krizhevsky, A., Hinton, G., et al.: Learning multiple layers of features from tiny images (2009)
13. Li, S., Xia, X., Ge, S., Liu, T.: Selective-supervised contrastive learning with noisy labels. In: Proceedings of the IEEE/CVF Conference on Computer Vision and Pattern Recognition, pp. 316–325 (2022)
14. Liu, D., Zhao, J., Wu, J., Yang, G., Lv, F.: Multi-category classification with label noise by robust binary loss. Neurocomputing **482**, 14–26 (2022)
15. Mai, Z., Li, R., Jeong, J., Quispe, D., Kim, H., Sanner, S.: Online continual learning in image classification: an empirical survey. Neurocomputing **469**, 28–51 (2022)
16. de Masson D'Autume, C., Ruder, S., Kong, L., Yogatama, D.: Episodic memory in lifelong language learning. In: Advances in Neural Information Processing Systems, vol. 32 (2019)
17. Ortego, D., Arazo, E., Albert, P., O'Connor, N.E., McGuinness, K.: Multi-objective interpolation training for robustness to label noise. In: Proceedings of the IEEE/CVF Conference on Computer Vision and Pattern Recognition, pp. 6606–6615 (2021)
18. Prabhu, A., Torr, P.H.S., Dokania, P.K.: GDumb: a simple approach that questions our progress in continual learning. In: Vedaldi, A., Bischof, H., Brox, T., Frahm, J.-M. (eds.) ECCV 2020. LNCS, vol. 12347, pp. 524–540. Springer, Cham (2020). https://doi.org/10.1007/978-3-030-58536-5_31
19. Ren, M., Zeng, W., Yang, B., Urtasun, R.: Learning to reweight examples for robust deep learning. In: International Conference on Machine Learning, pp. 4334–4343. PMLR (2018)
20. Riemer, M., et al.: Learning to learn without forgetting by maximizing transfer and minimizing interference. arXiv preprint arXiv:1810.11910 (2018)
21. Selvaraju, R.R., Cogswell, M., Das, A., Vedantam, R., Parikh, D., Batra, D.: Grad-CAM: visual explanations from deep networks via gradient-based localization. In: Proceedings of the IEEE International Conference on Computer Vision, pp. 618–626 (2017)
22. Song, H., Kim, M., Park, D., Shin, Y., Lee, J.G.: Learning from noisy labels with deep neural networks: a survey. IEEE Trans. Neural Netw. Learn. Syst. (2022)
23. Wang, H., et al.: Score-CAM: score-weighted visual explanations for convolutional neural networks. In: Proceedings of the IEEE/CVF Conference on Computer Vision and Pattern Recognition Workshops, pp. 24–25 (2020)
24. Wang, X., et al.: Transformer-based unsupervised contrastive learning for histopathological image classification. Med. Image Anal. **81**, 102559 (2022)
25. Wang, Y., Ma, X., Chen, Z., Luo, Y., Yi, J., Bailey, J.: Symmetric cross entropy for robust learning with noisy labels. In: Proceedings of the IEEE/CVF International Conference on Computer Vision, pp. 322–330 (2019)
26. Wei, H., Feng, L., Chen, X., An, B.: Combating noisy labels by agreement: a joint training method with co-regularization. In: Proceedings of the IEEE/CVF Conference on Computer Vision and Pattern Recognition, pp. 13726–13735 (2020)
27. Yao, Y., et al.: Jo-SRC: a contrastive approach for combating noisy labels. In: Proceedings of the IEEE/CVF Conference on Computer Vision and Pattern Recognition, pp. 5192–5201 (2021)
28. Yi, K., Wu, J.: Probabilistic end-to-end noise correction for learning with noisy labels. In: Proceedings of the IEEE/CVF Conference on Computer Vision and Pattern Recognition, pp. 7017–7025 (2019)

29. Zheng, M., You, S., Huang, L., Wang, F., Qian, C., Xu, C.: Simmatch: semi-supervised learning with similarity matching. In: Proceedings of the IEEE/CVF Conference on Computer Vision and Pattern Recognition, pp. 14471–14481 (2022)
30. Zhou, X., Liu, X., Zhai, D., Jiang, J., Ji, X.: Asymmetric loss functions for noise-tolerant learning: Theory and applications. IEEE Trans. Pattern Anal. Mach. Intell. (2023)

Author Index

Q. Liu et al. (Eds.): PRCV 2023, LNCS 14432, pp. 511–513, 2024.
https://doi.org/10.1007/978-981-99-8543-2

Printed in the United States
by Baker & Taylor Publisher Services